VOLUME NINETY

ADVANCES IN VIRUS RESEARCH

Control of Plant Virus Diseases: Seed-Propagated Crops

Edited by

GAD LOEBENSTEIN
Department of Plant Pathology
Volcani Center Bet Dagan,
Bet Dagan, Israel

NIKOLAOS KATIS
Faculty of Agriculture, Forestry and Natural Environment, School of Agriculture,
Plant Pathology Lab,
Aristotle University of Thessaloniki,
Thessaloniki, Greece

ELSEVIER

AMSTERDAM • BOSTON • HEIDELBERG • LONDON
NEW YORK • OXFORD • PARIS • SAN DIEGO
SAN FRANCISCO • SINGAPORE • SYDNEY • TOKYO
Academic Press is an imprint of Elsevier

Academic Press is an imprint of Elsevier
225 Wyman Street, Waltham, MA 02451, USA
525 B Street, Suite 1800, San Diego, CA 92101-4495, USA
32 Jamestown Road, London NW1 7BY, UK
The Boulevard, Langford Lane, Kidlington, Oxford OX5 1GB, UK

First edition 2014

ISBN: 978-0-12-801246-8
ISSN: 0065-3527

For information on all Academic Press publications
visit our website at store.elsevier.com

CONTENTS

CONTRIBUTORS

Yehezkel Antignus
Plant Pathology and Weed Research Department, ARO, The Volcani Center, Bet Dagan, Israel

Fabrizio Cillo
Istituto di Virologia Vegetale, CNR, Bari, Italy

Masarapu Hema
Department of Virology, Sri Venkateswara University, Tirupati, India

John H. Hill
Department of Plant Pathology and Microbiology, Iowa State University, Ames, Iowa, USA

Jacqueline d'A. Hughes
AVRDC—The World Vegetable Center, Shanhua, Tainan, Taiwan, P.R. China

Roger A.C. Jones
School of Plant Biology and Institute of Agriculture, University of Western Australia, Nedlands, and Department of Agriculture and Food, South Perth, Western Australia, Australia

Nikolaos Katis
Faculty of Agriculture, Forestry and Natural Environment, School of Agriculture, Plant Pathology Lab, Aristotle University of Thessaloniki, Thessaloniki, Greece

Lawrence Kenyon
AVRDC—The World Vegetable Center, Shanhua, Tainan, Taiwan, P.R. China

Sanjeet Kumar
AVRDC—The World Vegetable Center, Shanhua, Tainan, Taiwan, P.R. China

Safaa G. Kumari
International Centre for Agricultural Research in the Dry Areas (ICARDA), Tunis, Tunisia

Moshe Lapidot
Institute of Plant Sciences, Volcani Center, ARO, Bet Dagan, Israel

P. Lava Kumar
International Institute of Tropical Agriculture, Ibadan, Nigeria

Hervé Lecoq
INRA, UR407, Station de Pathologie Végétale, Montfavet Cedex, France

James P. Legg
International Institute of Tropical Agriculture, Dar es Salaam, Tanzania

Khaled M. Makkouk
National Council for Scientific Research, Beirut, Lebanon

Peter Palukaitis
Department of Horticultural Sciences, Seoul Women's University, Seoul, Republic of Korea

Basavaprabhu L. Patil
National Research Centre on Plant Biotechnology, IARI, Pusa Campus, New Delhi, India

Jane E. Polston
Department of Plant Pathology, University of Florida, Gainesville, Florida, USA

Dodla V.R. Reddy
Formerly Principal Virologist, ICRISAT, Patancheru, Hyderabad, India

Margaret G. Redinbaugh
USDA, Agricultural Research Service, Corn, Soybean and Wheat Quality Research Unit and Department of Plant Pathology, Ohio State University-OARDC, Wooster, Ohio, USA

Pothur Sreenivasulu
Formerly Professor of Virology, Sri Venkateswara University, Tirupati, India

Wen-Shi Tsai
AVRDC—The World Vegetable Center, Shanhua, Tainan, Taiwan, P.R. China

Joop A.G. van Leur
NSW Department of Primary Industries, Tamworth Agricultural Institute, New South Wales, Australia

Steven A. Whitham
Department of Plant Pathology and Microbiology, Iowa State University, Ames, Iowa, USA

William M. Wintermantel
USDA-ARS, Salinas, California, USA

José L. Zambrano
Instituto Nacional Autónomo de Investigaciones Agropecuarias (INIAP), Programa Nacional del Maíz, Quito, Ecuador

PREFACE

Crop plants suffer from a number of abiotic and biotic diseases with some of them causing high-yield losses. It is generally accepted that viral diseases rank second after fungal plant diseases in terms of economic losses they cause globally. So far, more a thousand viral diseases have been described worldwide and currently new viruses are characterized due to the new technologies. Viruses are intercellular parasites which depend on the host plant cell machinery for their replication. Therefore, they cannot (at present) be controlled by chemical pesticides as is practiced for fungal or bacterial plant diseases. Management of viral diseases relies on the prevention measures. The current methods for many important crops, both those propagated by seed and those propagated vegetatively, will be outlined, describing the most up-to-date state of knowledge. The chapters were written by experts who have been working for years with the respective crop(s). Each chapter dealing with a specific or a number of related crops (e.g., *Citrus*) summarizes the most important viruses infecting the crop, their epidemiology and ecology, and finally measures for their control. Also, chapters dealing with novel strategies concerning control of air-borne viruses in the field and in greenhouses will be outlined. Transgenic resistance a novel and promising approach will help to increase food production, once some public phobia will diminish.

The information on the control of plant viruses in the various crops necessitates splitting it into two volumes. The present volume contains two general chapters and chapters on control of crops propagated by seed, such as cucurbits, legumes, maize, and pepper. The following one will be centered on control in vegetative propagated crops, such as berry crops, cassava, citrus, grapevine, and pome and stone fruits.

We would like to express our gratitude to Prof. Karl Maramorosch who encouraged us to proceed with the subject and Ms. Helene Kabes and her staff who helped us with the technical details.

Finally, we hope that this book will be of interest to plant virologists, horticulturists, and practitioners and will further contribute to the control of viral diseases and minimize the losses they cause to the crops.

GAD LOEBENSTEIN AND NIKOLAOS KATIS
August 2014

CHAPTER ONE

Management of Air-Borne Viruses by "Optical Barriers" in Protected Agriculture and Open-Field Crops

Yehezkel Antignus*,[1]
*Plant Pathology and Weed Research Department, ARO, The Volcani Center, Bet Dagan, Israel
[1]Corresponding author: e-mail address: antignus@volcani.agri.gov.il

Contents

Abstract

The incurable nature of viral diseases and the public awareness to the harmful effects of chemical pest control to the environment and human health led to the rise of the integrated pest management (IPM) concept. Cultural control methods serve today as a central pivot in the implementation of IPM. This group of methods is based on the understanding of the complex interactions between disease agents and their vectors as well as the interactions between the vectors and their habitat. This chapter describes

Advances in Virus Research, Volume 90
ISSN 0065-3527
http://dx.doi.org/10.1016/B978-0-12-801246-8.00001-9

a set of cultural control methods that are based on solar light manipulation in a way that interferes with vision behavior of insects, resulting in a significant crop protection against insect pests and their vectored viruses.

1. INTRODUCTION

Insect-borne plant viruses may cause severe losses to many annual and perennial crops of a high economic value. Insect vectors of plant viruses are found in 7 of the 32 orders of the class Insecta and are therefore responsible for severe epidemics that form a threat to the world's agricultural industry. Insect vectors transmit plant viruses by four major transmission modes that are supported by a number of viral and insect proteins (Raccah & Fereres, 2009). The obligatory parasitism of plant viruses and their intimate integration within the plant cell requires an indirect approach for their control. This chapter will focus on the use of light manipulation to affect insects vision behavior in a way that interferes with their flight orientation, their primary landing on the crop, and the secondary dispersal within the crop. Manipulation of light signals simultaneously diminishes the insect immigration into the crop and reduces feeding contacts between the insect vector and the host plant, thus lowering significantly virus disease incidence.

2. THE INSECTS VISION APPARATUS

Insects perceive light through a single pair of compound eyes which facilitate a wide field of vision. The basic unit of the compound eyes is the ommatidium which rests on a basement membrane. The corneagen cells are located atop a long retinula formed by long neurons and secondary pigment cells. A crystalline cone lies within the corneagen cells. The dorsal surface of the ommatidium is covered with the corneal lens which is a specialized part of insect cuticle. Part of each retinula cell is a specialized area known as a rhabdomere. A nerve axon from each retinula cell projects through the basement membrane into the optic nerve. Ommatidia are functionally isolated because the retinula cells are surrounded by the secondary pigment cells (Diaz & Fereres, 2007).

Vision involves the transduction of light energy into a bioelectric signal within the nervous system. The first events in this process take place in the retinula cells. The fine structure of rhabdomeres consists of thousands of closely packed tubules (microvilli). The visual pigments occur mainly in these rhabdomeric microvilli. It has been suggested that the small diameter of each microvillus inhibits free rotation of visual pigments. This specific orientation may be the molecular basis of insects' sensitivity to polarized light. Photobiological processes in the insect eye occur in a narrow band of the electromagnetic spectrum between 300 and 700 nm. Visual pigments initiate vision by absorbing light in this spectral region. These pigments are a class of membrane-bound proteins known as opsins that are conjugated with a chromophore. Visual pigments whose chromophore is retinal are called rhodopsins. The visual pigments of all invertebrates, including insects, crustaceans, and squids, are all rhodopsins. According to which parameter of the light is being used or what information is extracted from the primary sensory data, vision is often divided into subcategories like polarization vision (Wehner & Labhart, 2006), color vision, depth perception, and motion vision (Borst, 2009). Polarization arises from the scattering of sunlight within the atmosphere enabling the insect to infer the location of the sun in the sky. The polarization plane is detected by an array of specialized photoreceptors (Heinze & Homberg, 2007). Many insects can discriminate between light wavelength (color) (Fukushi, 1990) its contrast and intensity. Motion signals are also part of vision cues that serve as a rich information source on the environment in which the insect is acting (Borst, 2009; Diaz & Fereres, 2007).

2.1. UV vision and insects behavior

In insects, the different visual pigments (opsins) are segregated into different subsets of cells that form the ommatidium. In the fruit fly *Drosophila*, seven genes encoding different opsins have been identified and sequenced (Hunt, Wilkie, Bowmaker, & Poopalasundaram, 2001). The ability of insects and mites (McEnrone & Dronka, 1966) to perceive light signals in the UV range (300–400 nm) is associated with the presence of specific photoreceptors within their compound eye. UV receptors of the greenhouse whitefly *Trialeurodes vaporariorum* (Westwood) as in other herbivorous insects are present in the dorsal eye region (Mellor, Bellingham, & Anderson, 1997; Vernon & Gillespie, 1990). Many insects have two rhodopsins, one with maximum absorption in ultraviolet wavelengths (365 nm) and one with maximum absorption in the green part of the spectrum (540 nm) (Borst, 2009; Matteson, Terry, Ascoli, & Gilbert, 1992). UV component of the light spectrum plays an important role in aspects of insect behavior, including orientation, navigation, feeding, and interaction between the sexes (Mazokhin-Porshnykov, 1969; Nguyen, Borgemeister, Max, & Poehling,

2009; Seliger, Lall, & Biggley, 1994). The involvement of UV rays in the flight behavior of some economically important insect pests has been studied by several workers (Coombe, 1982; Issacs, Willis, & Byrne, 1999; Kring, 1972; Matteson et al., 1992; Moericke, 1955; Mound, 1962; Vaishampayan, Kogan, Waldbauer, & Wooley, 1975; Vaishampayan, Waldbauer, & Kogan, 1975).

2.1.1 Effect of UV on insects dispersal and propagation

Whiteflies [*Bemisia tabaci* (Gennadius)] dispersal pattern under UV-absorbing films was examined using a release-recapture experiment. In "walk-in" tunnels covered with a UV-absorbing film and an ordinary film, a grid of yellow-sticky traps was established forming two concentric circles: an inner and an external. Under UV-absorbing films, significantly higher numbers of whiteflies were captured on the internal circle of traps than that on the external circle. The number of whiteflies that were captured on the external circle was much higher under regular covers, when compared with UV-absorbing covers, suggesting that filtration of UV light hindered the ability of whiteflies to disperse in a UV-deficient environment (Antignus, Nestel, Cohen, & Lapidot, 2001).

Following artificial infestation of pepper plants with the peach aphid [*Myzus persicae* (Sulzer)] in commercial tunnels, covered with a UV-absorbing film, aphid population growth and spread were significantly lower compared to tunnels covered with an ordinary film. In laboratory experiments, no differences in development time (larvae to adult) were observed when aphids were maintained in a UV-deficient environment. However, propagation was faster in cages covered with the regular film. The numbers of aphids was 1.5–2 times greater in cages or commercial tunnels covered with an ordinary film. In all experiments, the number of trapped winged aphids was significantly lower under UV-absorbing films. It was suggested that elimination of UV from the light spectrum reduces flight activity and dispersal of the alate aphids (Chyzik, Dobrinin, & Antignus, 2003).

Mazza, Izaguirre, Zavala, Scopel, and Ballaré (2002) reported that in choice situations *Caliothrips phaseoli* (Hood) (Thysanoptera: Thripidae), favored areas with ambient UV-A (320–400 nm) radiation compared with areas where this part of the light spectrum was blocked. This type of behavior was explained by the relatively broad gap between the peak sensitivities of the photoreceptors that are responsible for sensing the UV range (365 nm) and the visible light (540 nm). It was assumed that under UV-deficient environment formed by the photoselective film, UV receptors are not stimulated by the ambient light, lacking the short wavelength (<400 nm) and thus did not trigger the dispersal flight of thrips. Moreover, it was hypothesized that if only the 540-nm receptor is activated, thrips should be unable to discriminate colors but only light brightness because at least stimulation of two receptors is essential for color vision, and blue and UV are of special importance regarding color opponency mechanism (Doring & Chittka, 2007). In large-scale dispersal experiments with *Frankliniella occidentalis* [western flower thrips (WFTs)], plants or blue sticky cards were arranged in concentric circles around a source plant at the release point. Dispersal of the WFT tended to exhibit reduced dispersal from source plants under UV-deficient conditions (Kigathi & Poehling, 2012).

2.1.2 UV stimulated phototaxis of insects

The UV range (360–400 nm) forms a strong stimulus for whiteflies to fly; e.g., the greenhouse whitefly, *T. vaporariorum*, took off more readily and walked faster when exposed to light of wavelengths under 400 nm than when exposed to that between 400 and 500 nm (Coombe, 1981). Similar photosystems and effects of UV light were suggested for aphids and thrips (Kring, 1972; Matteson et al., 1992). Monochromatic UV lamps in a flight chamber served to study the attraction of *B. tabaci* to distinct wave lengths in the UV-A and UV-C parts of the spectrum. Whiteflies were attracted similar to light source emitting at 366 and 254 nm (Antignus, Mor, Ben Joseph, Lapidot, & Cohen, 1996). Preference for richer UV environments has been shown for the whiteflies *B. argentifolii* (Bellows & Perring) and *T. vaporariorum*, when tested in choice situations (Costa & Robb, 1999; Costa, Robb, & Wilen, 2002; Doukas & Payne, 2007b; Mutwiwa et al., 2005). A striking demonstration of the effect of UV phototaxis on aphids and whiteflies occurs in greenhouses with roof arches that are covered alternately, with a UV-transmitting film and a UV-absorbing film. In cases where such greenhouses are invaded by virus born insects, all the plants under the UV-transmitting film become infected, while plants under the UV-absorbing film remain totally virus free (Y. Antignus, unpublished) (Fig. 1.1). This dramatic scenario is a consequence of the fact that when an insect is in a choice situation, it will be attracted always to a UV-rich environment.

In choice situations, *Ceratothripoides claratris* (Shumsher) thrips exhibited a clear preference to fly toward rooms enlightened with higher UV intensities, while it avoids greenhouse constructions with low internal UV radiation, as initially shown by Kumar and Poehling (2006). Costa and Robb (1999)

Figure 1.1 A demonstration of aphids phototaxis toward UV irradiation and the formation of the "two-compartment effect" in greenhouses with roof arches covered alternately with UV-absorbing (films with a bluish hue) and ordinary films (A). The massive immigration of aphids into greenhouse sections with a rich UV environment resulted in total infection with the aphid-borne nonpersistent *Zucchini yellow mosaic virus* (ZYMV) (B Left hand). All the stunted plants show the typical mosaic symptoms of the virus (C). None of the plants grown under the UV-absorbing film were infected (A Right hand). (See the color plate.)

reported a distinct flight preference of the WFT, to higher UV levels in choice experiments carried out in plastic tunnels. Likewise, Kigathi (2005) reported similar findings for WFT in choice experiments, in the laboratory, and in small greenhouses.

2.1.3 Effect of UV reflection

Moller (2002) assumed that UV reflection from plant surfaces plays an important role in making plants visible to herbivorous insects, thus directing their take off and initial orientation. It was assumed that stimulation of UV receptors by incoming UV mainly controls initiation of take off and directs the route of dispersion flight of the thrips *C. claratris* (Nguyen et al., 2009). However, high reflectance of UV rays seems to disrupt the host-finding behavior of insect pests. For example, in the case of thrips, *F. occidentalis*, the strong reflection of UV caused an attractively colored surface to become repellent (Vernon & Gillespie, 1990), and the use of UV-reflecting mulches has been proposed to reduce populations of *Frankliniella* spp. and the associated incidence of tomato spotted wilt disease in open-field tomato (Stavisky, Funderburk, Brodbeck, Olson, & Andersen, 2002).

3. USE OF UV-ABSORBING CLADDING MATERIALS FOR GREENHOUSE PROTECTION AGAINST THE SPREAD OF INSECT PESTS AND VIRUS DISEASES

The first report on the effect of UV-absorbing films on insects came from a Japanese experimental station (Onuma & Nakagaki, 1982). Few years later, this effect was discovered independently by Antignus et al. (1996) through a field experiment in which an anti-Botrytis UV-absorbing film (Raviv & Antignus, 2004; Reuveni & Raviv, 1992) was tested for its ability to protect cucumbers against the spread of downy mildew. Surprisingly, 100% of the plants grown in tunnels covered by an ordinary non-UV-absorbing film were infected by the whitefly-borne *Cucumber yellow stunting disorder virus* (CYSDV), while those that were grown under the UV-absorbing film remained healthy. Further studies have shown the blocking effect of UV-absorbing films on the invasion of a wide range of insect pests into structures covered with different UV-absorbing cladding materials. Since then, an extensive research was carried out in different laboratories worldwide enabling a better understanding of the interactions among UV light, insects, and their host plants.

3.1. Spectral transmission properties of UV-absorbing cladding materials

Improved greenhouse film technology has given growers access to plastic covers that last up to twice as long, allow a better light transmission, and are engineered to lift crop production. Unique among these covering materials are multilayered polyethylene films, often referred to as "smart films" that include UV inhibitors and colorants to filter or enhance light quality, and additives that offer anti-drip and UV blocking, as well as cooling and thermal benefits (Espí, Salmerón, Fontecha, García, & Real, 2006).

Photosynthesis active radiation, 400–700 nm, the main source of energy for plants, should be transmitted through the covering material at a high rate to enhance yields. According to environmental photobiologists, the UV portion of the spectrum is subdivided into three regions, defined as UV-C (200–290 nm), UV-B (290–320 nm), and UV-A (320–400 nm). The near-UV spectrum (200–400 nm)-absorbing qualities that plastics naturally have can be manipulated by the addition of additive materials (Edser, 2002).

The introduction of UV-blocking compounds into polyethylene films determines their UV-absorbing capacity. UV films block the transmission of UV irradiation in the range of 280–380 nm, while most of visible part of the solar light spectrum is transmitted (Fig. 1.2). Only 7% of the solar light spectrum between 250 and 400 nm is transmitted through UV-absorbing films compared to 39% transmission by the ordinary films (Chyzik et al., 2003).

UV-absorbing screens permit the influx of a mixture of filtered and unfiltered light, resulting in transmission of less than 40% of the natural UV radiation. Ordinary nets allow more than 75% of incident light to reach the crop (Legarrea, Karnieli, Fereres, & Weintraub, 2010).

The level of the UV-blocking capacity of the plastic is determined by the level of the UV-blocking additive that it contains or by the molecular properties of the material. Enhanced UV-blocking capability increases the protection efficiency of the cladding material (Antignus et al., 1996; Doukas, 2002). The activity of UV blockers is time limited due to chemical degradation. Polyvinyl chloride (PVC) films on the other hand act as efficient UV filters due to their molecular properties which increase both the UV-absorbing capacity and stability (Antignus et al., 1996).

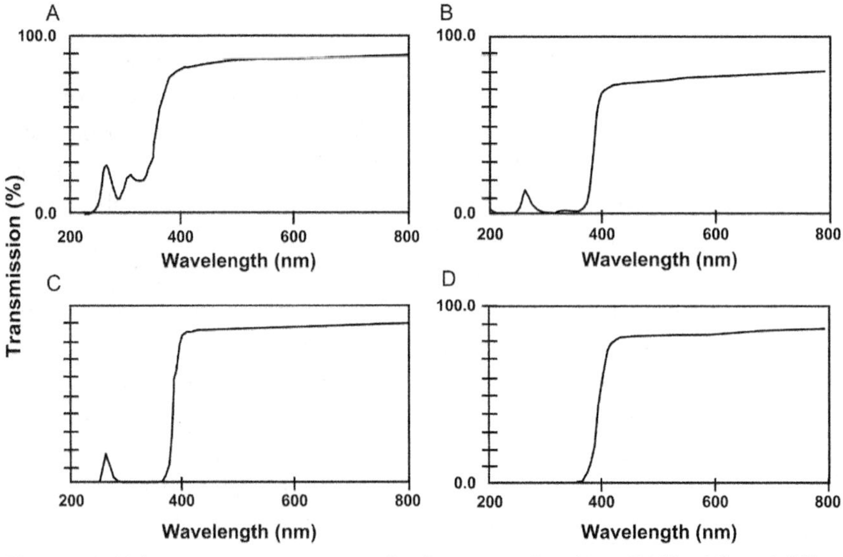

Figure 1.2 Light transmission spectra of ordinary nonabsorbing UV film (A), and different brands of UV-absorbing films: Solarig (B), IR-Veradim (C), PVC sheet, Rav-Hozek (D).

3.2. Effect of UV-absorbing films on the immigration of insect pests into greenhouses

Photoselective greenhouse cladding materials can serve as mega filters to eliminate parts of the UV light spectrum, thus inhibiting insect development and virus epidemics (Antignus et al., 1996). UV-absorbing polyethylene films were highly efficient in protection of greenhouse crops against infestation by different insect pests and viral diseases. Tomato crops grown in "walk- in" tunnels covered with a UV-absorbing polyethylene were highly protected against the immigration of *B. tabaci* (Fig. 1.3). UV-absorbing plastic sheets were also highly efficient in protecting cucumbers from infestation by the WFTs, *F. occidentalis* (Fig. 1.4), and the cotton aphid, *Aphis gossypii* (Glover) (Fig. 1.5) (Antignus et al., 1996; Raviv & Antignus, 2004). Similar results were obtained by other workers showing that these protection effects seem to be universal and are not species specific. UV-absorbing films were found effective against the invasion of thrips *Ceratothripoides claratris* (Kumar & Poehling, 2006), *Thrips tabaci* (Lindenman) (Doukas & Payne, 2007a), the peach aphid *M. persicae* (Chyzik et al., 2003), the whitefly *T. vaporariorum* (Doukas & Payne, 2007a; Gonzales, Rodriguez, Bafion, Franco, & Fernandez, 2001), *Macrosiphum euphorbiae* (Thomas), *Acyrthosiphum lactucae* (Passerini), and *Aphis fabae* (Scopoli) (Diaz, Biurrun, Moreno, Nebreda, & Fereres, 2006; Doukas & Payne, 2007a). Some reports

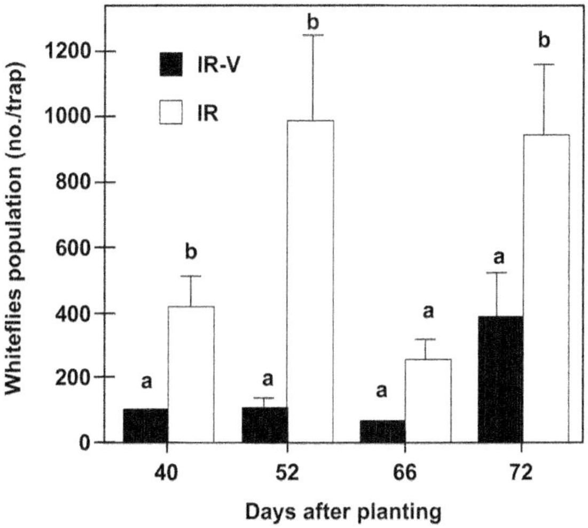

Figure 1.3 Protection of UV-absorbing films from whiteflies immigration. Trapping of *Bemisia tabaci* in "walk-in" tunnels covered with either nonabsorbing polyethylene film (IR) or a UV-absorbing polyethylene film (IR-V).

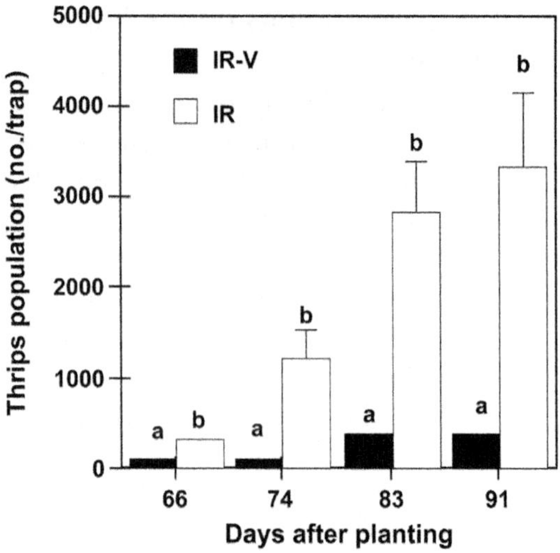

Figure 1.4 Protection of UV-absorbing films from thrips immigration. Trapping of *Frankliniella occidentalis* in "walk-in" tunnels covered with either nonabsorbing polyethylene film (IR) or a UV-absorbing polyethylene film (IR-V).

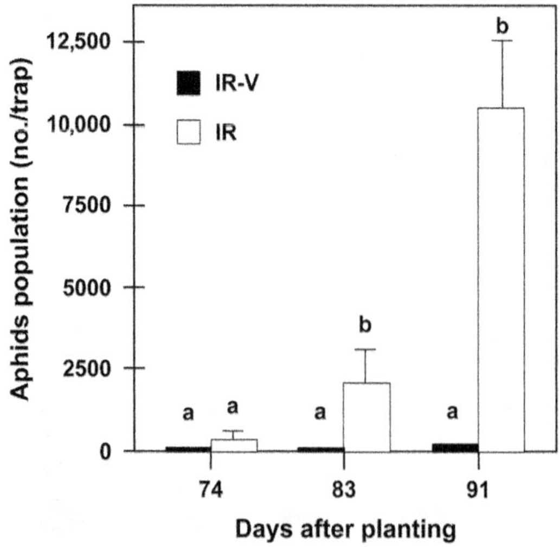

Figure 1.5 Protection of UV-absorbing films from aphids' immigration. Trapping of *Aphis gossypii* in "walk-in" tunnels covered with either a nonabsorbing polyethylene film (IR) or a UV-absorbing polyethylene film (IR-V).

indicate also inhibiting effects on immigration of the leafhopper *Hauptida maroccana* (Melichar 1907) (Doukas & Payne, 2007a). Lower numbers of *Lyriomyza* spp. were found in structures covered with UV-blocking cladding materials (Antignus, Lapidot, Hadar, Messika, & Cohen, 1998; Costa, Newman, & Robb, 2003). The following mite species were deterred by UV-absorbing films: red mites (*Tetranychus telarius*) (Antignus et al., 1998), the tomato rust mite (*Aculpas lycopersici*), and the broad mite (*Polyphagotarsonemus latus*) (Antignus et al., 1998). Lower numbers of caterpillars of the nocturnal moth *Spodoptera exigua* were found in mint grown in commercial walk-in tunnels covered with a UV-absorbing film (Messika et al., 1999). Population density of the Lepidopteran pest *Autographa gamma* was reduced in lettuce grown in a UV-deficient environment (Diaz et al., 2006). The significant reduction in insect pest populations in commercial greenhouses protected with UV-absorbing films enabled a dramatic reduction in the number of insecticide applications during the growing season. In field experiments carried out in commercial structures (Messika et al., 1999; Messika, Lapidot, & Antignus, 1997; Messika, Nishri, Gokkes, Lapidot, & Antignus, 1998), only three applications were given to the cut flower Lisianthus grown in a greenhouse protected by a UV-absorbing film, compared with 20 insecticide applications given to control leafminers, thrips, and whiteflies in the same crop grown under ordinary polyethylene films.

3.3. Effect of UV filtration on the spread of insect-vectored virus diseases

The inhibitory effect of UV-absorbing films on the migration of insect pests into plastic protected structures had a drastic effect on the spread and disease incidence of insect-vectored virus diseases. Thus, TYLCV spread rate and disease incidence under UV-absorbing films were 5- to 10-folds lower compared to that in control structures covered with an ordinary polyethylene film (Fig. 1.6) (Antignus et al., 1996; Kumar & Poehling, 2006; Rapisarda & Tropea-Garzia, 2002). Kumar and Poehling (2006) reported on the attenuation effect of UV-absorbing films on symptoms induced in tomato by *Capsicum chlorosis virus*. Blocking effects on virus spread were reported for potyviruses and the tospovirus *Tomato spotted wilt virus* (Diaz et al., 2006). High protection against the spread of *Zucchini yellow mosaic virus* was recorded in a commercial greenhouse covered with a UV-absorbing plastic (Y. Antignus, unpublished). Cucumber and melon crops grown in commercial "walk-in" tunnels or greenhouses covered with UV-blocking films were highly protected against the whitefly-borne CYSDV. CYSDV

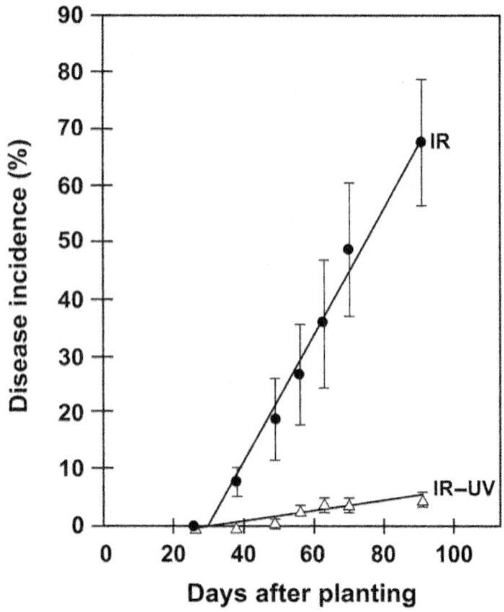

Figure 1.6 The spread dynamics of Tomato yellow leaf curl virus (TYLCV) in tomato plants grown in "walk-in" tunnels covered with either regular (IR) or UV-absorbing films (IR-UV).

incidence under the UV-absorbing films was 2% compared to 14% under an ordinary film (Mizrahi et al., 1998). Infection rates of the nonpersistent viruses *Cucumber mosaic virus* (CMV) and *Lettuce mosaic virus* were lower under UV-absorbing screens, probably due to the lower population density and dispersal rate of the aphid vector *M. euphorbiae* (Legarrea, Betancourtb, et al., 2012; Legarrea, Weintraub, Plaza, Viñuela, & Fereres, 2012).

3.4. Effect of UV filtration on crop plants

The effect of UV filtration on plants was reviewed in the past (Antignus & Ben-Yakir, 2004; Raviv & Antignus, 2004). Although UV-A radiation (315–400 nm) is less potent per photon than UV-B (280–315 nm), its damaging and inhibiting effects on growth and photosynthesis of aquatic and terrestrial plants can also be considerable, since its flux rate is higher than that of UV-B radiation. UV-A radiation may also induce increased amounts of UV-absorbing pigments (Rozema et al., 2002). UV-exclusion studies on cucumber (Krizek, Mirecki, & Britz, 1997) and a red-pigmented lettuce (Krizek, Britz, & Mirecki, 1998) indicate that ambient UV-A radiation greatly inhibits leaf enlargement, stem elongation, and biomass production. UV-B radiation is known to affect the secondary metabolism of plants via the activation of UV-B photoreceptor, the upregulation of genes of the phenylpropanoid pathway, and the accumulation of flavonoids, anthocyanins, alkaloids, waxes, and polyamines (Marcel, Jansen, Gaba, & Greenberg, 1998). UV-B may also have a negative impact on the plant including damage to DNA, reduction in the activity of several enzymes, reduction in the levels of chlorophyll and carotenoids, downregulation of photosynthetic genes, and changes in the chloroplast ultra structure. It is obvious that elimination of the UV-B from light by greenhouse cladding material may affect plants in different ways (Fiscus, Philbeck, & Britt, 1999; Marcel et al., 1998; Robson et al., 2003). In some species (e.g., cucumber, mung bean, New Zealand spinach, and "New Fire" lettuce), growth is inhibited by solar UV-B (Adamse, Reed, Krizek, Britz, & Mirecki, 1997; Krizek et al., 1998, 1997). In some plants (e.g., tomato), growth is promoted (Cybulski & Peterjohn, 1999; Krizek et al., 1997), whereas others (e.g., cotton, oats) remain unaffected (Adamse et al., 1997; Krizek et al., 1997). No significant differences were found in growth, yield, maturing time, fresh, or dry weights of plant parts in tomatoes grown in greenhouses under standard and UV-blocking films. The yield and quality of pepper and cucumbers were not affected by blocking UV (Onuma & Nakagaki, 1982). No differences were found in pigment intensity and total soluble sugars of tomato and pepper fruits grown under regular and UV-blocking films (Pressman, Moshkovitz, Rsenfeld, & Shaked, 1996). The type of polyethylene film covering did not affect the percentage of viable pollen grains of tomato and pepper plants (Pressman et al., 1996) or on the firmness and shelf life of tomato fruits (Antignus et al., 1999). UV blocking caused an improper pigmentation in the violet cultivar of Lisianthus flowers, thus reducing its marketing quality (Messika et al., 1998). Changes in quality, such as pigmentation and taste, were observed in lettuce when plants were grown under UV-opaque film, which absorbed 50% of UV-A and 95% of UV-B light (Paul, Jacobson, Taylor, Wargent, & Moore, 2005). These effects on pigmentation may be explained by the requirement of UV irradiation for the synthesis and accumulation of anthocyanins and flavonoids (Marcel et al., 1998). Therefore, UV-blocking films may not be used to protect crops in which anthocyanin pigmentation is a determinant of quality. The implementation of UV-blocking cladding materials in greenhouses is feasible only if it will not affect negatively plant development, yield level, and quality. Experimental results indicate that UV-absorbing films can be used safely in most crop plants. UV-absorbing films were successfully used to protect tomato (*Lycopersicon esculentum*;

Antignus et al., 1996; Kumar & Poehling, 2006; Rapisarda & Tropea-Garzia, 2002), pepper (Capsicum annuum; Chyzik et al., 2003; Mizrahi et al., 1998), cucumbers (*Cucumis sativus*; Antignus et al., 1996), melons (*Cucumis melo*; Antignus, unpublished), chive (*Allium schoenoprasum*), sage (*Salvia fruticosa*), basil (*Ocimum basilicum*), mint (*Mentha* sp.), chervil (*Anthriscus cerefolium*; Messika et al., 1997, 1999), lettuce (*Lactuca sativa*; Diaz et al., 2006), and eggplant (*Solanum melongena*; Kittas, Tchamitchian, Katsoulas, Karaiskou, & Papaioannou, 2006). The growth and performance of the following organic crops were tested in tunnels covered with UV-blocking films compared to non-UV-blocking films: chard, chicory, winter parsley, endive, winter lettuce, winter spinach, winter radish, rocket, Chinese cabbage, and spring cabbage. The crops under the UV-blocking film appeared to grow quicker and also reached a greater final size (Leigh, 2004).

3.5. Effect of UV filtration on pollinators

Pollination improves the yield and increases the quantity of most crop species, thus contributing to one-third of global crop production. More than 75% of the 115 leading crop species worldwide are dependent on or at least benefit from animal pollination, whereas wind and self-pollination are sufficient for only 28 crop species (Klatt et al., 2014).

UV-poor environments might have an influence on pollinator behavior in two ways: first, the overall flight activity may be diminished due to deficient light conditions, and second, the light conditions might change the color perception of the crop flowers by the pollinators so that they will have difficulty in localizing the flowers among the leaf mass (van der Blom, 2010).

Bumblebees [*Bombus terrestris* (Linnaeus)] are important pollinators of angiosperms. The pollination of tomato flowers requires the agitation of flower anther cones to enable an efficient pollination, and bumblebees are widely used in tomato greenhouses (Kevan, Straver, Offer, & Laverty, 1991). Studies carried out under laboratory conditions have shown that bumblebees perceive when ultraviolet radiation is either removed or added to an illumination source, and are capable of using their visual system to forage efficiently in a UV-deficient environment. Thus, their forage efficiency is not affected by the type of greenhouse covering (Dyer & Chittka, 2004). A delay in the hive start up of the bumblebee *B. terrestris* (Bio-Bee, Ltd., Israel) was observed in experimental mini greenhouses covered with UV-blocking films (Steinberg et al., 1997; van der Blom, 2010). Later, this problem was solved by placing the hives near the greenhouse walls, where they were exposed to unfiltered light (Y. Antignus, unpublished). In a field study, no significant differences were found in bumblebee activity or in the numbers of flowers visited, under standard or UV-blocking films (Fig. 1.7) (Antignus & Ben-Yakir, 2004). Studies in commercial tomato greenhouses have demonstrated that biomass and size of hives were not significantly affected, whether the greenhouses were covered with standard or UV-blocking films (Antignus & Ben-Yakir, 2004; Hefez, Izikovitch, & Dag, 1999; Seker, 1999). No differences were found in the numbers of workers that foraged nor in the final harvest in field trials where the pollination activity of bumblebees, under UV-absorbing and -ordinary films, was compared in both tomato and watermelon crops (van der Blom, 2010). Contrary to the bumblebees, honeybees did show significant behavioral changes under the UV-blocking plastic. Two trials were carried out in watermelon, and one in melon using honey bees for pollination. In all three cases, a reduced foraging activity was observed under the UV-blocking material, resulting in a significantly lower fruit yield. This reduction was seen in the number of workers leaving and entering the hive, so it seems to be the result of deficient general light conditions, more than of the difficulty to localize the flowers once foraging (van der Blom, 2010).

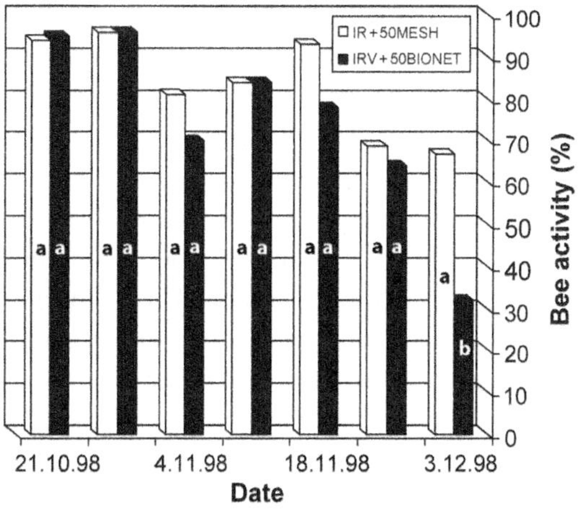

Figure 1.7 Bumblebee (*Bombus terrestris*) pollination activity in tomatoes grown in a greenhouse with ordinary cladding materials (polyethylene film + 50 mesh screen) versus a greenhouse covered with UV-absorbing cladding materials (polyethylene film + 50 mesh "Bionet" screen). Pollination activity is expressed as percentage of visited flowers identified by typical brown ring on the flower stamens cone.

In Canada, bumblebees' activity was 94% greater under standard films than under UV-blocking films (expressed as the number of entrances and exits to and from the hive). No relationship was found there between bumblebees' activity and the amount of solar radiation or the humidity in the greenhouse (Morandin, Laverty, Kevan, Khosla, & Shipp, 2001). The differences between the results from Canada and Israel may be explained by the differences in sun light intensities and temperatures between these two locations. A positive correlation exists between the rate of bumblebees' activity and temperature. Higher temperatures (in the range 5–25 °C) may compensate for the inhibitory effect of reduced UV radiation (Morandin et al., 2001; Morandin, Laverty, Kevan, Khosla, & Shipp, 2002). Areas within the greenhouse that have relatively high levels of UV radiation (normally the southern wall side) were found as optimal sites for placing bumblebees' hives in greenhouses covered with UV-blocking films (Y. Antignus, unpublished).

3.6. Effect of UV filtration on insect natural enemies

The effect of UV-absorbing plastic sheets on the host location ability of three commercially available parasitoids—*Aphidius colemani* (Viereck), *Diglyphus isaea* (Walker), and *Eretmocerus mundus* (Mercet)—was tested in the laboratory as well as in field trials. The parasitoids preference for natural versus UV-filtered light was tested under laboratory conditions using Y-tube system. Approximately 90% of the tested insects, regardless of species, chose natural light. The parasitoid's ability to locate a host-infested plant from a distance (~10 m) was also tested in field trials (Chiel, Messika, Steinberg, & Antignus, 2006). Host location by *A. colemani* (Fig. 1.8) and *D. isaea* as monitored by parasitism rates was not affected by greenhouse covering plastic type whether standard or UV-absorbing plastic was used. *E. mundus*, on the other hand, was unable to locate the host-infested plant when the latter was placed in the center of the UV-absorbing plastic covered greenhouses. Also, parasitism rates were lower under UV-absorbing plastic than under regular plastic when the host-infested plants were located in the corners of the greenhouse and the wasps were released in the center. It was therefore recommended that the number of release points must be increased to facilitate host location when releasing *E. mundus* in greenhouses covered with UV-absorbing plastic, whereas no modification was necessary for *D. isaea* and *A. colemani* (Chiel et al., 2006). Kajita (1986) found that parasitism of whiteflies by *Encarsia formosa* (Gahan) was similar under both standard and UV-blocking films. In choice experiments, significantly more

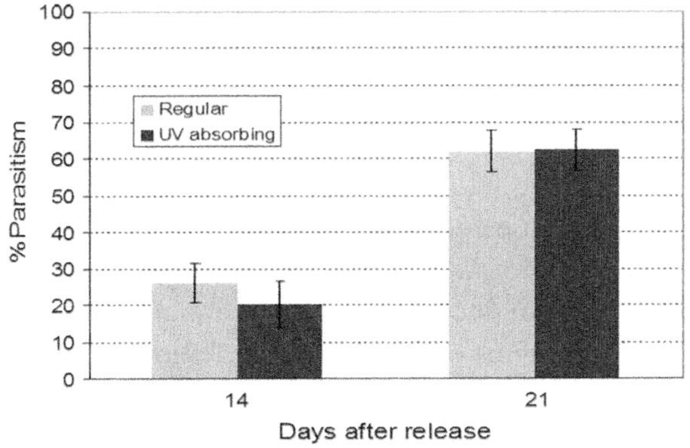

Figure 1.8 Parasitism level of the green peach aphid (*Myzus persicae*) by *Aphidius colemani* in greenhouses covered with regular cladding materials (plastic film roof and 50 mesh wall screens) or with UV-absorbing cladding materials (plastic film roof and 50 mesh Bionet® screened walls). Parasitoids were released in the center or at the greenhouse perimeter.

(two to three times) *E. formosa* individuals were trapped under standard rather than under UV-blocking films. It seemed that the parasitoids—like their hosts—oriented more toward an environment with high UV radiation. However, when they had no other choice, they performed well in a UV-deficient environment (Doukas & Payne, 2007c). Results from experiments carried out in cages covered with UV-absorbing nets showed that visual cues in *Orius laevigatus* (Fieber) (Hemiptera: Anthocoridae) may be disturbed under UV-absorbing covers, inducing a reduction in dispersal. UV-deficient environments formed under photoselective screens seem to be attractive for *Amblyseius swirskii* (Athias-Henriot) (Acari: Phytoseiidae) (Legarrea, Weintraub, et al., 2012). It was suggested that the predatory mite attempts to avoid UV-B radiation that may reduce survival, egg laying, and hatching, as has been found for other predatory mite species (Onzo, Sabelis, & Hanna, 2010).

3.7. Effect of UV-absorbing screens on the immigration of insect pests into greenhouses

Fifty-mesh screens installed as greenhouse walls were implemented in Israeli greenhouses as a protection mean against the spread of the whitefly-borne TYLCV (Berlinger et al., 1991; Cohen & Berlinger, 1986). However, these dense screens prevent adequate ventilation, especially during summer when temperatures are high. The negative effect of the resulting heat stress can be decreased, however, by using different types of cooling systems. The protection efficiency of 50 mesh screens was dramatically increased by introduction of a UV-absorbing additive into the polyethylene used for the production of the screens. The first UV-absorbing screens (BioNet®) were developed and reported by Antignus et al. (1998). These screens are characterized by a double insect–exclusion mechanism based on both their physical and optical properties. When compared to ordinary 50 mesh screens, the UV-absorbing screens reduced whitefly penetration and spread of TYLCV by a factor of 4 (Antignus et al., 1998). The "Bionets" were significantly more effective than the conventional 50-mesh screens in protecting tomato from infestation with *B. tabaci*, red spider mites, and leafminers [*Lyriomyza trifolii* (Burgess)]. "Bionets" also protected cucumbers against aphids (*A. gossypii*) (Antignus et al., 1998) and leafhoppers (Weintraub, Pivonia, & Gera, 2008). However, 50-mesh "Bionet" screens failed to prevent the ingress and build up of *F. occidentalis* in the protected structures and these results were later confirmed by others (Ben-Yakir, Hadar, Offir, Chen, & Tregerman, 2008; Diaz & Fereres, 2007; Legarrea, Betancourtb, et al., 2012; Legarrea, Diaz, Morales, Vinuela, & Fereres, 2008; Legarrea et al., 2010; Legarrea, Weintraub, et al., 2012; Weintraub et al., 2008). However, later, a different brand of photoselective screens designated OptiNet® were developed which were also able to protect against thrips. Forty and fifty mesh OptiNet® screens reduced thrips infestations on cucumber, tomato, and chive plants by three- to fourfolds compared with standard 50 mesh screen (Ben-Yakir et al., 2008).

3.8. Mode of action of UV-absorbing greenhouse cladding materials

A putative twofold mechanism is suggested to explain the defense impact of UV filtration on the immigration and spread of insect pests and virus diseases in greenhouses covered with UV-absorbing cladding materials. Initially, insects are excluded from the greenhouse wall due to the lack of UV emission from the inner space of the greenhouse. This effect was well demonstrated by Antignus et al. (2001) and Nguyen et al. (2009) who monitored lower numbers of whiteflies and thrips, respectively, on sticky traps placed on the outer sidewalls of greenhouses covered with UV-absorbing cladding materials. Hypothetically, a "two-compartment effect" is formed by the photoselective cladding materials: the greenhouse ambient environment has a normal level of UV irradiation, thus representing a UV-rich compartment, while a second UV-deficient compartment is formed within the greenhouse due to UV filtration. Insects that approach the greenhouse wall from the external environment exhibit a positive UV phototactic behavior, and as they lose contact with UV, near the walls of the protected greenhouse, they become diverted toward the UV-rich environment, away from the UV-deficient greenhouse compartment (Antignus, 2010).

While the first step of the defense mechanism is interfering with the primary infestation of the crop, the second step is involved in the secondary spread by altering the insects' normal behavior paradigm. As described in paragraphs 2.1.1, 3.2 and 3.3, the reduced flight activity that was observed in greenhouses covered with UV-absorbing cladding materials is leading to a lower rate of secondary spread of the invading insect and as a consequence insect-vectored virus diseases are hindered (Antignus, 2010; Antignus et al., 2001; Chyzik et al., 2003; Diaz & Fereres, 2007).

4. STICKY TRAPS FOR MONITORING AND INSECTS MASS TRAPPING

Mound (1962) suggested that *B. tabaci* is attracted by two groups of wavelengths of transmitted light, the blue/ultraviolet, and the yellow parts of the spectrum. He correlated the reaction to ultraviolet to the induction of migratory behavior, whereas yellow radiation induces vegetative behavior, which may be part of the host selection mechanism. The attraction of whiteflies to yellow has been utilized as an important instrument in sampling and monitoring of whiteflies populations (Gerling & Horowitz, 1984; Gonzalez & Rawlin, 1968). Aphid-landing preference is determined by the color of the background, mainly by the degree of contrast between the green plant and the color of the soil (background color) (A'Brook, 1968). When horizontal sticky yellow traps are installed on stands at different heights above ground, their whitefly (*B. tabaci*) trapping efficiency is directly correlated to the distance between the trap plane and soil level. A gradient of number of trapped insects is obtained, with a maximum on traps that are placed directly on soil providing the highest contrast between the yellow plate and the soil background (Fig. 1.9). However, when the yellow-sticky cards are placed on a large yellow plate which drastically diminish contrast, very few insects are trapped (Fig. 1.9) (Y. Antignus, unpublished).

This vision cue was used by Cohen & Marco (1973) to protect pepper (*C. annuum*, L.) from nonpersistently transmitted aphid-borne viruses in the open field. A "protection belt" consisting of polyethylene sheets, 2–3 m long, hung on stakes 70 cm above ground and covered with transparent glue, was erected around the field 6–7 m beyond the end rows. Disease incidence in the

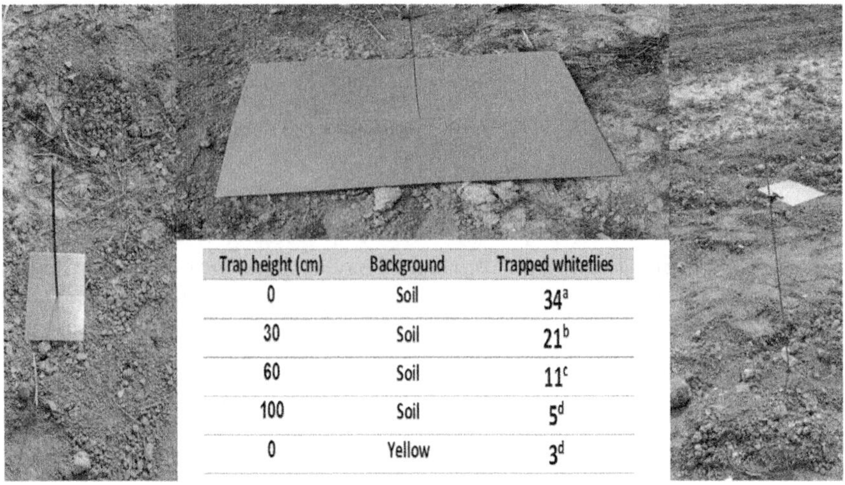

Trap height (cm)	Background	Trapped whiteflies
0	Soil	34[a]
30	Soil	21[b]
60	Soil	11[c]
100	Soil	5[d]
0	Yellow	3[d]

Figure 1.9 The contrast effect on whiteflies (*Bemisia tabaci*) trapping by yellow-sticky cards. The highest trapping efficiency was obtained when traps were placed directly on bare soil. A gradient of numbers of trapped insects was formed according to the distance of the yellow cards' plane from ground level. Lower trapping numbers are correlated with lower levels of contrast between the yellow color and background formed by the brown soil. When the yellow cards were placed over a large yellow "poligal" plate trapping was zero or near zero, indicating again the importance of high contrast between the trap and its background. (See the color plate.)

protected plots was one-half that in unprotected ones, where the incidence reached 52%. This method was used to protect seed potato crops against the persistently transmitted *Potato leafroll virus* (PLRV) (Zimmerman-Gries, 1979). However, such success cannot be assumed for other systems as there have been failures using this practice (Harpaz, 1982; Raccah, 1986). A predominance of olfactory and gustatory cues over those of vision may explain some of the failures. Mass trapping of insects in greenhouses by yellow-sticky cards was reported by Van de Veire & Vacante (1984). Vertical sticky traps consisting of blue and yellow plastic strips (Poly-traps) are used routinely in Israeli greenhouses and they are positioned around the lower part of the inner walls of greenhouses to trap invading insects.

5. SOIL MULCHES

Mulching consists of spreading a thin plastic film directly over the ground. Mulching films usually have a thickness between 12 and 80 mm and a width of up to 3 m. Usually, they are designed for a lifetime of 2–4 months. In order to avoid the expense of collection after use, they are sometimes photo- or biodegradable. The insulating effect of the mulching film helps to maintain or increase temperature and humidity of the ground, minimizing the seed time and enhancing crop growth. Mulching also helps to maintain the structure of the ground, avoiding the erosion and improving the management of water, an important issue in areas with limited water resources. Mulching films can be transparent or pigmented: black films minimize the growth of weeds, reducing the use of agrochemicals; aluminized or white films increase the reflection of light toward the low parts of the plants. Light reflection affects plant physiology as well as attracts or repels certain insects (Espí et al., 2006). The use of soil mulch to protect tomato plants from infestation by whiteflies was reported by Avidov (1956), who used sawdust or whitewash spray to mulch the crop seedbeds. Similar results were obtained from straw mulches that not only markedly reduced whitefly population but also delayed the spread of *Cucumber vein yellowing virus* and TYLCV both vectored by *B. tabaci* (Cohen, 1982). Later on, Cohen and Melamed-Madjar (1978) tested yellow, aluminum, and blue polyethylene film, demonstrating the high efficiency of the yellow polyethylene in delaying infection of tomatoes by TYLCV. Loebenstein, Alper, Levy, and Menagem (1975) found that aluminum foil was highly efficient in protecting pepper crops from aphid-transmitted viruses such as CMV and PVY. Corn plants grown in early spring on transparent polyethylene mulch to obtain an early yield were highly protected from *Maize dwarf mosaic virus* due to repelling effect of the reflected light on winged aphids that vector the virus (Y. Antignus, unpublished). Transparent polyethylene mulches reduced both whitefly and aphid populations and virus incidence (Orozco-Santos, Perez-Zamora, & Lopez-Arriaga, 1995). Similar protective effects against whiteflies, aphids, and their vectored viruses were reported later by others (Csizinszky, Schuster, & Kring, 1995; Summers, Mitchell, & Stapleton, 2005; Suwwan, Akkawi, Al-Musa, & Mansour, 1988). Yellow and silver polyethylene mulches efficiently protected zucchini plants from the spread of the whitefly-borne Squash leaf curl virus (SLCV). A delay in SLCV spread was observed 2 weeks after planting, when disease incidence was 10–20% in plants grown over yellow or silver mulches, respectively, compared with 50% disease incidence in the unmulched plots (Fig. 1.10). The landing rate of whiteflies on plants grown over silver and yellow soil mulches was five- to sevenfolds lower than that on plants grown over bare soil (Antignus, 2012; Antignus, Lachman, et al., 2004).

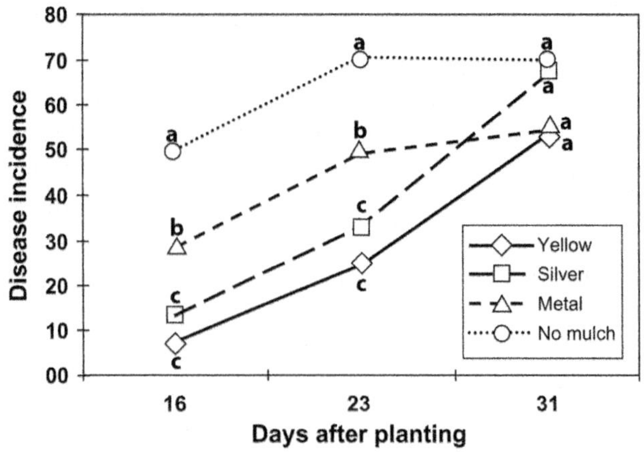

Figure 1.10 Delay in the spread of *Squash leaf curl geminivirus* (SLCV) infection in zucchini crops grown on different polyethylene mulches. A disease incidence of 70% was found 20 days after planting when plants grew on bare soil, while 25%, 32%, and 50% disease incidence was recorded in zucchini plots grown on yellow, silver gray, and metal polyethylene mulching films, respectively. SLCV disease delay by reflective mulches resulted in a double increase of high-grade fruit.

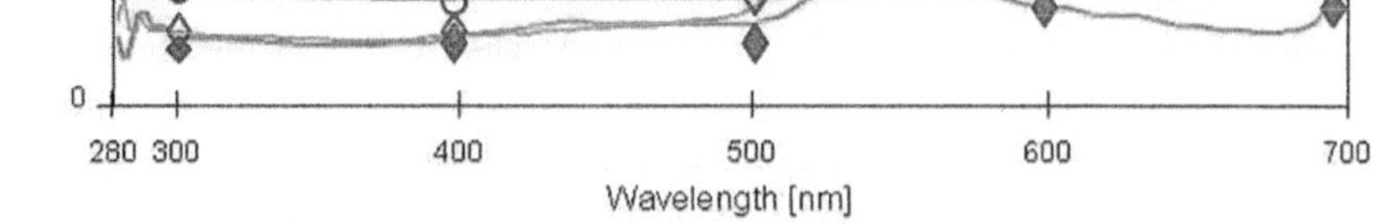

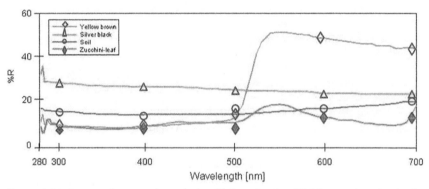

Figure 1.11 Spectrophotometric analysis of light reflection (*R*%) from colored polyethylene films, soil surface, and leaf surface. A reflection peak at 540 nm (50 *R*%) measured from yellow mulching polyethylene film (Ginegar Plastic, Israel) (◊). A constant reflection level at 30 *R*% at both the UV and visible light ranges measured from silver black mulching polyethylene film (Ginegar Plastic, Israel) (▽). A constant reflection level at ca.18 *R*% at both the UV and visible light ranges measured from bare soil (○). A reflection peak at 540 nm (19 *R*%) measured from zucchini leaf (♦). The contrast between the plant canopy and soil is highest when plants are grown on bare soil. Plant image is less visible for landing insects when plants are grown on highly reflective yellow or silver black mulches blurring the plant image.

Spectrophotometric analysis of light reflection from yellow and silver polyethylene mulches, soil surfaces, and plant canopy (Fig. 1.11) has demonstrated a relatively high level of light reflection from the plastic mulches in the range of 300–700 nm, compared with low levels of light reflection from bare soil. The plant foliage had a distinct reflection peak at 550 nm, considerably higher than that of the reflection from the bare soil. Under these circumstances, the contrast between the soil background and the plant canopy was maximal, enabling insects to detect the crop for landing. On the other hand, when the background of the plant was formed by yellow or silver mulches, the amount of reflected light in the visible range was considerably higher than the reflection of the soil and plant canopy. The poor contrast resulting from the reflection of the plastic interfered with the ability of the insect to detect the plant image and perceive a landing signal (Antignus, 2012; Antignus et al., 2005).

6. REFLECTIVE AND COLORED SHADING NETS

Cohen (1981) used white coarse nets to get a highly efficient protection of pepper crops from aphids and their vectored viruses: CMV and PVY. Recently, photoselective shade nettings were designed to screen various light spectral components of the solar radiation and/or transform direct light into diffused light. These spectral manipulations are utilized to promote desired physiological responses in ornamental plants, vegetables, and fruit trees. Thus, growing vegetables, fruits, and ornamental crops under certain colored shading nets can increase their yields and improve their quality (Shahak, Gal, Offir, & Ben-Yakir, 2008; Shahak et al., 2009).

Although shading nets permit free passage of small size pests, the infestation levels of aphids and whiteflies in "walk-in" tunnels covered by either the yellow or pearl nets were consistently two- to threefolds lower than in tunnels covered by the black or red nets. The reduced level of insect infestation resulted in 2- to 10-folds lower infection rates of pepper plants by CMV under yellow or pearl nets compared to disease incidence under black or red nets, where infection level ranged between 35% and 89%. Similarly, the incidence of necrotic PVY in tomato grown under black or red nets ranged between 42% and 50%, two- to threefolds higher than under the yellow or pearl nets. Yellow or pearl nets also provided efficient protection against the whitefly-borne, TYLCV in tomato. Disease incidence under these nets was two- to fourfolds lower compared to tomato plants grown under the black or red nets

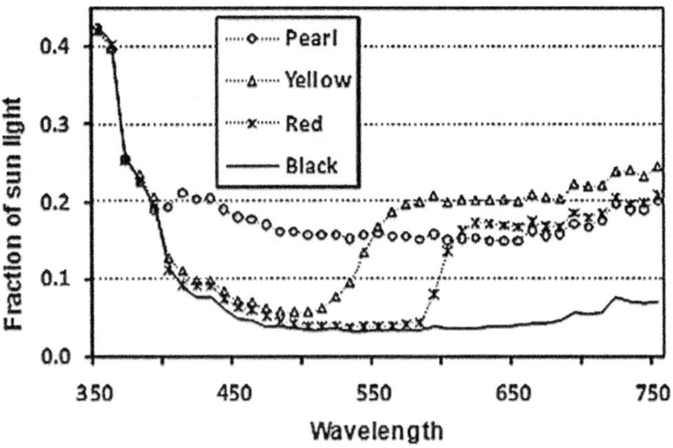

Figure 1.12 Sunlight reflectance profile from colored 35% shading nets in the UV and visible ranges.

where disease incidence ranged between 15% and 50% (Antignus et al., 2009; Ben-Yakir, Antignus, Offir, & Shahak, 2012; Shahak et al., 2008).

Apparently, the protection mechanism of shading nets is associated with insect repellence by the intensive light reflection from the nets. Indeed, spectral analysis of the reflection profiles of the different nets showed clearly the high light reflection by the yellow and pearl nets, in the visible range of the spectrum, compared to the low reflection characteristics the ordinary black net and the red one (Fig. 1.12) (Ben-Yakir et al., 2012). The protection against nonpersistent viruses (PVY and CMV) is probably a result of both the deterring effect on invading aphids due to light reflection and the mechanical effect formed by the nets in delaying aphid landing on the crop beyond the relatively short retention time of these viruses.

7. REFLECTIVE FILMS FORMED BY WHITEWASHES

Whitewash spray has been shown to be effective in preventing the colonization of plants by aphids (Bar Joseph & Frenkel, 1983) and preventing vector-borne virus and spiroplasma diseases (Yokomi, Bar-Joseph, Oldfield, & Gumpf, 1981). Marco (1986) showed that weekly sprays with 15% whitewashes of Loven and Yalbin reduced the number of landing aphids on field grown potatoes by 30–50%. PLRV and PVY incidence in tubers harvested from whitewashed plants was 0–61% and 0–68% lower, respectively, compared to untreated plots. However, whitewashes reduced the tuber yield by about 30%.

Kaolin particle film technology has been proposed as a new measure against several arthropod and disease pests (Glenn, Puterka, van der Zwet, Byers, & Feldhake, 1999). Hydrophilic processed kaolin, composed mainly of kaolinite (Al_2 [$(OH)_2$, Si_2O_5]), was developed as a commercial product named "surround WP" crop protectant (Engelhard Corporation, Iselin, NJ, USA). It was approved by the US Food and Drug Administration (FDA) as a human food additive. "Surround WP" was registered by the US Environmental Protection Agency (EPA) in 2000. Additionally, the Organic Materials Review Institute (OMRI) in the United States has listed it for use in organic production. This processed kaolin is a nonabrasive, nontoxic sprayable particle barrier which effectively controlled psyllid *Cacopsylla pyricola* (Foerster) on pear; *Diaphorina citri* (Kuwayama) on citrus (Hall, Lapointe, & Wenninger, 2007; Puterka, Glenn, & Pluta, 2005); codling moth *Cydia pomonella* (L.) on pear (Unruh, Knight, Upton, Glenn, & Puterka, 2000); the oblique banded leafroller moth Choristoneura roseceana (Knight, Unruh, Christianson, Puterka, & Glenn, 2000) Mediterranean fruit fly *Ceratitis capitata* (Wiedemann) on apple (Mazor & Erez, 2004); thrip *T. tabaci* (Lindeman) on onions (Larentzaki, Shelton, & Plate, 2008); aphids *M. persicae* on peach (Karagounis, Kourdoumbalos, Margaritopoulos, Nanos, & Tsitsipis, 2006); and the spirea aphid *Aphis citricola* (van der Goot) (Bar Joseph & Frenkel, 1983). Kaolin has also horticultural benefits of reducing sunburns and heat stress in fruit trees (Glenn, Prado, Erez, McFerson, & Puterka, 2002). Kaoline film also affects some biological characteristics of the insect, as oviposition rate of *T. tabaci* and hatch rate in onions were significantly reduced on kaolin-treated plants. The time required for development of larval stages was significantly increased, and mortality was significantly higher on kaolin than on water-treated onion leaves. Feeding choice was influenced by the presence of the kaolin treatment and in choice assays both larvae and adults fed significantly less on kaolin-treated than on water-treated leaves. In a field study, significantly more adults were captured in the beginning of the season on control than on kaolin-treated plots, and at population peaks, significantly more larvae and adults were harbored in control plots (Larentzaki et al., 2008).

The mode of action of whitewashes can be explained by the increased light reflectance from the leaf surface, 130–250% higher for leaves sprayed by Loven and Yalbin whitewashes (Marco, 1986). Interestingly, different aphids responded differently to the whitewashes treatment. *A. citricola* (van der Goot) was repelled to a greater extent compared to other aphid species, while *A. gosypii* was attracted to treated leaves (Marco, 1986). Another mechanism that may work in insect repellency by whitewashes is the contact between the insect and the whitewash particles that are present on the leaf surface. The whitewash film or powder may irritate the sensilla located at the insects' tarsus and induce a takeoff response from the treated plant canopy (Glenn et al., 1999). The use of whitewashes is relevant only to dry climates where the whitewash film is not washed off by frequent rains.

8. PROSPECTS AND OUTLOOKS

This review summarized the research and international efforts to develop and implement alternatives for chemical control of insect pests and virus diseases. The array of optical barriers described are based on insects ability to perceive light signals that drastically affect their flight orientation and landing as well as their interaction with the host plants in terms of feeding and propagation rates. Light plays also a central role in the metabolism, physiology, and development of plants. A comprehensive study is required to shed light on how different parts of the solar light spectrum affect the physiology of plants and insects and how it affects their interaction. More research should be directed to understand how light manipulation is affecting the ecology of other organisms that occupy the crop habitat. A better understanding of these parameters may help to design light filters and reflectors to manipulate light signals in order to improve crop yields and better environmental friendly, plant protection technologies.

REFERENCES

A'Brook, J. (1968). The effect of plant spacing on the numbers of aphids trapped over the groundnut crop. *Annals of Applied Biology, 61*, 289–294.

Adamse, P., Reed, H. E., Krizek, D. T., Britz, S. J., & Mirecki, R. M. (1997). An inexpensive setup for assessing the impact of ambient solar ultraviolet radiation on seedlings. *Journal of Natural Resources and Life Science Education, 26*, 139–144.

Antignus, Y. (2010). Optical manipulations block the spread of Bemisia tabaci in greenhouses and the open field. In P. A. Stansly, & S. E. Naranjo (Eds.), *Bemisia: Bionomics and management of a global pest* (pp. 349–356). Dordrecht Heidelberg London New York: Springer Science+Business Media BV.

Antignus, Y. (2012). Control methods of virus diseases in the Mediterranean Basin. In G. Loebenstein, & H. Lecoq (Eds.), *Advances in virus research. Viruses and virus diseases of vegetables in the Mediterranean Basin* (pp. 349–356). Oxford, UK: Elsevier.

Antignus, Y., & Ben-Yakir, D. (2004). Greenhouse photoselective cladding materials serve as an IPM tool to control the spread of insect pests and their vectored viruses. In R. Horowitz, & Y. Ishaya (Eds.), *Insect pest management* (pp. 319–335). Berlin: Springer.

Antignus, Y., Ben-Yakir, D., Offir, Y., Messika, Y., Dombrovsky, A., Chen, M., et al. (2009). Colored shade nets form optical barrier protecting pepper and tomato crops against aphid-borne non-persistent viruses. *Sade Va'Yerek, 12*, 60–62 (in Hebrew).

Antignus, Y., Lachman, O., Leshem, Y., Matan, E., Yehezkel, H., & Messika, Y. (1999). Protection efficiency of UV-absorbing films in greenhouses with vertical walls. In O. Zeidan (Ed.), *Bulletin of the Israeli extension service. Summary of research projects and field experiments in tomato crops for 1999* (pp. 29–39). Bet Dagan: Israeli Ministry of Agriculture.

Antignus, Y., Lachman, O., Pearlsman, M., Koren, A., Matan, E., Tregerman, M., et al. (2004). Development of an IPM system to reduce the damage of squash leaf curl begomovirus in zucchini squash crops. (Abstract), *Compendium, 2nd European Whitefly Symposium*, Cavtat, Croatia.

Antignus, Y., Lapidot, M., Hadar, D., Messika, Y., & Cohen, S. (1998). UV absorbing screens serve as optical barriers to protect vegetable crops from virus diseases and insect pests. *Journal of Economic Entomology, 91*, 1401–1405.

Antignus, Y., Mor, N., Ben Joseph, R., Lapidot, M., & Cohen, S. (1996). Ultraviolet-absorbing plastic sheets protect crops from insect pests and from virus diseases vectored by insects. *Environmental Entomology, 25*, 919–924.

Antignus, Y., Nestel, D., Cohen, S., & Lapidot, M. (2001). Ultraviolet-deficient greenhouse environment affects whitefly attraction and flight behavior. *Environmental Entomology, 30*, 394–399.

Antignus, Y., Lachman, O., & Pearlsman, M. (2005). Light manipulation by soil mulches protect crops from the spread of Begomoviruses. In *Abstracts of the IX International Plant virus Epidemiology Symposium*. Lima: Peru.

Avidov, Z. (1956). Bionomics of the tobacco whitefly (*Bemisia tabaci* Gennad.) in Israel. *Ktavim*, 7, 25–41.

Bar Joseph, M., & Frenkel, H. (1983). Spraying citrus plants with kaoline suspensions reduces colonization by the Spirea aphid (*Aphis citricola* van der Goot). *Crop Protection, 2*, 371–374.

Ben-Yakir, D., Antignus, Y., Offir, Y., & Shahak, Y. (2012). Colored shading nets impede insect invasion and decrease the incidences of insect-transmitted viral diseases in vegetable crops. *Entomologia Experimentalis et Applicata, 144*, 249–257.

Ben-Yakir, D., Hadar, M. D., Offir, Y., Chen, M., & Tregerman, M. (2008). Protecting crops from pests using OptiNet® and ChromatiNet® shading nets. *Acta Horticulturae, 770*, 205–212.

Berlinger, M. J., Dahan, R., Mordechi, S., Liper, A., Katz, J., & Levav, N. (1991). The use of nets to prevent the penetration of *Bemisia tabaci* into greenhouse. *Hassadeh, 71*, 1579–1583 (in Hebrew).

Borst, A. (2009). Drosophila's view. *Insect Vision Current Biology, 19*, 36–47. http://dx.doi.org/10.1016/j.cub.2008.11.001, Elsevier Ltd All rights reserved.

Chiel, E., Messika, Y., Steinberg, S., & Antignus, Y. (2006). The effect of UV-absorbing plastic sheet on the attraction and host location ability of three parasitoids: *Aphidius colemani, Diglyphus isaea* and *Eretmocerus mundus*. *Biocontrol, 51*, 65–78.

Chyzik, R., Dobrinin, S., & Antignus, Y. (2003). Effect of a UV-deficient environment on the biology and flight activity of *Myzus persicae* and its hymenopterous parasite *Aphidius matricariae*. *Phytoparasitica, 31*, 467–477.

Cohen, S. (1981). Reducing the spread of aphid transmitted viruses in peppers by coarse-net cover. *Phytoparasitica, 9*, 69–76.

Cohen, S. (1982). Control of whitefly vectors of viruses by color mulches. In K. F. Harris, & K. Maramorosch (Eds.), *Pathogens, vectors and plant diseases, approaches to control* (pp. 45–56). New York: Academic Press.

Cohen, S., & Berlinger, M. J. (1986). Transmission and cultural control of whitefly-borne viruses. *Agriculture and Ecosystem Environment, 17*, 89–97.

Cohen, S., & Marco, S. (1973). Reducing the spread of aphid transmitted viruses in pepper by trapping the aphids on sticky yellow polyethylene sheets. *Phytopathology, 63*, 1207–1209.

Cohen, S., & Melamed-Madjar, V. (1978). Prevention by soil mulching of the spread of tomato yellow leaf curl virus transmitted by *Bemisia tabaci* (Gennadius) (Hemiptera: Aleyrodidae). *The Israeli Bulletin of Entomological Research, 68*, 465–470.

Coombe, P. E. (1981). Wave length specific behaviour of the whitefly *Trialeurodes vaporariorum* (Homoptera: Aleyrodidae). *Journal of Comparative Physiology, 144*, 83–90.

Coombe, P. E. (1982). Visual behavior of the greenhouse whitefly, Trialeurodes vaporariorum. *Physiological Entomology*, 7, 243–251.

Costa, H. S., Newman, J., & Robb, K. L. (2003). Ultraviolet blocking greenhouse plastic films for management of insect pest. *HortScience, 38*, 465.

Costa, H. S., & Robb, K. (1999). Effects of ultraviolet-absorbing greenhouse plastic films on flight behavior of *Bemisia argentifolii* (Homoptera: Aleyrodidae) and *Frankliniella occidentalis* (Thysanoptera: Thripidae). *Journal of Economic Entomology, 92*, 557–562.

Costa, H. S., Robb, K. L., & Wilen, C. A. (2002). Field trials measuring the effects of ultraviolet-absorbing greenhouse plastic films on insect populations. *Journal of Economic Entomolology, 95*, 113–120.

Csizinszky, A. A., Schuster, D. J., & Kring, J. B. (1995). Color mulches influence yield and insect pest populations in tomatoes. *Journal of the American Society of Horticultural Science, 120*, 778–784.

Cybulski, W. J. I., & Peterjohn, W. T. (1999). Effects of ambient UV-B radiation on the above-ground biomass of seven temperate-zone plant species. *Plant Ecology, 145*, 175–181.

Diaz, B. M., Biurrun, R., Moreno, A., Nebreda, M., & Fereres, A. (2006). Impact of ultraviolet-blocking plastic films on insect vectors of virus diseases infecting crisp lettuce. *HortScience, 41*, 711–716.

Diaz, B. M., & Fereres, A. (2007). Ultraviolet-blocking materials as a physical barrier to control insect pests and plant pathogens in protected crops. *Pest Technology, 1*, 85–95.

Doring, T. F., & Chittka, L. (2007). Visual ecology of aphids—A critical review on the role of colors in host finding. *Arthropod-Plant Interactions, 1*, 3–16.

Doukas, D. (2002). Impact of spectral cladding materials on the behavior of glasshouse whitefly *Trialeurodes vaporariorum* and *Encarsia formosa*, its Hymenopteran parasitoid. In *British crop protection conference, pests and diseases 2002, Brighton* (pp. 773–776).

Doukas, D., & Payne, C. C. (2007a). The use of ultraviolet blocking films in insect pest management in the UK; effect of naturally occurring arthropod pest and natural enemy populations in a protected cucumber crop. *Annals of Applied Biology, 151*, 221–231.

Doukas, D., & Payne, C. C. (2007b). Greenhouse whitefly (Homoptera: Alerodidae) dispersal under different UV environments. *Journal of Economic Entomology, 100*, 389–397.

Doukas, D., & Payne, C. C. (2007c). Effects of UV-blocking films on the dispersal behavior of *Encarsia formosa* (Hymenoptera: Aphelinidae). *Journal of Economic Entomology, 100*, 110.

Dyer, A. G., & Chittka, L. (2004). Bumblebee search time without ultraviolet light. *Journal of Experimental Biology, 207*, 1683–1688.

Edser, C. (2002). Light manipulation additives extend opportunities for agricultural plastic films. *Plastic Additive Compounds, 4*, 20–24.

Espí, E., Salmerón, A., Fontecha, A., García, Y., & Real, A. I. (2006). Plastic films for agricultural applications. *Journal of Plastic Film and Sheeting, 22*, 85–101.

Fiscus, E. L., Philbeck, R., & Britt, A. B. (1999). Growth of Arabidopsis flavonoid mutants under solar radiation and UV filters. *Environmental and Experimental Botany, 41*, 231–245.

Fukushi, T. (1990). Colour discrimination from various shades of grey in the trained blowfly, Lucilia cuprina. *Journal of Insect Physiology, 36*, 69–75.

Gerling, D., & Horowitz, A. R. (1984). Yellow traps for evaluating the population levels and dispersal patterns of *Bemisia tabaci* (Gennadius) (Homoptera: Aleyrodidae). *Annual Meeting of the Entomological Society of America, 77*, 753–759.

Glenn, D. M., Prado, E., Erez, A., McFerson, J., & Puterka, G. J. (2002). A reflective, processed-kaolin particle film affects fruit temperature, radiation reflection, and solar injury in apple. *Journal of the American Society of Horticultural Science, 127*, 188–193.

Glenn, D. M., Puterka, F., van der Zwet, J. T., Byers, R. E., & Feldhake, C. (1999). Hydrophobic particle films: A new paradigm for suppression of arthropod pests and plant diseases. *Journal of Economic Entomolology, 92*, 759–771.

Gonzales, A., Rodriguez, R., Bafion, S., Franco, J., & Fernandez, J. A. (2001). The influence of photoselective plastic films as a greenhouse cover on sweet pepper yield and on insect pest levels. *Acta Horticulturae, 559*, 233–238.

Gonzalez, D., & Rawlin, W. A. (1968). Aphid sampling efficiency of Moericke traps affected by height and background. *Journal of Economic Entomology, 61*, 109–114.

Hall, D. G., Lapointe, S. L., & Wenninger, E. J. (2007). Effects of a particle film on biology and behavior of *Diaphorina citri* (Hemiptera: Psyllidae) and its infestations in citrus. *Journal of Economic Entomology, 100*, 847–854.

Harpaz, I. (1982). Non-pesticidal control of vector-borneviruses. In K. F. Harris, & K. Maramorosch (Eds.), *Pathogens, vectors, and plant diseases: Approaches to control* (pp. 1–21). New York: Academic Press.

Hefez, A., Izikovitch, D., & Dag, A. (1999). Effects of UV-absorbing films on the activity of pollinators (honey bees and bumble bees). *A report to the chief scientist. Israeli Ministry of Agriculture*, Bet-Dagan ID code 891-0117-96.

Heinze, S., & Homberg, U. (2007). Map like representation of celestial E-vector orientations in the brain of an insect. *Science, 315*, 995–997.

Hunt, D. M., Wilkie, S. E., Bowmaker, J. K., & Poopalasundaram, S. (2001). Vision in the ultraviolet. *Cellular and Molecular Life Sciences, 58*, 1583–1598.

Issacs, R., Willis, M. A., & Byrne, D. N. (1999). Modulation of whitefly take-off and flight orientation by wind speed and visual cues. *Physiological Entomology, 28*, 311–318.

Kajita, H. (1986). Parasitism of the greenhouse whitefly, *Trialeurodes vaporariorum* (Westwood) (Homoptera: Aleurodidae) by *Encarsia Formosa* (Hymenoptera: Aphelinidae) in a greenhouse covered with near-ultraviolet absorbing vinyl film. *Proceedings of Plant Protection, Kyushu, 32*, 155–157 (in Japanese).

Karagounis, C., Kourdoumbalos, A. K., Margaritopoulos, J. T., Nanos, G. D., & Tsitsipis, J. A. (2006). Organic farming-compatible insecticides against the aphid *Myzus persicae* (Sulzer) in peach orchards. *Journal of Applied Entomology, 130*, 150–154.

Kevan, P. G., Straver, W. A., Offer, O., & Laverty, T. M. (1991). Pollination of greenhouse tomatoes by bumblebees in Ontario. *Proceedings of the Entomological Society of Ontario, 122*, 15–19.

Kigathi, R. (2005). Effect of UV absorbing greenhouse covering materials and UV reflecting mulches on the immigration and dispersal of Western flower thrips. *IPP Int. Msc, 18*.

Kigathi, R., & Poehling, H. M. (2012). UV-absorbing films and nets affect the dispersal of western thrips, *Frankliniella occidentalis*. *Journal of Applied Entomology, 136*, 761–771.

Kittas, C., Tchamitchian, M., Katsoulas, N., Karaiskou, P., & Papaioannou, C. H. (2006). Effect of two UV-absorbing greenhouse-covering films on growth and yield of an eggplant soilless crop. *Scientia Horticulturae, 110*, 30–37.

Klatt, B. K., Holzschuh, A., Westphal, C., Clough, Y., Smit, I., Pawelzik, E., et al. (2014). Bee pollination improves crop quality, shelf life and commercial value. *Proceedings of the Royal Society, 281*, 2013–2440. http://dx.doi.org/10.1098/rspb.2013.2440.

Knight, A. L., Unruh, T. R., Christianson, B. A., Puterka, G. J., & Glenn, D. M. (2000). Effects of a kaolin-based particle film on oblique banded leafroller (Lepidoptera: Tortricidae). *Journal of Economic Entomolology, 93*, 744–749.

Kring, J. B. (1972). Flight behavior of aphids. *Annual Review of Entomology, 17*, 461–492.
Krizek, D. T., Britz, S. J., & Mirecki, R. M. (1998). Inhibitory effects of ambient levels of solar UV-A and UV-B radiation on growth of cv. New red fire lettuce. *Plant Physiology, 103*, 1–7.
Krizek, D. T., Mirecki, R. M., & Britz, S. J. (1997). Inhibitory effects of ambient levels of solar UV-A and UV-B radiation on growth of cucumber. *Plant Physiology, 100*, 886–893.
Kumar, P., & Poehling, H. M. (2006). UV-blocking plastic films and nets influence vectors and virus transmission on greenhouse tomatoes in the humid tropics. *Environmental Entomology, 35*, 1069–1082.
Larentzaki, E., Shelton, A. M., & Plate, J. (2008). Effect of kaolin particle film on *Thrips tabaci* (Thysanoptera: Thripidae), oviposition, feeding and development on onions: A lab and field case study. *Crop Protection, 27*, 727–734.
Legarrea, S., Betancourtb, M., Plazaa, M., Fraile, A., García-Arena, F., & Fereres, A. (2012). Dynamics of nonpersistent aphid-borne viruses in lettuce crops covered with UV-absorbing nets. *Virus Research, 165*, 1–8.
Legarrea, S., Diaz, B. M., Morales, I., Vinuela, E., & Fereres, A. (2008). Effect of UV absorbing nets in the spread and population growth of the potato aphid. *HortScience, 43*, 1249.
Legarrea, S., Karnieli, A., Fereres, A., & Weintraub, P. G. (2010). Comparison of UV absorbing nets in pepper crops: Spectral properties, effects on plants and pest control. *Photochemistry & Photobiology, 86*, 324–330.
Legarrea, S., Weintraub, P. G., Plaza, M., Viñuela, E., & Fereres, A. (2012). Dispersal of aphids, whiteflies and their natural enemies under photoselective nets. *BioControl, 57*, 523–532. http://dx.doi.org/10.1007/s10526-011-9430-2.
Leigh, M. (2004). *Ultraviolet blocking greenhouse polyethylene covers for insect pest crops: May 2003–September 2004*. www.organic.aber.ac.uk/library/UVBlockingpolytunnels.pdf.
Loebenstein, G., Alper, M., Levy, S., & Menagem, E. (1975). Protecting peppers from aphid-borne viruses with aluminium foil or plastic mulch. *Phytoparasitica, 3*, 43–53.
Marcel, A. K., Jansen, M. A. K., Gaba, V., & Greenberg, B. M. (1998). Higher plants and UV-B radiation: Balancing damage, repair and acclimation. *Trends in Plant Science, 3*, 131–135.
Marco, S. (1986). Incidence of aphid transmitted virus infections reduced by whitewashes sprays on plants. *Phytopathology, 76*, 1344–1348.
Matteson, N., Terry, I., Ascoli, C. A., & Gilbert, C. (1992). Spectral efficiency of the western flower thrips, *Frankliniella occidentalis*. *Journal of Insect Physiology, 38*, 453–459.
Mazokhin-Porshnykov, G. A. (1969). Structure of faceted eyes and visual centers. In T. H. Goldsmith (Ed.), *Insect vision* (pp. 1–19). New York: Plenum Press.
Mazor, M., & Erez, A. (2004). Processed kaolin protects fruits from Mediterranean fruit fly infestations. *Crop Protection, 23*, 47–51.
Mazza, C. A., Izaguirre, M. M., Zavala, J., Scopel, A. L., & Ballaré, C. L. (2002). Insect perception of ambient ultraviolet-B radiation. *Ecological Letters, 5*, 722–726.
McEnrone, W. D., & Dronka, K. (1966). Color vision in the adult female two-spotted spider mite. *Science, 154*, 782–784.
Mellor, H. E., Bellingham, J., & Anderson, M. (1997). Spectral efficiency of the glasshouse whitefly *Trialeurodes vaporariorum* and *Encarsia formosa* its Hymenopteran parasitoid. *Entomologia Experimentalis et Applicata, 83*, 11–20.
Messika, Y., Antignus, Y., Lapidot, M., Ben Yakir, D., Chen, M., & Zimmerman, C. (1999). Effect of UV filtration on the spread of insect pests and spraying regime in green herbs grown in 'walk-in' tunnels. *Gan Sadeh Vameshek*, 53–55 (in Hebrew).
Messika, Y., Lapidot, M., & Antignus, Y. (1997). The effect of UV-absorbing sheets on the infestation of green spice crops with insect pests. *Hassadeh, 77*, 27–31 (in Hebrew).

Messika, Y., Nishri, Y., Gokkes, M., Lapidot, M., & Antignus, Y. (1998). UV-absorbing films and Aluminet screens—An efficient control means to block the spread of insect and viral pests in Lisianthus. *Dapey Meyda, the Flower Grower Magazine, 13*, 55–57 (in Hebrew).

Mizrahi, S., Sacs, Y., Mor, N., Elad, Y., Reuveni, R., & Antignus, Y. (1998). Comparative study on the protection effects of commercial polyethylene films with different absorption spectra against insect, fungal and viral pests. *Gan Sadeh Vameshek*, 33–37 (in Hebrew).

Moericke, V. (1955). Neue untersuchungen uber das farbensehen der homopteran. In *Proceedings of conference on potato diseases 2nd meeting, Lisse-Wageningen, Netherlands, 1954* (pp. 55–69).

Moller, R. (2002). Insects could exploit UV-green contrast for landmark navigation. *Journal of Theoretical Biology, 214*, 619–631.

Morandin, L. A., Laverty, T. M., Kevan, P. G., Khosla, S., & Shipp, L. (2001). Bumble bee (Hymenoptera: Apidae) activity and loss in commercial tomato greenhouses. *Canadian Journal of Entomology, 133*, 883–893.

Morandin, L. A., Laverty, T. M., Kevan, P. G., Khosla, S., & Shipp, L. (2002). Effect of greenhouse polyethylene covering on activity level and photo-response of bumble bees. *Canadian Journal of Entomology, 134*, 539–549.

Mound, L. A. (1962). Studies on the olfaction and color sensitivity of *Bemisia tabaci* (GENN.) (Homoptera, Aleurodidae). *Entomologia Experimentalis et Applicata, 5*, 99–104.

Mutwiwa, U. N., Brogemeister, C., Von Elsner, B., & Tantu, H. (2005). Effects of UV absorbing plastic films on greenhouse whitefly (Homoptera: Aleyrodidae). *Journal of Economical Entomology, 98*, 1221–1228.

Nguyen, T. H. N., Borgemeister, C., Max, J., & Poehling, H.-M. (2009). Manipulation of ultraviolet light affects immigration behavior of *Ceratothripoides claratris* (Thysanoptera: Thripidae). *Journal of Economic Entomology, 102*, 1559–1566.

Onuma, K., & Nakagaki, S. (1982). The growth of vegetable crops and establishment of insect and mite pests in a plastic greenhouse treated to exclude near UV radiation. (1) The growth of pepper and cucumber. *Bulletin of Ibaraki-Ken Horticultural Experimental Station, 10*, 31–38 (in Japanese).

Onzo, A., Sabelis, M. W., & Hanna, R. (2010). Effects of ultraviolet radiation on predatory mites and the role of refugees in plant structures. *Environmental Entomology, 39*, 695–701.

Orozco-Santos, M., Perez-Zamora, O., & Lopez-Arriaga, O. (1995). Effect of transparent mulch on insect populations, virus diseases, soil temperature, and yield of cantaloupe in a tropical region. *New Zealand Journal of Crop and Horticultural Science, 23*, 199–204.

Paul, N. D., Jacobson, R. J., Taylor, A., Wargent, J. J., & Moore, J. P. (2005). The use of wavelength-selective plastic cladding materials in horticulture: Understanding of crop and fungal responses through the assessment of biological spectral weighting functions. *Photochemistry and Photobiology, 81*, 1052–1060.

Pressman, E., Moshkovitz, A., Rsenfeld, K., & Shaked, R. (1996). *The effects of UV-blocking films on quality parameters of tomato and pepper fruits*. Final report submitted to Ginegar plastic industries LTD, Israel.

Puterka, G. J., Glenn, D. M., & Pluta, R. C. (2005). Action of particle films on the biology and behavior of pear psylla (Homoptera: Psyllidae). *Journal of Economic Entomology, 98*, 2079–2088.

Raccah, B. (1986). Non-persistent viruses: Epidemiology and control. In K. Maramorosch, F. A. Murphy, & A. J. Shatkin (Eds.), *Advances in virus research* (pp. 387–429). New York: Academic Press.

Raccah, B., & Fereres, A. (March 2009). Plant Virus Transmission by Insects. In *Encyclopedia of Life Sciences (ELS)* Chichester: John Wiley & Sons, Ltd. http://dx.doi.org/10.1002/9780470015902.A0021525.a0000760.pub2.

Rapisarda, C., & Tropea-Garzia, G. (2002). Tomato yellow leaf curl Sardinia virus and its vector *Bemisia tabaci* in Sicilia (Italy): Present status and control. *OEPP/EPPO Bulletin, 32*, 25–29.

Raviv, M., & Antignus, Y. (2004). UV radiation effects on pathogens and insect pests of greenhouse-grown crops. *Photochemistry & Photobiology, 79*, 219–226.

Reuveni, R., & Raviv, M. (1992). The effect of spectrally-modified polyethylene films on development of *Botrytis cinerea* in greenhouse grown tomato plants. *Biology and Agricultural Horticulture, 9*, 77–89.

Robson, T. M., Pancotto, V. A., Flint, S. D., Ballare, C. L., Sala, O. E., Scopel, A. L., et al. (2003). Six years of solar UV-B manipulations affect growth of sphagnum and vascular plants in a Tierra del Fuego peat land. *New Phytology, 160*, 379–389.

Rozema, J., Björn, L. O., Bornman, J. F., Gaberscik, A., Häder, D. P., Trost, T., et al. (2002). The role of UV-B radiation in aquatic and terrestrial ecosystems an experimental and functional analysis of the evolution of UV-absorbing compounds. *Journal of Photochemistry & Photobiology, 66*, 2–12.

Seker, I. (1999). The use of UV-blocking films to reduce insect pests damage in tomato greenhouses and the effect of UV-filtration on the pollination activity of bumble bees. *Gan Sadeh Va'Meshek, 12*, 55–59 (in Hebrew).

Seliger, H. H., Lall, A. B., & Biggley, W. H. (1994). Blue through UV polarization sensitivities in insects. Optimizations for the range of atmospheric polarization conditions. *Journal of Comperative Physiology, 175*, 475–486.

Shahak, Y., Gal, E., Offir, Y., & Ben-Yakir, D. (2008). Photoselective shade netting integrated with greenhouse technologies for improved performance of vegetable and ornamental crops. *Acta Horticulturae, 797*, 75–80.

Shahak, Y., Ratner, K., Zur, N., Offir, Y., Matan, E., Yehezkel, H., et al. (2009). Photoselective netting: An emerging approach in protected agriculture. *Acta Horticulturae, 807*, 79–84.

Stavisky, J., Funderburk, J., Brodbeck, B. V., Olson, S. M., & Andersen, P. C. (2002). Population dynamics of *Frankliniella* spp. and tomato spotted wilt incidence as influenced by cultural management tactics in tomato. *Journal of Economic Entomology, 95*, 1216–1221.

Steinberg, S., Prag, H., Gouldman, D., Antignus, Y., Pressman, E., Asenheim, D., et al. (1997). The effect of ultraviolet-absorbing plastic sheets on pollination of greenhouse tomatoes by bumblebees. In *Proceedings of the international congress for plastics in agriculture (CIPA)*. Tel-Aviv: Israel.

Summers, C. G., Mitchell, J. P., & Stapleton, J. J. (2005). Mulches reduce aphid-borne viruses and whiteflies in cantaloupe. *California Agriculture, 59*, 90–94.

Suwwan, M. A., Akkawi, M. A., Al-Musa, M., & Mansour, A. (1988). Tomato performance and incidence of tomato yellow leaf curl (TYLC) virus as affected by type of mulch. *Scientia Horticulturae, 37*, 39–45.

Unruh, T. R., Knight, A. L., Upton, J., Glenn, D. M., & Puterka, G. J. (2000). Particle films for suppression of codling moth (Lepidoptera: Tortricidae) in apple and pear orchards. *Journal of Economic Entomology, 93*, 737–743.

Vaishampayan, S. M., Kogan, M., Waldbauer, G. P., & Wooley, J. T. (1975). Spectral specific responses in the visual behavior of the greenhouse whitefly, *Trialeurodes vaporariorum* (Homoptera: Aleurodidae). *Entomologia Experimentalis et Applicata, 18*, 344–356.

Vaishampayan, S. M., Waldbauer, G. P., & Kogan, M. (1975). Visual and olfactory responses in orientation to plants by the greenhouse whitefly, *Trialeurodes vaporariorum* (Homoptera: Aleurodidae). *Entomologia Experimentalis et Applicata, 18*, 412–422.

Van de Veire, M., & Vacante, V. (1984). Greenhouse whitefly control through the combined use of color attraction system with the parasite wasp *Encarsia formosa* (Hym. Aphelinidae). *Entomophaga, 29*, 303–310.

van der Blom, Jan. (2010). Applied entomology in Spanish greenhouse horticulture. *Proceedings of The Netherland Entomological Society Meeting*, *21*, 9–17.
Vernon, R. S., & Gillespie, D. R. (1990). Spectral responsiveness of *Frankliniella occidentalis* (Thysanoptera: Thripidae) determined by trap catches in greenhouses. *Environmental Entomology*, *19*, 1229–1241.
Wehner, R., & Labhart, T. (2006). Polarisation vision. In E. Warrent, & D. E. Nilsson (Eds.), *Invertebrate vision* (pp. 291–293). Cambridge University Press: Cambridge.
Weintraub, P. G., Pivonia, S., & Gera, A. (2008). Physical control of leafhoppers. *Journal of Economic Entomology*, *101*, 1337–1340.
Yokomi, R. H., Bar-Joseph, M., Oldfield, G. N., & Gumpf, J. J. (1981). A preliminary report of reduced infection by *Spiroplasma citri* and virescence in whitewash-treated periwinkle. *Phytopathology*, *81*, 914 (abstract).
Zimmerman-Gries, S. (1979). Reducing the spread of potato leaf roll virus, alfalfa mosaic virus and potato virus Y in seed potatoes by trapping aphids on sticky yellow polyethylene sheets. *Potato Research*, *22*, 123–131.

CHAPTER TWO

Transgenic Resistance

Fabrizio Cillo*, Peter Palukaitis[†,1]
*Istituto di Virologia Vegetale, CNR, Bari, Italy
[†]Department of Horticultural Sciences, Seoul Women's University, Seoul, Republic of Korea
[1]Corresponding author: e-mail address: scripath1@yahoo.co.uk

Contents

Abstract

Transgenic resistance to plant viruses is an important technology for control of plant virus infection, which has been demonstrated for many model systems, as well as for the most important plant viruses, in terms of the costs of crop losses to disease, and also for many other plant viruses infecting various fruits and vegetables. Different approaches have been used over the last 28 years to confer resistance, to ascertain whether particular genes or RNAs are more efficient at generating resistance, and to

Advances in Virus Research, Volume 90
ISSN 0065-3527
http://dx.doi.org/10.1016/B978-0-12-801246-8.00002-0

take advantage of advances in the biology of RNA interference to generate more efficient and environmentally safer, novel "resistance genes." The approaches used have been based on expression of various viral proteins (mostly capsid protein but also replicase proteins, movement proteins, and to a much lesser extent, other viral proteins), RNAs [sense RNAs (translatable or not), antisense RNAs, satellite RNAs, defective-interfering RNAs, hairpin RNAs, and artificial microRNAs], nonviral genes (nucleases, antiviral inhibitors, and plantibodies), and host-derived resistance genes (dominant resistance genes and recessive resistance genes), and various factors involved in host defense responses. This review examines the above range of approaches used, the viruses that were tested, and the host species that have been examined for resistance, in many cases describing differences in results that were obtained for various systems developed in the last 20 years. We hope this compilation of experiences will aid those who are seeking to use this technology to provide resistance in yet other crops, where nature has not provided such.

1. INTRODUCTION

The year 2014 marks 30 years since the regeneration of the first transgenic plants (De Block, Herrera-Estrella, Van Montagu, Schell, & Zambryski, 1984; Horsch et al., 1984) and 28 years since the first report of transgenic plants expressing plant viral sequences and showing disease protection (Powell Abel et al., 1986). Since the latter report, many types of sequences of different viruses have been used to confer resistance to various extents against plant viruses; many successful, but not all. These include expression of proteins (capsid proteins, replication-related proteins, movement proteins, and other viral-encoded proteins) and various RNAs (satellites of plant viruses, defective-interfering RNAs, noncoding RNAs, antisense RNAs, ribozymes, double-stranded RNAs (dsRNAs), inverted repeat RNAs (irRNAs), and artificial microRNAs (amiRNAs)). In addition, nonviral sequences also have been introduced into transgenic plants to obtain resistance or protection against infection by plant viruses. These include RNases, antiviral inhibitor proteins, plantibodies, plant defense response elicitors, and various host-derived genes, including natural resistance genes. The use of various sequences and the mechanisms of resistance (Castel & Martienssen, 2013; Csorba, Pantaleo, & Burgyán, 2009; Dietzgen & Mitter, 2006; Ding, 2010; Duan, Wang, & Guo, 2012; Eames, Wang, Smith, & Waterhouse, 2008; Pumplin & Voinnet, 2013; Vanderschuren, Stupak, Fütterer, Gruissem, & Zhang, 2007; Wang, Masuta, Smith, & Shimura, 2012), as well as the successful application plus various biosafety-related and political issues arising from use of this technology have been considered in detail in a number of reviews (Collinge, Jørgensen, Lond, & Lyngkjaer, 2010; Fuchs et al., 2007; Fuchs & Gonsalves, 2007; Gottula & Fuchs, 2009; Prins et al., 2008; Ready, Sudarshana, Fuchs, Rao, & Thottappilly, 2009; Thompson & Tepfer, 2010). However, there has not been a review that has covered all of the technical approaches used to date, in light of the current status of knowledge of the mechanisms of RNA silencing. In this review, we will not focus on the mechanisms of resistance, which have been covered adequately, but will consider these various systems and the shifts in approaches that have occurred in the last 28 years, in terms of the technologies being used and those under consideration at present, to obtain resistance to virus infection in model systems and in various crops. Although there are many approaches that have been developed, few have made it to the field for evaluation, and even fewer have been deployed. The reasons for this will be not considered here, as they relate largely to various perceived and imagined risks, as well as political issues, which have been discussed in other recent reviews (Collinge et al., 2010; Fuchs et al., 2007; Fuchs & Gonsalves, 2007; Gottula & Fuchs, 2009; Prins et al., 2008; Ready et al., 2009; Thompson & Tepfer, 2010).

It should be noted that historically many transgenic plants were generated to express viral-encoded proteins, in the expectations that any resistance obtained would be protein-mediated. In some cases, these expectations were met, while in many other cases the resistance was shown to be RNA-mediated, or a combination of protein-mediated and RNA-mediated; however, in many cases, the basis of the inhibition was not established conclusively. Hence, the various sections and subsection below relate to what was expressed in the transgenic plants and not the nature of the specific resistance mechanism(s).

2. VIRAL PROTEIN-MEDIATED RESISTANCE

2.1. Coat protein-mediated resistance

The conception of pathogen-derived resistance has several fathers (Hamilton, 1980; Sanford & Johnston, 1985; Sequeira, 1984), but only one mother (Powell Abel et al., 1986). The pioneering work of the laboratory of Roger Beachy, using the cutting-edge technology of the time available from the Monsanto Corporation, generated transgenic tobacco (*Nicotiana tabacum*) plants expressing the capsid protein encoding sequences of *Tobacco mosaic virus* (TMV) and demonstrated a partial resistance to TMV, largely manifested as a delayed infection (Powell Abel et al., 1986). Subsequent transformation of tomato with a similar construct resulted in stronger resistance to infection by TMV in the field (Nelson et al., 1988). Interestingly, this was not the first report demonstrating that the TMV capsid protein could be expressed transgenically in tobacco. That report (Bevan, Mason, & Goelet, 1985) demonstrated capsid protein expression, albeit at a very low level, but did not assess resistance to infection by TMV. However, Bevan et al. (1985) did demonstrate the first use of the *Cauliflower mosaic virus* (CaMV) 35S promoter for expression of other viral sequences which has since been used for expression of genes and RNA sequences in the overwhelming majority of the transgenic plants generated and assessed for virus resistance over the last 29 years.

Soon after the initial report by Powell Abel et al. (1986), numerous publications appeared expressing capsid protein genes of various viruses in tobacco and demonstrating protection or resistance (Table 2.1). In at least six cases, the capsid protein gene of a potyvirus was expressed in a nonhost plant (tobacco) of that virus, conferring resistance to other potyviruses (Dinant, Blaise, Kusiak, Astier-Manifacier, & Albouy, 1993; Fang & Grumet, 1993; Ling, Namba, Gonsalves, Slightom, & Gonsalves, 1991; Ravelonandro, Minsion, Delbos, & Dunez, 1993; Sohn et al., 2004; Stark & Beachy, 1989). In one case, the nucleocapsid (N) proteins of three tospoviruses [*Tomato spotted wilt virus* (TSWV), *Tomato chlorotic spot virus* (TCSV), and *Groundnut ringspot virus* (GRSV)] were expressed together in transgenic tobacco conferring resistance to all three viruses (Prins et al., 1995).

In some cases, *Nicotiana benthamiana* was the preferred plant for assessment, especially if the viruses assessed did not infect tobacco efficiently, or at all (Table 2.2). Transgenic *Nicotiana debneyii* plants expressing the capsid protein gene of *Potato virus S* (PVS) also were shown to be resistant to infection by PVS (Mackenzie & Tremaine, 1990), as were transgenic *Nicotiana clevelandii* plants expressing the capsid protein coding sequences of *Plum pox virus* (PPV) to infection by PPV (da Câmara Machado et al., 1994; Ravelonandro et al., 1993; Regner et al., 1992). Additionally, transgenic *Nicotiana occidentalis* plants expressing the capsid protein genes of *Grapevine virus A*, *Grapevine virus B* (Minafra et al., 1998), or *Apple chlorotic leaf spot virus* (ACLSV) to infection by the corresponding viruses (Yoshikawa et al., 2000).

With regard to what was referred to as either coat protein-mediated protection or coat protein-mediated resistance in those early years, as technology advanced and allowed transformation and the regeneration of many

Table 2.1 Coat protein-mediated resistance to viruses in transgenic tobacco (*Nicotiana tabacum*)

Virus	Capsid protein gene status	References
Alfalfa mosaic virus (AMV)	Translatable	Tumer et al. (1987)
	Translatable	Van Dun, Bol, and Van Vloten-Doting (1987)
	Translatable	Xu, Collins, Hunt, and Nielsen (1998)
	Translatable	Loesch-Fries et al. (1987)
	Translatable/ nontranslatable	Van Dun, Overduin, Van Vloten-Doting, and Bol (1988)
	Translatable/ antisense	Jayasena, Ingham, Hajimorad, and Randles (1997)
Andean potato mottle virus (APMoV)	Translatable/ nontranslatable	Neves-Borges et al. (2001)
Arabis mosaic virus (ArMV)	Translatable	Bertioli, Cooper, Edwards, and Hawes (1992)
	Translatable	Cooper, Edwards, Rosenwasser, and Scott (1994)
Cherry leaf roll virus (CLRV)	Nontranslatable	Cooper et al. (1994)
Cucumber mosaic virus (CMV)	Translatable	Namba, Ling, Gonsalves, Gonsalves, and Slightom (1991)
	Translatable	Quemada, Gonsalves, and Slightom (1991)
	Translatable	Nakajima, Hayakawa, Nakamura, and Suzuki (1993)
(+Satellite RNA)	Translatable	Yie et al. (1992) and Yie et al. (1995)
	Translatable	Okuno, Nakayama, Yoshida, Furusawa, and Komiya (1993)
	Translatable	Ryu, Lee, Park, Lee, and Park (1998)
	Translatable	Jacquemond, Teycheney, Carrère, Navas-Castillo, and Tepfer (2001)
	Nontranslatable	Rezaian, Skene, and Ellis (1988)
	Nontranslatable	Tan et al. (2012)

Continued

Table 2.1 Coat protein-mediated resistance to viruses in transgenic tobacco (*Nicotiana tabacum*)—cont'd

Virus	Capsid protein gene status	References
(+Movement protein gene)	Antisense	Kim, Kim, and Paek (1995)
	Translatable/ antisense	Cuozzo et al. (1988)
Grapevine chrome mosaic virus (GCMV)	Translatable	Brault et al. (1993)
Pepper mild mottle virus (PMMoV)	Translatable	Lim et al. (1997)
Pepper severe mosaic virus (PepSMV)	Translatable	Rabinowicz et al. (1998)
Physalis mottle virus (PhMV)	Translatable	Ranjith-Kumar et al. (1999)
Potato aucuba mosaic virus (PAMV)	Translatable/ antisense	Leclerc and AbouHaidar (1995)
Potato virus X (PVX)	Translatable	Wang, Yang, Pan, and Chen (1993)
	Translatable	Spillane, Verchot, Kavanagh, and Baulcombe (1997)
	Translatable	Bazzini, Hopp, Beachy, and Asurmendi (2006)
(+Movement protein gene)	Nontranslatable	Liu, Zhang, et al. (2005)
	Translatable/ antisense	Hemenway, Fang, Kamiewski, Chua, and Tumer (1988)
Potato virus Y (PVY)	Translatable	Kollár, Thole, Dalmay, Salamon, and Balázs (1993)
	Translatable	Malnoë, Farinelli, Collet, and Reust (1994)
	Translatable	Sudarsono et al. (1995)
	Translatable	Han et al. (1999)
	Translatable	Józsa, Stasevski, and Balázs (2002)
	Nontranslatable	Guo et al. (2001)

Table 2.1 Coat protein-mediated resistance to viruses in transgenic tobacco (*Nicotiana tabacum*)—cont'd

Virus	Capsid protein gene status	References
	Nontranslatable	Masmoudi, Yacoubi, Hassairi, Elarbi, and Ellouz (2002)
	Nontranslatable	Pourrahim, Ahoonmanesh, Hashemi, Zeinali, and Farzadfar (2006)
	Translatable/ nontranslatable	Van der Vlugt, Ruiter, and Goldbach (1992)
	Translatable/ nontranslatable	Van der Vlugt and Goldbach (1993)
	Trans/ nontrans/ antisense[a]	Farinelli and Malnoë (1993)
	Trans/ nontrans/ antisense	Smith, Swaney, Parks, Wernsman, and Dougherty (1994)
PVX and PVY	Translatable	Song et al. (1994)
	Nontranslatable	Bai et al. (2005)
Strawberry latent ringspot virus (SLRV)	Translatable	Kreiah et al. (1996)
Tobacco etch virus (TEV)	Translatable	Lindbo, Silva-Rosales, Proebsting, and Dougherty (1993)
	Translatable	Silva-Rosales, Lindbo, and Dougherty (1994)
	Nontranslatable	Doughterty et al. (1994)
	Trans/ nontrans/ antisense	Lindbo and Dougherty (1992a) and Lindbo and Dougherty (1992b)
Tobacco mosaic virus (TMV)	Translatable	Powell Abel et al. (1986)
	Antisense	Powell, Stark, Sanders, and Beachy (1989)
Tobacco necrosis virus (TNV)	Translatable/ antisense	Hackland, Coetzer, Rybicki, and Thomson (2000)

Continued

Table 2.1 Coat protein-mediated resistance to viruses in transgenic tobacco (*Nicotiana tabacum*)—cont'd

Virus	Capsid protein gene status	References
Tobacco rattle virus (TRV)	Translatable	Van Dun and Bol (1988)
	Translatable	Angenent, van den Ouweland, and Bol (1990)
	Translatable	Ploeg, Mathis, Bol, Brown, and Robinson (1993)
Tobacco ringspot virus (TRSV)	Translatable	Zadeh and Foster (2004)
Tobacco streak virus (TSV)	Translatable	Van Dun et al. (1988)
Tobacco vein mottling virus (TVMV)	Translatable	Maiti, Murphy, Shaw, and Hunt (1993) and Maiti, Von Lanken, Hong, Dey, and Hunt (1999)
	Translatable	Xu, Collins, Hunt, and Nielsen (1997)
Tomato black ring virus (TBRV)	Translatable	Pacot-Hiriart, Le Gall, Candresse, Delbos, and Dunez (1999)
Tomato mosaic virus (ToMV)	Translatable	Yeh and Pon (1996)
Tomato mottle virus (ToMoV)	Translatable	Sinisterra, Polston, Abouzid, and Hiebert (1999)
Tomato ringspot virus (ToRSV)	Translatable	Yepes, Fuchs, Slightom, and Gonsalves (1996)
Tomato spotted wilt virus (TSWV)[b]	Translatable	Gielen et al. (1991)
	Translatable/nontranslatable	de Haan et al. (1992)
	Translatable	Mackenzie and Ellis (1992)
	Translatable	Pang, Nagpala, Wang, Slightom, and Gonsalves (1992)
	Antisense	Prins et al. (1996)

[a]Translatable/nontranslatable/antisense RNA.
[b]Nucleocapsid protein.

Table 2.2 Coat protein-mediated resistance to viruses in transgenic *Nicotiana benthamiana*

Virus	Capsid protein gene status	References
Arabis mosaic virus (ArMV)	Translatable	Spielmann, Krastanova, Douet-Orhant, and Gugerli (2000)
Bean yellow mosaic virus (BYMV)	Translatable	Nakamura, Honkura, Ugaki, Ohshima, and Ohaski (1994)
Cassava brown streak Uganda virus (CBSUV)	Translatable	Patil et al. (2011)
Cassava brown streak virus (CBSV)	Translatable	Patil et al. (2011)
Cowpea aphid-borne mosaic virus (CABMV)	Translatable	Mundembe, Matibiri, and Sithole-Niang (2009)
Cucumber mosaic virus (CMV)	Nontranslatable	Tan et al. (2012)
Cymbidium mosaic virus (CymMV)	Translatable	Chia, Chan, and Chua (1992)
Cymbidium ringspot virus (CymRSV)	Translatable	Rubino, Capriotti, Lupo, and Russo (1993)
Grapevine fanleaf virus (GFLV)	Translatable	Bardonnet, Hans, Serghini, and Pinck (1994)
Grapevine leafroll associated virus-2 (GFLRaV-2)	Translatable	Ling, Zhu, and Gonsalves (2008)
Grapevine virus A (GVA)	Translatable	Minafra et al. (1998)
	Translatable	Radian-Sade et al. (2000)
Grapevine virus B (GVB)	Translatable	Minafra et al. (1998)
Passionfruit woodiness virus (PWV)	Translatable	Yeh and Chu (1996)
PWV and CMV	Translatable	Sokhandan-Bashir, Gillings, and Bowyer (2012)
Peanut stripe virus (PStV)	Translatable/ nontranslatable	Cassidy and Nelson (1995)

Continued

Table 2.2 Coat protein-mediated resistance to viruses in transgenic *Nicotiana benthamiana*—cont'd

Virus	Capsid protein gene status	References
Plum pox virus (PPV)	Translatable	Regner et al. (1992)
	Translatable	Ravelonandro et al. (1993)
	Translatable	da Câmara Machado, Katinger, da Câmara, and Machado (1994)
	Translatable	Jacquet, Ravelonandro, Bachelier, and Dunez (1998)
Poplar mosaic virus (PopMV)	Trans/nontrans/antisense[a]	Edwards, Liu, Hawes, Henderson, and Cooper (1997)
Potato mop-top virus (PMTV)	Translatable	Reavy, Arif, Kashiwazaki, Webster, and Barker (1995)
Potato virus A (PVA)	Translatable	Savenkov and Valkonen (2001)
Prune dwarf virus (PDV)	Trans/nontrans/antisense	Raquel, Lourenço, Moita, and Oliveira (2008)
Sweet potato feathery mottle virus (SPFMV)	Translatable	Sonada and Nishiguchi (1999)
Tobacco etch virus (TEV)	Translatable	Voloudakis, Aleman-Verdaguer, Padgett, and Beachy (2005)
Tomato ringspot virus (ToRSV)	Translatable/antisense	Yepes et al. (1996)
Tomato spotted wilt virus (TSWV)[b]	Translatable	Pang, Bock, Gonsalves, Slightom, and Gonsalves (1994)
	Translatable	Vaira et al. (1995)
Turnip crinkle virus (TCV)	Translatable	Vasudevan et al. (2008)
Turnip mosaic virus (TuMV)	Translatable	Jan, Pang, Fagoaga, and Gonsalves (1999)
Vanilla necrosis virus (VNV)	Translatable/antisense	Wang, Gardner, and Pearson (1997)
Watermelon mosaic virus (WMV)	Translatable	Namba, Ling, Gonsalves, Slightom, and Gonsalves (1992)
Zucchini yellow mosaic virus (ZYMV)	Translatable	Namba et al. (1992)

[a]Translatable/nontranslatable/antisense RNA.
[b]Nucleocapsid protein.

other crop species, such plants were generated subsequently to express viral capsid protein genes and were assessed for resistance to their eponymous viruses (Table 2.3).

The modes of resistance operating in these numerous transgenic events were not always characterized completely, and even when characterized, the state of knowledge available concerning the existence or expression of RNA silencing was such that the exact cause that may have been ascribed to the resistance may have been incorrect. It is clear that in the case of transgenic tobacco plants expressing the TMV capsid protein the resistance was protein-mediated interference (Asurmendi, Berg, Koo, & Beachy, 2004; Asurmendi, Berg, Smith, Bendahmane, & Beachy, 2007; Bendahmane et al., 2007; Bendahmane, Szecsi, Chen, Berg, & Beachy, 2002; reviewed by Bendahmane & Beachy, 1999). The transgenic events described using the capsid protein of *Alfalfa mosaic virus* (AMV) (Loesch-Fries et al., 1987; Taschner, Van Marle, Brederode, Tumer, & Bol, 1994; Tumer et al., 1991, 1987) probably also reflect protein-mediated interference, as does, to some extent, the coat protein-mediated protection described for *Potato virus X* (PVX) (Bazzini et al., 2006; Hemenway et al., 1988; Hoekema et al., 1989; Spillane et al., 1997). However, the other examples given above probably were due to either RNA silencing or a mixture of RNA silencing and capsid protein-mediated interference. By a mixture, we mean that some plants transformed with particular construct exhibited RNA silencing, while other plants transformed with the same construct showed protein-mediated interference. It is also conceivable that the same plants could show both forms of resistance. A general characteristic of this type of protein-mediated resistance was the ability to be broken either with higher concentrations of virus inoculum or by using viral RNA as the inoculum (Loesch-Fries et al., 1987; Okuno et al., 1993; Powell Abel et al., 1986; Van Dun et al., 1987), although this was not always true. This may indicate that this type of resistance is more easily broken; however, since presumably the inoculum dose per inoculation event in the field would be much less than in manual inoculations, is not clear whether the above observations also would be true for plants in the field. Moreover, the mode of resistance of the plants in which field assessments have been made was not always known, let alone the mechanism.

2.2. Replicase-mediated resistance

Although originally designed to generate a system for examining the function of the TMV 54-kDa protein, transgenic tobacco plants expressing

Table 2.3 Coat protein-mediated resistance to viruses in other transgenic plant species

Host/virus species	CP gene status	References
Alfalfa (Lucerne)		
Alfalfa mosaic virus (AMV)	Translatable	Hill et al. (1991)
Arabidopsis (Thale cress)		
Turnip mosaic virus (TuMV)	Translatable	Nomura, Ohshima, Anai, Uekusa, and Kita (2004)
Barley		
Barley yellow dwarf virus (BYDV)	Translatable	McGrath et al. (1997)
Barrel medic		
AMV	Translatable/ antisense	Jayasena et al. (2001)
Chinese cabbage		
TuMV	Translatable	Zhu, Song, Zhang, Guo, and Wen (2001)
Chrysanthemum		
Cucumber mosaic virus (CMV)	Translatable	Kumar, Raj, Sharma, and Varma (2012)
Citrus		
Citrus mosaic virus (CiMV)	Translatable	Iwanami, Shimizu, Ito, and Hirabayashi (2004)
Citrus psorosis virus (CPsV)	Translatable	Reyes, De Francesco, et al. (2011)
Citrus tristeza virus (CTV)	Translatable	Domínguez et al. (2002)
	Trans/nontrans/ antisense[a]	Febres, Lee, and Moore (2008)
	Translatable	Loeza-Kuk et al. (2011)
Cucumber		
CMV	Translatable	Gonsalves, Chee, Provvidenti, Seem, and Slightom (1992)
	Translatable	Nishibayashi, Hayakawa, Nakajima, Suzuki, and Kaneko (1996)
Zucchini yellow mosaic virus (ZYMV)	Translatable	Wako, Terami, Hanada, and Tabei (2001)

Table 2.3 Coat protein-mediated resistance to viruses in other transgenic plant species—cont'd

Host/virus species	CP gene status	References
Eggplant (Aubergene; Brinjal)		
CMV	Translatable	Pratap, Kumar, Raj, and Sharma (2011)
Gladiolus		
Bean yellow mosaic virus (BYMV)	Translatable/ antisense	Kamo et al. (2005)
CMV	Translatable	Kamo, Jordan, Guaragna, Hsu, and Ueng (2010)
Lettuce		
Lettuce mosaic virus (LMV)	Translatable	Dinant et al. (1997)
Lettuce big-vein associated virus (LBVaV)	Translatable/ antisense	Kawazu, Fujiyama, Sugiyama, and Sasaya (2006)
Tomato spotted wilt virus (TSWV)[b]	Translatable	Pang et al. (1996)
Maize (Corn)		
Maize dwarf mosaic virus (MDMV)	Translatable	Murry et al. (1993)
	Translatable	Liu, Song, et al. (2005)
Sugarcane mosaic virus (SCMV)	Nontranslatable	Liu, Tan, Li, Zhang, and He (2009)
Melon		
CMV	Translatable	Yoshioka, Hanada, Harada, Minobe, and Oosawa (1993)
	Translatable	Gonsalves et al. (1994)
	Translatable	Xu, Shi, and Xue (2005)
Watermelon mosaic virus (WMV)	Translatable	Wang, Zhao, and Zhou (2000)
ZYMV	Translatable/ antisense	Fang and Grumet (1993)
WMV and ZYMV	Translatable	Clough and Hamm (1995)
CMV, WMV, and ZYMV	Translatable	Fuchs et al. (1997)

Continued

Table 2.3 Coat protein-mediated resistance to viruses in other transgenic plant species—cont'd

Host/virus species	CP gene status	References
Melon, oriental		
ZYMV	Translatable	Wu, Yu, Raja, Wang, and Yeh (2009)
ZYMV and PRSV-W	Nontranslatable	Wu et al. (2010)
Oat		
BYDV	Translatable	McGrath et al. (1997)
Oilseed rape		
TuMV	Translatable	Lu et al. (1996)
	Translatable/ nontranslatable	Lehmann, Jenner, Kozubek, Greenland, and Walsh (2003)
Orchid—*Dendrobium*		
Cymbidium mosaic virus (CymMV)	Translatable	Chang et al. (2005)
Orchid—*Phalaenopsis*		
CymMV	Translatable	Liao et al. (2004)
Osteospermum		
TSWV	Translatable	Vaira, Berio, Accotto, Vecchiati, and Allavena (2000)
Papaya		
Papaya ringspot virus (PRSV)	Translatable	Fitch, Manshardt, Gonsalves, Slightom, and Sandford (1992)
	Translatable	Tennant et al. (1994)
	Translatable	Ferreira et al. (2002)
	Translatable	Bau, Cheng, Yu, Yang, and Yeh (2003)
	Translatable	Fermin et al. (2004)
	Translatable	Kertbundit et al. (2007)
	Nontranslatable	Lines, Persley, Dale, Drew, and Bateson (2002)
	Translatable/ nontranslatable	Souza, Nickel, and Gonsalves (2005)

Table 2.3 Coat protein-mediated resistance to viruses in other transgenic plant species—cont'd

Host/virus species	CP gene status	References
PRSV and *Papaya leaf-distortion mosaic virus* (PLDMV)	Nontranslatable	Kung et al. (2009)
	Nontranslatable	Kung et al. (2010)
Passionflower		
Cowpea aphid-borne mosaic virus (CABMV)	Translatable	Trevisan et al. (2006) Monteiro-Hara et al. (2011)
Patchouli		
Patchouli mild mosaic virus (PaMMV)	Translatable	Kadotani and Ikegami (2002)
Pea		
AMV	Translatable	Timmerman-Vaughan et al. (2001)
Pea enation mosaic virus (PEMV)	Translatable	Chowrira, Cavileerm, Gupta, Lurquin, and Berger (1998)
Peanut (groundnut)		
Peanut stripe virus (PStV)	Translatable/ nontranslatable	Higgins, Hall, Mitter, Cruickshank, and Dietzgen (2004)
Tobacco streak virus (TSV)	Translatable	Mehta et al. (2013)
TSWV	Translatable	Li, Jarret, and Demski (1997)
	Antisense	Magbanua et al. (2000)
Pepper (Capsicum)		
CMV	Translatable	Zhang et al. (1994)
	Translatable	Li, Hu, Wang, and Fan (2000)
	Translatable	Balázs et al. (2008)
	Translatable	Lee et al. (2009)
	Translatable	Pack et al. (2012)
CMV and *Pepper mild mottle virus* (PMMoV)	Translatable	Shin, Han, Lee, and Peak (2002)
CMV and *Tobacco mosaic virus* (TMV)	Translatable	Bi, Shan, Wang, and Xu (1999)
	Translatable	Xu, Shang, and Wang (2002)
	Translatable	Cai et al. (2003)

Continued

Table 2.3 Coat protein-mediated resistance to viruses in other transgenic plant species—cont'd

Host/virus species	CP gene status	References
CMV and *Tomato mosaic virus* (ToMV)	Translatable	Shin, Han, et al. (2002)
Plum		
PRSV	Translatable	Scorza et al. (1995)
Plum pox virus (PPV)	Translatable	Ravelonandro et al. (1997)
	Translatable	Scorza et al. (2001)
Potato		
Potato leafroll virus (PLRV)	Translatable	Kawchuk, Martin, and McPherson (1990)
	Translatable/ antisense	Kawchuk, Martin, and McPherson (1991)
	Translatable	Van der Wilk, Posthumus-Lutke Willink, Huisman, Huttinga, and Goldbach (1991)
	Translatable	Barker, Reavy, Kumar, Webster, and Mayo (1992)
	Translatable	Presting, Smith, and Brown (1995)
	Translatable/ antisense	Palucha, Zagórski, Chrzanowska, and Hulanicka (1998)
Potato mop-top virus (PMTV)	Translatable/ nontranslatable	Barker, Reavy, and McGeachy (1998)
Potato virus S (PVS)	Translatable	Mackenzie, Tremaine, and McPherson (1991)
Potato virus X (PVX)	Translatable	Hoekema, Huisman, Molendijk, Van den Elzen, and Cornelissen (1989)
	Translatable	Van den Elzen et al. (1989)
	Translatable	Jongedijk, de Schutter, Stolte, Van den Elzen, and Cornelissen (1992)

Table 2.3 Coat protein-mediated resistance to viruses in other transgenic plant species—cont'd

Host/virus species	CP gene status	References
(+8-kDa movement protein gene)	Translatable	Xu, Khalilian, Eweida, Squire, and Abouhaidar (1995)
	Translatable	Spillane, Baulcombe, and Kavanagh (1998)
Potato virus Y (PVY)	Translatable	Van den Heuvel, Van der Vlugt, Verbeek, de Haan, and Huttinga (1994)
	Translatable	Józsa, Stasevski, Wolf, Horváth and Balázs (2002)
	Translatable	Gargouri-Bouzid et al. (2005)
	Translatable/ antisense	Sokolova, Pugin, Shulga, and Skryabin (1994)
	Translatable/ nontranslatable	Rachman et al. (2001)
	Trans/nontrans/ antisense	Smith, Powers, Swaney, Brown, and Dougherty (1995)
PLRV and PVY	Translatable	Zhang, Peng, Song, and Li (1997)
PVX and PVY	Translatable	Kaniewski et al. (1990)
	Translatable	Lawson et al. (1990)
	Translatable	Zhang et al. (1996)
Red clover		
AMV	Translatable	Zhao and Mu (2004)
Rice		
Rice stripe virus (RSV)	Translatable	Hayakawa, Zhu, Itoh, Kimura, and Izawa (1992)
	Translatable	Yan, Wang, Qiu, and Tien (1997)
	Antisense	Park et al. (2012)
Rice tungro bacilliform virus (RTBV)	Translatable	Ganesan, Suri, Rajasubramaniam, Rajam, and Dasgupta (2009)

Continued

Table 2.3 Coat protein-mediated resistance to viruses in other transgenic plant species—cont'd

Host/virus species	CP gene status	References
Rice tungro spherical virus (RTSV)	Translatable	Sivamani et al. (1999)
	Nontranslatable	Verma, Sharma, Devi, Rajasubramaniam, and Dasgupta (2012)
Rice yellow mottle virus (RYMV)	Nontranslatable/ antisense	Kouassi et al. (2006)
Ryegrass		
Ryegrass mosaic virus (RgMV)	Nontranslatable	Xu, Schubert, and Altpeter (2001)
Soybean		
Bean pod mottle virus (BPMV)	Translatable	Di et al. (1996) and Reddy et al. (2001)
Soybean dwarf virus (SDV)	Translatable	Tougou et al. (2007)
Soybean mosaic virus (SMV)	Translatable	Wang, Eggenberger, Nutter, and Hill (2001)
	Translatable	Furutani, Hidaka, Kosaka, Shizukawa, and Kanematsu (2006) and Furutani et al. (2007)
Squash (Marrow; Courgette)		
Squash mosaic virus (SqMV)	Translatable	Pang et al. (2000)
WMV and ZYMV	Translatable	Clough and Hamm (1995)
WMV; ZYMV; WMV and ZYMV	Translatable	Fuchs and Gonsalves (1995)
CMV; WMV; ZYMV; CMV and WMV; WMV and ZYMV; CMV, WMV, and ZYMV	Translatable	Tricoli et al. (1995)
CMV; WMV; ZYMV; WMV and ZYMV; CMV, WMV, and ZYMV	Translatable	Fuchs et al. (1998)

Table 2.3 Coat protein-mediated resistance to viruses in other transgenic plant species—cont'd

Host/virus species	CP gene status	References
Subterranean clover		
BYMV	Translatable	Chu, Anderson, Khan, Shukla, and Higgins (1999)
Sugar beet		
Beet necrotic yellow vein virus (BNYVV)	Translatable	Mannerlöf, Lennerfors, and Tenning (1996)
Sugarcane		
Sorghum mosaic virus (SrMV)	Nontranslatable	Ingelbrecht, Ervine, and Mirkov (1999)
SCMV	Translatable	Guo et al. (2008)
Sugarcane yellow leaf virus (SCYLV)	Nontranslatable	Zhu et al. (2011)
Sweet potato		
Sweet potato feathery mottle virus (SPFMV)	Translatable	Okada et al. (2001)
	Translatable	Okada and Saito (2008)
	Translatable	Okada and Yoshinaga (2010)
Tomato		
AMV	Translatable	Tumer et al. (1987)
CMV	Translatable	Xue et al. (1994)
	Translatable	Gielen et al. (1996)
	Translatable	Cheng, Wu, Wang, Pan, and Chen (1997)
	Translatable	Kaniewski et al. (1999)
	Translatable	Shi and Xue (1999)
	Translatable	Tomassoli, Ilardi, Barba, and Kaniewski (1999)
	Translatable	Pratap et al. (2012)
Physalis mottle virus (PhMV)	Translatable	Vidya, Manoharan, Kumar, Savithra, and Sita (2000)
Tobacco mosaic virus (TMV)	Translatable	Nelson et al. (1988)
	Translatable	Sanders et al. (1992)

Continued

Table 2.3 Coat protein-mediated resistance to viruses in other transgenic plant species—cont'd

Host/virus species	CP gene status	References
Tomato leaf curl virus (ToLCV)	Translatable	Raj, Singh, Pandey, and Singh (2005)
Tomato leaf curl Taiwan virus (ToLCTWV)	Translatable	Sengoda et al. (2012)
ToMV	Translatable	Sanders et al. (1992)
TSWV	Translatable	Ultzen et al. (1995)
	Translatable	Gubba, Gonsalves, Stevens, Tricoli, and Gonsalves (2002)
Tomato yellow leaf curl virus (TYLCV)	Translatable	Kunik et al. (1994)
Watermelon		
Cucumber green mottle mosaic virus (CGMMV)	Translatable	Park et al. (2005)
CMV and WMV	Nontranslatable	Lin, Ku, et al. (2012)
WMV	Translatable	Wang, Zhao, Xu, Zhao, and Zhang (2003)
WMV (and replicase genes of CMV and ZYMV)	Translatable	Niu et al. (2005)
ZYMV and PRSV-W	Nontranslatable	Yu et al. (2011)
Wheat		
BYDV (GPV)	Translatable	Cheng et al. (2000)
Wheat streak mosaic virus (WSMV)	Translatable/ nontranslatable	Sivamani et al. (2002)
	Translatable	Li, Liu, and Berger (2005)
Wheat yellow mosaic virus (WYMV)	Translatable	Dong et al. (2002)
White clover		
AMV	Translatable	Zhao, Mu, and Chu (2005)
	Translatable	Panter et al. (2012)

[a]Translatable/nontranslatable/antisense RNA.
[b]Nucleocapsid protein.

RNA for the gene encoding the 54-kDa protein exhibited a very strong resistance to TMV, but not to other viruses (Golemboski, Lomonossoff, & Zaitlin, 1990). Either this observation, or other considerations, prompted a number of other laboratories to generate and test transgenic plants expressing intact, partial, or modified viral polymerase genes. The other considerations included the desire to test other plant viral genes for dominant-negative interference effects on viral replication, or an earlier report showing that mutation of the G residue in the conserved GDD motif in the polymerase of the RNA phage Qβ resulted in dominant-negative interference of Qβ replication (Inokuchi & Hirashima, 1987). As was the case for coat protein-mediated resistance, the first examples were assessed using either tobacco or *N. benthamiana*, but subsequently were assessed in crop plant species (Table 2.4). All those examined for the various aspects of resistance exhibited a narrow spectrum of resistance (vis-à-vis virus strains and related viruses), but showed strong resistance against very high levels of virus inocula, and also against high levels of viral RNAs used as inocula. By contrast, transgenic tobacco expressing one particular truncated form of the TMV replicase protein gave a very broad-spectrum resistance also against high levels of inocula of other tobamoviruses (Donson et al., 1993). In addition, transgenic tobacco expressing both *Cucumber mosaic virus* (CMV) RNA1 and RNA2, produced by crossing plants expressing either RNA1 or RNA2, gave a strong resistance to infection by CMV RNAs, but a much weaker resistance to infection by CMV virions, while the parental transgenic plants showed susceptibility (Suzuki, Masuta, Takanami, & Kuwata, 1996).

Some studies suggested that the nature of the resistance was protein-mediated, while other studies suggested that the resistance was RNA-mediated. This was even the case for the same transgenic plants. For example, in the case of the TMV *54-kDa* gene-mediated resistance (Golemboski et al., 1990), one study provided evidence supporting the resistance being protein-mediated (Carr, Marsh, Lomonossoff, Sekiya, & Zaitlin, 1992), while another study showed evidence suggesting an RNA-mediated component was involved (Marano & Baulcombe, 1998). A different approach also indicated that both RNA and protein components could be involved in resistance to TMV mediated through replicase gene sequences (Goregaoker, Eckhardt, & Culver, 2000). In the case of tobacco expressing a defective form of the CMV 2a replicase protein, both protein-mediated and RNA-mediated aspects of the nature of the resistance could be demonstrated, with the protein-mediated resistance being more important to the high level of resistance (Hellwald & Palukaitis, 1995; Wintermantel & Zaitlin, 2000). In the case of resistance conferred by

Table 2.4 Replicase-mediated resistance to viruses in various transgenic plant species

Host species	Virus species	References
Barley	*Barley yellow dwarf virus* (BYDV-PAV)	Koev et al. (1998)
Citrus	*Citrus tristeza virus* (CTV)	Febres et al. (2008)
Cucumber	*Cucumber fruit mottle mosaic virus* (CFMoMV)	Gal-On et al. (2005)
Gladiolus	*Cucumber mosaic virus* (CMV)	Kamo et al. (2010)
Lily	CMV	Azadi et al. (2011)
N. benthamiana	*Cowpea mosaic virus* (CPMV)	Sijen, Wellink, Hendriks, Verver, and van Kammen (1995)
	Cymbidium ringspot virus (CymRSV)	Rubino, Lupo, and Russo (1993)
	Pea early browning virus (PEBV)	MacFarlane and Davies (1992)
	Peanut stripe virus (PStV)	Cassidy and Nelson (1995)
	Pepper mild mottle virus (PMMoV)	Tenllado, García-Luque, Serra, and Díaz-Ruíz (1995)
	Plum pox virus (PPV)	Guo and García (1997) Guo, Cervera, and García (1998)
Papaya	*Papaya ringspot virus* (PRSV)	Chen, Ye, Huang, Yu, and Li (2001)
Pea	*Pea seed-borne mosaic virus* (PSbMV)	Jones, Johansen, Bean, Bach, and Maule (1998)
Potato	*Potato leafroll virus* (PLRV)	Thomas, Lawson, Zalewski, Reed, and Kaniewski (2000) Lawson, Weiss, Thomas, and Kaniewski (2001) Vazquez Rovere, Asurmendi, and Hopp (2001)
	Potato virus Y (PVY-N)	Schubert, Matoušek, and Mattern (2004)
	Tobacco rattle virus (TRV)	Melander (2006)

Table 2.4 Replicase-mediated resistance to viruses in various transgenic plant species—cont'd

Host species	Virus species	References
Rice	*Maize dwarf mosaic virus* (MDMV)	Du et al. (2011)
	Rice tungro spherical virus (RTSV)	Huet et al. (1999)
	Rice yellow mottle virus (RYMV)	Pinto, Kok, and Baulcombe (1999)
Tobacco	*Alfalfa mosaic virus* (AMV)	Brederode, Taschner, Posthumus, and Bol (1995)
	CMV	Anderson, Palukaitis, and Zaitlin (1992)
	Potato virus X (PVX)	Braun and Hemenway (1992) Longstaff, Brigneti, Boccard, Chapman, and Baulcombe (1993)
	PVY	Audy, Palukaitis, Slack, and Zaitlin (1994)
	Tobacco mosaic virus (TMV)	Golemboski et al. (1990) Donson et al. (1993)
	TRV	Vassilakos et al. (2008)
Tomato	CMV	Gal-On et al. (1998) Wu, Chen, Liang, and Wang (2006)
Watermelon	CMV, *Zucchini yellow mosaic virus* (ZYMV), and the WMV *CP* gene	Niu et al. (2005)
Wheat	*Wheat streak mosaic virus* (WSMV)	Sivamani, Brey, Dyer, Talbert, and Qu (2000)
	Wheat yellow mosaic virus (WYMV)	Wu, Zhang, Gao, Xu, and Cheng (2006)

expression of the replicase gene of *Pea early browning virus*, the initial data suggested that the resistance was protein-mediated (MacFarlane & Davies, 1992). However, a subsequent study indicated that the resistance was RNA-mediated (van den Boogaart, Wen, Davies, & Lomonossoff, 2001). Similarly, in the case of replicase-mediated resistance to *Potato virus Y* (PVY) conferred by the NIb encoding sequences, the initial data indicated the resistance was probably protein-mediated (Audy et al., 1994). However, a more recent analysis of these same transgenic lines indicates that the resistance is based on RNA silencing (B.N. Chung & P. Palukaitis, unpublished data). Other studies in which transgenic expression of nontranslatable genes showed resistance (see Section 3.1) demonstrated that RNA-mediated inhibition can be involved in the resistance process, but do not exclude the possibility that there could also be a protein-mediated supplemental or alternative resistance process. In many cases where protein-mediated interference was established, ruled out, or considered the most/least likely basis for the resistance, the number of transgenic lines generated was insufficient to draw such definite conclusions. This was exemplified by the situation with the PVX replicase gene, where the ability of the ADD mutant and the inability of the GAD or GED mutants of the GDD motif in the replicase protein to convey resistance led to an initial conclusion that the resistance probably was protein-mediated (Longstaff et al., 1993). Subsequently, in a follow-up study, in which more transgenic lines were generated, including lines with the wild-type GDD control, all constructs were shown to yield some transgenic lines conveying resistance, and the resistance was considered RNA-mediated (Mueller, Gilbert, Davenport, Brigneti, & Baulcombe, 1995).

2.3. Movement protein-mediated resistance

Based on the results observed from a limited set of systems examined, there do not seem to be any general rules established for this type of resistance. For example, transgenic expression of defective-mutant 30-kDa movement protein of TMV in tobacco (Lapidot et al., 1993; Malyshenko et al., 1993), 12-kDa movement protein of PVX in potato (Seppänen et al., 1997), and 13-kDa movement protein of *White clover mosaic virus* (WClMV) in *N. benthamiana* (Beck et al., 1994), resulted in the inhibition of eponymous virus infection, as well as infection by viruses in other genera and sometimes families (Beck et al., 1994; Cooper, Lapidot, Heick, Dodds, & Beachy, 1995). On the other hand, transgenic expression of the PVX 12-kDa protein in transgenic tobacco conferred resistance only to PVX and not to either TMV or PVY (Kobayashi, Cabral, Calamante, Maldonado, & Mentaberry, 2001). Transgenic expression of the wild-type movement proteins of *Cowpea mosaic virus* (CPMV) in *N. benthamiana* showed a narrow spectrum of resistance against CPMV strains, with no resistance against another comovirus (Sijen et al., 1995), while transgenic expression of the wild-type *Potato leafroll virus* (PLRV) movement protein in potato showed broad-spectrum resistance to PLRV, PVY, and PVX (Tacke, Salamini, &

Rohde, 1996). Transgenic tobacco expressing the NS_M gene of TSWV showed complete resistance to TSWV in a small number of lines in the R2 generation (Prins et al., 1997), while transgenic expression of a defective-mutant BC1 protein of the geminivirus *Tomato mottle virus* (ToMoV) showed resistance to the eponymous virus as well as to *Cabbage leaf curl virus* (CaLCuV) (Duan, Powell, Webb, Purcifull, & Hiebert, 1997). Transgenic expression of the wild-type 24-kDa movement protein of PVX in tobacco inhibited movement of TMV and Ob TMV (Ares et al., 1998), while transgenic expression of the wild-type movement protein of *Brome mosaic virus* in tobacco (a nonhost plant) inhibited TMV movement (Malyshenko et al., 1993). Transgenic expression of a defective-mutant 15-kDa movement protein of *Beet necrotic yellow vein virus* (BNYVV) in sugar beet also showed resistance to infection by the eponymous virus; other viruses were not tested (Lauber et al., 2001). By contrast, transgenic expression of functional movement proteins of TMV and PVX in tobacco either enhanced or showed no affect on virus movement (Kaplan et al., 1995; Ziegler-Graff, Guilford, & Baulcombe, 1991), while transgenic expression of the 50-kDa movement protein of ACLSV enhanced infection of the eponymous virus in transgenic *N. occidentalis*, although those plants also inhibited infection by another trichovirus, *Grapevine berry inner necrosis virus*, but not *Apple stem grooming virus* or *Apple stem pitting virus* (Yoshikawa et al., 2000). However, the situation with CMV was more complicated: Sanz, Serra, and García-Luque (2000) reported that transgenic expression of the wild-type movement protein of CMV also had no effect on virus movement, while Nadarajah, Hanafi, and Tan (2009) reported some resistance to disease and reduced viral accumulation. The latter group also saw a similar response with transgenic expression of movement protein mutants of CMV, whereas we did not see any effect on disease or interference with the movement of the wild-type virus in transgenic tobacco expressing CMV movement protein mutants (M.H. Shintaku & P. Palukaitis, unpublished data). Transgenic plants expressing a mutant version of the gene encoding the 13-kDa movement protein of *Potato mop-top virus* (PMTV) in potato were assessed for resistance in the field. There was a reduced incidence of infection by PMTV in the tubers, and the virus titers were also up to 79% lower in the most resistant line (Melander, Lee, & Sandgren, 2001). Transgenic melon expressing tandem duplicate copies of the *Cucumber green mottle mosaic virus* (CGMMV) gene encoding the movement protein conferred resistance against the virus, but only in lines where the transgene expression was silenced (Emran, Tabei, Kobayashi, Yamaoka, & Nishiguchi, 2012). To the best of our knowledge, such analyses of movement protein mutants in transgenic plants do not seem to have been extended to other viruses. Since the modes of action were not studied in detail, it cannot be concluded conclusively that they are all protein-mediated; however, given that several of the systems operated against viruses unrelated by sequence, it is highly likely that those at least are protein mediated.

2.4. Other viral protein-mediated resistance

2.4.1 Rep protein-mediated resistance

In the case of geminiviruses, which contain a DNA genome, three approaches to transgenic pathogen-derived resistance were examined early after the first reports of pathogen-derived resistance to RNA viruses. Two of these were the use of antisense RNA (see Section 3.1) and the use of defective-interfering (DI) sequences (see Section 3.3). Since some of the sense and antisense sequences covered the gene encoding the replication initiator protein (Rep) of particular geminiviruses, transgenic expression of partial, or complete *Rep* genes also were assessed for resistance in transgenic plants. These included the *Rep* genes of *Tomato golden mosaic virus* (TGMV) (Hong & Stanley, 1996), *Tomato yellow leaf curl virus* (TYLCV) (Antignus et al., 2004; Brunetti et al., 1997; Lucioli et al., 2003; Noris et al., 1996; Yang, Sherwood, Patte, Hieber, & Polston, 2004), *African cassava mosaic virus* (ACMV) (Chellappan, Masona, Vanitharani, Taylor, & Fauquet, 2004; Sangaré, Deng, Fauquet, & Beachy, 1999), *Bean golden mosaic virus* (BGMV) (Faria et al., 2006), *Maize streak virus* (Shepherd, Mangwende, Martin, Bezuidenhout, Kloppers, et al., 2007; Shepherd, Mangwende, Martin, Bezuidenhout, Thomson, et al., 2007), *Cotton leaf curl virus* (CLCuV) (Hashmi, Zafar, Arshad, Mansoor, & Asad, 2011), and *Tomato leaf curl Taiwan virus* (ToLCTWV) (Lin, Tsai, Ku, & Jan, 2012).

2.4.2 NIa protease-mediated resistance

The NIa protease coding sequences, engineered to express the NIa protein of *Tobacco vein mottling virus* (TVMV) in transgenic tobacco plants, were shown to provide resistance to TVMV, but not to two other potyviruses, *Tobacco etch virus* (TEV) and PVY (Maiti et al., 1993). Similarly, two transgenic tobacco lines engineered to express most of the NIa coding sequences of PVY from an added initiation codon, as well as part of the NIb coding sequences, showed resistance in some plants to infection by PVY. Some plants of two tobacco lines expressing the same NIa coding sequences along with those of the complete NIb, the capsid protein sequences of PYV and part of the 3′-nontranslated region (NTR) also showed resistance

(Vardi et al., 1993). Transgenic tobacco plants expressing the NIa protease coding sequences of TEV and PVY also were evaluated for resistance to the corresponding viruses, with both showing infection, although the plants subsequently recovered from disease in the new growth (Fellers, Collins, & Hunt, 1998). In addition, transgenic tobacco lines expressing paired NIa protease coding sequences for two viruses (TEV–PVY, TEV–TVMV, and TVMV–PVY) also were evaluated for resistance. Those plants expressing the NIa sequences of TEV–PVY or TEV–TVMV either did not develop symptoms or showed a recovery from infection by TEV, while plants expressing the NIa sequences of TEV–TVMV and TVMV–PVY only showed the recovery phenotype after infection by TVMV, and plants expressing the NIa sequences of TEV–PVY and TVMV–PVY only showed the recovery response after infection by PVY (Fellers et al., 1998). In the case of transgenic *N. benthamiana* expressing the coding sequences of the VPg component of the NIa protein of *Potato virus A* (PVA), but not the VPg protein, complete resistance to the virus was observed in only one of eight lines, with some other lines eventually recovering from infection (Germundsson & Valkonen, 2006). In transgenic *Arabidopsis thaliana* plants expressing the VPg coding sequence of PVY-O, two of three lines showed a high level of resistance to infection by PVY-NTN with no virus detected (Wojtal et al., 2011).

The question of whether expression of multiple potyvirus genes of the same virus gave better resistance than just the expression of a single gene was examined by Maiti et al. (1999), who expressed a TVMV segment encoding the NIa/NIb/capsid protein, and compared this to expression of just the TVMV capsid protein coding sequences in transgenic tobacco. They found that lines carrying the former construct gave less resistance than lines expressing the capsid protein gene alone. In addition, lines expressing the three NIa/NIb/capsid protein coding regions showed no resistance to TEV, which was exhibited by most of the lines expressing only the TVMV capsid protein gene (Maiti et al., 1999). Similarly, transgenic expression of a segment of the PVA genome containing translatable sequences of the CI/NIa/capsid protein coding region did not show any resistance among 10 *N. benthamiana* lines (Germundsson, Savenkov, Ala-Poikela, & Valkonen, 2007). By contrast, transgenic tobacco expressing a segment of PVY-N consisting of sequences of the 3′-part of the NIb coding sequence, the capsid protein coding sequence and part of the 3′-NTR showed resistance to PNY-N and PVY-O in some lines, and resistance to disease but not virus accumulation for PVA or *Potato virus V* (Farinelli & Malnoë, 1993).

2.4.3 P1 protein-mediated resistance

Transgenic potato plants expressing the P1 coding sequences of PVY-O were resistant to infection by PVY-O, but not by PVY-N (Mäki-Valkama, Valkonen, Lehtinen, & Pehu, 2001; Pehu et al., 1995). Transgenic *N. benthamiana* expressing part or all of the P1 coding sequence of PPV showed either complete resistance or a recovery form of resistance after inoculation with the virus. The resistant plants also were resistant to infection by other strains of PPV (Tavert-Roudet, Ravelonandro, Bachelier, & Dunez, 1998). Transgenic expression of the P1 coding sequences of TVMV in tobacco resulted in only a recovery-type of resistance in one line and different ratios of recovery-type resistance versus complete resistance in seven lines, with susceptible plants in two of those seven lines. Only one line showed any resistance to a more distantly related strain of TVMV (only 82% identical in sequence in the P1 coding region) and no lines showed any resistance to another potyvirus, PVY (Moreno, Bernal, Jimenez, & Rodríguez-Cerezo, 1998). In the case of transgenic *N. benthamiana* expressing the coding sequences for the PVA P1 protein, complete resistance to the virus was observed in only one of seven lines, with some other lines eventually recovering from infection (Germundsson & Valkonen, 2006).

2.4.4 HCPro-mediated resistance

Transgenic *N. benthamiana* plants expressing the HC-Pro of PVA showed no initial resistance to infection, but a recovery phenotype developed in which the upper leaves contained no virus and were resistant to superinfection by the same virus. Methylation of the transgene followed recovery (Savenkov & Valkonen, 2002). Similarly, transgenic *N. benthamiana* expressing the HC-Pro protein of *Cowpea aphid-borne mosaic virus* (CABMV) showed no resistance to infection by CABMV, but showed enhanced symptoms after infection by CABMV, as well as after infection by PVY or CPMV. This was not observed in transgenic plants expressing CABMV HC-Pro in which the central domain of the protein was deleted. However, the highly expressing HC-Pro lines underwent a brief recovery from infection followed by reinfection, while one line expressing a low level of HC-Pro RNA was highly resistant to infection by CABMV, but not PVY (Mlotshwa et al., 2002). Transgenic *N. benthamiana* plants expressing the HC-Pro coding sequence of PPV showed resistance to disease and virus accumulation in those lines in which the transgene transcript accumulated to a low level, but not in those in which the translatable transgene transcript accumulated at a high level (Barajas et al., 2004). Transgenic soybean expressing the HC-Pro coding region of *Soybean mosaic virus* (SMV) did not show complete resistance to infection by SMV in any lines, but those lines expressing low levels of the transgene RNA showed mild symptoms. The line expressing high levels of the transgene RNA showed high levels of disease initially followed by recovery from disease in new leaves with lower virus titers in that line (Lim et al., 2007). The extent and duration of recovery may depend on the level of expression of the HC-Pro protein versus the suppression of its synthesis by RNA silencing of the transgene producing the HC-Pro protein, which also would be affected by the level of HC-Pro versus siRNAs produced from the invading virus.

Transgenic expression of HC-Pro also had the effect of enhancing some host defense responses resulting in enhanced resistance to unrelated viruses: TMV or *Tomato black ring virus* (TBRV) (Pruss et al., 2004), as well as *Tobacco rattle virus* (TRV), but not CMV (Shams-Bakhsh, Canto, & Palukaitis, 2007).

2.4.5 Other viral gene-mediated resistance

The 13- and 16-kDa nonstructural proteins of TRV strain PLB (found in the common 3′ coterminal sequences of both RNAs) and the 29-kDa protein of TRV strain TCM (found on RNA2) were expressed in transgenic tobacco plants, but no plants tested showed any resistance to infection by TRV (Angenent et al., 1990). Transgenic tobacco plants expressing the CI coding sequence of TVMV did not show any resistance to infection by TVMV, TEV, or PVY (Maiti et al., 1993), nor did transgenic *N. benthamiana* expressing the CI coding sequences of PPV (Wittner, Palkovics, & Balázs, 1998). However, transgenic *N. benthamiana* expressing the coding sequences of a mutant of the PPV CI did show complete resistance to PPV infection in one line (Wittner et al., 1998). Expression of the 6-kDa/VPg coding region of TEV in transgenic tobacco plants resulted in some lines that were highly resistant to infection and more lines that showed recovery from infection (Swaney, Powers, Goodwin, Rosales, & Dougherty, 1995). The P3 coding sequences of TVMV were expressed in transgenic tobacco plants and all five lines showed different proportions of complete resistance versus a recovery-type of resistance with two lines also showing a few susceptible plants (Moreno et al., 1998). Transgenic tobacco expressing RNA1 of CMV (encoding a component of the viral replicase containing the methyl transferase and helicase domains) showed resistance to systemic infection of CMV, as well as complementing infection by

CMV RNAs 1 and 3 (Canto & Palukaitis, 1998), while transgenic tobacco plants expressing the sequences encoding the TMV methyl transferase domain of the 126-kDa protein were completely resistant to infection by TMV (Song, Koh, Kim, & Lee, 1999). Transgenic *N. benthamiana* plants expressing the wild-type or either of two mutant forms of the VPg-protease coding region of *Tomato ringspot virus* (ToRSV) were assessed for resistance to infection. More plants expressing the wild-type transgene were resistant to infection (46–89%) than plants expressing either mutant gene (22–55% and 0–57%). These plants were all susceptible to another ToRSV isolate with 21% difference in sequence with the transgene (Sun, Xiang, & Sanfaçon, 2001). Transgenic *N. benthamiana* expressing sequences encoding the 54-kDa capsid protein readthrough domain of BNYVV in a translatable form showed either high resistance or a recovery-type resistance after foliar inoculation, although the plants could be infected through inoculation of the roots by viruliferous *Polymyxa betae* (Andika, Kondo, & Tamada, 2005). Some transgenic Mexican lime plant lines silenced for the expression the p23 silencing suppressor protein of *Citrus tristeza virus* (CTV) were resistant to infection by CTV (Fagoaga et al., 2006). Transgenic tobacco plants expressing the wild-type or mutant CMV 2b protein showed some level of resistance to disease, decreasing with increasing virus concentration, and only limited effects on virus accumulation (Nadarajah et al., 2009).

3. VIRAL RNA-MEDIATED RESISTANCE

In parallel with their work on what was expected at the outset to be protein-mediated resistance, a number of laboratories also developed transgenic plants expressing viral antisense RNAs, viral sense RNAs derived from noncoding regions, and later, nontranslatable sense RNAs (i.e., sense RNAs in sequence altered to prevent translation), in an effort either to understand the basis of the resistance or to determine whether other forms of resistance could be established. In addition, other forms of interference were examined based on RNA-mediated approaches, such as the use of natural or artificial DI RNAs/DNAs, and the use of satellite RNAs (satRNAs). As the development of ribozyme technology paralleled the early development of transgenic plants, it was only natural that artificial ribozymes also would be considered as a method of conferring resistance to plant viruses (Edgington, 1992). And, as a better understanding of the process of RNA silencing developed (under various names including posttranscriptional gene silencing, homology-dependent gene silencing, and RNA interference), modifications to the expression of the transgenically expressed viral RNA were made, including the use of dsRNAs and hairpin RNAs (hpRNAs), leading to the use of short hpRNAs and amiRNAs. The viruses and plant species examined for resistance using these various approaches are described below.

3.1. Noncoding single-stranded RNAs

This subsection deals with those RNA segments expressed in transgenic plants that derive from (i) noncoding regions of the virus genome, such as 5′ and 3′ NTRs, or intergenic regions (Section 3.1.1); (ii) (+)-sense open reading frames that have been rendered nontranslatable (Section 3.1.2); or (iii), noncoding (−)-sense RNAs or viral DNA sequences complementary to the viral sequences containing the open reading frames (Section 3.1.3).

3.1.1 Noncoding regions of viral genomes

Expression of the 3′ terminal 100-nt 3′-NTR of *Turnip yellow mosaic virus* (TYMV) in transgenic oilseed rape plants, as a fusion with a *CAT* gene, resulted in a delay of infection, which was dose dependent, with 20–40% of the plants not becoming infected at lower levels of inocula (Zaccomer, Cellier, Boyer, Haenni, & Tepfer, 1993). Similar results were obtained by transgenic expression of antisense RNA, but not sense RNA, corresponding to a 52-nt fragment of the TMV 5′-NTR (nucleotides 7–58), but not to a second 52-nt fragment (nucleotides 64–115) starting near the beginning of the first open reading frame. Among the resistant lines, one line showed good resistance at low inoculum levels, while other lines showed less resistance with lower virus accumulation detected (Nelson, Roth, & Johnson, 1993). The 505-nt 3′-NTR of *Andean potato mottle virus* (APMoV) was expressed transgenically in tobacco and while there was some resistance to the virus in some lines of the R0 generation, this was lost in all of the R1-generation plants. However, a construct containing this 505 nt and the adjoining 425 nt of the smaller 591-nt capsid protein coding region gave excellent resistance in several lines (Vaskin, Vidal, Alves, Farinelli, & de Oliveira, 2001). The 197-nt intergenic region of PLRV was expressed in transgenic potato plants and different lines were evaluated for resistance. The transgenic plants showed little or no symptoms, with those plants expressing (+)-polarity PLRV RNAs showing a reduction in virus titer ranging from 43% to 72%, and those plants expressing (−)-polarity PLRV RNAs showing a reduction in virus accumulation of 74–86% (Dong, Li, Hasi, & Zhang, 1999). Transgenic grapefruit plants expressing the

3′-NTR of CTV showed resistance to CTV accumulation in one of seven lines tested over 3 years (Febres et al., 2008). In many other cases, sequences of the 3′NTR, or sometimes the 5′NTR, were expressed in either the sense or antisense orientation along with various (often adjacent) gene sequences in the same orientation. In these situations, the contribution of each part of the RNA to the resistance level observed could not be ascertained separately.

3.1.2 Nontranslatable sense RNAs

Transgenic expression of a nontranslatable form of the TEV capsid protein coding sequence in tobacco produced plants most of which were resistant to infection by TEV (Doughterty et al., 1994; Lindbo & Dougherty, 1992a, 1992b). Transgenic tobacco expressing a nontranslatable capsid protein coding sequence of PVY-N also showed resistance to infection by PVY-N (Van der Vlugt et al., 1992), while transgenic tobacco expressing a segment of PVY-N consisting of sequences of the 3′-part of the NIb coding sequence, the capsid protein coding sequence and part of the 3′-NTR showed resistance to PNY-N and PVY-O in some lines (Farinelli & Malnoë, 1993). In those constructs, the capsid protein was potentially translatable, although only lines without detectable capsid protein were resistant. However, resistance also was obtained in constructs with a frameshift preventing translation of the capsid protein (Farinelli & Malnoë, 1993). This was also true in other studies of PVY in tobacco (Smith et al., 1994) and in potato (Smith et al., 1995). Expression of frameshifted, capsid protein coding sequences of PVY-NTN in transgenic potato gave rise to either completely resistant plants in two lines or partially resistant plants with low virus titers and delayed symptoms appearance in three lines (Rachman et al., 2001). In transgenic tobacco plants expressing a nontranslatable version of the capsid protein coding region of PVY-N, complete resistance was obtained to infection by PVY-N in seven lines, which were also resistant to infection by PVY-O (Guo et al., 2001). By contrast, in another study expressing similar nontranslatable sequences of PVY-N, resistance was seen in some lines to PVY-N isolates, but not to a PVY-O isolate (Pourrahim et al., 2006), while in transgenic tobacco expressing similar sequences derived from PVY-O, two lines showed no resistance, three lines showed partial resistance (delayed and milder infection, for most plants), and one line showed complete resistance (Masmoudi et al., 2002). In transgenic tobacco expressing nontranslatable fragments of the 3′-part of the PVY capsid protein encoding sequences, resistance to PVY varied with the length of the RNA: fragments of 607 and 602 nt both conferred resistance, but a fragment of 202 nt did not (Zhu, Zhu, Wen, Bai, et al., 2004).

Transgenic expression in tobacco of a nontranslatable form of the nucleoprotein gene of TSWV also provided resistance to infection by TSWV (de Haan et al., 1992), while transgenic tobacco expressing a nontranslatable form of the NS_M coding region of TSWV showed complete resistance to virus infection in several lines (Prins et al., 1996, 1997). Expression of a nontranslatable form of the TEV 6-kDa/VPg encoding region in transgenic tobacco plants resulted in some lines that were highly resistant to infection and more lines that showed recovery from infection (Swaney et al., 1995). The transgenic expression of a nontranslatable form of the capsid protein coding region of *Sorghum mosaic virus* in sugarcane gave rise to plants with two forms of resistance: complete resistance or recovery (Ingelbrecht et al., 1999). Expression of nontranslatable RNA derived from the large (42 kDa) capsid-protein coding region of APMoV in transgenic tobacco resulted in inhibition of systemic infection by the virus to a high extent in two lines, with two other lines showing partial resistance; all lines showed virus accumulation in the inoculated leaves (Neves-Borges et al., 2001). Transgenic ryegrass expressing nontranslatable RNA derived from the capsid protein coding sequences of *Ryegrass mosaic virus* conferred different levels of resistance in several lines: three lines contained very low levels of virus, while 10 lines showed a two- to fivefold reduction in virus levels (Xu et al., 2001). *N. benthamiana* transgenic for expression of a nontranslatable form of the capsid protein coding region of *Peanut stripe virus* (PStV) showed either resistance to infection by PStV or a delay in the appearance of symptoms, but no recovery-type resistance (Cassidy & Nelson, 1995). Transgenic peanut expressing a nontranslatable RNA derived from the capsid protein coding sequence of PStV gave excellent resistance in several lines (Higgins et al., 2004). Transgenic papaya lines showed good resistance to a nontranslatable form of the capsid protein coding region of *Papaya ringspot virus* (PRSV) (Lines et al., 2002; Souza et al., 2005). Nontranslatable forms of the capsid protein gene or 25-kDa movement protein gene of PVX expressed in transgenic tobacco plants showed excellent resistance in some lines to PVX infection, with more lines expressing nontranslatable movement protein gene (39.7%) than capsid protein gene (13.3%) showing resistance (Liu, Zhang, et al., 2005). Six transgenic lines of tobacco plants expressing a fusion of nontranslatable sequences of the capsid protein coding sequences of PVY-N and PVX showed high resistance to infection by either or both viruses, with no detectable virus in the plants (Bai et al., 2005). Transgenic rice expressing a nontranslatable form of the *Rice yellow mottle virus* (RYMV) capsid protein gene showed only a delay in virus accumulation (Kouassi et al., 2006). Transgenic *N. benthamiana* expressing nontranslatable sequences of *Prune dwarf virus* (PDV) capsid protein gene showed only a recovery form of resistance in one line (Raquel et al., 2008). Transgenic grapefruit expressing nontranslatable RNAs of the capsid protein gene of CTV only showed resistance to virus accumulation in two of four plants in one of six lines evaluated (Febres et al., 2008). Transgenic papaya of various cultivars were generated expressing nontranslatable RNAs of either the capsid protein encoding sequences of *Papaya leaf-distortion mosaic virus* (PLDMV) or a fusion of parts of the capsid protein coding sequences of PRSV and PLDMV. Several lines of each construct were highly resistant to infection by the corresponding viruses with no detectable virus accumulation (Kung et al., 2009, 2010). One of 23 transgenic oriental melon lines expressing a nontranslatable RNA derived from fused parts of the capsid protein encoding sequences of *Zucchini yellow mosaic virus* (ZYMV) and PRSV-W showed complete resistance to both viruses with no virus accumulation (Wu et al., 2010). Similarly, two of 10 transgenic watermelon lines expressing RNA from the same construct were completely resistant to infection by the corresponding viruses with no detectable virus accumulation (Yu et al., 2011). Transgenic sugarcane expressing a nontranslatable capsid protein gene conferred resistance to *Sugarcane yellow leaf virus* accumulation; symptoms did not occur until after a cold treatment, by which only one line was evaluated and showed a resistance to disease as well (Zhu et al., 2011). No transgenic rice plants expressing nontranslatable RNA of the capsid protein coding region of *Rice tungro spherical virus* (RTSV) showed complete resistance, but some showed a delay in infection with lower virus titers, which prevented transmission by the leafhopper vector (Verma et al., 2012).

3.1.3 Antisense RNAs

3.1.3.1 DNA viruses

Transgenic tobacco plants expressing antisense RNA to the *AL1* (*Rep-TrAP-REn*) gene of TGMV showed resistance to disease, with the symptomless plants containing lower levels of virus (Day, Bejarano, Buck, Burrel, & Lichtenstein, 1991). By contrast, antisense RNA to the TYLCV *C1* (*Rep*) gene expressed in transgenic *N. benthamiana* conferred either little-to-no resistance (Noris et al., 1996), or resistance in which some plants were symptomless and the inoculated virus was almost completely suppressed

(Bendahmane & Gronenborn, 1997), while transgenic *N. benthamiana* plants expressing both sense and antisense RNAs to a truncated *Rep* gene were all susceptible to the same virus (Brunetti et al., 1997). Transgenic beans expressing antisense RNAs to the BGMV *Rep-TrAP-REn* genes and also to the *BC1* gene showed attenuated or delayed symptoms in two lines, and the DNA levels were either undetectable in the attenuated plants, or reduced in the plants with delayed infection (Aragão et al., 1998). Transgenic tobacco expressing antisense RNAs to various regions of DNA A of CLCuV showed resistance to infection in many lines with no viral DNA accumulation (Asad et al., 2003), while transgenic cotton expressing antisense RNA to the *Rep* gene of CLCuV were said to be resistant to infection by the virus, although no data were presented to substantiate this (Amudha et al., 2010). Transgenic cassava expressing antisense RNAs of ACMV *Rep*, *TrAP*, and *REn* genes separately showed delayed and attenuated symptom development with reduced virus accumulation (Zhang, Biu, & Zhou, 2005). Transgenic tomato expressing antisense RNAs of the *Rep* gene of *Tomato leaf curl virus* (ToLCV) showed an attenuation of disease symptoms in ~80% of the plants in two lines (Praveen, Kushwaha, et al., 2005), while ToLCV-infected tomato plants transformed with the same construct resulted in a recovery of the plants from infection (Praveen, Mishra, & Dasgupta, 2005). Transgenic tobacco expressing antisense RNAs of the *AV2* gene of ToLCV New Delhi variant showed resistance to the virus, with no symptoms and little-to-no virus accumulation (Mubin, Mansoor, Hussain, & Zafar, 2007). Transgenic soybean expressing antisense RNA to the *Rep* gene of *Mungbean yellow mosaic India virus* showed resistance to disease in some lines and resistance to infection by the virus in other lines (Singh, Haq, & Malathi, 2013).

3.1.3.2 RNA viruses

Transgenic tobacco plants expressing antisense RNAs to the capsid protein coding region of CMV were protected from infection only at very low doses of inoculum (Cuozzo et al., 1988; Rezaian et al., 1988), while transgenic tobacco plants expressing antisense RNA to part of the CMV capsid protein gene and movement protein gene showed only a slight attenuation in disease development (Kim, Kim, et al., 1995). Protection only at low doses of inocula also was observed for transgenic tobacco plants expressing antisense RNA to the capsid protein gene of PVX (Hemenway et al., 1988) or *Potato aucuba mosaic virus* (Leclerc & AbouHaidar, 1995). Transgenic tobacco plants expressing antisense RNA to the TMV capsid protein coding region and

3′-NTR showed resistance to infection only at low levels of virus inocula, while similar plants expressing antisense RNA against only the capsid protein coding region but not the 3′-NTR showed no resistance to infection by TMV (Powell et al., 1989). Transgenic potato expressing antisense RNAs to the capsid protein gene of PLRV were shown to be highly resistant to the virus (Kawchuk et al., 1991; Palucha et al., 1998). Expression of antisense RNAs to the capsid protein coding sequences of ZYMV in transgenic melon resulted in a delay in symptom development and a reduction in virus levels (Fang & Grumet, 1993). Transgenic tobacco expressing antisense RNAs to the capsid protein coding region of TEV showed differences in the results in two studies. Only one of seven lines showed any resistance, and either more than half the plants inoculated showed good resistance (Lindbo & Dougherty, 1992a) or some inoculated plants occasionally showed attenuated symptoms (Lindbo & Dougherty, 1992b). Antisense RNA to the PVY-N capsid protein coding region and flanking sequences of parts of the NIb coding region and 3′-NTR expressed in transgenic tobacco offered only partial resistance to infection by PVY-N and no resistance to PVY-O (Farinelli & Malnoë, 1993). However, resistance to PVY accumulation was observed in transgenic potato plants expressing antisense RNAs to the capsid protein gene of PVY (Sokolova et al., 1994). Transgenic expression of antisense RNA against the *Potato spindle tuber viroid* (PSTVd) resulted in only a delay of infection for most potato plants of all lines tested (Matoušek et al., 1994). Antisense RNA of a segment of the *Bean yellow mosaic virus* genome corresponding to the 3′ end of the capsid protein coding sequences, the 3′-NTR and part of the polyA tract was expressed in transgenic *N. benthamiana* and showed varying degrees of resistance including partial protection (most lines), recovery from infection, and complete resistance to infection and virus accumulation; however, none of the lines were resistant to infection by *Pepper mottle virus* (PepMoV) or *Turnip mosaic virus* (TuMV) (Hammond & Kamo, 1995). A similar construct expressed in transgenic gladiolus showed resistance in some lines, which broke down the following season (Kamo et al., 2005). Expression of antisense RNA to the TEV 6-kDa/VPg encoding region in transgenic tobacco plants resulted in some lines that were highly resistant to infection and more lines that showed recovery from infection (Swaney et al., 1995). Transgenic *N. tabacum* and *N. benthamiana* plants expressing antisense sequences of the capsid protein coding sequences of ToRSV were evaluated for resistance to infection by ToRSV. Three out of four R2-generation lines of the transgenic *N. benthamiana* showed complete resistance while the fourth line showed

30–60% resistance, but the transgenic tobacco plants showed a lower level of complete resistance, with only one of five lines showing complete resistance (Yepes et al., 1996). Expression of antisense RNA to the NS_M coding sequence of TSWV in transgenic tobacco showed complete resistance to the virus in several lines (Prins et al., 1997). Transgenic expression of antisense RNA to the AMV capsid protein gene in tobacco showed resistance to infection by AMV to a few plants in three out if six transgenic lines (Jayasena et al., 1997), while the same construct expressed in barrel medic (*Medicago truncatula*) did not yield resistance to infection by AMV, although infection was delayed with lower virus accumulation (Jayasena et al., 2001). Transgenic *N. occidentalis* plants expressing antisense RNA to the *Cymbidium mosaic virus* (CymMV) capsid protein gene showed a delay in infection by CymMV, as well as milder symptoms and tolerance to systemic infection (Lim, Ko, Lee, La, & Kim, 1999). Some transgenic tobacco lines expressing antisense RNA to the *Tobacco necrosis virus* (TNV) capsid protein gene showed a reduction in local lesions after infection by TNV (Hackland et al., 2000). Transgenic peanuts expressing antisense RNAs against the TSWV nucleocapsid gene in the field showed good, but not complete, protection against the virus (Magbanua et al., 2000). Antisense RNA generated against the P1 coding sequence of PVY-O and expressed in transgenic potato plants showed complete resistance to PVY-O, but no resistance to PVY-N (Mäki-Valkama et al., 2000, 2001). Expression of antisense RNA to the 505-nt 3′-NTR of APMoV in transgenic tobacco plants resulted in segregating susceptibility, resistance, or partial protection in several lines (Vaskin et al., 2001). Transgenic lettuce expressing antisense RNA against the capsid protein gene of *Lettuce big-vein associated virus* (LBVaV) showed resistance to both LBVaV and to *Mirafiori lettuce virus* (MiLV; Kawazu et al., 2006). Transgenic rice expressing antisense RNA of the RYMV capsid protein gene showed only a delay in virus accumulation (Kouassi et al., 2006). Transgenic Chinese cabbage expressing antisense RNA to the NIb coding sequence of TuMV showed resistance in most lines to infection by TuMV (Yu, Zhao, & He, 2007). Transgenic *N. benthamiana* expressing antisense RNA to capsid protein gene of PDV showed either resistance or a recovery form of resistance (Raquel et al., 2008). Transgenic maize expressing antisense RNAs to the capsid protein coding region and 3′-NTR of *Sugarcane mosaic virus* (SCMV) showed variable resistance to infection by the virus (Liu et al., 2009). Antisense RNA to the capsid protein gene of *Rice stripe virus* (RSV) expressed in transgenic rice provided 0–63% resistance to disease in different lines to infection by RSV, and all but one line of plants showing resistance to disease had virus levels similar to the nontransformed plants (Park et al., 2012). No transgenic rice plants expressing antisense RNA to the replicase gene of RTSV showed complete resistance, but some showed a delay in infection with lower virus titers, which prevented transmission by the leafhopper vector (Verma et al., 2012).

3.2. Satellite RNA

Some viruses exhibit a supernumerary RNA component that does not show any apparent homology to the viral RNA genome, but depends on its helper virus for replication, encapsidation and transmission, and is defined as a satRNA (Simon, Roossinck, & Havelda, 2004). In certain virus–host plant interactions, satRNA variants can sometimes affect viral replication, pathogenesis, and symptom expression. As for the latter, satRNA variants have been described with hugely different characteristics, ranging from attenuated, to unaffected, to intensified disease symptoms. Variants with attenuating effects (benign satRNAs) have been studied extensively as useful biocontrol agents both in cross-protection experiments (Kaper, 1993; Sayama, Sato, Kominato, Natsuaki, & Kaper, 1993; Tien & Wu, 1991) and in transgenic plants. Two seminal papers paved the way to the subsequent search for satRNA-mediated transgenic resistance in different virus–host systems, by demonstrating that an attenuating CMV-associated satRNA expressed from its cDNA in transgenic tobacco plants could be rescued from basal mRNA expression, replicated at high levels during CMV infection, decreasing helper virus replication, and suppressing infection symptoms (Baulcombe, Saunders, Bevan, Mayo, & Harrison, 1986; Harrison, Mayo, & Baulcombe, 1987). Replication of satRNA, and the correlated suppression of symptoms, occurred when virus infection was transmitted by aphids (Jacquemond, Amselem, & Tepfer, 1988) and also could originate from negative-sense transcripts (Tousch, Jacquemond, & Tepfer, 1994).

The capability of transgenically expressed satRNA sequences to conferring tolerance to CMV infection also was demonstrated in other solanaceous host such as petunia and pepper (Kim, Lee, Kim, & Paek, 1997; Kim, Paek, & Kim, 1995). The availability of this new resistance approach coincided with destructive outbreaks in several countries of CMV isolates containing necrogenic satRNA variants in commercial tomato fields (Gallitelli, 2000; García-Arenal & Palukaitis, 1999; Jacquemond & Lot, 1981; Kaper & Waterworth, 1977; Palukaitis, 1988). Thus, a direct application of the satRNA technology was developed in transgenic tomato plants as a protection strategy against CMV. The first applications in tomato demonstrated, as already shown in tobacco, that transgenic satRNA was replicated by the satRNA-free helper virus to levels that were able to suppress both viral replication and symptom expression (Saito et al., 1992; Stommel, Tousignant, Wai, Pasini, & Kaper, 1998; Yie et al., 1995, 1992). Quantitative assays showed that replication of transcript satRNA induced a 10-fold decrease in CMV accumulation in the transgenic tomato plants compared to controls. When the same plants were challenged with *Tomato aspermy virus*, another cucumovirus, they showed satRNA replication but neither helper virus downregulation nor significant tolerance to virus-induced symptoms (McGarvey, Montasser, & Kaper, 1994). Later, it was shown that upon inoculation of a necrogenic satRNA/CMV combination, the transgenic satRNA was able to subvert initial cell death, to suppress significantly both the necrogenic satRNA and the helper virus replication and to codetermine an ameliorative symptomless phenotype. Thus, a double mechanism involving both downregulation of CMV replication and transgene-guided satRNA silencing was recognized as the molecular basis for resistance (Cillo, Finetti-Sialer, Papanice, & Gallitelli, 2004). A demonstration of a similar double protection mechanism that also involved silencing of both the transgenic and the natural satRNA was reported in *N. benthamiana* plants transformed for the expression of the *Groundnut rosette virus* (GRV)-associated satRNA. RNA silencing also was the mechanism determining resistance in transgenic lines transformed with sequences representing only the 5′ terminal one-third of the mild GRV satRNA (Taliansky, Ryabov, & Robinson, 1998).

Nepovirus is another plant virus genus where satRNA variants have been described, some of which coinduce mild symptoms associated to helper viruses. *Tobacco ringspot virus* (TRSV) is the type member of the genus. As seen in the case of CMV, when a satRNA-free isolate of TRSV was inoculated on transgenic tobacco plants expressing a mild variant of its satRNA in either positive or negative orientation, the satRNA was replicated and amplified at high levels, while the severity of symptom expression was remarkably reduced and delayed (Gerlach, Llewellyn, & Haseloff, 1987).

In *Arabidopsis*, the satRNA-C variant satRNA, when in association with its helper virus *Turnip crinkle virus* (TCV), usually coinduces intensification of viral symptoms. However, substitution of the TCV capsid protein open reading frame by that of the related species *Cardamine chlorotic fleck virus* generated a chimeric virus the disease phenotype of which was milder in the presence of satRNA-C. Transgenic expression of the satRNA-C sequence in *Arabidopsis* also resulted in chimeric TCV attenuated symptoms and in a 70% reduction of viral RNA accumulation levels. These results indicated that the capsid protein is a viral determinant for symptom suppression by satRNA (Kong, Oh, & Simon, 1995).

The only satRNA known to be dependent on a potexvirus is that associated with *Bamboo mosaic virus* (BaMV) infection in bamboo. In transgenic *N. benthamiana* and *A. thaliana* plants expressing BaMV satRNA, high levels of resistance to the helper virus were observed, which did not depend on either RNA silencing or salicylic acid-mediated signaling, but positively correlated with the transcript level of the transgene (Lin, Hsu, Chen, & Lin, 2013).

3.3. Defective-interfering RNAs/DNAs

Defective DNAs and RNAs are produced in significant amounts during the replication of some viral species. These defective nucleic acids retain sequences indispensable for replication, while showing deletions in other particular regions. In the course of replication, the generation of these shorter replicons can reduce substantially full-length genomic accumulation, hence the denomination of DI nucleic acids, associated with the modulation of symptom expression. Both ssDNA geminiviruses like ACMV and ssRNA tombusviruses like *Cymbidium ringsport virus* (CymRSV) and *Tomato bushy stunt virus* (TBSV) are known to form DIs, which have been exploited as transgenic resistance strategies. When a cloned copy of a naturally occurring subgenomic ACMV DNA B was expressed in *N. benthamiana*, it interfered with viral replication and coinduced milder symptoms upon ACMV infection. Tolerance correlated with accumulation of the episomally replicating subgenomic DNA B and was extended to another ACMV isolate, but not to the related TGMV that did not mediate transgenic DI DNA accumulation (Stanley, Frischmuth, & Ellwood, 1990). Similarly, when DI DNAs of *Beet curly top virus* (BCTV) were expressed in transgenic *N. benthamiana* the plants showed delayed and attenuated symptoms, with reduced virus levels, and the effect was strain specific (Frischmuth & Stanley, 1994; Stenger, 1994). In the case of the two tombusviruses, DI RNAs were cloned and expressed in *N. benthamiana*, where they replicated to high levels upon helper virus infection. Transgenic plants infected with the homologous virus showed tolerance and were protected from cell death symptoms. Plants transformed with TBSV DI RNA also exhibited broad-spectrum protection against related tombusviruses, *Cucumber necrosis virus* (CNV), *Carnation Italian ringspot virus*, and CymRSV, which share high sequence homology within their genomic terminal regions containing *cis*-acting elements necessary for replication (Kollàr, Dalmay, & Burgyàn, 1993; Rubio et al., 1999).

3.4. Ribozymes

Self-cleaving RNA (ribozymes), structures typically observed in some viroids and some satRNA that induce autocatalysis at specific positions, were used for achieving resistance in transgenic plants. Ribozyme modules embedded within sequences complementary to target viral RNA were predicted to cleave RNA in a sequence-specific manner and suppress virus accumulation in host tissues. However, earlier examples of transgenically expressed ribozymes targeting either *Citrus exocortis viroid* (Atkins et al., 1995) or TMV (De Feyter, Young, Schroeder, Dennis, & Gerlach, 1996) RNAs were ineffective and the partially tolerant phenotypes observed were likely due to antisense RNA. Partial resistance was also observed when a ribozyme construct was designed for targeting either the movement protein gene on CMV RNA3 (Nakamura, Honkura, Ugaki, Ohshima, & Ohaski, 1995) or the conserved leader sequences of CMV RNA1 and RNA2 (Kwon, Chung, & Paek, 1997). Unlike earlier studies, Huttner et al. (2001) generated polyribozymes to target two pathogenic viruses, WMV and ZYMV. Transgenic melon plants contained some degree of resistance to the viruses and plants of one line expressing a ribozyme directed against WMV were tested in the field under natural infection pressure and were found to be immune to WMV.

3.5. dsRNAs and hpRNAs

Expressing a combination of sense and antisense transcripts in the same transgenic plants greatly increased the percentage of transformed and regenerated lines that exhibited resistance (Waterhouse, Graham, & Wang, 1998). This approach has been used in a few laboratories to achieve resistance to virus disease to various extents, such as resistance to *Rice tungro bacilliform virus* (RTBV) (Tyagi, Rajasubramaniam, Rajam, & Dasgupta, 2008). To achieve this, either a rather complicated vector construction was necessary to introduce viral sequences expressing both transcripts in the same plants, or one had to cross plants that were expressing sense and antisense sequences from individual transformation events. Therefore, Waterhouse, Wang, and colleagues further improved this approach to produce a single RNA transcript of both polarities, with sense and antisense sequences separated by introns or other spacer sequences, to generate dsRNAs with loops, which stabilized the inverted repeat DNA sequences in *Escherichia coli* (Smith et al., 2000; Wesley et al., 2001). Then, these dsRNAs could be digested by Dicer-like nucleases *in planta* and produce the siRNAs that would flow into the rest of the RNA silencing mechanism, resulting in resistance to the invading virus, in most instances (reviewed by Castel & Martienssen, 2013; Csorba et al., 2009; Ding, 2010; Eames et al., 2008; Pumplin & Voinnet, 2013). The initial products of this dsRNA-producing strategy became known as hpRNAs, intron hpRNAs (ihpRNAs), or irRNAs; terms that are all still used to describe the strategy. This would be particularly useful for crops in which the transformation and regeneration frequencies were much lower than in model crops, reducing the number of lines that needed to be generated and screened to find good sources of resistance. In addition, as the use of increasingly shorter sequences expressed from these various ihpRNAs, irRNAs, or hpRNAs (the last will be referred to here) were assessed for resistance, it became possible to generate chimeric, fused hpRNAs expressing sequences of several viruses from the same transformation vector. The first demonstration of the proof of principle with hpRNA was to the NIa protease coding sequences of PVY in transgenic tobacco (Smith et al., 2000) and in transgenic barley expressing hpRNAs against the polymerase coding sequences of BYDV (PAV strain) (Wang, Abbott, et al., 2000). Subsequently, many laboratories have utilized this approach and reported resistance against numerous viruses in a variety of crop species. Resistance has been conferred using numerous gene segments and to various extents, but in general, complete resistance could be achieved in many transgenic lines.

3.5.1 Cassava expressing hpRNAs

Several transgenic cassava lines expressing hpRNAs to the *Rep* gene of ACMV showed complete resistance to the virus by different inoculation procedures (Vanderschuren, Alder, Zhang, & Gruissem, 2009). Transgenic cassava plants expressing hpRNAs to the capsid protein encoding sequences of *Cassava brown streak Uganda virus* and graft inoculated with cassava infected by the same virus were completely resistant to infection (Yadav et al., 2011); resistance also extended to *Cassava brown streak virus* (CBSV), which caused the same disease (Ogwok et al., 2012). Transgenic cassava plants with natural resistance to ACMV expressing hpRNA sequences of the capsid protein encoding sequence of CBSV showed complete resistance to both viruses with no virus accumulation (Vanderschuren, Moreno, Anjanappa, Zainuddin, & Gruissem, 2012).

3.5.2 Citrus expressing hpRNAs

Transgenic sweet orange expressing hpRNAs derived from the capsid protein gene on RNA3 of *Citrus psorosis virus* (CPsV) showed excellent resistance to infection, as well as little-to-no virus accumulation. By contrast, transgenic sweet orange plants expressing hpRNAs from the *54-kDa* gene on RNA2 showed little-to-no resistance and variable effects on virus accumulation (Reyes et al., 2009), as did transgenic sweet orange expressing hpRNAs of the *p24* gene on RNA1 (Reyes, Zanek, et al., 2011). Transgenic citrus plants expressing CTV hpRNA sequences of either part of the *p23* gene and 3′-NTR (López et al., 2010) or capsid protein gene (Muniz et al., 2012) were all infected by CTV, although some propagated clones of particular lines showed very low levels of virus accumulation. By contrast, transgenic citrus plants expressing hpRNA sequences of parts of three, fused, RNA silencing suppressor coding regions of CTV provided complete resistance to the virus (Soler et al., 2012).

3.5.3 Cucurbits expressing hpRNAs

Transgenic cantaloupe plants expressing hpRNAs to the capsid protein coding sequences of PRSV were resistant to infection by the virus in all four lines (Krubphachaya, Juříček, & Kertbundit, 2007). Transgenic cucumber expressing hpRNAs from the capsid protein gene of CGMMV showed 70–90% resistance to infection in five of seven transgenic lines of the R1 generation, while two lines showed 86% resistance in the R2 generation (Kamachi, Mochizuki, Nishiguchi, & Tabei, 2007). Transgenic cucumber and melon expressing hpRNAs against HC-Pro encoding sequences of ZYMV showed complete resistance against ZYMV, while plants of one transgenic cucumber line that also showed enhanced expression of *RDR1* and *AGO1* exhibited very strong resistance to both WMV and PRSV (Leibman et al., 2011).

3.5.4 Legumes expressing hpRNAs

Transgenic bean plants expressing hpRNAs to the *C1* gene of BGMV showed a high level of resistance in one of 18 lines tested, with infection occurring in only 7% of the plants in the resistant line (Bonfim, Faria, Nogueira, Mendes, & Aragão, 2007). Plants of several transgenic white clover lines expressing hpRNAs to part of the replicase gene of WClMV provided complete resistance to infection by WClMV (Ludlow, Mouradov, & Spangenberg, 2009).

3.5.5 Maize expressing hpRNAs

Transgenic maize expressing hpRNAs to the NIb coding sequences of SCMV showed 50–70% resistance to the virus in the field in the R1 generation, while plants in the R2 generation showed 85–90% resistance to infection (Bai et al., 2008). Transgenic maize expressing hpRNAs of the P1 coding sequence (Zhang, Fu, Gou, Wang, & Li, 2010) or the capsid protein coding sequence (Zhang, Li, et al., 2011) of *Maize dwarf mosaic virus* (MDMV) showed resistance in several lines. In these lines, the disease incidence was 28–45% for the P1 encoding hpRNA sequences and 38–54% for the capsid protein encoding hpRNA sequences, with the resistant plants exhibiting much less symptoms and low levels of virus. Transgenic maize with hpRNA sequences derived from the P1 encoding region of MDMV showed moderate resistance with 33–62% of the plants infected with orders of magnitude lower virus accumulation in the resistant plants (Zhang, Wang, et al., 2013).

3.5.6 N. benthamiana *expressing hpRNAs*

Transgenic *N. benthamiana* expressing hpRNAs derived from 197 bp of the 5′-NTR of PPV showed resistance in 80% of the inoculated plants and no virus was detected, while 20% showed mild symptoms and virus in upper leaves (Pandolfini, Molesini, Avesani, Spena, & Polverari, 2003). In two studies involving transgenic *N. benthamiana* plants expressing hpRNAs to PPV, a high level of resistance to infection by PPV was achieved, whether the hpRNAs were derived from the 5′-NTR/P1 coding region, the P1/HC-Pro coding region, the HC-Pro coding region, or the HC-Pro/P3 coding region (Di Nicola-Negri, Brunetti, Tavazza, & Ilardi, 2005), or separate P1 and HC-Pro coding regions (Zhang et al., 2006). In the former study, more than 90% of all lines tested were completely resistant or nearly so to PPV infection (Di Nicola-Negri et al., 2005), while in the later study, more than half of the lines tested showed complete resistance to infection by PPV (Zhang et al., 2006). A line expressing hp RNAs to the 5′-NTR/P1 coding region of PPV also was resistant to all PPV strains tested (Di Nicola-Negri, Tavazza, Salandri, & Ilardi, 2010). Transgenic *N. benthamiana* expressing hpRNAs of different regions of the PPV genome (5′-NTR/P1, P1/HC-Pro, HC-Pro, and two lines of HC-Pro/P3) showed excellent resistance to infection by seven of nine isolates of PPV (representing three strains with 96.2–98.5% sequence identity to the transgene). Only the plants expressing the 5′-NTR/P1 hpRNA sequences showed resistance to two other strains of PPV (with 76.1% and 78.6% sequence identity to the transgene) (Di Nicola-Negri et al., 2010). *N. benthamiana* plants transgenically expressing hpRNAs to sequences encoding the capsid protein of PPV also showed resistance to infection by PPV, with those constructs containing longer based-paired arms showing a higher percentage of resistant plants (Hily et al., 2007). Transgenic *N. benthamiana* expressing hpRNAs to various part of RNA2 (replicase gene or 3′-NTR), or the capsid protein gene and part of the 3′-NTR of CMV showed resistance to infection by the virus in all cases. The resistance varied from 32% (3′-NTR of RNA2) to 75% (replicase gene) in the R0 generation, and 100% resistance in most lines of the R2 generation, although only to same virus isolate, and with no resistance to a CMV strain in a different subgroup (Chen, Lohuis, Goldbach, & Prins, 2004). In the first demonstration of multiple virus resistance using hpRNAs, 15 of 32 transgenic *N. benthamiana* lines expressing fused, 150-bp hpRNA sequences of four tospoviruses [TSWV, GRSV, TCSV, and *Watermelon silver mottle virus* (WSMoV)] showed 100% resistance to a mixture of the four viruses, 10 lines showed some level of resistance (three lines with 33%, two lines with 66%, and five lines with 80% or 83%), and seven of 32 lines showed no resistance (Bucher et al., 2006). Complete resistance to infection occurred in about half the transgenic *N. benthamiana* plants from two of five lines expressing a 938-bp hpRNA of the geminivirus *Tomato chlorotic mottle virus*, encompassing the intergenic regions and flanking sequences covering parts of the *AC1*, *AC4*, *AC5*, and *AV1* genes (Ribiero, Lohuis, Goldbach, & Prins, 2007). Resistance, but not immunity to infection, with lower virus titers present, was observed after inoculation with PPV, PDV, and ToRSV in transgenic *N. benthamiana* expressing hpRNAs from a cassette of six fruit tree viruses (Liu, Scorza, Hily, Scott, & James, 2007); the other viruses could not be tested. Transgenic *N. benthamiana* expressing hpRNA against capsid protein encoding sequences of PStV showed complete resistance to the virus in three transgenic lines of the R2 generation, and less resistance in other lines of earlier generations (Yan, Xu, Chen, Goldbach, & Prins, 2007). Transgenic *N. benthamiana* plants expressing hpRNAs of the movement protein coding sequence of *Grapevine fanleaf virus* (GFLV) exhibited a range of responses, even within the same line, including susceptibility, delayed infection, recovery, and complete resistance; the last was observed in 7.1–66.7% in six of seven transgenic lines (Jardak-Jamoussi et al., 2009). Transgenic *N. benthamiana* expressing hpRNAs derived from the capsid protein gene on RNA3 of the negative-stranded ophiovirus, CPsV, showed excellent resistance to infection, as well as little-to-no virus accumulation, while transgenic plants expressing hpRNAs from the *54K* gene on RNA2 showed little-to-no resistance and variable effects on CPsV accumulation (Reyes et al., 2009). Transgenic *N. benthamiana* expressing hpRNAs of the part of the *Rep* gene/the intergenic region/and part of the movement protein gene of the geminivirus *Chickpea chlorotic dwarf Pakistan virus* exhibited complete resistance to infection by the virus (Nahid, Amin, Briddon, & Mansoor, 2011). Expression of hpRNAs against the 2b coding region of CMV in *N. benthamiana* gave rise to some highly resistant plants, but the percentages of resistant plants differed significantly, depending on whether a gene encoding a translatable green fluorescent protein (GFP) was fused upstream of the hpRNAs or not, and also the nature of the plasmid vector (Tan et al., 2012).

3.5.7 Potato expressing hpRNAs

Transgenic potato plants expressing 605-bp hpRNA sequences derived from the 3′ part of the PVY-N capsid protein coding sequences showed resistance to the virus in 11 of 15 lines, and the plants of one line tested further were resistant to infection by another PVY-N isolate, as well as isolates of PVY-NTN and PVY-O (Missiou et al., 2004). Plants of four transgenic potato cultivars expressing hpRNAs of the capsid protein gene of PVY-NTN showed between 29% and 80% resistance to infection with no detectable virus accumulation (Bukovinszki, Divéki, Csányi, Palkovics, & Balázs, 2007). Two lines of transgenic potato expressing 930-bp hpRNAs to parts of the fused PVX capsid protein and PVY NIb-encoding region were resistant to infection by PVX, PVY-O, PVY-N, and PVY-C, and the viruses could not be detected in those plants (Bai, Guo, Wang, Bai, & Zhang, 2009). Some transgenic potato lines expressing a chimeric hpRNAs containing fused sequences from PVX, PVY, and PLRV were completely resistant to infection by all three viruses (Arif et al., 2012). Transgenic potato expressing hpRNA sequences of PVY capsid protein showed complete resistance to infection by PVY-O and PVY-NTN in most of the transgenic lines (McCue et al., 2012). Transgenic potato expressing hpRNA sequences of the replicase gene of CMV showed complete resistance in three of six lines, with high resistance and low virus levels in three other lines (Ntui et al., 2013). Transgenic potato plants expressing fused hpRNA sequences of three viruses (PVA, PVY, and PLRV) were shown to be resistant to these various viruses, including both PVY-O and PVY-N, although resistance to PLRV was never 100% in any line (Chung, Yoon, & Palukaitis, 2013).

3.5.8 Prunus *species expressing hpRNAs*

All seven plants of one transgenic plum line expressing hpRNAs to the entire capsid protein coding region of PPV exhibited strong resistance to infection by PPV (Hily et al., 2007). Two transgenic plum lines expressing hpRNA sequences of the 5′-NTR/P1 coding sequence of PPV and grafted onto PPV-infected plum plants showed high levels of resistance to infection and virus accumulation (Monticelli, Di Nicola-Negri, Gentile, Damiano, & Ilardi, 2012). Transgenic cherry plants expressing hpRNAs to the capsid protein of *Prunus necrotic ringspot virus* showed resistance to disease and virus accumulation in a number of lines, with reduced virus accumulation in other lines (Song et al., 2013).

3.5.9 Rice expressing hpRNAs

Transgenic rice expressing hpRNAs to 627 bp of segment 8 of *Rice dwarf virus* (RDV) showed either a high level of resistance or delayed and attenuated symptoms to infection by RDV (Ma, Yang, Wang, & Tien, 2005). Similarly, transgenic rice expressing hpRNAs to parts of either the *Pns4* or *Pns12* genes of RDV showed excellent resistance to the virus in some lines for each of two hpRNAs target sites for each gene. However, most lines expressing either hpRNA of the *Pns4* gene also showed some plants with a delayed infection as well as some plants with no resistance, demonstrating that the target RNA or gene on a multicomponent virus also was an important factor to consider (Shimizu, Yoshii, Wei, Hirochika, & Omura, 2009). Transgenic rice expressing hpRNAs to parts of each of the seven genes (distributed over the four genomic RNAs) of RSV showed several types of results when they were assessed for resistance in infection by RSV (Shimizu, Nakazono-Nagaoka, Uehara-Ichiki, Sasaya, & Omura, 2011). Those plants expressing hpRNA sequences of *pC2* (*NSvc2*) and *p4* (disease-specific protein) showed no resistance. Those plants expressing hpRNA sequences of *p2* (*NS2*) and *p3* (*NS3*) showed moderate resistance with a delayed symptom response in the case of *p2* plants and lower virus accumulation for both types of plants. And those plants expressing hpRNA sequences of *pC1* (replicase), *pC3* (nucleocapsid), and *pC4* (movement protein) showed high levels of resistance to symptoms, with *pC1* plants showing lower virus accumulation and *pC3* and *pC4* plants showing no virus accumulation (Shimizu et al., 2011). In another study, transgenic rice expressing hpRNAs to the RSV nucleocapsid protein gene, the disease-specific protein gene, or a fusion of both viral genes showed excellent resistance (with 0–6% of the plants becoming infected) in 25%, 29%, and 39% of the lines, respectively (Ma et al., 2011). Similarly, transgenic rice of two cultivars expressing hpRNAs derived from either the nucleocapsid protein gene- or disease-specific protein gene of RSV showed 96% and 90% resistance, respectively (Zhou et al., 2012). In a second similar study, 96% resistance in two RSV nucleocapsid protein gene hpRNA lines and 100–96% resistance in three RSV disease-specific protein gene hpRNA lines was observed

(Jiang et al., 2013). Transgenic rice expressing hpRNA sequences of RSV nucleocapsid protein showed 62–100% resistance, with no virus detectable in most plants (Park et al., 2012). In another study, involving transgenic rice expressing hpRNAs of the RSV nucleocapsid protein gene, 16–29% resistance to infection by RSV was observed; the differences in the extent of resistance being due in part to the type of promoter used (Zhang et al., 2012). Transgenic rice expressing hpRNA sequences of a nonstructural gene (*P9-1*) of the Fijivirus *Rice black streaked dwarf virus* (RBSDV) showed excellent resistance to the virus in various transgenic lines. Similarly, transgenic rice expressing hpRNA sequences of a nonstructural gene (*Pns9*) of *Rice gall dwarf virus* showed excellent resistance to the virus in various transgenic lines, but no resistance to another phytoreovirus, RDV (Shimizu et al., 2012). Transgenic rice expressing hpRNAs against either the nucleocapsid protein or movement protein gene of *Rice ragged stunt virus* showed excellent resistance to the virus in various transgenic lines, but no resistance to RSV (Shimizu et al., 2013).

3.5.10 Soybean expressing hpRNAs

Transgenic soybean expressing hpRNAs to the capsid protein gene of SDV showed high levels of resistance to the virus, although in a minority of the R2 lines (Tougou et al., 2006). Transgenic soybean expressing hpRNAs to the capsid protein coding sequence of SMV showed either complete resistance in plants and seeds or a milder foliar infection with lower virus levels, as well as milder symptoms on the seeds versus susceptible plants (Kim et al., 2013; Zhang, Xie, et al., 2013).

3.5.11 Sugar beet expressing hpRNAs

Transgenic sugar beets, transformed using *Agrobacterium tumefaciens* and expressing 400-bp hpRNAs of the BNYVV replicase gene, showed excellent resistance in several lines to the A, B, and P types of BNYVV with little virus accumulation in the plant (Lennefors et al., 2006). The resistance to BNYVV was not affected by infection with other beet viruses (Lennefors, van Roggen, Flemming, Savenkov, & Valkonen, 2008). By contrast, transgenic sugar beets roots, transformed using *A. rhizogenes* expressing hpRNAs of three different segments of the BNYVV replicase gene, conferred resistance to the rhizomania disease, but showed a delayed systemic infection in the nontransformed shoots and leaves (Pavli, Panopoulos, Goldbach, & Skaracis, 2010).

3.5.12 Sweet potato expressing hpRNAs

Transgenic sweet potato expressing the sequences encoding the capsid protein and replicase, as well as hpRNA to the capsid protein of *Sweet potato feathery mottle virus* (SPFMV) were assessed for resistance to SPFMV and *Sweet potato chlorotic stunt virus* (SPCSV), two viruses that interact synergistically in sweet potato, and seven of 597 transgenic lines showed resistance to both viruses (Nyaboga, Ateka, Gichuki, & Bulimo, 2008). Transgenic sweet potato plants expressing hpRNAs to parts of the replicase coding sequences of both SPFMV and SPCSV did not show complete resistance to SPCSV, but 10 of 20 transgenic events showed lower virus accumulation and some plants showed milder symptoms. However, following inoculation with both SPCSV and SPFMV, the plants all showed synergistic disease caused by infection with both viruses (Kreuze et al., 2008). Transgenic sweet potato expressing fused hpRNA sequences of four viruses [SPFMV, SPCSV, *Sweet potato virus G* (SPVG), and *Sweet potato mild mottle virus*] were graft inoculated with sweet potato infected by all four viruses, but only two viruses (SPFMV and SPVG) could be transmitted by the grafts. The transgenic lines did not show complete resistance to these viruses, but showed delayed infections with milder symptoms and lower viruses levels (Sivparsad & Gubba, 2014).

3.5.13 Tobacco expressing hpRNAs

Transgenic tobacco expressing hpRNAs to part of the capsid protein gene of CMV showed resistance to virus infection in 17% of the transgenic lines and recovery from infection in 11% of the transgenic lines (Kalantidis, Psaradakis, Tabler, & Tsagris, 2002). Transgenic tobacco expressing hpRNAs against part of the 3′-NTR of CMV largely became infected followed by a recovery, although a low percentage of plants (3.7–9.1%) in various lines showed resistance to any disease (Duan, Wang, & Guo, 2008). Transgenic tobacco plants expressing hpRNA sequences of the *2b* gene of CMV provided strong resistance to CMV in about 30% of the transgenic lines and a delayed infection in another 30% of the transgenic lines (Kavosipour, Niazi, Izadpanah, Afsharifar, & Yasaie, 2012). Elsewhere, expression of hpRNAs against the 2b coding region of CMV in tobacco gave rise to some highly resistant plants, but the percentages of resistant plants differed significantly, depending on whether a gene encoding a translatable GFP was fused upstream of the hpRNAs or not, and also the nature of the plasmid vector (Tan et al., 2012). Transgenic tobacco plants expressing a 1793-bp fragment of the replicase gene from CMV stain O showed complete protection to infection by CMV strains O and Y in three of four lines, while the fourth line showed mild symptoms only in the inoculated leaves with no virus detected in upper leaves (Ntui et al., 2014). Transgenic tobacco plants expressing hpRNA to part of the capsid protein coding region of PVY gave similar high levels of resistance to plants expressing a direct repeat of the same capsid protein coding region, although more resistant transgenic lines were obtained from the hpRNA transformants than for the direct repeat RNA transformants (Zhu, Zhu, Wen, & Song, 2004). Transgenic tobacco plants expressing hpRNAs of different sequences of PVY gave similar results overall: plants expressing hpRNAs of the 3′ region of the capsid protein coding sequences of PVY showed 82.8% resistance to infection by PVY (Zhu et al., 2006). Similarly, plants expressing hpRNAs of the NIb coding sequences of PVY showed complete resistance to five PVY potato strains sharing 99.5–88.1% sequence identity with the transgene. However, only a delayed infection occurred with two tomato strains and a pepper strain sharing 86.8–86.3% with the transgene (Gaba et al., 2010). Transgenic tobacco plants expressing hpRNAs of different regions of the PVY-N genome (the 3′ end 400 nt of the coding sequences for HC-Pro, CI, NIb, or capsid protein) showed resistance levels to infection by the virus of 55.3%, 73.7%, 51.5%, or 84.2%, respectively (Liu et al., 2010). Transgenic tobacco expressing hpRNA sequences derived from each of eight functional genes of PVY showed excellent resistance in most lines regenerated from each construct (Chen et al., 2010). Transgenic tobacco expressing hpRNA sequences of PVY capsid protein showed 24–83% resistance to PVY (Zhang et al., 2012); the differences in the extent of resistance due in part to the type of promoter used. Transgenic tobacco expressing short hpRNAs against various protein coding regions of PVY-N in tobacco were resistant to both PVY-N and TEV (Zhang, Li, Hu, Shan, & Tang, 2013). Transgenic tobacco expressing hpRNA to part of the movement protein gene of TMV demonstrated immunity to infection, but in only about 10% of the transgenic lines (Zhang, Vanderschuren, Fütterer, & Gruissem, 2005). Transgenic tobacco lines expressing hpRNAs of part of the TMV movement protein gene or part of the CMV replicase gene were completely resistant to infection by the corresponding virus (Hu, Niu, Zhang, Liu, & Zhou, 2011; Niu, Wang, Yao, Yan, & You, 2011). Transgenic tobacco expressing fused hpRNA sequences of three viruses (PVA, PVY, and PLRV) or five viruses (PVA, PVY, PLRV, TRV, and PMTV) were shown to be resistant to the various viruses, including both PVY-O and PVY-N, although resistance to PLRV was variable in different lines and only occurred when inoculation was by aphid transmission and not by agroinoculation (Chung & Palukaitis, 2011). Transgenic tobacco expressing hpRNA of the *Tobacco streak virus* (TSV) capsid protein gene showed excellent resistance against the virus, with no symptoms and little-to-no virus detectable (Pradeep et al., 2012). Transgenic tobacco with sense or hpRNA against the AC4 and *Rep* gene of *Mungbean yellow mosaic virus* showed either enhancement of virus infection (sense RNA) or resistance to infection (hpRNA) (Sunitha, Shanmugapriya, Balamani, & Veluthambi, 2013).

3.5.14 Tomato expressing hpRNAs

Transgenic tomato plants expressing 726-bp hpRNAs of the *C1* (*Rep*) gene of TYLCV showed complete resistance to infection in one in six lines, and partial resistance with lower virus accumulation in a second line (Fuentes et al., 2006). Transgenic tomato plants expressing hpRNAs targeted against the TYLCV capsid protein gene showed a reduction in the extent of infection, with 41.5–55.5% of the plants in four transgenic lines showing no symptoms and much lower virus titers (Zrachya et al., 2007). Transgenic tomato expressing hpRNAs of the *Rep* gene of TYLCV (Sardinia variant) were assessed for resistance to TYLCV, and some lines were completely resistant to the virus (Ben Tamarzizt et al., 2009). Transgenic tomato plants expressing hpRNAs derived from PSTVd expressed complete resistance to infection by PSTVd in two of three transgenic lines (Schwind et al., 2009).

3.5.15 Wheat expressing hpRNAs

Transgenic wheat expressing hpRNAs of sequences encoding the replicase gene of BYDV (GPV strain), with an antisense RNA loop of part of the capsid protein gene, showed a high level of resistance in 28.5% of 21 lines, while the other lines showed no symptoms when infected by the virus at a lower dose (Yan, Zhang, Xiao, Li, & Cheng, 2007). Transgenic wheat expressing hpRNAs of the NIa coding sequence of *Wheat streak mosaic virus* (WSMV) showed complete resistance to infection in several lines (Fahim, Ayala-Vavarrete, Millar, & Larkin, 2010). Transgenic wheat expressing hpRNAs against the polymerase gene of BYDV (PAV) did not show symptoms after inoculation with the virus (Yassaie, Afsharifar, Niazi, Salehzadeh, & Izadpanah, 2011).

3.5.16 Other plant species expressing hpRNAs

Transgenic *Arabidopsis* expressing hpRNAs against the capsid protein gene of TuMV showed excellent resistance to the virus (Zeng et al., 2013).

Transgenic banana expressing hpRNAs to the master replication initiation protein gene of *Banana bunchy top virus* gave complete resistance to the virus (Elayabalan et al., 2013; Shekawat, Ganapathi, & Hadapad, 2012).

Transgenic lettuce plants expressing hpRNAs of the capsid protein gene of the ophiovirus MiLV showed good, but not complete, resistance against the virus, with 100%, 93%, and 96% of the plants showing resistance to disease in the R3, R4, and R5 generations, respectively, and 15%, 11%, and 4% of the plants showing the presence of virus in the same respective generations (Kawazu, Fujiyama, & Noguchi, 2009; Kawazu et al., 2010). Even in resistant R3- and R4-generation plants, where virus could not be detected in the leaves, virus was found in the roots (Kawazu et al., 2009).

Transgenic poinsettia plants expressing hpRNAs of sequences encoding the replicase gene (two different constructs) or the capsid protein of *Poinsettia mosaic virus* showed complete resistance to the virus in the two lines expressing the capsid protein gene hpRNAs and the two lines of one construct expressing replicase gene hpRNAs, but only in two of five lines of the other construct expressing replicase gene hpRNAs (Clarke et al., 2008).

Transgenic sunflower expressing hpRNA of the TSV capsid protein gene showed excellent against the virus, with no symptoms and little-to-no virus detectable (Pradeep et al., 2012).

3.5.17 Various parameters affecting resistance

Various hpRNA sequences less than 100 bp have been used successfully to confer high or complete resistance to infection. In one study where 50-bp segments all along the capsid protein coding region of PVY were used (Jiang, Wu, et al., 2011), it was found that resistance was not observed for the 200-bp region encoding the N-terminus of the capsid protein, and only occurred in 6% up to 78% of the plants for the central and C-terminal coding regions. This mirrored and extended a previous study in which transgenic tobacco expressing 50-bp hpRNA sequences corresponding to the 5′ end of the capsid protein coding sequence of PVY-N showed no resistance to infection in 166 transformed plants. At the same time, plants expressing hpRNA sequences corresponding to the 3′ end of the capsid protein coding region gave resistance in 69% of 126 transformed plants (Li, Song, Liu, Zhu, & Wen, 2007). A similar observation was made in transgenic tobacco expressing fused hpRNA sequences of part of the capsid protein encoding sequences of PVY, TMV, and CMV, which were assessed for resistance to all three viruses (Zhu, Song, & Wen, 2008). When hpRNA segments of 100, 100, and 200 bp, respectively, derived from the 3′ regions of the capsid protein encoding sequences of the above three viruses were used, 23% of the transgenic plants were completely resistant to infection by the three viruses (Zhu et al., 2008). However, when hpRNA segments of 200, 200, and 250 bp, respectively, derived from the 5′ regions of the capsid protein encoding sequences of the above three viruses, were used, only 18% of the transgenic plants were completely resistant to all three viruses (Zhu, Song, Yin, & Wen, 2009). Similarly, in two studies involving transgenic tobacco lines expressing one-of-five 51-bp hpRNAs covering different regions of the 1557-nt NIb coding sequences of PVY, differences in the level of resistance to infection by the homologous virus were noted (Xu et al., 2009). When the transgenic lines were derived from a PVY-NTN strain, the levels of resistance were highest in the central region (51.5%) and right-of-center region (67.1%), while the 5′ region (31.1%), left-of-center region (10.3%), and 3′ region (16.1%) showed much lower levels of resistance (Xu et al., 2009). By contrast, when the transgenic lines were derived from a PVY-O strain, the levels of resistance corresponding to the same five regions were 26.2% (5′ region), 22.7% (left-of center-region), 36.3% (central region), 20.3% (right-of-center region), and 21.7% (3′ region). By contrast, resistance to the PVY-NTN strain [with 98%, 96%, 96%, 84%, and 90% sequence identity to the PVY-O strain for the five (5′ to 3′) regions], was 2.4%, 3.0%, 15.9%, and 0% for the last two regions, respectively (Xu et al., 2009). Furthermore, the transgenic lines expressing hpRNAs from the five regions of the PVY-O NIb encoding sequence showed no resistance at all to a PVY-N strain, with 82.4–88.8% sequence identity to the five regions of the PVY-O strain, or to a TEV isolate, with 65–84.3% sequence identity to the PVY-O strain (Xu et al., 2009). This indicates that a high sequence identity, while a prerequisite up to some level to confer resistance, is not by itself sufficient to predict the occurrence of resistance. Moreover, not all regions are equally efficient at generating siRNAs that are able to activate RNA silencing. In some cases, this may be due to secondary structure of the viral RNA and access of the siRNA to its binding site, as has been seen for some amiRNAs (see Section 3.6).

Most studies have used sequences in the loops of the hpRNA-expressing constructs that derive from introns, but many have used loops derived from adjacent viral sequences. In the initial demonstration of using hpRNAs to obtain resistant plants, Smith et al. (2000) showed that the use of introns increased the frequency of PVY-resistant tobacco plants to nearly 100%, while the use of noncleavable introns or other spacer sequences increased the frequency to between 58% and 65%. By contrast,

Chen et al. (2004) found no differences in the frequency of resistant plants for cleavable versus noncleavable introns, for three different constructs with frequencies of resistance of ~30%, ~45%, and 75%. Also in contrast to Smith et al. (2000), another study from the same lab (Wang, Abbott, et al., 2000), using hpRNA with a loop derived from adjacent BYDV sequences, found only 36% resistance in the R0-generation plants. There were a number of differences between the two studies from the same lab, including the host plant, the virus, the promoter, and the size of the hpRNA, as well as the ratio of the hpRNA stem to loop. Does the size of the loop in the hairpin structure affect the extent of resistance? In the study by Smith et al. (2000) involving PVY hpRNA and a noncleavable intron, the ratio of the stem to loop was 0.875:1, while in the work by Wang, Abbott, et al. (2000), the ratio of BYDV stem to loop was 1.86:1. In the report by Chen et al. (2004), the ratios of stems to loops were 3.4:1, 1.8:1, and 1.1:1, with the resistance frequencies observed at 75%, ~45%, and ~30%, respectively. In a study involving a 50-bp hpRNA stem derived from the 3′ region of the PVY-N capsid protein encoding region and different size loops derived from a pUC plasmid, Li, Song, Zhu, and Wen (2008) found that as the ratio of the stem to loop decreased initially from 4:1 to 1:1, the percentage of resistant lines did not change; transgenic lines with a 4:1 to 1:1 ratio showed about 60% resistant plants, as was observed by Smith et al. (2000). However, as the ratio decreased further (1:2, 1:4, and 1:8), the percentage resistant plants decreased to 50%, 44%, and 9%, respectively. Thus, the length of the stem and, if the loop is not an intron, the ratio of the stem-to-loop size can potentially affect the efficiency of generation of resistant transgenic plants. This may be particularly the case for shorter hpRNAs.

It also is not clear what the maximum length is of the viral sequence cassette that can be used to produce hpRNA transcripts and whether this is limited at the stage of stability in the transformation plasmid while in bacteria (Kohli et al., 1999), or at some other stage, such as during transfer of the DNA into the plant genome. In one report, transgenic plants were regenerated that produced hpRNAs of 2500 bp (encompassing ~400 to ~500 bp of each of six fruit tree viruses) and siRNAs to the various viral hpRNAs could be detected. Resistance was observed against three of the viruses that could be tested in *N. benthamiana* (Liu et al., 2007). Although there was no comment with regard to reduced recovery of transformants in that report, in a report by Song et al. (2013) using the same vector construct in two cherry genotypes, no transformants out of 2600 explants could be obtained, while using a vector that produced a 414-bp hpRNA against one virus only, 22 transformants could be obtained from 900 explants. Similarly, in another report, fewer transgenic lines were regenerated in tobacco from a vector in which the viral sequence cassette would produce hpRNAs of 1000 versus 600 bp (Chung & Palukaitis, 2011). This was also the case for transgenic potato expressing the same hpRNAs (Chung et al., 2013; B.N. Chung. & P. Palukaitis, unpublished data). Whether increasing the length of the hpRNA region will increase the deletion frequency of the hpRNA-coding sequences, as has already been noted in at least one case (Sunitha, Shivaprasad, Sujata, & Veluthambi, 2012), remains to be established generally. In response to this possibility, Zhang, Sato, et al. (2011) generated transgenic soybean in which the three hpRNAs (of 109, 123, and 147 bp) to the replicase genes of three viruses examined (AMV, *Bean pod mottle virus*, and SMV) were separated from each other by ssRNA regions. These plants also showed complete resistance to infection by all three viruses. Thus, overall, the use of hpRNAs has shown very high to complete resistance against the target viruses, and resistance to multiple viruses has been achieved by combining (or pyramiding) sequences from up to six viruses (to date) into one vector in one crop.

3.6. Artificial microRNAs

The endogenous microRNA (miRNA) pathway is a powerful mechanism by which eukaryotic cells regulate gene expression, either at the posttranscriptional or at the translational level (Bartel, 2004). miRNA precursors could be engineered to express siRNAs whose sequences were unrelated to the corresponding mature miRNAs (Zeng, Wagner, & Cullen, 2002). The so-called amiRNA technology was then demonstrated to act in plants, guiding the silencing of target genes with high specificity (Schwab, Ossowski, Riester, Warthmann, & Weigel, 2006). The technology of amiRNAs was quickly adapted to produce virus-resistant plants, a strategy that since then has provided numerous successful examples (Table 2.5). Niu et al. (2006) designed expression vectors where, under the transcriptional control of a CaMV 35S promoter, the native pre-miRNA backbone of Ath-miR159a was modified with the insertion of 21 nucleotide sequences, complementary to each other and forming a stem-loop structure, targeting the RNA genomes of TYMV and TuMV. The transformed *A. thaliana* lines

Table 2.5 Artificial microRNAs mediated resistance to viruses in various transgenic plant species

Host	Virus	Genome	Target region in the viral genome	Pre-miRNA backbone	Degrees of resistance	References
Arabidopsis thaliana	TYMV	ssRNA	P69	Ath-miR159a	Complete	Niu et al. (2006)
	TuMV	ssRNA	HC-Pro, CP	Ath-miR159a	Complete	
	TYMV + TuMV				Double, complete	
	TuMV (same as in Niu et al., 2006)	ssRNA	HC-Pro	Ath-miR159a	Susceptible to TuMV resistance-breaking mutants	Lafforgue et al. (2011)
					Susceptible to TuMV when TRV, CaMV, or CMV are preinfected	Martinez, Elena, and Daros (2013)
Nicotiana tabacum	CMV	ssRNA	2b	Ath-miR171a	Various from complete to susceptible	Qu, Ye, and Fang (2007)
Arabidopsis thaliana, Nicotiana tabacum	CMV (isolates of different subgroups)	ssRNA	RISC-accessible RNA3 3′-UTR	Ath-miR159a	Various from complete to susceptible	Duan, Wang, Fang, et al. (2008)
Nicotiana tabacum	PVY	ssRNA	HC-Pro	Ath-miR159a, Ath-miR167b, Ath-miR171a	Various from complete to susceptible	Ai, Zhang, Gao, Zhu, and Guo (2011)
	PVX	ssRNA	p25		Various from complete to susceptible	
	PVY + PVX				Double, complete	

Nicotiana tabacum	PVY	ssRNA	CP	Ath-miR319a	Moderate to low	Jiang, Song, et al. (2011)
Solanum lycopersicum	CMV	ssRNA	2a, 2b, 3′-UTR	Ath-miR159a	Complete	Zhang, Yang, et al. (2011)
Triticum aestivum	WSMV	ssRNA	5′-UTR, P1, HC-Pro, P3, pipo-P3	Osa-miR395 (five amiRNAs in one polycistronic precursor)	Various from complete to susceptible	Fahim, Millar, Wood, and Larkin (2012)
Nicotiana benthamiana	WSMoV	(−) Sense ssRNA	RdRp (L-RNA), six amiRNAs expressed individually, or as triple constructs	Ath-miR159a	Various from complete (triple construct) to susceptible	Kung et al. (2012)
Solanum lycopersicum	ToLCNDV	Circular ssDNA	AV1, AV1/AV2 overlapping region	Ath-mir-159a (highest levels), sly-mir-319a, and sly-mir-168a (low levels)	Various from highly tolerant to susceptible	Vu, Choudhury, and Mukherjee (2013)
Nicotiana benthamiana	CLCuBuV	Circular ssDNA	V2	*Gossypium* mir-169a	Highly tolerant, susceptible to related virus species	Ali, Amin, Briddon, and Mansoor (2013)

accumulated high levels of virus-specific amiRNAs and showed complete resistance to TYMV, TuMV, or to both viruses when they were coinoculated in plants expressing both amiRNAs in a dimeric construct.

The amiRNA strategy for obtaining virus-resistant transgenic genotypes has been extended since then in several different directions. It has been tested positively against other RNA viruses, mainly with positive-sense, ssRNA genome such as CMV, PVY, and PVX, but also including WSMoV (genus *Tospovirus*), a negative-sense, ssRNA virus (Table 2.5). The amiRNA approach also provided resistance to two ssDNA in the genus *Begomovirus*: ToLCV New Delhi variant (Vu et al., 2013) and *Cotton leaf curl Burewala virus* (Ali et al., 2013). In these instances, amiRNAs guided specific cleavage of viral RNA transcripts produced during the viral gene expression processes. Monocot hosts such as wheat have been successfully transformed for amiRNA expression and resistance to WSMV (Fahim et al., 2012).

The level of plant protection from virus infection conferred by amiRNA-based transgenes varied from complete to low, depending on several factors. First of all, the sequence of the pre-miRNA backbone of choice has a major influence on the levels of amiRNA expression and consequently on the efficiency of posttranscriptional cleavage. Comparative studies have shown that pre-miRNAs based on Ath-miR159a ensure higher expression levels than other pre-miRNAs in different host plants (Ai et al., 2011; Vu et al., 2013). Other reports indicate that the levels of resistance depend upon the levels of complementarity between the amiRNA and the target sequence, which may or may not allow cross-protection against related viruses (Ali et al., 2013). Moreover, different target sequences within the virus genome have different effects on virus suppression; for instance, amiRNA sequences targeting sequences coding for viral suppressors of RNA silencing are usually more efficient than amiRNAs directed against capsid protein or other viral genes (Vu et al., 2013). The important issue of whether amiRNA-mediated transgenic approaches to produce virus-resistant genotypes may be durable in time and under different conditions is being addressed appropriately. It was shown that TuMV developed a resistance-breaking mutant population when replicating in transgenic hosts expressing the amiRNA transgene at sub-optimal levels, and that this drawback can be avoided by the expression of amiRNAs complementary either to independent targets in the viral genome or to highly conserved RNA motifs in the viral genome (Lafforgue et al., 2013, 2011). Finally, preinfection of transgenic plants by another virus (e.g., CMV, but also TRV or CaMV) allowed TuMV to overcome the amiRNA-mediated resistance (Martinez et al., 2013).

4. NONVIRAL-MEDIATED RESISTANCE

4.1. Nucleases

The transgenic expression of components of the dsRNA-activated mammalian 2–5A and RNase L system was tested in several laboratories for resistance to plant viruses in transgenic plants. This approach utilizes the interferon-induced antivirus system in which dsRNA activates the synthesis of 2′–5′ oligoadenylate synthetase to produce 2′,5′-linked oligoadenylate (2–5A) that is required to activate RNase L (Lengyel, 1981). In the first example from plants, transgenic potatoes were engineered for expression of 2′–5′ oligoadenylate synthetase and showed some level of protection against PVX in some but not all plants in the field (Truve et al., 1993). However, two subsequent studies did not observe any protection by transgenic expression of 2′–5′ oligoadenylate synthetase alone, but only in combination with the transgenic expression of RNase L (Mitra et al., 1996; Ogawa, Hori, & Ishida, 1996). In the study by Mitra et al. (1996), transgenic tobacco showed good resistance to AMV, but less resistance to TMV and TEV. In all three cases, virus infection produced necrotic lesions on the inoculated leaves and the virus remained confined to the inoculated leaves, but at higher inoculums doses, both TMV and TEV moved systemically, resulting in veinal necrosis of upper and lower leaves. In the study by Ogawa et al. (1996), CMV infection also was limited to necrotic lesions on the inoculated leaves, while PVY infection resulted in systemic necrosis and plant death. Thus, the 2–5 A system appears to have a major drawback for use in controlling potyviruses in particular. This system was soon abandoned and both laboratories focused their attention on two different systems based on transgenic expression of dsRNA-specific RNases in plants.

The yeast-derived dsRNase, Pac1, has been expressed transgenically in a number of studies to examine its ability to provide broad-spectrum protection against plant virus infection. In transgenic tobacco expressing Pac1, a delay in infection and reduced virus titer were observed with CMV, *Tomato mosaic virus* (ToMV), and PVY, as well as a reduced incidence of infection by CMV (Watanabe et al., 1995). Similarly, transgenic potato plants expressing Pac1 were shown to be protected from infection by PSTVd, with a reduction in both the incidence of infection and the viroid accumulation, in both the inoculated plants and the plants generated from tubers (Sano, Nagayama, Ogawa, Ishida, & Okada, 1997). In transgenic chrysanthemum plants, expression of Pac1 resulted in both reduction in infection frequency and reduced titers of TSWV and *Chrysanthemum stunt viroid* (Ogawa et al., 2005). In transgenic potato plants, expression of Pac1 resulted in reduced levels of accumulation of PVX and PVY, but there was no mention of effects on disease development (Colalongo, Contaldo, Faccioli, & Pagan, 2005). Using other dsRNases, including the *E. coli* dsRNase (RNase III), several laboratories demonstrated some protection against infection in transgenic plants to various viruses. Langenberg, Zhang, Court, Guinchedi, and Mitra (1997) used both a wild-type *E. coli* RNase III (rnc) and its cleavage-defective mutant (rnc70), to examine protection against TMV and TEV, which was poor and did not extend either to systemic leaves (TMV) or beyond the R0 generation (TEV). In the case of AMV, a lettuce isolate of TSWV, or *Impatiens necrotic spot virus* (INSV), the protection was quite variable, but generally low, and only in the case of a tomato isolate of TSWV was the protection to a level high (Langenberg et al., 1997). In the case of the mutant RNase, rnc70, in transgenic wheat, the protection against *Barley stripe mosaic virus* improved from transgenic generation R1 to R3, and was manifested by a lack of symptoms and lower virus titer (Zhang, French, Langenberg, & Mitra, 2001). In transgenic potato expressing RNase III, the plants showed only limited protection to PVS, with a lower infection rate (Matoušek et al., 2001). Transgenic expression of an *E. coli* dsRNase gene in maize gave some protection against the Fijivirus RBSDV, one of several viruses that cause maize rough dwarf disease (Cao et al., 2013). Transgenic expression in tobacco of either bovine pancreatic RNase (Trifonova et al., 2007) or an inducible extracellular RNase from *Zinnia elegans* (Trifonova et al., 2012) led to some level of protection against infection by TMV, with a delay in infection or some plants not showing any symptoms. Thus, none of these systems has been able to demonstrate the type of broad-spectrum resistance against a range of viruses, in which all of the plants are completely protected from infection and the resistance is stably maintained.

Interestingly, a novel nuclease-mediated resistance has been described recently. In this case, both a monoclonal antibody against DNA and a single-chain variable fragment (scFv) derived from the monoclonal antibody were shown to have catalytic properties of a sequence-nonspecific nuclease, able to digest ssDNA, dsDNA, as well as ssRNA, and dsRNA (Kim et al., 2006). When expressed in transgenic tobacco plants, this nuclease was able to prevent disease and virus accumulation in all of the plants tested against two geminiviruses, BCTV and *Beet severe curly top virus* (BSCTV) (Lee et al., 2013a), as well as four tobamoviruses [TMV, ToMV, *Pepper mild mottle virus* (PMMoV), and *Tobacco mild green mosaic virus*] and CMV (Lee et al., 2013b), with no apparent effects on plant growth. However, while extension of this system to TuMV in Chinese cabbage provided some initial resistance, but only in a few plants, this resistance was not stable, due to methylation of the transgene, reducing transcription levels of the transgene with loss of the resistance (Zhao, An, Lee, Kim, & Kang, 2013).

4.2. Antiviral inhibitor proteins

A number of plant-derived antiviral proteins have been described in the literature. However, only a few of them have been expressed transgenically and have been shown to have some resistance to virus infection.

4.2.1 Ribosome-inactivating proteins

Several ribosome-inhibiting proteins or ribosome-inactivating proteins (RIPs), such as the pokeweed (*Phytolacca americana*) antiviral protein (PAP), have been examined for their ability to control virus infection in transgenic plants. Transgenic expression of one of three PAPs from *P. americana* was used to assess broad-spectrum resistance to virus infection in several plant species (Lodge, Kaniewski, & Tumer, 1993), showing that good-to-excellent resistance to PVX in tobacco and potato, and to PVY in tobacco, potato, and *N. benthamiana* could be obtained, while moderate resistance to CMV in tobacco also was obtained. PAP also was introduced into transgenic *Brassica napus*, where it provided resistance to infection by TuMV (Zhang et al., 1999). Since wild-type PAP affected the transformation frequency and plant development, a mutant variant of PAP, showing less effects on the hosts, was used in most transgenic events examined (Lodge et al., 1993). This variant PAP was still able to depurinate ribosomes, and exhibit plant toxicity, but a C-terminal deletion of the wild-type PAP did not, although it still retained the ability to reduce the incidence of infection by PVX to between 30% and 60% (Tumer, Hwang, & Bonness, 1997). Transgenic expression of a second *P. americana* PAP (PAPII) in tobacco containing the *N* gene for resistance to TMV also exhibited some protection against PVX and TMV infection (Wang, Zoubenko, & Tumer, 1998), as did RIP I from *Phytolacca heterotepala* expressed transgenically in tobacco and challenged with PVX (Corrado et al., 2008). However, the extent of protection against systemic infection could not be evaluated in either study. Similarly, a RIP from *Phytolacca insularis* expressed in transgenic potato plants also showed some level of protection against infection by PVX, PVY, and PLRV (Moon, Song, Choi, & Lee, 1997). A RIP from *Dianthus caryophyllus*, dianthin, expressed in transgenic *N. benthamiana* from an ACMV AV1 promoter (which is only active after transactivation by the ACMV *AC2* gene product), reduced infection by ACMV, with lower viral DNA accumulation in the infected tissues, attenuated systemic symptoms, and recovery. However, the plants were not resistant to infection by other geminiviruses (Hong, Saunders, Hartley, & Stanley, 1996). In another system, the RIP trichosanthin isolated from *Trichosanthes kirilowii*, was expressed transgenically in tobacco and showed no symptoms of infection by TuMV, the only virus assayed (Lam et al., 1996). However, as tobacco is not a systemic host of TuMV, it is difficult to determine the extent of the resistance in this system. A trichosanthin RIP expressed transgenically in tomato showed some level of protection against infection by TMV, CMV, and TBRV (Jiang, Jin, Weng, Guo, & Wang, 1999, Jiang, Weng, Jin, & Wang, 1998). Similarly, a trichosanthin RIP expressed transgenically in tobacco showed a long delay in infection by TMV and CMV, resulting in milder symptoms, and much lower virus accumulation was detected for TMV (Krishnan, McDonald, Dandekar, Jackman, & Falk, 2002). Two types of RIPs from *Iris hollandica* expressed in transgenic plants showed protection against TMV and TEV in the inoculated leaves, but not against systemic infection (Vandenbussche et al., 2004). Thus, while earlier results appeared promising, variation in expression and the extent of protection in different transgenic lines have not provided the hoped for high level of broad-spectrum protection. In addition, the mechanism of action of RIPs is not clearly understood, but may represent both direct effects on the viral RNAs, as well as induction of host defense responses (reviewed by Wang & Tumer, 2000).

4.2.2 Virus replication inhibitor protein

An inhibitor of virus replication (IVR), isolated originally from TMV infected *N*-gene tobacco (reviewed by Loebenstein, 2009) and having previously been shown to inhibit replication of several viruses (TMV, CMV, and PVX) to various extents when applied ectopically (Gera & Loebenstein, 1983), also was shown to be able to inhibit partially replication of TMV in transgenic plants expressing IVR protein (Akad et al., 2005). Resistance to other viruses was not assessed in this transgenic system, but the inheritance of the resistance was not stable, and when expressed from transgenic tomato plants, no resistance to TMV was observed (A. Gal-On, personal communication), while partial resistance to infection by *Botrytis cinerea* occurred in both plant species (Akad et al., 2005; Loebenstein et al., 2010). The mode of action of IVR against either viruses or fungi is not known.

4.2.3 Artificial zinc finger protein

An artificial zing finger protein (AZP), containing six 28-amino acid, near-repeated, peptide sequences was designed, based on the 19 nt, Rep binding site of the geminivirus BSCTV, which contains two 9-nt direct repeats (Sera, 2005). The AZP was shown to bind to the 19-nt Rep binding site and to interfere with the binding of the Rep protein. R3 transgenic *Arabidopsis* plants expressing the AZP (fused to a nuclear localization signal and a FLAG epitope trag) were evaluated for resistance to infection by BSCTV, with 97 of 116 plants (84%) showing no symptoms and no virus accumulation. These plants were derived from two transgenic R1 lines that showed either no symptoms or slight curling of one inflorescence after agroinfection with BSCTV. The other R3 plants showed either curling of the tops of some inflorescences (14 plants) or deformed or curled tops of all inflorescences, but no effects on plants growth (5 plants). Virus was detected at a low level in the symptomed areas, but not in other parts of the plants (Sera, 2005).

4.2.4 Peptide aptamers

Peptide aptamers are short polypeptides fused to a carrier protein and which bind to and inactivate a target protein. The carrier protein constrains the structure of the peptide aptamer allowing for a higher binding specificity of the peptide (reviewed by Colas, 2000, 2008; Crawford, Woodman, & Ferrigno, 2003). Screening of the initial binding of short peptides (containing ~20–30 amino acids) to a target protein is usually done in the yeast two-hybrid system. Subsequently, tests are done either directly in transgenic plants, or first in plant cells, before testing in transgenic plants. In the first use of this system to examine resistance to plant viruses in transgenic plants, Rudolph, Schreier, and Uhrig (2003) screened a library of random DNase I-generated fragments of the TSWV nucleoprotein gene for expressed and fused peptides that could bind to the TSWV nucleoprotein in the yeast two-hybrid system. (The nucleoprotein contained two mutations that inactivated the C-terminal self-interaction domain.) One identified 29-amino acid peptide was them fused to GUS and tested again in the yeast two-hybrid system with the nucleoproteins of other tospoviruses, all containing C-terminal deletions, and showed interaction with five of eight of the tospovirus nucleoproteins. This peptide was expressed transgenically (as a fusion to GUS) in *N. benthamiana*, and the plants showed resistance to disease development in all four transgenic lines against TSWV, GRSV, and *Chrysanthemum stem necrosis virus*, while plants of only one line showed 50% resistance to TCSV, and none showed resistance to INSV (Rudolph et al., 2003).

The levels of virus accumulation were not assessed in the symptomless plants. In another study, Lopez-Ochoa, Ramirez-Prado, and Handley-Bowdoin (2006) screened a yeast two-hybrid library of random peptides embedded in the active site of thioredoxin A for interaction with full-length or truncated Rep protein of TGMV and identified 88 peptides that bound to both Rep proteins. When expression cassettes for these 88 peptides were cotransfected into tobacco protoplasts along with a replicon cassette of DNA A of TGMV, 31 peptides showed significant reduction in replication of DNA A. Fourteen of these peptides were equal to or better than a trans-dominant-negative Rep mutant described previously (Orozco, Kong, Batts, Elledge, & Hanley-Bowdoin, 2000); i.e., showing 70% or better inhibition of virus replication. The 31 peptides also showed the ability to interact with the Rep protein of another geminivirus, CaLCuV (Lopez-Ochoa et al., 2006). In a follow-up study, 16 peptide aptamers were found to bind to all (5 peptides) or nearly all (11 peptides) Rep proteins of nine geminiviruses in the yeast two-hybrid system (Reyes, Nash, Dallas, Ascencio-Ibáñez, & Hanley-Bowdoin, 2013). In transgenic tomato plants, expressing two of these (20-amino acid) peptide aptamers, which bound strongly to the Rep protein, resistance was observed against disease development induced by either TYLCV or ToMoV, although usually virus accumulation only was delayed. It was suggested that combining peptide aptamer-based resistance with conventional resistance could provide a stronger form of resistance, which might also show a broader spectrum than conventional resistance alone (Reyes et al., 2013).

4.2.5 Cationic peptides

More than 800 antimicrobial cationic peptides have been described. One of these, indolicidin, a 13-amino acid long cationic peptide rich in tryptophan and proline, isolated from cytoplasmic granules of bovine neutrophil cells (Selsted et al., 1992), was modified in sequence and expressed in transgenic tobacco plants also containing the *N* gene for resistance to TMV (Bhargava et al., 2007). Detached leaves from these transgenic plants showed fewer necrotic lesions and lower virus accumulation titers than for either nontransformed plants or GUS-transgenic plants. However, as these plants already contained a resistance gene conferring a hypersensitive response (HR), limiting infection to the inoculated leaves, and whole plants also were not tested, it is difficult to extrapolate this situation to what would happen in whole plants without a natural resistance gene present.

4.3. Plantibodies

The first demonstration of antibody-mediated resistance to a plant virus, *Artichoke mottled crinkle virus* (AMCV), in transgenic *N. benthamiana* plants expressing an scFv against AMCV (Tavladoraki et al., 1993), occurred soon after it was shown that polyclonal antibodies (Düring, Hippe, Kreuzaler, & Schell, 1990; Hiatt, Cafferkey, & Bowdish, 1989) and then scFv antibodies (Firek et al., 1993; Owen, Gandecha, Cockburn, & Whitelam, 1992) could be expressed stably in transgenic tobacco plants. This technology was not that widespread at the time and a number of issues needed to be resolved concerning choosing the appropriate target in the virus, the levels of scFv protein expression, and stabilizing the expressed proteins by directing the proteins to various subcellular compartments (endoplasmic reticulum or plasma lemma; reviewed by Nölke, Fischer, & Schillberg, 2004). Thus, it was some years before further examples of the use of plant-expressed antibodies, also referred to as plantibodies, appeared in the literature. These other examples of transgenic expression of plantibodies against plant viruses and obtaining some measure of resistance to infection have since been reported for 15 plant viruses, not only largely in tobacco and *N. benthamiana* but also in crops such as Chinese cabbage, citrus, gladiolus, potato, and tomato (Table 2.6). Only one of these studies involved a full-size antibody (Voss et al., 1995) and one study expressed a heavy chain variable region (VH) domain (Bouaziz et al., 2009), with the other studies utilized scFv antibodies. The earlier transgenic events did not result in strong resistance, but rather produced delays in infection, or reduced levels of infection. The level of resistance increased in many later studies, when improvements in plantibody targeting, and stabilization were made, and also when plantibodies were generated against nonstructural viral proteins, such as replication-associated/protease proteins (Gargouri-Bouzid et al., 2006; Nickel et al., 2008) and viral replicases (Boonrod et al., 2004; Gil et al., 2011). For example, scFvs targeted against the TBSV replicase and expressed in transgenic *N. benthamiana* were able to confer resistance not only to TBSV and two other members of the family *Tombusviridae* (CNV and TCV) but also to the dianthovirus *Red clover necrotic mosaic virus* (Boonrod et al., 2004). Nevertheless, it is rare that the level of resistance observed using plantibody-mediated systems ever approaches that observed with hpRNA-mediated resistance or most of the earlier examples of coat protein-, movement protein-, or replicase-mediated resistance.

Table 2.6 Plantibody-mediated resistance to viruses in various transgenic plant species

Virus species	Plant species	References
Arabis mosaic virus (ArMV)	*Nicotiana benthamiana*	Nölke et al. (2009)
Artichoke mottled crinkle virus (AMCV)	*N. benthamiana*	Tavladoraki et al. (1993)
Beet necrotic yellow vein virus (BNYVV)	*N. benthamiana*	Fecker, Koenig, and Obermeier (1997)
Citrus tristeza virus (CTV)	Citrus (Mexican lime)	Cervera et al. (2010)
Clover yellow vein virus (CYVV)	Tobacco	Xiao et al. (2000)
Cucumber mosaic virus (CMV)	Gladiolus	Kamo et al. (2012)
	Tomato	Villani, Roggero, Bitti, Benvenuto, and Franconi (2005)
Grapevine fanleaf virus (GFLV)	*N. benthamiana*	Nölke et al. (2009)
Plum pox virus (PPV)	*N. benthamiana*	Gil, Esteban, García, Peña, and Cambra (2011)
Potato leafroll virus (PLRV)	Potato	Nickel et al. (2008)
	Tobacco	Nickel et al. (2008)
Potato virus Y (PVY)	Potato	Gargouri-Bouzid et al. (2006)
	Potato	Bouaziz et al. (2009)[a]
	Potato	Ayadi et al. (2012)
	Tobacco	Xiao et al. (2000)
Tobacco mosaic virus (TMV)	Tobacco	Voss et al. (1995)[a] Zimmermann, Schillberg, Liao, and Fischer (1998) Schillberg, Zimmermann, Findlay, and Fischer (2000) Bajrovic, Erdag, Atalay, and Cirakoclu (2001)
Tomato bushy stunt virus (TBSV)	*N. benthamiana*	Boonrod, Galetzka, Nagy, Conrad, and Krczal (2004)

Table 2.6 Plantibody-mediated resistance to viruses in various transgenic plant species—cont'd

Virus species	Plant species	References
Tomato spotted wilt virus (TSWV)	*N. benthamiana*	Prins, Lohuis, Schots, and Goldbach (2005)
	Tobacco	Zhang, Zimmermann, Fischer, and Schillberg (2008)
Tomato yellow leaf curl virus (TYLCV)	*N. benthamiana*	Safarnejad, Fischer, and Commandeur (2009)
Turnip mosaic virus (TuMV)	Chinese cabbage	Zhao et al. (2013)

[a]Bouaziz et al. (2009) expressed a heavy chain variable region (VH) domain antibody and Voss et al. (1995) expressed genes for a full-length monoclonal antibody; all others expressed genes for single-chain variable fragment (scFv) antibodies.

5. HOST-DERIVED RESISTANCE

Plants respond to their pathogens by deploying a range of molecular interaction mechanisms, some of which orchestrate the first basal defense responses, while others may lead to specific, effective, and durable resistance (Dangl & Jones, 2001). Many genes and different pathways have been demonstrated to have a role in these mechanisms. Some of these genes have been overexpressed in transgenic plants, as a mean for introducing or reinforcing genetic defense barriers to pathogens in otherwise susceptible hosts. Plant genes that have been used as heterologous source of resistance to viral diseases fall principally into three distinct categories: dominant resistance (*R*) genes, recessive resistance genes, and transcription factors (TFs). [See Maule, Caranta, and Boulton (2007) and Palukaitis and Carr (2008) for lists of dominant and recessive *R* genes that have been isolated and characterized.]

5.1. Dominant resistance genes

R genes code for proteins that recognize specific pathogen effectors, known as avirulence proteins, in a specific gene-for-gene fashion. They are classified, according to their domain organization, in nucleotide-binding leucine-rich repeat (NB-LRR) and the extracellular LRR resistance proteins. The NB-LRR class is the most abundant, and members can possess amino-terminal coiled-coil (CC) or Toll/interleukin-1 receptor (TIR) domains.

Extensive reviews on *R* genes and their protein structures and functions have been published previously (Belkhadir, Subramaniam, & Dangl, 2004; Chisholm, Coaker, Day, & Staskawicz, 2006; Dangl & Jones, 2001).

The approach of conferring resistance to viruses by transgenic expression of *R* genes was first described with the *N* gene, which codes for a TIR-NB-LRR class protein that mediates the resistance to TMV and other tobamoviruses in *Nicotiana glutinosa*, where it was originally identified, and in tobacco. The resistant phenotype is accompanied by a HR, consisting of necrotic spots on inoculated leaves. In tomato plants overexpressing the *N* gene, the infection of different strains of TMV and ToMV was confined to limited areas of inoculated leaves, where HR also was observed, whereas systemic host invasion was not observed. The ability of the ectopically expressed *N* gene to reproduce in tomato the resistant phenotype naturally occurring in *Nicotiana* spp. demonstrated that all of the components required by *N* for both TMV recognition and signal transduction are conserved in tomato, another member of the same *Solanaceae* family (Whitham, McCormick, & Baker, 1996). The existence of shared defense pathways between *Nicotiana* and *Solanum* spp. were confirmed by further evidence that *R* genes isolated from wild tomato species and introgressed by breeding in cultivated genotypes were able to confer resistance against viruses in transgenic tobacco. Two *R* genes, $Tm2^2$ and *Sw5*, both encoding proteins belonging to the CC-NBS-LRR class of proteins, have been used extensively in commercial tomato varieties for resistances to ToMV and TSWV, respectively. When transferred to transgenic tobacco lines, both genes mediated a resistant phenotype not different in efficiency and specificity from that observed in the source host (Lanfermeijer et al., 2004; Spassova et al., 2001). The *Rx* gene mediates extreme resistance against PVX. Rx protein is different in structure to the other above mentioned *R* gene-encoded proteins, since it contains a leucine zipper domain at its N-terminus. Another important difference is that resistance response to PVX in potato does not involve a necrotic HR at the site of initial infection and that the suppression of viral infection occurs at the single cell level. This form of resistance has been referred to as extreme resistance. Interestingly, transgenic *N. benthamiana* carrying the *Rx* gene exhibited an identical extreme resistance to avirulent strains of PVX, which was not associated with any cell death mechanism (Bendahmane, Kanyuka, & Baulcombe, 1999).

The concept of transferring *R* genes to a different host species has not always been confirmed in such a straightforward way. For instance, when the *HRT* gene, controlling the resistance to TCV in *A. thaliana*, was overexpressed in the susceptible Col-0, transgenic plants developed an HR but generally remained susceptible to TCV, because a second gene, *RRT*, is also needed for regulating resistance to TCV. Only some plants when inoculated with TCV were resistant to infection (Cooley, Pathirana, Wu, Kachroo, & Klessig, 2000). Seo et al. (2006) demonstrated that a TIR-NBS-LRR gene from common bean (*Phaseolus vulgaris*), *CMR1*, was in fact a *R* gene controlling CMV, by transferring its coding sequence to *N. benthamiana*, a nonrelated host. However, *CMR1* in *N. benthamiana* did not induce resistance, but did activate a resistance-related response (systemic necrosis) to several strains of CMV.

In at least one case, transgenic expression of a nonviral *R* gene led to virus resistance. Overexpression of the *Prf R* gene against *Pseudomonas syringae pathovar tomato* in tomato led to reduced infection by a range of bacterial pathogens, as well as by TMV, through activation of systemic acquired resistance (Oldroyd & Staskawicz, 1998). Whether this system functions against other viruses is not known.

5.2. Recessive resistance genes

Susceptibility genes code for proteins that are indispensable for pathogen attack. Their mutant alleles can confer resistance if they are incompetent for functions that assist the pathogen invasion. This type of resistance is recessive because the plant must be homozygous relatively to the negative mutant allele for the trait to be exhibited phenotypically. Recessive *R* genes against plant viruses have been mostly correlated to components of the eukaryotic translation initiation complex. Eukaryotic translation initiation factors (eIFs), and particularly the eIF4E and eIF4G protein families, were found to be essential determinants in the outcome of infections by some groups of RNA viruses, notably potyviruses (reviewed by Kang, Yeam, & Jahn, 2005; Maule et al., 2007; Palukaitis & Carr, 2008). Potyviruses rely on host eIFs for different functions in their replication and translation, some of which are yet to be identified (Robaglia & Caranta, 2006). The hypothesis that transgenic expression of mutant eIFs alleles derived from resistant plants could lead to loss of viral susceptibility in a heterologous system was first tested in tomato Micro-Tom lines transformed to express *pvr1*, an eIF4E homologue naturally occurring in wild pepper (*Capsicum chinense*). The recessive allele *pvr1* is known to confer resistance to a number of pepper-infecting potyviruses including PVY, PepMoV, and most TEV strains. Transgenic tomato progenies that overexpressed *pvr1* exhibited dominant resistance, and a range protection against different potyviruses similar to that observed in pepper with *pvr1*-mediated resistance (Kang, Yeam, Li, Perez, & Jahn, 2007). When the pepper eIF4E resistance allele *pvr1*^{2} was expressed transgenically in potato, the resulting transgenic plants were found to be completely resistant to mechanical inoculation of multiple PVY strains. Resistance was not observed in the case of transgenic overexpression of the *pvr1* allele (Cavatorta, 2010). In addition, the endogenous potato *eIF4E* cDNA was isolated and mutated at sites homologous to resistance-related mutations found in pepper *pvr1*^{2}. When these artificial mutant alleles were overexpressed in susceptible potato cultivars, they conferred virus resistance (Cavatorta et al., 2011). Recently, a different strategy has been employed to achieve eIF-mediated resistance to potyviruses. PPV infection in plum requires eIF(iso)4E. Thus, when transgenic plum (*Prunus domestica*) plants expressing short hpRNAs specifically targeting eIF(iso)4E and showing RNA silencing were evaluated for resistance to PPV, 82% of the plants were highly resistant to PPV infection. By contrast, eIF4E-silenced transgenic plum plants did not show PPV resistance, confirming the specificity of a potyviral species for only one eIF isoform in a particular plant–virus interaction (Wang et al., 2013).

5.3. Defense response factors

A third category of plant genes that have been successfully employed for transgenic resistance to viruses includes TFs. TFs are master regulators that control the expression of single genes or gene clusters, through binding to the *cis*-acting element in the promoters of respective targets. They have a major role in fine-tuning host responses to environmental stresses, and therefore several attempts have been made to reinforcing resistance to viral pathogens by engineering plants for TF overexpression.

Ethylene response factors (ERFs) have been typically associated to defense response pathways mediated by ethylene production. Transgenic plants overexpressing different ERF members showed consistently, in different laboratories, resistance to infections of tobamoviruses like TMV and ToMV in tobacco and PMMoV in hot pepper (Cao, Wu, Zheng, & Song, 2005; Fischer & Dröge-Laser, 2004; Shin, Park, et al., 2002; Zhang et al., 2009). Shin, Park, et al. (2002) achieved TF-mediated transgenic resistance for the first time by expressing in pepper *Tsi1*, encoding a tobacco ethylene-responsive element-binding protein/APETALA 2 (EREBP/AP2)-type TF, and observed resistance to both PMMoV and

CMV, but also increased resistance to a bacterial pathogen, *Xanthomonas campestris pv. vesicatoria* and an oomycete pathogen, *Phytophthora capsici*. In another example, constitutive expression in transgenic tobacco plants of the *NtERF5* gene, coding for a TF induced during the *N* gene-mediated HR, and inoculation of those plants with TMV, resulted in a reduced infection, with smaller lesions, less virus accumulation (10–30% of the wild-type levels) and impaired systemic spread at temperatures at which the *N* gene is inactive (Fischer & Dröge-Laser, 2004). In some cases, overexpression of the TF of the ERF family brought broad-spectrum resistance or tolerance not only to diverse pathogens but also to different abiotic stresses. For instance, overexpression of rice *OsBIERF3*, which encodes a protein belonging to the EREBP family, in transgenic tobacco, mediated enhanced disease resistance against infections by ToMV and the bacterial wild fire pathogen, *Pseudomonas syringae pv. tabaci*, and increased tolerance to salt stress (Cao et al., 2005). Similarly, the expression of the soybean *GmERF3* gene in transgenic tobacco plants induced enhanced resistance against TMV and infections by *Ralstonia solanacearum* and *Alternaria alternata*, and provided tolerance to salt and dehydration stresses (Zhang et al., 2009).

In two cases, non-ERF TFs also have been employed for transgenic resistance to viruses. A MYB TF, encoded by the *Osmyb4* rice gene, involved in cold acclimation, was overexpressed in *A. thaliana*. Overexpression of *Osmyb4* induced the modulation of many host genes, including other stress-responsive TFs such as ERF-, WRKY-, and NAM-like proteins. This improved tolerance not only to cold but also to other abiotic adverse conditions like drought, salt, and oxidative stresses and gave tolerance/resistance to TNV and other bacterial and fungal pathogens (Vannini et al., 2006). RTBV, a plant pararetrovirus with a circular dsDNA genome, is the causal agent of rice "tungro" disease. In its host, RTBV relies on two basic leucine zipper (bZIP)-type TFs, RF2a, and RF2b, for activating transcription from the RTBV promoter. Transgenic rice lines overexpressing RF2a and RF2b suppressed the accumulation of both viral RNA and DNA and remained free of tungro disease symptoms in laboratory and greenhouse trials (Dai et al., 2008).

A number of genes encoding other proteins involved in defense responses in plants have been overexpressed transgenically and shown to inhibit infection by diverse pathogens. For example, constitutive expression of a gene encoding a small GTP-binding protein in transgenic tobacco containing the *N* gene enhanced resistance to TMV, but also increased cytokinin levels and induced salicylic acid over-production (Sano et al., 1994).

Another system utilized a human dsRNA-dependent protein kinase, which is part of the interferon system in mammals and inhibits virus replication after dsRNAs-induction by inactivation of initiation translation factor eIF-2α. A gene coding for this kinase was fused to an *Arabidopsis* promoter, which is induced by wounding, and was introduced into transgenic tobacco plants. Some lines of the transgenic plants expressing the kinase showed either delayed and milder symptoms or no symptoms after inoculation with CMV, PVY, or TEV. Those plants with milder symptoms contained lower levels of accumulated virus, while the asymptomatic plants contained no detectable virus (Lim et al., 2002). A TMV-induced gene in pepper (*Capsicum annum*), designated *CaTin2*, was shown to encode a protein that was cell wall-associated and probably a pathogenesis-related protein. Constitutive expression of *CaTin2* in transgenic tobacco resulted in partial protection against infection by CMV and TMV, with lower viruses titers at 2 weeks after inoculation, but not by 1 month after inoculation (Shin, Park, An, & Paek, 2003). Overexpression in transgenic tobacco of a germin-like protein from *C. chinense* (CchGLP), which was identified as a superoxide dismutase, resulted in delayed and milder symptoms of infection by the geminiviruses *Pepper huasteco yellow vein virus* and *Pepper golden mosaic virus*, as well as reduced virus accumulation (Guevara-Olvera et al., 2012). To our knowledge, these various systems have not been tested against other viruses.

Broad resistance to infection by plant viruses was achieved by antisense inhibition of a host gene encoding an *S*-adenosylhomocysteine hydrolase (SAHH). This enzyme is important for a number of *S*-adenosylmethionine-dependent reactions, including the capping of the 5′-end of viral RNAs during replication. Transgenic expression of antisense RNAs to the *SAHH* gene showed greatly reduced infection by TMV, CMV, and PVX, and some reduction in PVY accumulation (Masuta et al., 1995). However, many of the transgenic plants also showed abnormal growth, in the absence of virus infection, due to enhanced cytokinin activity.

6. CONCLUSIONS AND PERSPECTIVES

The above compilation is not comprehensive. We did not include a number of reports that we did not have access too, or those that only appeared in meeting abstracts. Similarly, we did not include those reports of transgenic plants that were generated and never tested for virus resistance, those in which no form of resistance occurred at all, those in which resistance was tested only in protoplasts, cell cultures, or detached leaves, or those for which resistance was not evaluated against the eponymous virus donating the transgene, but rather against either only another virus or PVX expressing the segment of the transgene. Finally, in the interests of space and sanity (ours), we did not include all of the many reports that were published recapitulating the results obtained by the first five-or-so laboratories demonstrating resistance to TMV, CMV, PVX, or PVY in tobacco, CMV in tomato, and PVX or PVY in potato. *Mea culpa*!

The results presented here demonstrate that pathogen-derived resistance against viruses is feasible for most viruses, as long as one can regenerate the plant species involved. The differences in results obtained in different laboratories for the same viral sequences in the same plant species is most likely a combination of particular vectors used, or bacterial strain compatibility for the transformation, and generating sufficient plants to obtain data that are statistically meaningful regarding the conclusions drawn. This is especially true for those studies done before the use of expressing dsRNAs/hpRNAs, where the probability of obtaining resistant plants was much lower, and hence often due to chance. There were at least two studies where the effect of different promoters also showed differences in the numbers of resistant plants generated, and several studies found that the same vector would work in one species of plant but not another, as far as either generating transgenic plants or obtaining resistance in the regenerated plants. Thus, there are many factors that need to be considered when attempting to develop this technology in a new laboratory, especially for a crop species that had not been transformed previously.

Which approach appears to be the most promising? We would say the use of hpRNAs, since this technology has proven to give the highest frequency of resistant transgenic lines that show complete resistance, rather than simply recovery forms of resistance. The use of amiRNAs also is becoming more popular, but there has already been one study indicating that in some cases, virus mutation may readily overcome the ability of the amiRNA to bind to the virus (Lafforgue et al., 2011), with follow-up demonstrations that it could be improved by using amiRNAs against two independent targets in the virus (Kung et al., 2012; Lafforgue et al., 2013). In addition, as has been shown for amiRNAs, and applies as well to the use of very short hpRNAs, in the absence of knowledge of the *in situ* secondary structure of nonencapsidated viral RNA, one must choose several target sites, since many target sites may be too structured for the small RNAs to access their target sites (Duan, Wang, Fang, et al., 2008; Duan, Wang, et al., 2008; Jiang, Song, et al., 2011). One of the supposed advantages of either amiRNAs or short hpRNAs is that they are less likely to have "off-target effect"; i.e., interacting with a host RNA causing some form of developmental defect. However, there is at least one report that short hpRNAs can induce off-target effects not observed with longer hpRNAs (Praveen, Ramesh, Mishra, Koundal, & Palukaitis, 2010). Finally, it appears that the amiRNAs do not confer resistance to grafted, nontransgenic parts plants (Zhang, Li, et al., 2011), which would have implications for those situations where the rootstock was transgenic but the scion was not, or vice versa. However, time and more examples of applications will determine to what extent these concerns are just aberrations. It must also be remembered that the original approaches applied in the early 1990s have conferred resistance and continue to do so after many years in both vegetable crops and fruit trees.

What of the future? There will undoubtedly be more examples of new plant species with resistance developed against regional viruses, as well as against new viruses that have yet to be described. New technologies may come along to enhance the efficiency of development of resistant plants still further. The application of cisgenesis, the use of genes from the host or related species but introduced by biolistics or some other mechanisms that does not involves introduction of microbial sequences (Espinoza et al., 2013; Holme, Wendt, & Holm, 2013; Manoj & Shekhawat, 2014; Telem et al., 2013), may aid in removing some of the hostility to the application of this technology. The technologies we have at present will see further deployment in the Developing World, where it is most needed to assist in reducing the losses incurred due to virus infection. However, in the foreseeable future, it is unlikely that Europe will see further deployment of this technology. Western Europe is self-sufficient for food and has chosen to accept political rather than scientific bases for rejection of this technology. Eastern Europe appears to be following the same nonscientific arguments and thus those nations will also not see use of this technology in the near future. Hopefully, the rest of the world where this technology is really needed will be more enlightened.

ACKNOWLEDGMENTS

We would like to thank various colleagues for their unpublished information regarding transgenic resistance. F. C. was supported with a grant from the Ministry of Economy and Finance to the CNR, "Integrate knowledge for Sustainability and Innovation of Made in Italy Agro-Food" (CISIA, l. 191/2009), and a grant to the Public-Private Laboratory GenoPom "Integrating post-genomic platforms to enhance the tomato production chain" (GenoPOMpro, Cod. PON02_00395_3082360). P. P. was supported by grant number RDA-PJ007984 from the Next Generation BioGreen21 Program of the Rural Development Administration, Republic of Korea, and grant number NRF-2013R1A2A2A01016282 from the Korean National Research Foundation.

REFERENCES

Ai, T., Zhang, L., Gao, Z., Zhu, C. X., & Guo, X. (2011). Highly efficient virus resistance mediated by artificial microRNAs that target the suppressor of PVX and PVY in plants. *Plant Biology*, *13*, 304–316.

Akad, A., Teverovsky, E., Gidoni, D., Elad, Y., Kirschner, B., Rav-David, D., et al. (2005). Resistance to *Tobacco mosaic virus* and *Botrytis cinerea* in tobacco transformed with the complementary DNA encoding an inhibitor of viral replication-like protein. *Annals of Applied Biology*, *147*, 89–100.

Ali, I., Amin, I., Briddon, R. W., & Mansoor, S. (2013). Artificial microRNA-mediated resistance against the monopartite begomovirus Cotton leaf curl Burewala virus. *Virology Journal*, *10*, 231.

Amudha, J., Balasubramani, G., Malathi, V. G., Monga, D., Bansal, K. C., & Kranthi, K. R. (2010). Cotton transgenics with antisense AC1 gene for resistance against cotton leaf curl virus. *Electronic Journal of Plant Breeding*, *1*, 360–369.

Anderson, J. M., Palukaitis, P., & Zaitlin, M. (1992). A defective replicase gene induces resistance to cucumber mosaic virus in transgenic tobacco plants. *Proceedings of the National Academy of Sciences of the United States of America*, *89*, 8759–8763.

Andika, I. B., Kondo, H., & Tamada, T. (2005). Evidence that RNA silencing-mediated resistance to *Beet necrotic yellow vein virus* is less effective in roots than in leaves. *Molecular Plant-Microbe Interactions*, *18*, 194–204.

Angenent, G. C., van den Ouweland, J. M. W., & Bol, J. F. (1990). Susceptibility to virus infection of transgenic tobacco plants expressing structural and nonstructural genes of tobacco rattle virus. *Virology*, *175*, 191–198.

Antignus, Y., Vunsh, R., Lachman, O., Pearlsman, M., Maslenin, L., Hananya, U., et al. (2004). Truncated Rep gene originated from *Tomato yellow leaf curl virus-Israel* (Mild) confers strain-specific resistance in transgenic tomato. *Annals of Applied Biology*, *144*, 39–44.

Aragão, F. J. L., Ribeiro, S. G., Barros, L. M. G., Brasileiro, A. C. M., Maxwell, D. P., Rech, E. L., et al. (1998). Transgenic beans (*Phaseolus vulgaris* L.) engineered to express viral antisense RNAs show delayed and attenuated symptoms to bean golden mosaic geminivirus. *Molecular Breeding*, *4*, 491–499.

Ares, X., Calamante, G., Cabral, S., Lodge, J., Hemenway, P., Beachy, R. N., et al. (1998). Transgenic plants expressing the potato virus X ORF2 protein (p24) are resistant to tobacco mosaic virus and Ob tobamoviruses. *Journal of Virology*, *72*, 731–738.

Arif, M., Azhar, U., Arshad, M., Zafar, Y., Mansoor, S., & Asad, S. (2012). Engineering broad-spectrum resistance against RNA viruses in potato. *Transgenic Research*, *21*, 303–311.

Asad, S., Haris, W. A. A., Bashir, A., Zafar, Y., Malik, K. A., Malik, N. N., et al. (2003). Transgenic tobacco expressing geminiviral RNAs are resistant to the serious viral pathogen causing cotton leaf curl disease. *Archives of Virology*, *148*, 2341–2352.

Asurmendi, S., Berg, R. H., Koo, J. C., & Beachy, R. N. (2004). Coat protein regulates formation of replication complexes during *tobacco mosaic virus* infection. *Proceedings of the National Academy of Sciences of the United States of America*, *101*, 1415–1420.

Asurmendi, S., Berg, R. H., Smith, T. J., Bendahmane, M., & Beachy, R. N. (2007). Aggregation of TMV CP plays a role in CP functions and in coat-protein-mediated resistance. *Virology*, *366*, 98–106.

Atkins, D., Young, M., Uzzell, S., Kelly, L., Fillatti, J., & Gerlach, W. L. (1995). The expression of antisense and ribozyme genes targeting citrus exocortis viroid in transgenic plants. *Journal of General Virology*, *76*, 1781–1790.

Audy, P., Palukaitis, P., Slack, S. A., & Zaitlin, M. (1994). Replicase-mediated resistance to potato virus Y in transgenic tobacco plants. *Molecular Plant-Microbe Interactions*, 7, 15–22.

Ayadi, M., Bouaziz, D., Nouri-Ellouz, O., Rouis, S., Drira, N., & Gargouri-Bouzid, R. (2012). Efficient resistance to Potato virus Y infection conferred by cytosolic expression of anti-viral protease single-chain variable fragment antibody in transgenic potato plants. *Journal of Plant Pathology*, *94*, 561–569.

Azadi, P., Otang, N. V., Supaporn, H., Khan, R. S., Chin, D. P., Nakamura, I., et al. (2011). Increased resistance to cucumber mosaic virus (CMV) in *Lilium* transformed with a defective CMV replicase gene. *Biotechnology Letters*, *33*, 1242–1255.

Bai, Y., Guo, Z., Wang, X., Bai, D., & Zhang, W. (2009). Generation of double-virus-resistant marker-free transgenic potato plants. *Progress in Natural Science*, *19*, 543–548.

Bai, Y., Yang, H., Qu, L., Zheng, J., Zhang, J., Wang, M., et al. (2008). Inverted-repeat transgenic maize plants resistant to sugarcane mosaic virus. *Frontiers of Agriculture in China*, *2*, 125–130.

Bai, Q., Zhu, J., Liu, X., Zhu, C., Song, Y., & Wen, F. (2005). Production of transgenic tobacco plants resistant to two viruses via an RNA-mediated virus resistance. *Acta Phytopathologica Sinica*, *35*, 148–154.

Bajrovic, K., Erdag, B., Atalay, E. O., & Cirakoclu, B. (2001). Full resistance to tobacco mosaic virus infection conferred by the transgenic expression of a recombinant antibody in tobacco. *Biotechnology and Biotechnological Equipment*, *15*, 21–27.

Balázs, E., Bukovinszki, A., Csányi, M., Csillléry, G., Divéki, Z., Nagy, I., et al. (2008). Evaluation of a wide range of pepper genotypes for regeneration and transformation with an *Agrobacterium tumefaciens* shooter strain. *South African Journal of Botany*, *74*, 720–725.

Barajas, D., Tenllado, F., González-Jara, P., Martínez-García, B., Atencio, F. A., & Díaz-Ruíz, J. R. (2004). Resistance to *Plum pox virus* (PPV) in *Nicotiana benthamiana* plants transformed with the PPV HC-Pro silencing suppressor gene. *Journal of Plant Pathology*, *86*, 239–248.

Bardonnet, N., Hans, F., Serghini, M. A., & Pinck, L. (1994). Protection against virus infection in tobacco plants expressing the coat protein of grapevine fanleaf nepovirus. *Plant Cell Reports*, *13*, 357–360.

Barker, H., Reavy, B., Kumar, A., Webster, K. D., & Mayo, M. A. (1992). Restricted virus multiplication in potatoes transformed with the coat protein gene of potato leafroll luteovirus: Similarities with a type of host gene-mediated antigen resistance. *Annals of Applied Biology*, *120*, 55–64.

Barker, H., Reavy, B., & McGeachy, K. (1998). High level of resistance to potato mop-top virus induced by transformation with the coat protein gene. *European Journal of Plant Pathology*, *104*, 737–740.

Bartel, D. P. (2004). MicroRNAs: Genomics, biogenesis, mechanism, and function. *Cell*, *116*, 281–297.

Bau, H.-J., Cheng, Y.-H., Yu, T.-A., Yang, J.-S., & Yeh, S.-D. (2003). Broad spectrum resistance to different geographic strains of *Papaya ringspot virus* in coat protein gene transgenic plants. *Phytopathology*, *93*, 112–120.

Baulcombe, D. C., Saunders, G. R., Bevan, M. W., Mayo, M. A., & Harrison, B. D. (1986). Expression of biologically active viral satellite RNA from the nuclear genome of transformed plants. *Nature*, *321*, 446–449.

Bazzini, A. A., Hopp, H. E., Beachy, R. N., & Asurmendi, S. (2006). Posttranscriptional gene silencing does not play a significant role in *Potato virus X* coat protein-mediated resistance. *Phytopathology*, *96*, 1175–1178.

Beck, D. L., Van Dollenweerd, C. J., Lough, T. J., Balmori, E., Voot, D. M., Andersen, M. T., et al. (1994). Disruption of virus movement confers broad-spectrum resistance against systemic infection by plant viruses with a triple gene block. *Proceedings of the National Academy of Sciences of the United States of America*, *91*, 10310–10314.

Belkhadir, Y., Subramaniam, R., & Dangl, J. L. (2004). Plant disease resistance protein signaling: NBS-LRR proteins and their partners. *Current Opinion in Plant Biology*, *7*, 391–399.

Ben Tamarzizt, H., Chouchane, S. G., Lengliz, R., Maxwell, D. P., Marakchi, M., Fakhfakh, H., et al. (2009). Use of *Tomato leaf curl virus* (TYLCV) truncated Rep gene sequence to engineer TYLCV resistance in tomato plants. *Acta Virologia*, *53*, 99–104.

Bendahmane, M., & Beachy, R. N. (1999). Control of tobamovirus infections via pathogen-derived resistance. *Advances in Virus Research*, *53*, 369–386.

Bendahmane, M., Chen, I., Asurmendi, S., Bazzini, A. A., Szecsi, J., & Beachy, R. N. (2007). Coat protein-mediated resistance to TMV infection of *Nicotiana tabacum* involves multiple modes of interference by coat protein. *Virology*, *366*, 107–116.

Bendahmane, M., & Gronenborn, B. (1997). Engineering resistance against tomato yellow leaf curl virus (TYLCV) using antisense RNA. *Plant Molecular Biology*, *33*, 351–357.

Bendahmane, A., Kanyuka, K., & Baulcombe, D. C. (1999). The Rx gene from potato controls separate virus resistance and cell death responses. *Plant Cell*, *11*, 781–791.

Bendahmane, M., Szecsi, J., Chen, I., Berg, R. H., & Beachy, R. N. (2002). Characterization of mutant tobacco mosaic virus coat protein that interferes with virus cell-to-cell movement. *Proceedings of the National Academy of Sciences of the United States of America*, *99*, 3645–3650.

Bertioli, D. J., Cooper, J. I., Edwards, M. L., & Hawes, W. S. (1992). Arabis mosaic nepovirus coat protein in transgenic tobacco lessens disease severity and virus replication. *Annals of Applied Biology*, *120*, 47–54.

Bevan, M. W., Mason, S. E., & Goelet, P. (1985). Expression of tobacco mosaic virus coat protein by a cauliflower mosaic virus promoter in plants transformed by *Agrobacterium*. *The EMBO Journal*, *4*, 1921–1926.

Bhargava, A., Osusky, M., Hancock, R. E., Forward, B. S., Kay, W. W., & Misra, S. (2007). Antiviral indolicidin variant peptides: Evaluation from broad-spectrum disease resistance in transgenic *Nicotiana tabacum*. *Plant Science*, *172*, 515–523.

Bi, Y., Shan, L., Wang, X., & Xu, P. (1999). Regeneration of pepper plants resistant to TMV and CMV and its resistance assay. *Acta Agriculturae Boreali-Sinica*, *14*, 103–108.

Bonfim, K., Faria, J. C., Nogueira, E. O. P. L., Mendes, E. A., & Aragão, F. J. L. (2007). RNAi-mediated resistance to *Bean golden mosaic virus* in genetically engineered common bean (*Phaseolus vulgaris*). *Molecular Plant-Microbe Interactions*, *20*, 717–726.

Boonrod, K., Galetzka, D., Nagy, P. D., Conrad, U., & Krczal, G. (2004). Single-chain antibodies against a plant viral RNA-dependent RNA polymerase confer virus resistance. *Nature Biotechnology*, *22*, 856–862.

Bouaziz, D., Ayadi, M., Bidani, A., Rouis, S., Nouri-Ellouz, O., Jellouli, R., et al. (2009). A stable cytosolic expression of VH antibody fragment directed against PVY NIa protein in transgenic potato plants confers partial protection against the virus. *Plant Science*, *176*, 489–496.

Brault, V., Candresse, T., Le Gall, O., Delbos, R. P., Lanneau, M., & Dunez, J. (1993). Genetically engineered resistance against grapevine chrome mosaic nepovirus. *Plant Molecular Biology*, *21*, 89–97.

Braun, C. J., & Hemenway, C. L. (1992). Expression of amino-terminal portions or full-length viral replicase genes in transgenic plants confers resistance to potato virus X infection. *Plant Cell*, *4*, 735–744.

Brederode, F. Th., Taschner, P. E. M., Posthumus, E., & Bol, J. F. (1995). Replicase-mediated resistance to alfalfa mosaic virus. *Virology*, *207*, 467–474.

Brunetti, A., Tavazza, M., Noris, E., Tavazza, R., Caciagli, P., Ancora, G., et al. (1997). High expression of truncated viral Rep protein confers resistance to tomato yellow leaf curl virus in transgenic tomato plants. *Molecular Plant-Microbe Interactions*, *10*, 571–579.

Bucher, E., Lohuis, D., van Poppel, P. M. J. A., Geerts-Dimitriadou, C., Goldbach, R., & Prins, M. (2006). Multiple virus resistance at a frequency using a single transgene construct. *Journal of General Virology*, *87*, 3697–3701.

Bukovinszki, Á., Divéki, Z., Csányi, M., Palkovics, L., & Balázs, E. (2007). Engineering resistance to PVY in different potato cultivars in a marker-free transformation system using a 'shooter mutant' *A. tumefaciens*. *Plant Cell Reports*, *26*, 459–465.

Cai, W.-Q., Fang, R.-X., Shang, W.-S., Wang, X., Zhang, F.-L., Li, Y.-R., et al. (2003). Development of CMV- and TMV-resistant transgenic chili pepper: Field performance and biosafety assessment. *Molecular Breeding*, *11*, 25–35.

Canto, T., & Palukaitis, P. (1998). Transgenically expressed cucumber mosaic virus RNA 1 simultaneously complements replication of cucumber mosaic virus RNAs 2 and 3 and confers resistance to systemic infection. *Virology*, *250*, 325–336.

Cao, X., Lu, Y., Di, D., Zhang, Z., Liu, H., Tian, L., et al. (2013). Enhanced virus resistance in transgenic maize expressing a dsRNA-specific endoribonuclease gene from *E. coli*. *PLoS One*, *8*, e60829.

Cao, Y., Wu, Y., Zheng, Z., & Song, F. (2005). Overexpression of the rice EREBP-like gene OsBIERF3 enhances disease resistance and salt tolerance in transgenic tobacco. *Physiological and Molecular Plant Pathology*, *67*, 202–211.

Carr, J. P., Marsh, L. E., Lomonossoff, G. P., Sekiya, M. E., & Zaitlin, M. (1992). Resistance to tobacco mosaic virus induced by the 54-kDa gene sequence requires expression of the 54-kDa protein. *Molecular Plant-Microbe Interactions*, *5*, 397–404.

Cassidy, B. G., & Nelson, R. S. (1995). Differences in protection phenotypes in tobacco plants expressing coat protein genes from peanut stripe potyvirus with or without an engineered ATG. *Molecular Plant-Microbe Interactions*, *8*, 357–365.

Castel, S. E., & Martienssen, R. A. (2013). RNA interference in the nucleus: Roles for small RNAs in transcription, epigenetics and beyond. *Nature Reviews. Genetics*, *14*, 100–112.

Cavatorta, J. R. (2010). *Intragenic virus resistance in potato*. Doctoral dissertation, Cornell University.

Cavatorta, J., Perez, K. W., Gray, S. M., Van Eck, J., Yeam, I., & Jahn, M. (2011). Engineering virus resistance using a modified potato gene. *Plant Biotechnology Journal*, *9*, 1014–1021.

Cervera, M., Esteban, O., Gil, M., Gorris, M. T., Martínez, M. C., Peña, L., et al. (2010). Transgenic expression in citrus of single-chain antibody fragments specific to *Citrus tristeza virus* confers virus resistance. *Transgenic Research*, *19*, 1001–1015.

Chang, C., Chen, Y.-C., Hsu, Y.-H., Wu, J.-T., Hu, C.-C., Chang, W.-C., et al. (2005). Transgenic resistance to *Cymbidium mosaic virus* in *Dendrobium* expressing the viral capsid protein gene. *Transgenic Research*, *14*, 41–46.

Chellappan, P., Masona, M. V., Vanitharani, R., Taylor, N. J., & Fauquet, C. M. (2004). Broad spectrum resistance to ssDNA viruses associated with transgene-induced silencing in cassava. *Plant Molecular Biology*, *56*, 601–611.

Chen, X., Liu, J., Xu, L., Jiang, F., Xie, X., Zhu, C., et al. (2010). Inhibiting virus infection by RNA interference of the eight functional genes of the potato virus Y genome. *Journal of Phytopathology*, *158*, 776–784.

Chen, Y.-K., Lohuis, D., Goldbach, R., & Prins, M. (2004). High frequency induction of RNA-mediated resistance against *Cucumber mosaic virus* using inverted repeat constructs. *Molecular Breeding*, *14*, 215–226.

Chen, G., Ye, C. M., Huang, J. C., Yu, M., & Li, B. J. (2001). Cloning of the papaya ringspot virus (PRSV) replicase gene and generation of PRSV-resistant papayas through the introduction of the PRSV replicase gene. *Plant Cell Reports*, *20*, 272–277.

Cheng, Y., Wu, G., Wang, J., Pan, N., & Chen, Z. (1997). Resistance of transgenic tomatoes expressing cucumber mosaic virus coat protein to the virus. *Acta Botanica Sinica, 39*, 16–21.
Cheng, Z., Wu, M., Xia, G., Chen, H., He, X., & Zhou, G. (2000). Transgenic wheat resistant to BYDV were obtained by biolistics. *Acta Phytopathologica Sinica, 30*, 116–121.
Chia, T.-F., Chan, Y.-S., & Chua, N.-H. (1992). Characterization of cymbidium mosaic virus coat protein gene and its expression in transgenic tobacco plants. *Plant Molecular Biology, 18*, 1091–1099.
Chisholm, S. T., Coaker, G., Day, B., & Staskawicz, B. J. (2006). Host-microbe interactions: Shaping the evolution of the plant immune response. *Cell, 124*, 803–814.
Chowrira, G. M., Cavileerm, T. D., Gupta, S. K., Lurquin, P. F., & Berger, P. H. (1998). Coat protein-mediated resistance to pea enation mosaic virus in transgenic *Pisum sativum* L. *Transgenic Research, 7*, 265–271.
Chu, P. W. G., Anderson, B. J., Khan, M. R. I., Shukla, D., & Higgins, T. J. V. (1999). Production of *Bean yellow mosaic virus* resistant subterranean clover (*Trifolium subterraneum*) plants by transformation with the virus coat protein gene. *Annals of Applied Biology, 135*, 469–480.
Chung, B. N., & Palukaitis, P. (2011). Resistance to multiple viruses in transgenic tobacco expressing fused, tandem repeat, virus-derived double-stranded RNAs. *Virus Genes, 43*, 454–464.
Chung, B. N., Yoon, J.-Y., & Palukaitis, P. (2013). Engineering resistance in potato against potato leafroll virus, potato virus A and potato virus Y. *Virus Genes, 47*, 86–92.
Cillo, F., Finetti-Sialer, M. M., Papanice, M. A., & Gallitelli, D. (2004). Analysis of mechanisms involved in the Cucumber mosaic virus satellite RNA-mediated transgenic resistance in tomato plants. *Molecular Plant-Microbe Interactions, 17*, 98–108.
Clarke, J. L., Spetz, C., Haugslien, S., Xing, S., Dees, M. W., Moe, R., et al. (2008). *Agrobacterium tumefaciens*-mediated transformation of poinsettia, *Euphorbia pulcherrima*, with virus-derived hairpin RNA constructs confers resistance to *Poinsettia mosaic virus*. *Plant Cell Reports, 27*, 1027–1038.
Clough, G. H., & Hamm, P. B. (1995). Coat protein transgenic resistance to watermelon mosaic and zucchini yellows mosaic virus in squash and cantaloupe. *Plant Disease, 79*, 1107–1109.
Colalongo, C., Contaldo, N., Faccioli, G., & Pagan, P. (2005). The use of pac1 gene from *Schizosaccaromyces pombe* to protect potato from *Potato virus Y* (PVY) and *Potato virus X* (PVX) infections. *Phytopathologia Mediterranea, 44*, 226–231.
Colas, P. (2000). Combinational protein reagents to manipulate protein function. *Current Opinion in Chemical Biology, 4*, 54–59.
Colas, P. (2008). The eleven-year switch of peptide aptamers. *Journal of Biology, 7*, 2.
Collinge, D. B., Jørgensen, H. J. L., Lond, O. S., & Lyngkjaer, M. F. (2010). Engineering pathogen resistance in crop plants: Current trends and future prospective. *Annual Review of Phytopathology, 48*, 269–291.
Cooley, M. B., Pathirana, S., Wu, H.-J., Kachroo, P., & Klessig, D. F. (2000). Members of the Arabidopsis HRT/RPP8 family of resistance genes confer resistance to both viral and oomycete pathogens. *Plant Cell, 12*, 663–676.
Cooper, J. I., Edwards, M. L., Rosenwasser, O., & Scott, N. W. (1994). Transgenic resistance genes from nepoviruses: Efficacy and other properties. *New Zealand Journal of Crop and Horticultural Science, 22*, 129–137.
Cooper, B., Lapidot, M., Heick, J. A., Dodds, J. A., & Beachy, R. N. (1995). A defective movement protein of TMV in transgenic plants confers resistance to multiple viruses whereas the functional analog increases susceptibility. *Virology, 206*, 307–313.

Corrado, G., Scarpetta, M., Alioto, D., Di Maro, A., Polito, L., Parente, A., et al. (2008). Inducible antiviral activity and rapid production of the Ribosome-Inactivating Protein I from *Phytolacca heterotepala* in tobacco. *Plant Science, 174*, 467–474.

Crawford, M., Woodman, R., & Ferrigno, P. K. (2003). Peptide aptamers: Tools for biology and drug discovery. *Briefings in Functional Genomics and Proteomics, 2*, 72–79.

Csorba, T., Pantaleo, V., & Burgyán, J. (2009). RNA silencing: An antiviral mechanism. *Advances in Virus Research, 75*, 35–71.

Cuozzo, M., O'Connell, K. M., Kaniewski, W., Fang, R.-X., Chua, N.-H., & Tumer, N. E. (1988). Viral protection in transgenic tobacco plants expressing the cucumber mosaic virus coat protein or its antisense RNA. *Bio/Technology, 6*, 549–557.

da Câmara Machado, A., Katinger, H., da Câmara, Laimer, & Machado, M. (1994). Coat protein-mediated protection against plum pox virus in herbaceous model plants and transformation of apricot and plum. *Euphytica, 77*, 129–134.

Dai, S., Wei, X., Alfonso, A. A., Pei, L., Duque, U. G., Zhang, Z., et al. (2008). Transgenic rice plants that overexpress transcription factors RF2a and RF2b are tolerant to rice tungro virus replication and disease. *Proceedings of the National Academy of Sciences of the United States of America, 105*, 21012–21016.

Dangl, J. L., & Jones, J. D. (2001). Plant pathogens and integrated defence responses to infection. *Nature, 411*, 826–833.

Day, A. G., Bejarano, E. R., Buck, K. W., Burrel, M., & Lichtenstein, C. P. (1991). Expression of an antisense viral gene in transgenic tobacco confers resistance to the DNA virus tomato golden mosaic virus. *Proceedings of the National Academy of Sciences of the United States of America, 88*, 6721–6725.

De Block, M., Herrera-Estrella, L., Van Montagu, M., Schell, J., & Zambryski, P. (1984). Expression of foreign genes in regenerated plants and their progeny. *The EMBO Journal, 3*, 1681–1989.

De Feyter, R., Young, M., Schroeder, K., Dennis, E., & Gerlach, W. (1996). A ribozyme gene and an antisense gene are equally effective in conferring resistance to tobacco mosaic virus on transgenic tobacco. *Molecular and General Genetics, 250*, 329–338.

de Haan, P., Gielen, J. J. L., Prins, M., Wijkamp, I. G., Van Schepen, A., Peters, D., et al. (1992). Characterization of RNA-mediated resistance to tomato spotted wilt virus in transgenic tobacco plants. *Bio/Technology, 10*, 1133–1137.

Di Nicola-Negri, E., Brunetti, A., Tavazza, M., & Ilardi, V. (2005). Hairpin RNA-mediated silencing of *Plum pox virus* P1 and HC-Pro genes for efficient and predictable resistance to the virus. *Transgenic Research, 14*, 989–994.

Di Nicola-Negri, E., Tavazza, M., Salandri, L., & Ilardi, V. (2010). Silencing of *Plum pox virus* 5′UTR/P1 sequence confers resistance to a wide range of PPV strains. *Plant Cell Reports, 29*, 1435–1444.

Di, R., Purcell, V., Collins, G. B., & Ghabrial, S. A. (1996). Production of transgenic soybean lines expressing the bean pod mottle virus coat protein precursor gene. *Plant Cell Reports, 15*, 746–750.

Dietzgen, R. G., & Mitter, N. (2006). Transgenic gene silencing strategies for virus control. *Australasian Plant Pathology, 35*, 605–618.

Dinant, S., Blaise, F., Kusiak, C., Astier-Manifacier, S., & Albouy, J. (1993). Heterologous resistance to potato virus Y in transgenic tobacco plants expressing the coat protein gene of lettuce mosaic potyvirus. *Phytopathology, 83*, 818–824.

Dinant, S., Maisonneuve, B., Albouy, J., Chupeau, Y., Chupeau, M.-C., Bellec, Y., et al. (1997). Coat protein-mediated protection in *Lactuca sativa* against lettuce mosaic potyvirus strains. *Molecular Breeding, 3*, 75–86.

Ding, S.-W. (2010). RNA-based antiviral immunity. *Nature Reviews Immunology, 10*, 632–644.

Domínguez, A., Hermoso de Mendoza, A., Guerri, J., Cambra, M., Navarro, L., Moreno, P., et al. (2002). Pathogen-derived resistance to *Citrus tristieza virus* (CTV) in transgenic Mexican lime (*Citrus aurantifolia* (Christ.) Swing.) plants expressing its p25 coat protein gene. *Molecular Breeding, 10*, 1–10.

Dong, J., He, Z., Han, C., Chen, X., Zhang, L., Liu, W., et al. (2002). Generation of transgenic wheat resistant to wheat yellow mosaic virus and identification of gene silence induced by virus infection. *Chinese Science Bulletin, 47*, 1446–1450.

Dong, J., Li, T., Hasi, A., & Zhang, H. (1999). Resistance of transgenic potato expressing intergenic sequence of potato leafroll virus. *Virologica Sinica, 14*, 66–72.

Donson, J., Kearney, C. M., Turpen, T. H., Khan, I. A., Kurath, G., Turpen, A. M., et al. (1993). Broad resistance to tobamoviruses is mediated by a modified tobacco mosaic virus replicase gene. *Molecular Plant-Microbe Interactions, 6*, 635–642.

Dougherty, W. G., Lindbo, J. A., Smith, H. A., Parks, D. T., Swaney, S., & Proebsting, W. M. (1994). RNA-mediated virus resistance in transgenic plants: Exploitation of a cellular pathway possibly involved in RNA degradation. *Molecular Plant-Microbe Interactions, 7*, 544–552.

Du, J., Sun, Y., Wang, J., Hao, Y., Wang, Y., & Zhang, L. (2011). Transgenic maize plants with rice NibT gene and their MDMV-resistance. *Xi bei Zhi wu Xue bao, 31*, 893–901.

Duan, Y.-P., Powell, C. A., Webb, S. E., Purcifull, D. E., & Hiebert, E. (1997). Geminivirus resistance in transgenic tobacco expressing mutated BC1 protein. *Molecular Plant-Microbe Interactions, 10*, 617–623.

Duan, C.-C., Wang, C.-H., Fang, R.-X., & Guo, H.-S. (2008). Artificial microRNAs highly accessible to targets confer efficient virus resistance in plants. *Journal of Virology, 82*, 11084–11095.

Duan, C., Wang, C., & Guo, H. (2008). Delayed resistance to Cucumber mosaic virus mediated by 3′UTR-derived hairpin RNA. *Chinese Science Bulletin, 53*, 3301–3310.

Duan, C.-G., Wang, C.-H., & Guo, H.-S. (2012). Application of RNA silencing to plant disease resistance. *Silence, 3*, 5.

Düring, K., Hippe, S., Kreuzaler, F., & Schell, J. (1990). Synthesis and self assembly of a functional monoclonal antibody in transgenic *Nicotiana tabacum*. *Plant Molecular Biology, 15*, 281–293.

Eames, A., Wang, M.-B., Smith, N. A., & Waterhouse, P. M. (2008). RNA silencing in plants: Yesterday, today and tomorrow. *Plant Physiology, 147*, 456–468.

Edgington, S. M. (1992). Ribozymes: Stop making sense. *Bio/Technology, 10*, 256–262.

Edwards, M. L., Liu, Y. Y., Hawes, W. S., Henderson, J., & Cooper, J. I. (1997). Positive and negative sense coat protein gene-mediated protection against poplar mosaic carlavirus in *Nicotiana benthamiana*. *Annals of Applied Biology, 130*, 261–270.

Elayabalan, S., Kalaiponmani, K., Subramaniam, S., Selvarajan, R., Panchanathan, R., Muthuvelayoutham, R., et al. (2013). Development of *Agrobacterium*-mediated transformation of highly valued hill banana cultivar Virupakshi (AAB) for resistance to BBTV disease. *World Journal of Microbiology & Biotechnology, 29*, 589–596.

Emran, A. M., Tabei, Y., Kobayashi, K., Yamaoka, N., & Nishiguchi, M. (2012). Molecular analysis of transgenic melon plants showing virus resistance conferred by direct repeat of movement gene of *Cucumber green mottle mosaic virus*. *Plant Cell Reports, 31*, 1371–1377.

Espinoza, C., Schlechter, R., Herrera, D., Torres, E., Serrano, A., Medina, C., et al. (2013). Cisgenesis and intragenesis: New tools for improving crops. *Biological Research, 46*, 323–331.

Fagoaga, C., López, C., Hermosa de Mendoza, A., Moreno, P., Navarro, L., Flores, R., et al. (2006). Post-transcriptional gene silencing of the p23 silencing suppressor of *Citrus tristza virus* confers resistance to the virus in transgenic Mexican lime. *Plant Molecular Biology, 60*, 153–165.

Fahim, M., Ayala-Vavarrete, L., Millar, A. A., & Larkin, P. J. (2010). Hairpin RNA derived from viral *NIa* gene confers immunity to wheat streak mosaic virus infection in transgenic wheat plants. *Plant Biotechnology Journal, 8*, 821–834.

Fahim, M., Millar, A. A., Wood, C. C., & Larkin, P. J. (2012). Resistance to Wheat streak mosaic virus generated by expression of an artificial polycistronic microRNA in wheat. *Plant Biotechnology Journal, 10*, 150–163.

Fang, G., & Grumet, R. (1993). Genetic engineering of potyvirus resistance using constructs derived from the zucchini yellow mosaic virus coat protein gene. *Molecular Plant-Microbe Interactions, 6*, 358–367.

Faria, J. C., Albino, M. M. C., Dias, B. B. A., Cançado, L. J., da Cunha, N. B., Silva, L. de M., et al. (2006). Partial resistance to *Bean golden mosaic virus* in a transgenic common bean (*Phaseolus vulgaris* L.) line expressing a mutated *rep* gene. *Plant Science, 171*, 565–571.

Farinelli, L., & Malnoë, P. (1993). Coat protein gene-mediated resistance to potato virus Y in tobacco: Examination of the resistance mechanisms—Is the transgenic coat protein required for protection? *Molecular Plant-Microbe Interactions, 6*, 284–292.

Febres, V. J., Lee, R. J., & Moore, G. A. (2008). Transgenic resistance to *Citrus tristeza virus* in grapefruit. *Plant Cell Reports, 27*, 93–104.

Fecker, L. F., Koenig, R., & Obermeier, C. (1997). *Nicotiana benthamiana* plants expressing beet necrotic yellow vein virus (BNYVV) coat protein-specific scFv are partially protected against the establishment of the virus in the early stages of infection and the pathogenic effects in the late stages of infection. *Archives of Virology, 142*, 1857–1863.

Fellers, J. P., Collins, G. B., & Hunt, A. G. (1998). The NIa-proteinase of different plant potyviruses provides specific resistance to viral infection. *Crop Science, 38*, 1309–1319.

Fermin, G., Inglessis, V., Garboza, C., Rangel, S., Dagert, M., & Gonsalves, D. (2004). Engineered resistance against Papaya ringspot virus in Venezuelan transgenic papayas. *Plant Disease, 88*, 516–522.

Ferreira, S. A., Pitz, K. Y., Manschardt, R., Zee, F., Fitch, M., & Gonsalves, D. (2002). Viral coat protein transgenic papaya provides practical control of *Papaya ringspot virus* in Hawaii. *Plant Disease, 86*, 101–105.

Firek, S., Draper, J., Owen, M. R., Gandecha, A., Cockburn, B., & Whitelam, G. C. (1993). Secretion of a functional single-stranded Fv protein in transgenic tobacco plants and cell suspension cultures. *Plant Molecular Biology, 23*, 861–870.

Fischer, U., & Dröge-Laser, W. (2004). Overexpression of *NtERF5*, a new member of the tobacco ethylene response transcription factor family enhances resistance to *Tobacco mosaic virus*. *Molecular Plant-Microbe Interactions, 17*, 1162–1171.

Fitch, M. M. M., Manshardt, R. M., Gonsalves, D., Slightom, J. L., & Sandford, J. C. (1992). Virus resistant papaya plants derived from tissue bombarded with the coat protein gene of papaya ringspot virus. *Bio/Technology, 10*, 1466–1472.

Frischmuth, T., & Stanley, J. (1994). Beet curly top virus symptom amelioration in *Nicotiana benthamiana* transformed with a naturally occurring viral subgenomic DNA. *Virology, 200*, 826–830.

Fuchs, M., Cambra, M., Capote, N., Jelkmann, W., Kundu, J., Laval, V., et al. (2007). Safety assessment of transgenic plums and grapevines expressing viral coat protein genes: New insights into real environmental impact of perennial plants engineered for virus resistance. *Journal of Plant Pathology, 89*, 5–12.

Fuchs, M., & Gonsalves, D. (1995). Resistance of transgenic hybrid squash ZW-20 expressing the coat protein genes of zucchini yellow mosaic virus and watermelon mosaic virus 2 to mixed infections by both potyviruses. *Bio/Technology, 13*, 1466–1473.

Fuchs, M., & Gonsalves, D. (2007). Decades after their introduction: Lessons from realistic field risk assessment studies. *Annual Review of Phytopathology, 43*, 173–202.

Fuchs, M., McFerson, J. R., Tricoli, D. M., McMaster, J. R., Deng, R. Z., Boeshore, M. L., et al. (1997). Cantaloupe line CZW-30 containing coat protein genes of cucumber mosaic virus, zucchini yellow mosaic virus and watermelon mosaic virus-2 is resistant to these three viruses in the field. *Molecular Breeding*, *3*, 279–290.

Fuchs, M., Tricoli, D. M., Carney, K. J., Schesser, M., McFerson, J. R., & Gonsalves, D. (1998). Comparative virus resistance and fruit yield of transgenic squash wirth single and multiple coat protein genes. *Plant Disease*, *82*, 1350–1356.

Fuentes, A., Ramos, P. L., Fiallo, E., Callard, D., Sánchez, Y., Peral, R., et al. (2006). Intron-hairpin RNA derived from replication association protein C1 gene confers immunity to tomato yellow leaf curl virus infection in transgenic tomato plants. *Transgenic Research*, *15*, 291–304.

Furutani, N., Hidaka, S., Kosaka, Y., Shizukawa, Y., & Kanematsu, S. (2006). Coat protein gene-mediated resistance to soybean mosaic virus in transgenic soybean. *Breeding Science*, *56*, 119–124.

Furutani, N., Yamagishi, N., Hidaka, S., Shizukawa, Y., Kanematsu, S., & Kosaka, Y. (2007). Soybean mosaic virus resistance in transgenic soybean caused by post-transcriptional gene silencing. *Breeding Science*, *57*, 123–128.

Gaba, V., Rosner, A., Maslenin, L., Leibman, D., Singer, S., Kukurt, E., et al. (2010). Hairpin-based virus resistance depends on the sequence similarity between challenge virus and discrete, highly accumulating siRNA species. *European Journal of Plant Pathology*, *128*, 153–164.

Gallitelli, D. (2000). The ecology of Cucumber mosaic virus and sustainable agriculture. *Virus Research*, *71*, 9–21.

Gal-On, A., Wolf, D., Antignus, Y., Patlis, L., Ryu, K. H., Min, B. E., et al. (2005). Transgenic cucumbers harboring the 54-kDa putative gene of *Cucumber fruit mottle mosaic tobamovirus* are highly resistant to viral infection and protect non-transgenic scions from sol infection. *Transgenic Research*, *14*, 81–93.

Gal-On, A., Wolf, D., Wang, Y., Faure, J.-M., Pilowsky, M., & Zelcer, A. (1998). Transgenic resistance to cucumber mosaic virus in tomato: Blocking of long-distance movement of the virus in lines harboring a defective viral replicase gene. *Phytopathology*, *88*, 1101–1107.

Ganesan, U., Suri, S. S., Rajasubramaniam, S., Rajam, M. V., & Dasgupta, I. (2009). Transgenic expression of coat protein gene of *Rice tungro bacilliform virus* in rice reduces the accumulation of viral DNA in inoculated plants. *Virus Genes*, *39*, 113–119.

García-Arenal, F., & Palukaitis, P. (1999). Structure and functional relationships of satellite RNAs of cucumber mosaic virus. *Current Topics in Microbiology and Immunology*, *239*, 37–63.

Gargouri-Bouzid, R., Jaoua, L., Ben Mansour, R., Hathat, Y., Ayadi, M., & Ellouz, R. (2005). PVY resistant transgenic potato plants (cv Claustar) expressing the viral coat protein. *Journal of Plant Biotechnology*, *7*, 143–148.

Gargouri-Bouzid, R., Jaoua, L., Rouis, S., Saïdi, M. N., Bouaziz, D., & Ellouz, R. (2006). PVY-resistant transgenic potato plants expressing an anti-NIa protein scFv antibody. *Molecular Biotechnology*, *33*, 133–140.

Gera, A., & Loebenstein, G. (1983). Further studies of an inhibitor of virus replication from tobacco mosaic virus-infected protoplasts of a local lesion responding tobacco cultivar. *Phytopathology*, *73*, 111–115.

Gerlach, W. L., Llewellyn, D., & Haseloff, J. (1987). Construction of a plant disease resistance gene from the satellite RNA of tobacco ringspot virus. *Nature*, *328*, 802–804.

Germundsson, A., Savenkov, E. I., Ala-Poikela, M., & Valkonen, J. P. T. (2007). VPg of *Potato virus A* alone does not suppress RNA silencing but affects virulence of a heterologous virus. *Virus Genes*, *34*, 387–399.

Germundsson, A., & Valkonen, J. P. T. (2006). P1- and VPg-transgenic plants show similar resistance to *Potato virus A* and may compromise long distance movement of the virus in plant sections expressing RNA silencing-based resistance. *Virus Research*, *116*, 208–213.

Gielen, J. J. L., de Haan, P., Kool, A. J., Peters, D., van Grinsven, M. Q. J. M., & Goldbach, R. (1991). Engineering resistance to tomato spotted wilt virus, a negative-strand RNA virus. *Bio/Technology*, *9*, 1363–1367.

Gielen, J., Ultzen, T., Bontems, S., Loots, W., van Schepen, A., Westerbroek, W., et al. (1996). Coat protein-mediated protection to cucumber mosaic virus infections in cultivated cucumber. *Euphytica*, *88*, 139–149.

Gil, M., Esteban, O., García, J. A., Peña, L., & Cambra, M. (2011). Resistance to *Plum pox virus* in plants expressing cytosolic and nuclear single-chain antibodies against viral NIb replicase. *Plant Pathology*, *60*, 967–976.

Golemboski, D. B., Lomonossoff, G. P., & Zaitlin, M. (1990). Plants transformed with a tobacco mosaic virus nonstructural gene sequence are resistance to the virus. *Proceedings of the National Academy of Sciences of the United States of America*, *87*, 6311–6315.

Gonsalves, D., Chee, P., Provvidenti, R., Seem, R., & Slighton, J. L. (1992). Comparison of coat protein-mediated and genetically-derived resistance in cucumbers to infection by cucumber mosaic-virus under field conditions with natural challenge inoculations by vectors. *Bio/Technology*, *10*, 1562–1570.

Gonsalves, C., Xue, B., Yepes, M., Fuchs, M., Ling, K.-S., Namba, S., et al. (1994). Transferring cucumber mosaic virus-white leaf strain coat protein gene into *Cucumis melo* L. and evaluating transgenic plants for protection against infection. *Journal of the American Society for Horticultural Science*, *119*, 345–355.

Goregaoker, S. P., Eckhardt, L. G., & Culver, J. N. (2000). Tobacco mosaic virus replicase-mediated cross-protection: Contributions of RNA and protein-mediated mechanisms. *Virology*, *273*, 267–275.

Gottula, J., & Fuchs, M. (2009). Toward a quarter century of pathogen-derived resistance and practical approaches to plant virus disease control. *Advances in Virus Research*, *75*, 161–183.

Gubba, A., Gonsalves, C., Stevens, M. R., Tricoli, D. M., & Gonsalves, D. (2002). Combining transgenic and natural resistance to obtain broad resistance to tospovirus infection in tomato (*Lycopersicon esculentum* mill). *Molecular Breeding*, *9*, 13–23.

Guevara-Olvera, L., Ruíz-Nito, M. L., Rangel-Cano, R. M., Torres-Pacheco, I., Rivera-Bustamante, R. F., Muñoz-Sánchez, C. I., et al. (2012). Expression of a germin-like protein gene (CchGLP) from a geminivirus-resistant pepper (*Capsicum chinense* Jacq.) enhances tolerance to geminivirus infection in transgenic tobacco. *Physiological and Molecular Plant Pathology*, *78*, 45–50.

Guo, H. S., Cervera, M. T., & García, T. A. (1998). Plum pox potyvirus resistance to transgene silencing that can be stabilized after different number of plant generations. *Gene*, *206*, 263–272.

Guo, H. S., & García, T. A. (1997). Delayed resistance to plum pox potyvirus mediated by a mutated RNA replicase gene: Involvement of a gene-silencing mechanism. *Molecular Plant-Microbe Interactions*, *10*, 160–170.

Guo, X., Lu, S., Zhu, C., Song, Y., Meng, X., Zheng, C., et al. (2001). RNA-mediated viral resistance against potato virus Y (PVY) in transgenic tobacco plants. *Acta Phytopathologica Sinica*, *31*, 349–356.

Guo, Y., Ruan, M., Yao, W., Chen, L., Chen, R., & Zhang, M. (2008). Difference of coat protein mediated resistance to sugarcane mosaic virus between Badila and Funong 91-4621. *Journal of Fujian Agriculture and Forestry University*, *37*, 7–12.

Hackland, A. F., Coetzer, C. T., Rybicki, E. P., & Thomson, J. A. (2000). Genetically engineered resistance in tobacco against South African strains of tobacco necrosis and cucumber mosaic viruses. *South African Journal of Science*, *96*, 33–38.

Hamilton, R. I. (1980). Defenses triggered by previous invaders—Viruses. In J. G. Horsfall, & E. B. Cowling (Eds.), *How plants defend themselves: 5. Plant disease: An advanced treatise* (pp. 279–303). New York: Academic Press.

Hammond, J., & Kamo, K. K. (1995). Effective resistance to potyvirus infection conferred by expression of antisense RNA in transgenic plants. *Molecular Plant-Microbe Interactions, 8*, 674–682.

Han, S.-J., Cho, H. S., You, J.-S., Nam, Y.-W., Park, E. K., Shin, J.-S., et al. (1999). Gene silencing-mediated resistance in transgenic tobacco plants carrying potato virus Y coat protein gene. *Molecules and Cells, 9*, 376–383.

Harrison, B. D., Mayo, M. A., & Baulcombe, D. C. (1987). Virus resistance in transgenic plants that express cucumber mosaic virus satellite RNA. *Nature, 328*, 799–802.

Hashmi, J. A., Zafar, Y., Arshad, M., Mansoor, S., & Asad, S. (2011). Engineering cotton (*Gossypium hirsutum* L.) for resistance to cotton leaf curl disease using viral truncated AC1 DNA sequences. *Virus Genes, 42*, 286–296.

Hayakawa, T., Zhu, Y., Itoh, K., Kimura, Y., & Izawa, T. (1992). Genetically engineered rice resistant to rice stripe virus, an insect-transmissible virus. *Proceedings of the National Academy of Sciences of the United States of America, 89*, 9865–9869.

Hellwald, K.-H., & Palukaitis, P. (1995). Viral RNA as a potential target for two independent mechanisms of replicase-mediated resistance against cucumber mosaic virus. *Cell, 83*, 937–946.

Hemenway, C., Fang, R.-X., Kamiewski, W. K., Chua, N.-H., & Tumer, N. E. (1988). Analysis of the mechanism of protection in transgenic plants expressing the potato virus X coat protein or its antisense RNA. *The EMBO Journal, 7*, 1273–1280.

Hiatt, A., Cafferkey, R., & Bowdish, K. (1989). Production of antibodies in transgenic plants. *Nature, 342*, 76–78.

Higgins, C. M., Hall, R. M., Mitter, N., Cruickshank, A., & Dietzgen, R. (2004). Peanut stripe potyvirus resistance in peanut (*Arachis hypogaea* L.) plants carrying viral coat protein gene sequences. *Transgenic Research, 13*, 59–67.

Hill, K. K., Jarvis-Egan, N., Halk, E. L., Krahn, K. J., Liao, L. W., Mathewson, R. S., et al. (1991). The development of virus-resistant alfalfa, *Medicago sativa* L. *Bio/Technology, 9*, 373–377.

Hily, J.-M., Ravelonandro, M., Damsteegt, V., Bassett, C., Petri, C., Liu, Z., et al. (2007). Plum pox virus coat protein gene intron-hairpin-RNA (ihpRNA) constructs provide resistance to plum pox virus in *Nicotiana benthamiana* and *Prunus domestica*. *Journal of the American Society for Horticultural Science, 132*, 850–858.

Hoekema, A., Huisman, M. J., Molendijk, L., Van den Elzen, P. J. M., & Cornelissen, B. J. C. (1989). The genetic engineering of two commercial potato cultivars for resistance to potato virus X. *Bio/Technology, 7*, 273–278.

Holme, I. B., Wendt, T., & Holm, P. B. (2013). Intragenesis and cisgenesis as alternatives to transgenic crop development. *Plant Biotechnology Journal, 11*, 395–407.

Hong, Y., Saunders, K., Hartley, M. R., & Stanley, J. (1996). Resistance to geminivirus infection by virus-induced expression of dianthin in transgenic plants. *Virology, 220*, 119–127.

Hong, Y., & Stanley, J. (1996). Virus resistance in *Nicotiana benthamiana* conferred by *African cassava mosaic virus* replication-associated protein (AC1) transgene. *Molecular Plant-Microbe Interactions, 9*, 219–225.

Horsch, R. B., Fraley, R. T., Rogers, S. G., Sanders, P. R., Lloyd, A., & Hoffmann, N. (1984). Inheritance of functional foreign genes in plants. *Science, 223*, 496–498.

Hu, Q., Niu, Y., Zhang, K., Liu, Y., & Zhou, X. (2011). Virus-derived transgenes expressing hairpin RNA give immunity to *Tobacco mosaic virus* and *Cucumber mosaic virus*. *Virology Journal, 8*, 41.

Huet, H., Mahendra, S., Wang, J., Sivamani, E., Ong, C. A., Chen, L., et al. (1999). Near immunity to rice tungro spherical virus achieved in rice by a replicase-mediated resistance strategy. *Phytopathology, 89*, 1022–1027.

Huttner, E., Tucker, W., Vermeulen, A., Ignart, F., Sawyer, B., & Birch, R. (2001). Ribozyme genes protecting transgenic melon plants against potyviruses. *Current Issues in Molecular Biology*, *3*, 27–34.

Ingelbrecht, I. L., Ervine, J. E., & Mirkov, T. E. (1999). Posttranscriptional gene silencing in transgenic sugarcane. Dissection of homology-dependent virus resistance in a monocot that has a complex ployploid genome. *Plant Physiology*, *119*, 1187–1197.

Inokuchi, Y., & Hirashima, A. (1987). Interference with viral infection by defective RNA replicase. *Journal of Virology*, *61*, 3946–3949.

Iwanami, T., Shimizu, T., Ito, T., & Hirabayashi, T. (2004). Tolerance to Citrus mosaic virus in transgenic trifolita orange lines harboring capsid polyprotein gene. *Plant Disease*, *88*, 865–868.

Jacquemond, M., Amselem, J., & Tepfer, M. (1988). A gene coding for a monomeric form of cucumber mosaic virus satellite RNA confers tolerance to CMV. *Molecular Plant-Microbe Interactions*, *1*, 311–316.

Jacquemond, M., & Lot, H. (1981). L'ARN satellite du virus de la mosaïque du cocombre. I. Comparaison de l'aptitude à induire la nécrose létale de la tomate d'ARN satellites isolés de plusieurs souches du virus. *Agronomie*, *1*, 927–932.

Jacquemond, M., Teycheney, P.-Y., Carrère, I., Navas-Castillo, J., & Tepfer, M. (2001). Resistance phenotypes of transgenic tobacco plants expressing different cucumber mosaic virus (CMV) coat protein genes. *Molecular Breeding*, *8*, 85–94.

Jacquet, C., Ravelonandro, M., Bachelier, J.-C., & Dunez, J. (1998). High resistance to plum pox virus (PPV) in transgenic plants containing modified and truncated forms of PPV coat protein gene. *Transgenic Research*, *7*, 29–39.

Jan, F.-J., Pang, S.-Z., Fagoaga, C., & Gonsalves, D. (1999). Turnip mosaic virus resistance in *Nicotiana benthamiana* derived by post-transcriptional gene silencing. *Transgenic Research*, *8*, 203–213.

Jardak-Jamoussi, R., Winterhagen, P., Bouamama, B., Dubois, C., Mliki, A., Wetzel, T., et al. (2009). Development and evaluation of a GFLV inverted repeat construct for genetic transformation of grapevine. *Plant Cell, Tissue and Organ Culture*, *97*, 187–196.

Jayasena, K. W., Hajimorad, M. R., Law, E. G., Rehman, A.-U., Nolan, K. E., Zanker, T., et al. (2001). Resistance to *Alfalfa mosaic virus* in transgenic barrel medic lines containing the virus coat protein gene. *Australian Journal of Agricultural Research*, *52*, 67–72.

Jayasena, K. W., Ingham, B. J., Hajimorad, M. R., & Randles, J. W. (1997). The sense and antisense coat protein gene of alfalfa mosaic virus strain N20 confers protection in transgenic tobacco plants. *Australian Journal of Agricultural Research*, *48*, 503–510.

Jiang, G. Y., Jin, D. M., Weng, M. L., Guo, B. T., & Wang, B. (1999). Transformation and expression of trichosanthin gene in tomato. *Acta Botanica Sinica*, *41*, 334.

Jiang, F., Song, Y., Han, Q., Zhu, C., & Wen, F. (2011). The choice of target site is crucial in artificial miRNA-mediated virus resistance in transgenic *Nicotiana benthamiana*. *Physiological and Molecular Plant Pathology*, *76*, 208.

Jiang, Y., Sun, L., Jiang, M., Li, K., Song, Y., & Zhu, C. (2013). Production of marker-free and RSV-resistant transgenic rice using a twin T-DNA system and RNAi. *Journal of Biosciences*, *38*, 573–581.

Jiang, G. Y., Weng, M. L., Jin, D. M., & Wang, B. (1998). Characteristics of TCS transgenic tomato. *Acta Horticulturae Sinica*, *25*, 395–396.

Jiang, F., Wu, B., Zhang, C., Song, Y., An, H., Zhu, C., et al. (2011). Special origin of stem sequence influence the resistance of hairpin expressing plants against PVY. *Biologia Plantarum*, *55*, 528–535.

Jones, A. L., Johansen, I. E., Bean, S. J., Bach, I., & Maule, A. J. (1998). Specificity of resistance to pea seed-borne mosaic potyvirus in transgenic peas expressing the viral replicase (NIb) gene. *Journal of General Virology*, *79*, 3129–3137.

Jongedijk, E., de Schutter, A. A. J. M., Stolte, T., Van den Elzen, P. J. M., & Cornelissen, B. J. C. (1992). Increased resistance to potato virus X and preservation of cultivar properties in transgenic potato under field conditions. *Bio/Technology*, *10*, 422–429.

Józsa, R., Stasevski, Z., & Balázs, E. (2002). High level of field resistance of transgenic tobaccos induced by integrated potato virus Y coat protein gene. *Acta Phytopathologica et Entomologica Hungarica*, *37*, 311–312.

Józsa, R., Stasevski, Z., Wolf, I., Horváth, S., & Balázs, E. (2002). Potato virus Y coat protein gene induces resistance in valuable potato cultivars. *Acta Phytopathologica et Entomologica Hungarica*, *37*, 1–7.

Kadotani, N., & Ikegami, M. (2002). Production of patchouli mild mosaic virus resistant patchouli plants by genetic engineering of coat protein precursor gene. *Pest Management Science*, *58*, 1137–1142.

Kalantidis, K., Psaradakis, S., Tabler, M., & Tsagris, M. (2002). The occurrence of CMV-specific short RNAs in transgenic tobacco expressing virus-derived double-stranded RNA is indicative of resistance to the virus. *Molecular Plant-Microbe Interactions*, *15*, 826–833.

Kamachi, S., Mochizuki, A., Nishiguchi, M., & Tabei, Y. (2007). Transgenic *Nicotiana benthamiana* plants resistant to cucumber green mottle mosaic virus based on RNA silencing. *Plant Cell Reports*, *26*, 1283–1288.

Kamo, K., Aebig, J., Guaragna, M. A., James, C., Hsu, H. T., & Jordan, R. (2012). *Gladiolus* plants transformed with single-chain variable fragment antibodies to *Cucumber mosaic virus*. *Plant Cell, Tissue and Organ Culture*, *110*, 13–21.

Kamo, K., Gera, A., Cohen, J., Hammond, J., Blowers, A., Smith, F., et al. (2005). Transgenic *Gladiolus* plants transformed with the bean yellow mosaic virus coat-protein gene in either sense or antisense orientation. *Plant Cell Reports*, *23*, 654–663.

Kamo, K., Jordan, R., Guaragna, M. A., Hsu, H., & Ueng, P. (2010). Resistance to *Cucumber mosaic virus* in *Gladiolus* plants transformed with either a defective replicase or coat protein subgroup II gene from *Cucumber mosaic virus*. *Plant Cell Reports*, *29*, 695–704.

Kang, B.-C., Yeam, I., & Jahn, M. M. (2005). Genetics of plant virus resistance. *Annual Review of Phytopathology*, *43*, 581–621.

Kang, B. C., Yeam, I., Li, H., Perez, K. W., & Jahn, M. M. (2007). Ectopic expression of a recessive resistance gene generates dominant potyvirus resistance in plants. *Plant Biotechnology Journal*, *5*, 526–536.

Kaniewski, W., Ilardi, V., Tomassoli, L., Mitsky, T., Layton, J., & Barba, M. (1999). Extreme resistance to cucumber mosaic virus (CMV) in transgenic tomato expressing one or two viral coat proteins. *Molecular Breeding*, *5*, 111–119.

Kaniewski, W., Lawson, C., Sammons, B., Haley, L., Hart, J., Delannay, X., et al. (1990). Field resistance of transgenic Russet Burbank potato to effects of infection by potato virus X and potato virus Y. *Bio/Technology*, *8*, 750–754.

Kaper, J. M. (1993). Satellite-mediated symptom modulation: An emerging technology for the biological control of viral crop disease. *Microbial Releases*, *2*, 1–9.

Kaper, J. M., & Waterworth, H. E. (1977). Cucumber mosaic virus associated RNA 5: Causal agent for tomato necrosis. *Science*, *196*, 429–431.

Kaplan, I. B., Shintaku, M. H., Li, Q., Zhang, L., Marsh, L. E., & Palukaitis, P. (1995). Complementation of virus movement in transgenic tobacco expressing the cucumber mosaic virus 3a gene. *Virology*, *209*, 188–199.

Kavosipour, S., Niazi, A., Izadpanah, K., Afsharifar, A., & Yasaie, M. (2012). Induction of resistance to cucumber mosaic virus (CMV) using hairpin construct of 2b gene. *Iranian Journal of Plant Pathology*, *48*, 54–67.

Kawazu, Y., Fujiyama, R., & Noguchi, Y. (2009). Transgenic resistance to *Mirafiori lettuce virus* in lettuce carrying inverted repeats of the coat protein gene. *Transgenic Research*, *18*, 113–120.

Kawazu, Y., Fujiyama, R., Noguchi, Y., Kubota, M., Ito, H., & Fukuoka, H. (2010). Detailed characterization of *Mirafiora lettuce virus*-resistant transgenic lettuce. *Transgenic Research, 19*, 211–220.

Kawazu, Y., Fujiyama, R., Sugiyama, K., & Sasaya, T. (2006). A transgenic lettuce line with resistance to both lettuce big-vein associated virus and mirafiori lettuce virus. *Journal of the American Society for Horticultural Science, 131*, 760–763.

Kawchuk, L. M., Martin, R. R., & McPherson, J. (1990). Resistance in transgenic potato expressing the potato leafroll virus coat protein gene. *Molecular Plant-Microbe Interactions, 3*, 301–307.

Kawchuk, L. M., Martin, R. R., & McPherson, J. (1991). Sense and antisense RNA-mediated resistance to potato leafroll virus in Russet Burbank potato plants. *Molecular Plant-Microbe Interactions, 4*, 247–253.

Kertbundit, S., Pongtanom, N., Ruanjan, P., Chantasingh, D., Tanwanchai, A., Panyim, S., et al. (2007). Resistance of transgenic papaya plants to *Papaya ringspot virus*. *Biologia Plantarum, 51*, 333–339.

Kim, Y.-R., Kim, J.-S., Lee, S.-H., Lee, W.-R., Sohn, J.-N., Chung, Y.-C., et al. (2006). Heavy and light chain variable single domains of an anti-DNA binding antibody hydrolyze both double- and single-stranded DNAs without sequence specificity. *Journal of Biological Chemistry, 281*, 15287–15295.

Kim, S. J., Kim, B. D., & Paek, K. H. (1995). In-vitro translation inhibition and in-vivo viral symptom development attenuation by cucumber mosaic-virus RNA3 cDNA fragments. *Molecules and Cells, 5*, 65–71.

Kim, H.-J., Kim, M.-J., Pak, J. H., Jung, H. W., Choi, H. K., Lee, Y.-H., et al. (2013). Characterization of SMV resistance of soybean produced by genetic transformation of *SMV-CP* gene in RNAi. *Plant Biotechnology Reports, 7*, 425–433.

Kim, S. J., Lee, S. J., Kim, B. D., & Paek, K. H. (1997). Satellite-RNA-mediated resistance to cucumber mosaic virus in transgenic plants of hot pepper (*Capsicum annuum* cv. Golden Tower). *Plant Cell Reports, 16*, 825–830.

Kim, S. J., Paek, K. H., & Kim, B. D. (1995). Delay of disease development in transgenic Petunia plants expressing cucumber mosaic-virus I17N-satellite RNA. *Journal of the American Society for Horticultural Science, 120*, 353–359.

Kobayashi, K., Cabral, S., Calamante, G., Maldonado, S., & Mentaberry, A. (2001). Transgenic tobacco plants expressing the *Potato virus X* open reading frame 3 gene develop specific resistance and necrotic ring symptoms after infection with the homologous virus. *Molecular Plant-Microbe Interactions, 14*, 1274–1285.

Koev, G., Mohan, B. R., Dinesh-Kumar, S. P., Torbert, K. A., Somers, D. A., & Miller, W. A. (1998). Extreme reduction of oats transformed with the 5′ half of the barley yellow dwarf virus-PAV genome. *Phytopathology, 88*, 1013–1019.

Kohli, A., Griffiths, S., Palacios, N., Twyman, R. M., Vain, P., Laurie, D. A., et al. (1999). Molecular characterization of transforming plasmid rearrangements in transgenic rice reveals a recombination hotspot in the CaMV 35S promoter and confirms the predominance of microhomology mediated recombination. *Plant Journal, 17*, 591–601.

Kollàr, A., Dalmay, T., & Burgyàn, J. (1993). Defective interfering RNA-mediated resistance against cymbidium ringspot tombusvirus in transgenic plants. *Virology, 193*, 313–318.

Kollár, A., Thole, V., Dalmay, T., Salamon, P., & Balázs, E. (1993). Efficient pathogen-derived resistance induced by integrated potato virus Y coat protein gene in tobacco. *Biochemie, 75*, 623–629.

Kong, Q. Z., Oh, J. W., & Simon, A. E. (1995). Symptom attenuation by a normally virulent satellite RNA of Turnip crinkle virus is associated with the coat protein open reading frame. *Plant Cell, 7*, 1625–1634.

Kouassi, N. K., Chen, L., Siré, C., Bangratz-Reyser, M., Beachy, R. N., Fauquet, C. M., et al. (2006). Expression of rice yellow mottle virus coat protein enhances virus infection in transgenic rice. *Archives of Virology, 151*, 2111–2122.

Kreiah, S., Edwards, M. L., Hawes, W. S., Jones, A. T., Brown, D. J. F., McGavin, W. J., et al. (1996). Some coat protein constituents from strawberry latent ringspot virus expressed in transgenic tobacco protect plants from systemic invasion following root inoculation by nematode vectors. *European Journal of Plant Pathology, 102*, 297–303.

Kreuze, J. F., Samolski Klein, I., Untiveros Lazaro, M., Cuellar Chuquiyuri, W. J., Lajo Morgan, G., Cipriani Mejía, P. G., et al. (2008). RNA silencing-mediated resistance to a crinivirus (Costeroviridae) in cultivated sweetpotato (*Ipomoea batatas* L.) and development of sweetpotato virus disease following co-infection with a potyvirus. *Molecular Plant Pathology, 9*, 589–598.

Krishnan, R., McDonald, K. A., Dandekar, A. M., Jackman, A. P., & Falk, B. (2002). Expression of recombinant trichosanthin, a ribosome-inactivating protein, in transgenic tobacco. *Journal of Biotechnology, 97*, 69–88.

Krubphachaya, P., Juříček, M., & Kertbundit, S. (2007). Induction of RNA-mediated resistance to papaya ringspot virus type W. *Journal of Biochemistry and Molecular Biology, 40*, 404–411.

Kumar, S., Raj, S. K., Sharma, A. K., & Varma, H. N. (2012). Genetic transformation and development of *Cucumber mosaic virus* resistant transgenic plants of *Chrysanthemum morifolium* cv. Kundan. *Scientia Horticulturae, 134*, 40–45.

Kung, Y.-J., Bau, H.-J., Wu, Y.-L., Huang, C.-H., Chen, T.-M., & Yeh, S.-D. (2009). Generation of transgenic papaya with double resistance to *Papaya ringspot virus* and *Papaya leaf-distortion mosaic virus*. *Phytopathology, 99*, 1312–1320.

Kung, Y.-J., Lin, S.-S., Huang, Y.-L., Chen, T.-C., Harish, S. S., Cua, N.-H., et al. (2012). Multiple artificial microRNAs targeting conserved motifs of the replicase gene confer robust transgenic resistance to negative-sense single-stranded RNA plant virus. *Molecular Plant Pathology, 13*, 303–317.

Kung, Y.-J., Yu, T.-A., Huang, C.-H., Wang, H.-C., Wang, S.-L., & Yeh, S.-D. (2010). Generation of hermaphrodite transgenic papaya lines with virus resistance via transformation of somatic embryos derived from adventitious roots of in vitro shoots. *Transgenic Research, 19*, 621–635.

Kunik, T., Salomon, R., Zamir, D., Navot, N., Zeidan, M., Michelson, I., et al. (1994). Transgenic tomato plants expressing the *Tomato yellow leaf curl virus* capsid protein are resistant to the virus. *Bio/Technology, 12*, 500–504.

Kwon, C. S., Chung, W. I., & Paek, K. H. (1997). Ribozyme mediated targeting of cucumber mosaic virus RNA 1 and 2 in transgenic tobacco plants. *Molecules and Cells, 7*, 326–334.

Lafforgue, G., Martínez, F., Niu, Q. W., Chua, N. H., Daròs, J. A., & Elena, S. F. (2013). Improving the effectiveness of artificial microRNA (amiR)-mediated resistance against Turnip mosaic virus by combining two amiRNAs or by targeting highly conserved viral genomic regions. *Journal of Virology, 87*, 8254–8256.

Lafforgue, G., Martínez, F., Sardanyés, J., de la Iglesia, F., Niu, Q. W., Lin, S. S., et al. (2011). Tempo and mode of plant RNA virus escape from RNA interference-mediated resistance. *Journal of Virology, 85*, 9686–9695.

Lam, Y.-H., Wong, Y.-S., Wang, B., Wong, R. N.-S., Yeung, H.-W., & Shaw, P. C. (1996). Use of trichosanthin to reduce infection by turnip mosaic virus. *Plant Science, 114*, 111–117.

Lanfermeijer, F. C., Jiang, G., Ferwerda, M. A., Dijkhuis, J., de Haan, P., Yang, R., et al. (2004). The durable resistance gene Tm-2^2 from tomato confers resistance against ToMV in tobacco and preserves its viral specificity. *Plant Science, 167*, 687–692.

Langenberg, W. G., Zhang, L., Court, D. L., Guinchedi, L., & Mitra, A. (1997). Transgenic tobacco plants expressing the bacterial *rnc* gene resist virus infection. *Molecular Breeding, 3*, 391–399.

Lapidot, M., Gafny, R., Ding, B., Wolf, S., Lucas, W. J., & Beachy, R. N. (1993). A dysfunctional movement protein of tobacco mosaic virus that partially modifies the plasmodesmata and limits virus spread in transgenic plants. *Plant Journal, 4*, 959–970.

Lauber, E., Janssens, L., Weyens, G., Jonard, G., Richards, K. E., Lefèbvre, M., et al. (2001). Rapid screening for dominant negative mutations in the beet necrotic yellow vein virus triple gene block proteins P13 and P15 using a viral replicon. *Transgenic Research, 10*, 293–302.

Lawson, C., Kaniewski, W., Haley, L., Rozman, R., Newell, C., Sanders, P., et al. (1990). Engineering resistance to mixed infection in a commercial potato cultivar: Resistance to potato virus X and potato virus Y in transgenic Russet Burbank. *Bio/Technology, 8*, 127–134.

Lawson, E. C., Weiss, J. D., Thomas, P. E., & Kaniewski, W. K. (2001). NewLeaf Plus® Russet Burbank potatoes: Replicase-mediated resistance to potato leafroll virus. *Molecular Breeding, 7*, 1–12.

Leclerc, D., & AbouHaidar, M. G. (1995). Transgenic tobacco plants expressing a truncated form of the PAMV capsid protein (CP) gene show CP-mediated resistance to potato aucuba mosaic virus. *Molecular Plant-Microbe Interactions, 8*, 58–65.

Lee, Y. H., Jung, M., Shin, S. H., Lee, J. H., Choi, S. H., Her, N. H., et al. (2009). Transgenic peppers that are highly tolerant to a new CMV pathotype. *Plant Cell Reports, 28*, 223–232.

Lee, G., Shim, H.-K., Kwon, M.-H., Son, S.-H., Kim, K.-Y., Park, E.-Y., et al. (2013a). A nucleic acid hydrolyzing recombinant antibody confers resistance to curtovirus infection in tobacco. *Plant Cell, Tissue and Organ Culture, 115*, 179–187.

Lee, G., Shim, H.-K., Kwon, M.-H., Son, S.-H., Kim, K.-Y., Park, E.-Y., et al. (2013b). RNA virus accumulation is inhibited by ribonuclease activity of 3D8 scFv in transgenic *Nicotiana tabacum*. *Plant Cell, Tissue and Organ Culture, 115*, 189–197.

Lehmann, P., Jenner, C. E., Kozubek, E., Greenland, A. J., & Walsh, J. A. (2003). Coat protein-mediated resistance to *Turnip mosaic virus* in oilseed rape (*Brassica napus*). *Molecular Breeding, 11*, 83–94.

Leibman, D., Wolf, D., Saharan, V., Zelcer, A., Arazi, T., Shiboleth, Y., et al. (2011). A high level of transgenic viral small RNA is associated with broad potyvirus resistance in cucurbits. *Molecular Plant-Microbe Interactions, 24*, 1220–1238.

Lengyel, P. (1981). Enzymology of interferon action—A short survey. *Methods in Enzymology, 79*, 135–148.

Lennefors, B.-L., Savenkov, E. I., Bensefelt, J., Wremerth-Weich, E., van Roggen, P., Tuvesson, S., et al. (2006). dsRNA-mediated resistance to *Beet necrotic yellow vein virus* infections in sugar beet (*Beta vulgaris* L. spp. *vulgaris*). *Molecular Breeding, 18*, 313–325.

Lennefors, B.-L., van Roggen, P. M., Flemming, Y., Savenkov, E. I., & Valkonen, J. P. T. (2008). Efficient dsRNA-mediated transgenic resistance to *Beet necrotic yellow vein virus* in sugar beets is not affected by other soilborne and aphid-transmitted viruses. *Transgenic Research, 17*, 219–228.

Li, H. P., Hu, J. S., Wang, M., & Fan, H. Z. (2000). Studies on transgenic plants transferred with the coat protein gene of cucumber mosaic virus. *Chinese Journal of Virology, 16*, 276–278.

Li, A., Jarret, R. L., & Demski, J. W. (1997). Engineered resistance to tomato spotted wilt virus in transgenic peanut expressing the viral nucleocapsid gene. *Transgenic Research, 6*, 297–305.

Li, Z., Liu, Y., & Berger, P. H. (2005). Transgene silencing in wheat transformed with the WSMV-CP gene. *Biotechnology, 4*, 62–68.

Li, P., Song, Y., Liu, X., Zhu, C., & Wen, F. (2007). Study of virus resistance mediated by inverted repeats derived from 5′ and 3′ ends of coat protein gene of *Potato virus Y*. *Acta Phytopathologica Sinica*, *37*, 69–76.

Li, Y., Song, Y., Zhu, C., & Wen, F. (2008). Effect of stem-loop proportion in hpRNA on the RNA-mediated virus resistance. *Acta Phytopathologica Sinica*, *38*, 468–477.

Liao, L.-J., Pan, I.-C., Chan, Y.-L., Hsu, Y.-H., Chen, W.-H., & Chan, M.-T. (2004). Transgene silencing in *Phalaenopsis* expressing the coat protein of Cymbidium mosaic virus is a manifestation of RNA-mediated resistance. *Molecular Breeding*, *13*, 229–242.

Lim, H. S., Ko, T. S., Hobbs, H. A., Lambert, K. N., Yu, J. M., McCoppin, N. K., et al. (2007). *Soybean mosaic virus* helper component-protease alters leaf morphology and reduces seed production in transgenic plants. *Phytopathology*, *97*, 366–372.

Lim, S.-H., Ko, M. K., Lee, S. J., La, Y. J., & Kim, B.-D. (1999). Cymbidium mosaic virus coat protein gene in antisense confers resistance to transgenic *Nicotiana occidentalis*. *Molecules and Cells*, *9*, 603–608.

Lim, P. O., Lee, U., Ryu, J. S., Choi, J. K., Hovanessian, A., Kim, C. S., et al. (2002). Multiple virus resistance in transgenic plants conferred by the human dsRNA-dependent protein kinase. *Molecular Breeding*, *10*, 11–18.

Lim, P. O., Ryu, J. S., Lee, H. J., Lee, U., Park, J. S., Kwak, J. M., et al. (1997). Resistance to tobamoviruses in transgenic tobacco plants expressing the coat protein gene of pepper mild mottle virus (Korean isolate). *Molecules and Cells*, *7*, 313–319.

Lin, K. Y., Hsu, Y. H., Chen, H. C., & Lin, N. S. (2013). Transgenic resistance to Bamboo mosaic virus by expression of interfering satellite RNA. *Molecular Plant Pathology*, *14*, 693–707.

Lin, C.-Y., Ku, H.-S., Chiang, Y.-H., Ho, H.-Y., Yu, T.-A., & Jan, F.-J. (2012). Development of transgenic watermelon resistant to *Cucumber mosaic virus* and *Watermelon mosaic virus* by using a single chimeric transgene construct. *Transgenic Research*, *21*, 983–993.

Lin, C.-Y., Tsai, W.-S., Ku, H.-S., & Jan, F.-J. (2012). Evaluation of DNA fragments covering the entire genome of a monopartite begomovirus for induction of viral resistance in transgenic plants via gene silencing. *Transgenic Research*, *21*, 231–241.

Lindbo, J. A., & Dougherty, W. G. (1992a). Pathogen-derived resistance to a potyvirus: Immune and resistant phenotypes in transgenic tobacco expressing altered forms of a potyvirus coat protein nucleotide sequence. *Molecular Plant-Microbe Interactions*, *5*, 144–153.

Lindbo, J. A., & Dougherty, W. G. (1992b). Untranslatable transcripts of the tobacco etch virus coat protein gene sequence can interfere with tobacco etch virus replication in transgenic plants and protoplasts. *Virology*, *189*, 725–733.

Lindbo, J. A., Silva-Rosales, L., Proebsting, W. M., & Dougherty, W. G. (1993). Induction of highly specific antiviral state in transgenic plants: Implications for regulation of gene expression and virus resistance. *Plant Cell*, *5*, 1749–1759.

Lines, R. E., Persley, D., Dale, J. L., Drew, R., & Bateson, M. F. (2002). Genetically engineered immunity to *Papaya ringspot virus* in Australian papaya cultivars. *Molecular Breeding*, *10*, 119–129.

Ling, K., Namba, S., Gonsalves, C., Slightom, J. L., & Gonsalves, D. (1991). Protection against detrimental effects of potyvirus infections in transgenic tobacco plants expressing the papaya ringspot virus coat protein gene. *Bio/Technology*, *9*, 752–758.

Ling, K.-S., Zhu, H.-Y., & Gonsalves, D. (2008). Resistance to Grapevine leafroll associated virus-2 is conferred by post-transcriptional gene silencing in transgenic Nicotiana benthamiana. *Transgenic Research*, *17*, 733–740.

Liu, J., Chen, X.-M., Song, Y.-Z., Wu, B., Zhu, C.-X., & Wen, F.-J. (2010). The comparative study of *Potato virus Y HC-Pro*, *CI*, *NIb* and *CP* gene-mediated resistance. *Acta Phytopathologica Sinica*, *40*, 57–65.

Liu, Z., Scorza, R., Hily, J.-M., Scott, S. W., & James, D. (2007). Engineering resistance to multiple *Prunus* viruses through expression of chimeric hairpins. *Journal of the American Society for Horticultural Science*, *132*, 407–414.

Liu, X.-L., Song, Y.-Z., Liu, H.-M., Wen, F.-J., Zhu, C.-X., & Bai, Q.-R. (2005). Virus resistance mediated by the cDNAs encoding for the movement and coat proteins of potato virus X. *Acta Agronomica Sinica, 31*, 827–832.

Liu, X., Tan, Z., Li, W., Zhang, H., & He, D. (2009). Cloning and transformation of SCMV *CP gene* and regeneration of transgenic maize plants showing resistance to SCMV strain MDB. *African Journal of Biotechnology, 8*, 3747–3753.

Liu, X.-H., Zhang, H.-W., Liu, X., Liu, X.-J., Tan, Z.-B., & Rong, T.-Z. (2005). Isolation of the capsid protein gene of maize dwarf mosaic virus and its transformation in maize. *Chinese Journal of Biotechnology, 21*, 144–148.

Lodge, J. K., Kaniewski, W. J., & Tumer, N. E. (1993). Broad-spectrum virus resistance in transgenic plants expressing pokeweed antiviral protein. *Proceedings of the National Academy of Sciences of the United States of America, 90*, 7089–7093.

Loebenstein, G. (2009). Local lesions and induced resistance. *Advances in Virus Research, 75*, 73–117.

Loebenstein, G., Rav David, D., Leibman, D., Gal-On, A., Vunsh, R., Szosnek, H., et al. (2010). Tomato plants transformed with the inhibitor-of-virus-replication gene are partially resistant to *Botrytis cinerea*. *Phytopathology, 100*, 225–229.

Loesch-Fries, L. S., Merlo, D., Zinnen, T., Burhop, L., Hill, K., Krahn, K., et al. (1987). Expression of alfalfa mosaic virus RNA 4 in transgenic plants confers virus resistance. *The EMBO Journal, 6*, 1845–1851.

Loeza-Kuk, E., Gutiérrez-Espinosa, M. A., Ochoa-Martínez, D. L., Villegas-Monter, A., Mora-Agilera, G., Palacios-Torres, E. C., et al. (2011). Resistance analysis in grapefruit and Mexican lime transformed with the p25 *Citrus tristeza virus* gen. *Agrociencia, 45*, 55–65.

Longstaff, M., Brigneti, G., Boccard, F., Chapman, S., & Baulcombe, D. (1993). Extreme resistance to potato virus X infections in plants expressing a modified component of the putative viral replicase. *The EMBO Journal, 12*, 379–386.

López, C., Cervera, M., Fagoaga, C., Moreno, P., Navarro, L., Flores, R., et al. (2010). Accumulation of transgene-derived siRNAs is not sufficient for RNAi-mediated protection against *Citrus tristeza virus* in transgenic Mexican lime. *Molecular Plant Pathology, 11*, 33–41.

Lopez-Ochoa, L., Ramirez-Prado, J., & Handley-Bowdoin, L. (2006). Peptide aptamers that bind to a geminivirus replication protein interfere with viral replication in plant cells. *Journal of Virology, 80*, 5841–5853.

Lu, A., Chen, Z., Kong, L., Fang, R., Cun, S., & Man, K. (1996). Transgenic *Brassica napus* resistant to turnip mosaic virus. *Acta Genetica Sinica, 23*, 77–83.

Lucioli, A., Noris, E., Brunetti, A., Tavazza, R., Ruzza, V., Castillo, A. G., et al. (2003). *Tomato yellow leaf curl Sardinia virus* Rep-derived resistance to homologous and heterologous geminiviruses occurs by different mechanisms and is overcome if virus-mediated transgene silencing is activated. *Journal of Virology, 77*, 6785–6798.

Ludlow, E. J., Mouradov, A., & Spangenberg, G. C. (2009). Post-transcriptional gene silencing as an efficient tool for engineering resistance to white clover mosaic virus in white clover (*Trifolium repens*). *Journal of Plant Physiology, 166*, 1557–1567.

Ma, J., Song, Y., Wu, B., Jiang, M., Li, K., Zhu, C., et al. (2011). Production of transgenic rice new germplasm with strong resistance against two isolations of *Rice stripe virus* by RNA interference. *Transgenic Research, 20*, 1367–1377.

Ma, Z. L., Yang, H. Y., Wang, R., & Tien, P. (2005). Construct hairpin RNA to fight against rice dwarf virus. *Acta Botanica Sinica, 46*, 332–336.

MacFarlane, S. A., & Davies, J. W. (1992). Plants transformed with a region of the 201-kilodalton replicase gene form pea early browning virus RNA1 are resistant to virus infection. *Proceedings of the National Academy of Sciences of the United States of America, 89*, 5829–5833.

Mackenzie, D. J., & Ellis, P. J. (1992). Resistance to tomato spotted wilt virus infection in transgenic tobacco expressing the viral nucleocapsid gene. *Molecular Plant-Microbe Interactions, 5*, 34–40.
Mackenzie, D. J., & Tremaine, J. H. (1990). Transgenic *Nicotiana debneyii* expressing viral coat protein are resistant to potato virus S infection. *Journal of General Virology, 71*, 2167–2170.
Mackenzie, D. J., Tremaine, J. H., & McPherson, J. (1991). Genetically engineered resistance to potato virus S in potato cultivar Russet Burbank. *Molecular Plant-Microbe Interactions, 4*, 95–102.
Magbanua, Z. V., Wilde, H. D., Roberts, J. K., Chowdhury, K., Abad, J., Moyer, J. M., et al. (2000). Field resistance to Tomato spotted wilt virus in transgenic peanut (*Arachis hypogaea* L.) expressing an antisense nucleocapsid gene sequence. *Molecular Breeding, 6*, 227–236.
Maiti, I. B., Murphy, J. F., Shaw, J. G., & Hunt, A. G. (1993). Plants that express a potyvirus proteinase are resistant to virus infection. *Proceedings of the National Academy of Sciences of the United States of America, 90*, 6110–6114.
Maiti, I. B., Von Lanken, C., Hong, Y., Dey, N., & Hunt, A. G. (1999). Expression of multiple virus-derived resistance determinants in transgenic plants does not lead to additive resistance properties. *Journal of Plant Biochemistry and Biotechnology, 8*, 67–73.
Mäki-Valkama, T., Pehu, T., Santala, A., Valkonen, J. P. T., Koivu, K., Lehto, K., et al. (2000). High level of resistance to potato virus Y by expressing P1 sequence in antisense orientation in transgenic potato. *Molecular Breeding, 6*, 95–104.
Mäki-Valkama, T., Valkonen, J. P. T., Lehtinen, A., & Pehu, E. (2001). Protection against potato virus Y (PVY) in the field in potatoes transformed with the PVY P1 gene. *American Journal of Potato Research, 78*(3), 209–214.
Malnoë, P., Farinelli, L., Collet, G. F., & Reust, W. (1994). Small-scale field tests with transgenic potato, cv. Bintje, to test resistance to primary and secondary infections with potato virus y. *Plant Molecular Biology, 25*, 963–975.
Malyshenko, S. I., Kondakova, O. A., Nazarova, J. V., Kaplan, I. B., Taliansky, M. E., & Atabekov, J. G. (1993). Reduction of tobacco mosaic virus accumulation in transgenic plants producing nonfunctional viral transport proteins. *Journal of General Virology, 74*, 1149–1156.
Mannerlöf, M., Lennerfors, B.-L., & Tenning, P. (1996). Reduced titer of BNYVV in transgenic sugar beets expressing the BNYVV coat protein. *Euphytica, 90*, 293–299.
Manoj, K. R., & Shekhawat, N. S. (2014). Recent advances in genetic engineering for improvement of fruit crops. *Plant Cell, Tissue and Organ Culture, 116*, 1–5.
Marano, M. R., & Baulcombe, D. (1998). Pathogen-derived resistance targeted against the negative-strand RNA of tobacco mosaic virus: RNA strand-specific gene silencing? *Plant Journal, 13*, 537–546.
Martinez, F., Elena, S. F., & Daros, J. A. (2013). Fate of artificial microRNA-mediated resistance to plant viruses in mixed infections. *Phytopathology, 103*, 870–876.
Masmoudi, K., Yacoubi, I., Hassairi, A., Elarbi, L. N., & Ellouz, R. (2002). Tobacco plants transformed with an untranslatable form of the coat protein gene of *Potato virus Y* are resistant to viral infection. *European Journal of Plant Pathology, 108*, 285–292.
Masuta, C., Tanaka, H., Uehara, K., Kuwata, S., Koiwai, A., & Noma, M. (1995). Broad resistance to plant viruses in transgenic plants conferred by antisense inhibition of a host gene essential in *S*-adenosylmethionine-dependent transmethylation reactions. *Proceedings of the National Academy of Sciences of the United States of America, 92*, 6117–6121.
Matoušek, J., Schroder, A. R. M., Trnena, L., Reimers, I., Baumstark, T., Dědič, P., et al. (1994). Inhibition of viroid infection by antisense RNA expressed in transgenic plants. *Biological Chemistry Hoppe-Seyler, 375*, 765–777.

Matoušek, J., Schubert, J., Kuchař, M., Dědič, P., Ptáček, J., Vrbr, L., et al. (2001). The design and partial analysis of RNaseIII-anti-PVS antisense complex system to induce plant resistance. *Archiv für Phytopathologie und Pflanzenschutz, 33*, 381–394.

Maule, A. J., Caranta, C., & Boulton, M. I. (2007). Sources of natural resistance to plant viruses: Status and prospects. *Molecular Plant Pathology, 8*, 223–231.

McCue, K. F., Ponciano, G., Rockhold, D. R., Whitworth, J. L., Gray, S. M., Fofanov, Y., et al. (2012). Generation of PVY coat protein siRNAs in transgenic potatoes resistant to PVY. *American Journal of Potato Research, 89*, 374–383.

McGarvey, P. B., Montasser, M. S., & Kaper, J. M. (1994). Transgenic tomato plants expressing satellite RNA are tolerant to some strains of cucumber mosaic-virus. *Journal of the American Society for Horticultural Science, 119*, 642–647.

McGrath, P. F., Vincent, J. R., Leu, C.-H., Pawlowski, W. P., Torbert, K. A., Gu, W., et al. (1997). Coat protein-mediated resistance to isolates of barley yellow dwarf virus in oats and barley. *European Journal of Plant Pathology, 103*, 695–710.

Mehta, R., Radhakrishnan, T., Kumar, A., Yadav, R., Dobaria, J. R., Thirumalaisamy, P. P., et al. (2013). Coat protein-mediated transgenic resistance of peanut (*Arachis hypogaea* L.) to peanut stem necrosis disease through *Agrobacterium*-mediated genetic transformation. *Indian Journal of Virology, 24*, 205–213.

Melander, M. (2006). Potato transformed with a 57-kDa readthrough portion of the tobacco rattle virus replicase gene displays reduced tuber symptoms when challenged by viruliferous nematodes. *Euphytica, 150*, 123–130.

Melander, M., Lee, M., & Sandgren, M. (2001). Reduction of potato mop-top virus accumulation and incidence in tubers of potato transformed with a modified triple gene block gene of PMTV. *Molecular Breeding, 8*, 197–206.

Minafra, A., Golles, R., De Camara Machada, A., Saldarelli, P., Buzkan, N., Savino, V., et al. (1998). Expression of the coat protein gene of grapevine virus A and B in *Nicotiana* species and evaluation of the resistance conferred on transgenic plants. *Journal of Plant Pathology, 80*, 197–202.

Missiou, A., Kalantidis, K., Boutla, A., Tzortzakaki, S., Tabler, M., & Tsagris, M. (2004). Generation of transgenic potato plants highly resistant to potato virus Y (PVY) through RNA silencing. *Molecular Breeding, 14*, 185–197.

Mitra, A., Higgins, D. W., Langenberg, W. M., Nie, H., Sengupta, D. N., & Silverman, R. H. (1996). A mammalian 2-5A system functions as an antiviral pathway in transgenic plants. *Proceedings of the National Academy of Sciences of the United States of America, 93*, 6780–6785.

Mlotshwa, S., Verver, J., Sithole-Niang, I., Prins, M., Van Kammen, A., & Wellink, J. (2002). Transgenic plants expressing HC-Pro show enhanced virus sensitivity while silencing of the transgene results in resistance. *Virus Genes, 15*, 45–57.

Monteiro-Hara, A. C. B. A., Jadão, A. S., Mendes, B. M. J., Rezende, J. A. M., Trevesan, F., Mello, A. P. O. A., et al. (2011). Genetic transformation of passionflower and evaluation of R_1 and R_2 generations for resistance to *Cowpea aphid borne mosaic virus*. *Plant Disease, 95*, 1021–1025.

Monticelli, S., Di Nicola-Negri, E., Gentile, A., Damiano, C., & Ilardi, V. (2012). Production and *in vitro* assessment of transgenic plums for resistance to *Plum pox virus*: A feasible, environmental risk-free, cost-effective approach. *Annals of Applied Biology, 161*, 293–301.

Moon, Y. H., Song, S.-K., Choi, K. W., & Lee, J. S. (1997). Expression of a cDNA clone encoding *Phytolacca insularis* antiviral protein confers virus resistance on transgenic potato plants. *Molecules and Cells, 7*, 807–815.

Moreno, M., Bernal, J. J., Jimenez, I., & Rodríguez-Cerezo, E. (1998). Resistance in plants transformed with the P1 or P3 gene of tobacco vein mottling potyvirus. *Journal of General Virology, 79*, 2819–2827.

Mubin, M., Mansoor, S., Hussain, M., & Zafar, Y. (2007). Silencing of the AV2 gene by antisense RNA protects transgenic plants against a bipartite begomovirus. *Virology Journal, 4*, 10.

Mueller, E., Gilbert, J., Davenport, G., Brigneti, G., & Baulcombe, D. C. (1995). Homology-dependent resistance: transgenic resistance in plants related to homology-dependent gene silencing. *Plant Journal*, 7, 1001–1013.

Mundembe, R., Matibiri, A., & Sithole-Niang, I. (2009). Transgenic plants expressing the coat protein gene of cowpea aphid-borne mosaic potyvirus predominantly convey the delayed symptom development phenotype. *African Journal of Biotechnology, 8*, 2682–2690.

Muniz, F. R., De Souza, A. J., Stipp, L. C. L., Schinor, E., Freitas, W., Jr., Harakava, R., et al. (2012). Genetic transformation of *Citrus sinensis* with *Citrus tristeza virus* (CTV) derived sequences and reaction of transgenic lines to CTV infection. *Biologia Plantarum, 56*, 162–166.

Murry, L. E., Elliott, L. G., Capitant, S. A., West, J. A., Hanson, K. K., Scarafia, L., et al. (1993). Transgenic corn plants expressing MDMV strain-B coat protein are resistant to mixed infections of maize-dwarf mosaic-virus and maize chlorotic mottle virus. *Bio/Technology, 11*, 1559–1564.

Nadarajah, K., Hanafi, N. M., & Tan, S. L. (2009). Screening of transgenic tobacco for resistance against *Cucumber mosaic virus. Plant Pathology Journal, 8*, 42–52.

Nahid, N., Amin, I., Briddon, R. W., & Mansoor, S. (2011). RNA interference-based resistance against a legume mastrevirus. *Virology Journal, 8*, 499.

Nakajima, M., Hayakawa, T., Nakamura, I., & Suzuki, M. (1993). Protection against cucumber mosaic virus (CMV) strains O and Y and chrysanthemum mild mottle virus in transgenic tobacco plants expressing CMV-O coat protein. *Journal of General Virology, 74*, 319–322.

Nakamura, S., Honkura, R., Ugaki, M., Ohshima, M., & Ohaski, Y. (1994). Nucleotide sequence of the 3′-terminal region of bean yellow mosaic virus RNA and resistance to viral infection in transgenic *Nicotiana benthamiana* expressing its coat protein gene. *Annals of the Phytopathological Society of Japan, 60*, 295–304.

Nakamura, S., Honkura, R., Ugaki, M., Ohshima, M., & Ohaski, Y. (1995). Resistance to viral infection in transgenic plants expressing ribozymes designed against cucumber mosaic virus RNA3. *Annals of the Phytopathological Society of Japan, 61*, 53–55.

Namba, S., Ling, K., Gonsalves, C., Gonsalves, D., & Slightom, J. L. (1991). Expression of the gene encoding the coat protein of cucumber mosaic virus strain (CMV)-WL appears to provide protection to tobacco plants against infection by several different CMV strains. *Gene, 107*, 181–188.

Namba, S., Ling, K., Gonsalves, C., Slightom, J. L., & Gonsalves, D. (1992). Protection of transgenic plants expressing the coat protein gene of watermelon mosaic virus-II or zucchini yellow mosaic-virus against 6 potyviruses. *Phytopathology, 82*, 940–946.

Nelson, R. S., McCormick, S. M., Delannay, X., Dube, P., Layton, J., Anderson, E. J., et al. (1988). Virus tolerance, plant-growth, and field performance of transgenic tomato plants expressing coat protein from tobacco mosaic-virus. *Bio/Technology, 6*, 403–409.

Nelson, A., Roth, D. A., & Johnson, J. D. (1993). Tobacco mosaic virus infection of transgenic *Nicotiana tabacum* plants is inhibited by antisense constructs directed at the 5′ region of viral RNA. *Gene, 127*, 227–232.

Neves-Borges, A. C., Collares, W. M., Pontes, J. A., Breyne, P., Farinelli, L., & de Oliveira, D. E. (2001). Coat protein RNAs-mediated protection against *Andean potato mottle virus* in transgenic tobacco. *Plant Science, 160*, 699–712.

Nickel, H., Kawchuk, L., Twyman, R. M., Zimmermann, S., Junghans, H., Winter, S., et al. (2008). Plantibody-mediated inhibition of the potato leafroll virus P1 protein reduces virus accumulation. *Virus Research, 136*, 140–145.

Nishibayashi, S., Hayakawa, T., Nakajima, T., Suzuki, M., & Kaneko, H. (1996). CMV protection in transgenic cucumber plants with an introduced CMV-O cp gene. *Theoretical and Applied Genetics*, *93*, 672–678.

Niu, S.-N., Huang, X.-S., Wong, S.-M., Yu, J.-L., Zhao, F.-X., Li, D.-W., et al. (2005). Creation of trivalent transgenic watermelon resistant to virus infection. *Journal of Agricultural Biotechnology*, *13*, 10–15.

Niu, Q.-W., Lin, S.-S., Reyes, J. L., Chen, K.-C., Wu, H.-W., Yeh, S.-D., et al. (2006). Expression of artificial microRNAs in transgenic *Arabidopsis thaliana* confers virus resistance. *Nature Biotechnology*, *24*, 1420–1428.

Niu, Y.-B., Wang, D.-F., Yao, M., Yan, Z., & You, W.-X. (2011). Transgenic tobacco plants resistant for two viruses via RNA silencing. *Acta Agronomica Sinica*, *37*, 484–488.

Nölke, G., Cobanov, P., Uhde-Holzem, K., Reustle, G., Fischer, R., & Schillberg, S. (2009). *Grapevine fanleaf virus* (GFLV)-specific antibodies confer GFLV and *Arabis mosaic virus* (ArMV) resistance in *Nicotiana benthamiana*. *Molecular Plant Pathology*, *10*, 41–49.

Nölke, G., Fischer, R., & Schillberg, S. (2004). Antibody-based resistance in plants. *Journal of Plant Pathology*, *86*, 5–17.

Nomura, K., Ohshima, K., Anai, T., Uekusa, H., & Kita, N. (2004). RNA silencing of the introduced cat protein gene of *Turnip mosaic virus* confers broad-spectrum resistance to transgenic *Arabidopsis*. *Phytopathology*, *94*, 730–736.

Noris, E., Accotto, G. P., Tavazza, R., Brunetti, A., Crespi, S., & Tavazza, M. (1996). Resistance to tomato yellow leaf curl geminivirus in *Nicotiana benthamiana* plants transformed with a truncated viral C1 gene. *Virology*, *224*, 130–138.

Ntui, V. O., Kynet, K., Azadi, P., Khan, R. S., Chin, D. P., Nakamura, I., et al. (2013). Transgenic accumulation of a defective cucumber mosaic virus (CMV) replicase derived double stranded RNA modulates plant defence against CMV strains O and Y in potato. *Transgenic Research*, *22*, 1191–1205.

Ntui, V. O., Kynet, K., Khan, R. S., Ohara, M., Goto, Y., Watanabe, M., et al. (2014). Transgenic tobacco lines expressing defective CMV replicase-derived dsRNA are resistant to CMV-O and CMV-Y. *Molecular Biotechnology*, *56*, 50–63.

Nyaboga, E. N., Ateka, E. M., Gichuki, S. T., & Bulimo, W. D. (2008). Reaction of transgenic sweet potato (*Ipomoea batatas* L.) lines to virus challenge in the glasshouse. *Journal of Applied Biosciences*, *9*, 362–371.

Ogawa, T., Hori, T., & Ishida, I. (1996). Virus-induced cell death in plants expressing the mammalian 2′,5′ oligoadenylate system. *Nature Biotechnology*, *14*, 1566–1569.

Ogawa, T., Toguri, T., Kudoh, H., Okamura, M., Momma, T., Yoshioka, M., et al. (2005). Double-stranded RNA-specific ribonuclease confers tolerance against *Chrysanthemum stunt viroid* and *Tomato spotted wilt virus* in transgenic chrysanthemum plants. *Breeding Science*, *55*, 49–55.

Ogwok, E., Odipio, J., Halsey, M., Gaitán-Solís, E., Bua, A., Taylor, N. J., et al. (2012). Transgenic RNA interference (RNAi)-derived field resistance to cassava brown streak disease. *Molecular Plant Pathology*, *13*, 1019–1031.

Okada, Y., & Saito, A. (2008). Evaluation of resistance to complex infection of SPFMV in transgenic sweet potato. *Breeding Science*, *58*, 243–250.

Okada, Y., Saito, A., Nishiguchi, M., Kimura, T., Mori, M., Hanada, K., et al. (2001). Virus resistance in transgenic sweetpotato [*Ipomoea batatas* L. (Lam)] expressing the coat protein gene of sweet potato feathery mottle virus. *Theoretical and Applied Genetics*, *103*, 743–751.

Okada, Y., & Yoshinaga, M. (2010). Advanced resistance to *Sweet potato feathery mottle virus* (SPFMV) in transgenic sweetpotato. *Sweetpotato Research Front*, *23*, 4.

Okuno, T., Nakayama, M., Yoshida, S., Furusawa, I., & Komiya, T. (1993). Comparative susceptibility of transgenic tobacco plants and protoplasts expressing the coat protein gene of cucumber mosaic virus to infection with virions and RNA. *Phytopathology*, *83*, 542–547.

Oldroyd, G. E., & Staskawicz, B. J. (1998). Genetically engineered broad-spectrum disease resistance in tomato. *Proceedings of the National Academy of Sciences of the United States of America, 95*, 10300–10305.

Orozco, B. M., Kong, L. J., Batts, L. A., Elledge, S., & Hanley-Bowdoin, L. (2000). The multifunctional character of a geminivirus replication protein is reflected by its complex oligomerization properties. *Journal of Biological Chemistry, 275*, 6114–6122.

Owen, M., Gandecha, A., Cockburn, B., & Whitelam, G. (1992). Synthesis of a functional anti-phytochrome since-chain Fv protein in transgenic tobacco. *Bio/Technology, 10*, 790–794.

Pack, I. S., Kim, Y.-J., Youk, E. S., Lee, W. K., Yoon, W. K., Park, K. W., et al. (2012). A molecular framework for risk assessment of a virus-tolerant transgenic pepper line. *Journal of Crop Science and Biotechnology, 15*, 107–115.

Pacot-Hiriart, C., Le Gall, O., Candresse, T., Delbos, R. P., & Dunez, J. (1999). Transgenic tobaccos transformed with a gene encoding a truncated form of the coat protein of tomato black ring nepovirus are resistant to viral infection. *Plant Cell Reports, 19*, 203–209.

Palucha, A., Zagórski, W., Chrzanowska, M., & Hulanicka, D. (1998). An antisense coat protein gene confers immunity to potato leafroll virus in a genetically engineered potato. *European Journal of Plant Pathology, 104*, 287–293.

Palukaitis, P. (1988). Pathogenicity regulation by satellite RNAs of cucumber mosaic virus: Minor nucleotide sequence changes alter host responses. *Molecular Plant-Microbe Interactions, 1*, 175–181.

Palukaitis, P., & Carr, J. P. (2008). Plant resistance responses to viruses. *Journal of Plant Pathology, 90*, 153–171.

Pandolfini, T., Molesini, B., Avesani, L., Spena, A., & Polverari, A. (2003). Expression of self-complementary hairpin RNA under the control of the *rolC* promoter confers systemic disease resistance to plum pox virus without preventing local infection. *BMC Biotechnology, 3*, 7.

Pang, S.-Z., Bock, J. H., Gonsalves, C., Slightom, J. L., & Gonsalves, D. (1994). Resistance of transgenic *Nicotiana benthamiana* plants to tomato spotted wilt and impatiens necrotic spot tospoviruses: Evidence of involvement of the N protein and N gene RNA in resistance. *Phytopathology, 84*, 243–249.

Pang, S.-Z., Jan, F.-J., Carney, K., Stout, J., Tricoli, D. M., Quemada, H. D., et al. (1996). Post-transcriptional transgene silencing and consequent tospovirus resistance in transgenic lettuce are affected by transgene dosage and plant development. *Plant Journal, 9*, 899–909.

Pang, S.-Z., Jan, F.-J., Tricoli, D. M., Russell, P. F., Carney, K., Hu, J. S., et al. (2000). Resistance to squash mosaic comovirus in transgenic squash plants expressing its coat protein genes. *Molecular Breeding, 6*, 87–93.

Pang, S.-Z., Nagpala, P., Wang, M., Slightom, J. L., & Gonsalves, D. (1992). Resistance to heterologous isolates of tomato spotted wilt virus in transgenic tobacco expressing its nucleocapsid protein gene. *Phytopathology, 82*, 1223–1229.

Panter, S., Chu, P. G., Ludlow, E., Garrett, R., Kalla, R., Jahufer, M. Z. Z., et al. (2012). Molecular breeding of transgenic white clover (*Trifolium repens* L.) with field resistance to *Alfalfa mosaic virus* through expression of its coat protein gene. *Transgenic Research, 21*, 619–632.

Park, H.-M., Choi, M.-S., Kwak, D.-Y., Lee, B.-C., Lee, J.-H., Kim, M.-K., et al. (2012). Suppression of *NS3* and *MP* is important for the stable inheritance of RNAi-mediated *Rice stripe virus* (RSV) resistance obtained by targeting the fully complementary *RSV-CP* gene. *Molecules and Cells, 33*, 43–51.

Park, S. M., Lee, J. S., Jegal, S., Jeon, B. Y., Jung, M., Park, Y. S., et al. (2005). Transgenic watermelon rootstock resistant to CGMMV (cucumber green mottle mosaic virus) infection. *Plant Cell Reports, 24*, 350–356.

Patil, B. L., Ogwok, E., Wagaba, H., Mohammed, I. U., Yadav, I. S., Bagewadi, B., et al. (2011). RNAi-mediated resistance to diverse isolates belonging to two viruses involved in Cassava brown streak disease. *Molecular Plant Pathology*, *12*, 31–41.

Pavli, O. I., Panopoulos, N. J., Goldbach, R., & Skaracis, G. N. (2010). BNYVV-derived dsRNA confers resistance to rhizomania disease of sugar beet as evidenced by a novel transgenic hairy root approach. *Transgenic Research*, *19*, 915–922.

Pehu, T. M., Mäki-Valkama, T. K., Valkonen, J. P. T., Koivu, K. T., Lehto, K. M., & Pehu, E. P. (1995). Potato plants transformed with a potato virus Y P1 gene sequence are resistance to PVY-O. *American Journal of Potato Research*, *72*, 523–532.

Pinto, Y. M., Kok, R. A., & Baulcombe, D. C. (1999). Resistance to rice yellow mottle virus (RYMV) in cultivated African rice varieties containing RYMV transgenes. *Nature Biotechnology*, *17*, 702–707.

Ploeg, A. T., Mathis, A., Bol, J. F., Brown, D. J. F., & Robinson, D. J. (1993). Susceptibility of transgenic tobacco plants expressing tobacco rattle virus coat protein to nematode-transmitted and mechanically inoculated tobacco rattle virus. *Journal of General Virology*, *74*, 2709–2715.

Pourrahim, R., Ahoonmanesh, A., Hashemi, H., Zeinali, S., & Farzadfar, S. (2006). Assessment of virus resistance in transgenic *Nicotiana tabacum* cv. Samsun lines against three Iranian isolates of potato virus Y. *Applied Entomology and Phytopathology*, *73*, 5–9.

Powell Abel, P., Nelson, R. S., De, B., Hoffmann, N., Rogers, S. G., Fraley, R. T., et al. (1986). Delay of disease development in transgenic plants that express the tobacco mosaic virus coat protein gene. *Science*, *232*, 738–743.

Powell, P. A., Stark, D. M., Sanders, P. R., & Beachy, R. N. (1989). Protection against tobacco mosaic virus in transgenic plants that express tobacco mosaic virus antisense RNA. *Proceedings of the National Academy of Sciences of the United States of America*, *86*, 6949–6952.

Pradeep, K., Satya, V. K., Selvapriya, M., Vijayasamundeeswari, A., Ladhalakshmi, D., Paranidharan, V., et al. (2012). Engineering resistance against *Tobacco streak virus* in sunflower and tobacco using RNA interference. *Biologia Plantarum*, *56*, 735–741.

Pratap, D., Kumar, S., Raj, S. J., & Sharma, A. K. (2011). *Agrobacterium*-mediated transformation of eggplant (*Solanum melongena* L.) using cotyledon explants and coat protein gene of *Cucumber mosaic virus*. *Indian Journal of Biotechnology*, *10*, 19–24.

Pratap, D., Raj, S. K., Kumar, S., Snehi, S. K., Gautam, K. K., & Sharma, A. K. (2012). Coat protein-mediated transgenic resistance in tomato against a IB subgroup *Cucumber mosaic virus* strain. *Phytoparasitica*, *40*, 375–382.

Praveen, S., Kushwaha, C. M., Mishra, A. K., Singh, V., Jain, R. K., & Varma, A. (2005). Engineering tomato for resistance to tomato leaf curl disease using viral *rep* gene sequences. *Plant Cell, Tissue and Organ Culture*, *83*, 311–318.

Praveen, S., Mishra, A. K., & Dasgupta, A. (2005). Antisense suppression of replicase gene expression recovers tomato plants from leaf curl virus infection. *Plant Science*, *168*, 1011–1014.

Praveen, S., Ramesh, S. V., Mishra, A. K., Koundal, V., & Palukaitis, P. (2010). Silencing potential of viral derived RNAi constructs in *Tomato leaf curl virus*-AC4 gene suppression in tomato. *Transgenic Research*, *19*, 45–55.

Presting, G. G., Smith, O. P., & Brown, C. R. (1995). Resistance to potato leafroll virus in potato plants transformed with the coat protein gene or with vector control constructs. *Phytopathology*, *85*, 436–442.

Prins, M., de Haan, P., Luyten, R., van Veller, M., van Grinsven, M. Q. J. M., & Goldbach, R. (1995). Broad resistance to tospoviruses in transgenic tobacco plants expressing three tospoviral nucleoprotein gene sequences. *Molecular Plant-Microbe Interactions*, *8*, 85–91.

Prins, M., de Oliveira Resende, R., Anker, C., van Schepen, A., de Haan, P., & Goldbach, R. (1996). Engineering RNA-mediated resistance to tomato spotted wilt virus is sequence specific. *Molecular Plant-Microbe Interactions*, *9*, 416–418.

Prins, M., Kikkert, M., Ismayadi, C., de Graauw, W., de Haan, P., & Goldbach, R. (1997). Characterization of RNA-mediated resistance to tomato spotted wilt virus in transgenic tobacco plants expressing the NS_M gene sequences. *Plant Molecular Biology*, *33*, 235–243.

Prins, M., Laimer, M., Noris, E., Schubert, J., Wassenegger, M., & Tepfer, M. (2008). Strategies for antiviral resistance in transgenic plants. *Molecular Plant Pathology*, *9*, 73–83.

Prins, M., Lohuis, D., Schots, A., & Goldbach, R. (2005). Phage display-selected single-chain antibodies confer high levels of resistance against *Tomato spotted wilt virus*. *Journal of General Virology*, *86*, 2107–2113.

Pruss, G. J., Lawrence, C. B., Bass, T., Li, Q. Q., Bowman, L. H., & Vance, V. (2004). The potyviral suppressor of RNA silencing confers enhanced resistance to multiple pathogens. *Virology*, *320*, 107–120.

Pumplin, N., & Voinnet, O. (2013). RNA silencing suppression by plant pathogens: Defence, counter-defence, and counter-counter-defence. *Nature Reviews. Microbiology*, *11*, 745–760.

Qu, J., Ye, I., & Fang, R. (2007). Artificial microRNA-mediated virus resistance in plants. *Journal of Virology*, *81*, 6690–6699.

Quemada, H. D., Gonsalves, D., & Slightom, J. L. (1991). Expression of coat protein gene for cucumber mosaic virus strain C in tobacco: Protection against infection by CMV strains transmitted mechanically or by aphids. *Phytopathology*, *81*, 794–802.

Rabinowicz, P. D., Bravo-Almonacid, F. F., Lampasona, S., Rodriguez, F., Gracia, O., & Mentaberry, A. N. (1998). Resistance against pepper severe mosaic potyvirus in transgenic tobacco. *Journal of Phytopathology*, *146*, 315–319.

Rachman, D. S., McGeachy, K., Reavy, B., Štrukelj, B., Žel, J., & Barker, H. (2001). Strong resistance to potato tuber necrotic ringspot disease in potato induced by transformation with the coat protein gene sequences from an NTN isolate of *Potato virus Y*. *Annals of Applied Biology*, *139*, 269–275.

Radian-Sade, A., Perl, A., Edelbaum, O., Kunetsova, L., Gafny, R., Sela, I., et al. (2000). Transgenic *Nicotiana benthamiana* and grapevine plants transformed with grapevine virus A (GVA) sequences. *Phytoparasitica*, *28*, 79–86.

Raj, S. K., Singh, R., Pandey, S. K., & Singh, B. P. (2005). *Agrobacterium*-mediated tomato transformation and regeneration of transgenic lines expressing *Tomato leaf curl virus* coat protein gene for resistance against TLCV infection. *Current Science*, *88*, 1674–1679.

Ranjith-Kumar, C. T., Manoharan, M., Prasad, S. K., Cherian, S., Umashankar, M., Sita, G. L., et al. (1999). Engineering resistance against physalis mottle tymovirus by expression of the coat protein and 3′ noncoding region. *Current Science*, 77, 1542–1547.

Raquel, H., Lourenço, T., Moita, C., & Oliveira, M. M. (2008). Expression of prune dwarf *Ilarvirus* coat protein sequences in *Nicotiana benthamiana* plants interferes with PDV systemic proliferation. *Plant Biotechnology Reports*, *2*, 75–85.

Ravelonandro, M., Minsion, M., Delbos, R., & Dunez, J. (1993). Variable resistance to plum pox virus and potato virus Y infection in transgenic *Nicotiana* plants expressing plus pox virus coat protein. *Plant Science*, *91*, 157–169.

Ravelonandro, M., Scorza, R., Bachelier, J. C., Labonne, G., Levy, L., Damsteegt, V., et al. (1997). Resistance of *Prunus domestica* L. to plum pox virus infection. *Plant Disease*, *81*, 1231–1235.

Ready, D. V. R., Sudarshana, M. R., Fuchs, M., Rao, N. C., & Thottappilly, G. (2009). Genetically engineered virus-resistance plants in developing countries: Current status and future prospects. *Advances in Virus Research*, *75*, 185–220.

Reavy, B., Arif, M., Kashiwazaki, S., Webster, K. D., & Barker, H. (1995). Immunity to *Potato mop-top virus* in *Nicotiana benthamiana* plants expressing the coat protein gene is effective against fungal inoculation of the virus. *Molecular Plant-Microbe Interactions*, *8*, 286–291.

Reddy, M. S. S., Ghabrial, S. A., Redmond, C. T., Dinkins, R. D., & Collins, G. B. (2001). Resistance to *Bean pod mottle virus* in transgenic soybean lines expressing the capsid polyprotein. *Phytopathology*, *91*, 831–838.

Regner, F., da Câmara Machada, A., da Câmara, Laimer, Machada, M., Steinkellner, H., Mattanovich, D., et al. (1992). Coat protein mediated resistance to plum pox virus in *Nicotiana clevelandii* and *N. benthamiana*. *Plant Cell Reports*, *11*, 30–33.
Reyes, C. A., De Francesco, A., Peña, E. J., Costa, N., Plata, M. I., Sendin, L., et al. (2011). Resistance to Citrus psorosis virus in transgenic sweet orange plants is triggered by coat protein–RNA silencing. *Journal of Biotechnology*, *151*, 151–158.
Reyes, M. I., Nash, T. E., Dallas, M. M., Ascencio-Ibáñez, J. T., & Hanley-Bowdoin, L. (2013). Peptide aptamers that bind to geminivirus replication proteins confer resistance phenotype to *Tomato yellow leaf curl virus* and *Tomato mottle virus* infection in tomato. *Journal of Virology*, *87*, 9691–9706.
Reyes, C., Peña, E. J., Zanek, M. C., Sanchez, D. V., Grau, O., & García, M. L. (2009). Differential resistance to *Citrus psorosis virus* in transgenic *Nicotiana benthamiana* plants expressing hairpin RNA derived from the coat protein and 54K protein genes. *Plant Cell Reports*, *28*, 1817–1825.
Reyes, C. A., Zanek, M. C., Velázquez, K., Costa, N., Plata, M. I., & Garcia, M. L. (2011). Generation of sweet orange transgenic lines and evaluation of *Citrus psorosis virus*-derived resistance against psorosis A and psorosis B. *Journal of Phytopathology*, *159*, 531–537.
Rezaian, M. A., Skene, K. G. M., & Ellis, J. G. (1988). Any-sense RNAs of cucumber mosaic virus in transgenic plants assessed for control of virus. *Plant Molecular Biology*, *11*, 463–471.
Ribiero, S. G., Lohuis, H., Goldbach, R., & Prins, M. (2007). Tomato chlorotic mottle virus is a target of RNA silencing but the presence of specific short interfering RNAs does not guarantee resistance in transgenic plants. *Journal of Virology*, *81*, 1563–1573.
Robaglia, C., & Caranta, C. (2006). Translation initiation factors: A weak link in plant RNA virus infection. *Trends in Plant Science*, *11*, 40–45.
Rubino, L., Capriotti, G., Lupo, R., & Russo, M. (1993). Resistance to cymbidium ringspot tombusvirus infection in transgenic *Nicotiana benthamiana* plants expressing the virus coat protein gene. *Plant Molecular Biology*, *21*, 665–672.
Rubino, L., Lupo, R., & Russo, M. (1993). Resistance to cymbidium ringspot tombusvirus infection in transgenic *Nicotiana benthamiana* plants expressing a full-length viral replicase gene. *Molecular Plant-Microbe Interactions*, *6*, 729–734.
Rubio, T., Borja, M., Scholthof, H. B., Feldstein, P. A., Morris, T. J., & Jackson, A. O. (1999). Broad-spectrum protection against tombusviruses elicited by defective interfering RNAs in transgenic plants. *Journal of Virology*, *73*, 5070–5078.
Rudolph, C., Schreier, P. H., & Uhrig, J. (2003). Peptide-mediated broad-spectrum plant resistance to tospoviruses. *Proceedings of the National Academy of Sciences of the United States of America*, *100*, 4429–4434.
Ryu, K. H., Lee, G. P., Park, K. W., Lee, S. L., & Park, W. M. (1998). Transgenic tobacco expressing the coat protein of cucumber mosaic virus show different virus resistance. *Journal of Plant Biology*, *41*, 255–261.
Safarnejad, M. R., Fischer, R., & Commandeur, U. (2009). Recombinant-antibody-mediated resistance against *Tomato yellow leaf curl virus* in *Nicotiana benthamiana*. *Archives of Virology*, *154*, 457–467.
Saito, Y., Komari, T., Masuta, C., Hayashi, Y., Kumashiro, T., & Takanami, Y. (1992). Cucumber mosaic virus-tolerant transgenic tomato plants expressing a satellite RNA. *Theoretical and Applied Genetics*, *83*, 679–683.
Sanders, P. R., Sammons, B., Kaniewski, W., Haley, L., Layton, J., LaVallee, B. J., et al. (1992). Field resistance of transgenic tomatoes expressing the tobacco mosaic virus or tomato mosaic virus coat protein genes. *Phytopathology*, *82*, 683–690.
Sanford, J. C., & Johnston, S. A. (1985). The concept of parasite-derived resistance—Deriving resistance genes from the parasite's own genome. *Journal of Theoretical Biology*, *113*, 395–405.

Sangaré, A., Deng, D., Fauquet, C. M., & Beachy, R. N. (1999). Resistance to African cassava mosaic virus conferred by a mutant of the putative NTP-binding domain of the Rep gene (AC1) in *Nicotiana benthamiana*. *Molecular Breeding*, *5*, 95–102.

Sano, T., Nagayama, A., Ogawa, T., Ishida, I., & Okada, Y. (1997). Transgenic potato expressing a double-stranded RNA-specific ribonuclease is resistant to potato spindle tuner viroid. *Nature Biotechnology*, *15*, 1290–1294.

Sano, H., Seo, S., Orudgev, E., Youssefian, S., Ishizuka, K., & Ohashi, Y. (1994). Expression of the gene for a small GTP binding protein in transgenic tobacco elevates endogenous cytokinin levels, abnormally induces salicylic acid in response to wounding, and increases resistance to tobacco mosaic virus infection. *Proceedings of the National Academy of Sciences of the United States of America*, *91*, 10556–10560.

Sanz, A. I., Serra, M. T., & García-Luque, I. (2000). Altered expression and systemic spread of movement deficient virus in transgenic tobacco plants expressing the cucumber mosaic virus 3a protein. *Archives of Virology*, *145*, 2387–2401.

Savenkov, E. I., & Valkonen, J. P. T. (2001). Coat protein gene-mediated resistance to *Potato virus A* in transgenic plants is suppressed following infection with another potyvirus. *Journal of General Virology*, *82*, 2275–2278.

Savenkov, E. I., & Valkonen, J. P. T. (2002). Silencing of a viral RNA silencing suppressor in transgenic plants. *Journal of General Virology*, *83*, 2325–2335.

Sayama, H., Sato, T., Kominato, M., Natsuaki, T., & Kaper, J. M. (1993). Field testing of a satellite-containing attenuated strain of cucumber mosaic-virus for tomato protection in Japan. *Phytopathology*, *83*, 405–410.

Schillberg, S., Zimmermann, S., Findlay, K., & Fischer, R. (2000). Plasma membrane display of anti-viral single chain Fv fragments confers resistance to tobacco mosaic virus. *Molecular Breeding*, *6*, 317–326.

Schubert, J., Matoušek, J., & Mattern, D. (2004). Pathogen-derived resistance in potato to Potato virus Y—Aspects of stability and biosafety under field conditions. *Virus Research*, *100*, 41–50.

Schwab, R., Ossowski, S., Riester, M., Warthmann, N., & Weigel, D. (2006). Highly specific gene silencing by artificial microRNAs in *Arabidopsis*. *Plant Cell*, *18*, 1121–1133.

Schwind, N., Zwiebel, M., Itaya, A., Ding, B., Wang, M.-B., Krczal, G., et al. (2009). RNAi-mediated resistance to *Potato spindle tuber viroid* in transgenic tomato expressing a viroid hairpin RNA construct. *Molecular Plant Pathology*, *10*, 459–469.

Scorza, R., Callahan, A., Levy, L., Damsteegt, V., Webb, K., & Ravelonandro, M. (2001). Post-transcriptional gene silencing in plum pox virus resistant transgenic European plum containing the plum pox potyvirus coat protein gene. *Transgenic Research*, *10*, 201–209.

Scorza, R., Levy, L., Damsteegt, V., Yepes, L. M., Cordts, J., Hadidi, A., et al. (1995). Transformation of plum with the papaya ringspot virus coat protein gene and reaction of transgenic plants to plum pox virus. *Journal of the American Society for Horticultural Science*, *120*, 943–952.

Selsted, M. E., Novotny, M. J., Morris, W. L., Tang, Y. Q., Smith, W., & Cullor, J. S. (1992). Indolicidin, a novel bactericidal tridecapeptide amide from neutrophils. *Journal of Biological Chemistry*, *267*, 4292–4295.

Sengoda, V. K., Tsai, W.-S., de la Peña, R. C., Green, S. G., Kenyon, L., & Hughes, J. (2012). Expression of the full-length coat protein gene of *Tomato leaf curl Taiwan virus* is not necessary for recovery phenotype on transgenic tomato. *Journal of Phytopathology*, *160*, 213–219.

Seo, Y.-S., Rojas, M. R., Lee, J.-Y., Lee, S.-W., Jeon, J.-S., Ronald, P., et al. (2006). A viral resistance gene from common bean functions across plant families and is up-regulated in a non-virus-specific manner. *Proceedings of the National Academy of Sciences of the United States of America*, *103*, 11856–11861.

Seppänen, P., Puska, R., Honkanen, J., Tyulinka, L. G., Fedorkin, O., Morozov, S. Yu., et al. (1997). Movement protein-derived resistance to triple gene block-containing viruses. *Journal of General Virology*, *78*, 1241–1246.

Sequeira, L. (1984). Cross protection and induced resistance: Their potential for plant disease control. *Trends in Biotechnology*, *2*, 25–29.

Sera, T. (2005). Inhibition of virus DNA replication by artificial zinc finger proteins. *Journal of Virology*, *79*, 2614–2619.

Shams-Bakhsh, M., Canto, T., & Palukaitis, P. (2007). Enhanced resistance and neutralization of defense responses by suppressors of RNA silencing. *Virus Research*, *130*, 103–109.

Shekawat, U. K. S., Ganapathi, T. R., & Hadapad, A. B. (2012). Transgenic banana plants expressing small interfering RNAs targeted against viral replication initiation gene display high-level resistance to banana bunchy top virus infection. *Journal of General Virology*, *93*, 1804–1813.

Shepherd, D. N., Mangwende, T., Martin, D. P., Bezuidenhout, M., Kloppers, F. J., Carolissen, C. H., et al. (2007). Maize streak virus-resistant transgenic maize: A first for Africa. *Plant Biotechnology Journal*, *5*, 759–767.

Shepherd, D. N., Mangwende, T., Martin, D. P., Bezuidenhout, M., Thomson, J. A., & Rybicki, E. P. (2007). Inhibition of maize streak virus (MSV) replication by transient expression of MSV replication-associated protein mutants. *Journal of General Virology*, *88*, 325–556.

Shi, M., & Xue, B. (1999). Protection of transgenic tomato against CMV infection. *Journal of Zhejiang Agricultural Sciences*, *6*, 276–279.

Shimizu, T., Nakazono-Nagaoka, E., Akita, F., Wei, T., Sasaya, T., Omura, T., et al. (2012). Hairpin RNA derived from the gene for Pns9, a viroplasm matrix protein of *Rice gall dwarf virus*, confers string resistance to virus infection in transgenic rice plants. *Journal of Biotechnology*, *157*, 421–427.

Shimizu, T., Nakazono-Nagaoka, E., Uehara-Ichiki, U., Sasaya, T., & Omura, T. (2011). Targeting specific genes for RNA interference is crucial to the development of strong resistance to *Rice stripe virus*. *Plant Biotechnology Journal*, *9*, 503–512.

Shimizu, T., Ogamino, T., Hiraguri, A., Nakazono-Nagaoka, E., Uehara-Ichiki, U., Nakajima, M., et al. (2013). Strong resistance against *Rice grassy stunt virus* is induced in transgenic rice plants expressing double-stranded RNA of the viral genes for nucleocapsid or movement proteins as targets for RNA interference. *Phytopathology*, *103*, 513–519.

Shimizu, T., Yoshii, M., Wei, T., Hirochika, H., & Omura, T. (2009). Silencing by RNAi of the gene for Pns12, a viroplasm matrix protein of *Rice dwarf virus*, resulting in strong resistance of transgenic rice plants to the virus. *Plant Biotechnology Journal*, *7*, 24–32.

Shin, R., Han, J.-H., Lee, G.-J., & Peak, K.-H. (2002). The potential use of a viral coat protein gene as a transgene screening marker and multiple virus resistance in pepper plants coexpressing coat proteins of cucumber mosaic virus and tomato mosaic virus. *Transgenic Research*, *11*, 215–219.

Shin, R., Park, J. M., An, J.-M., & Paek, K.-H. (2002). Ectopic expression of Tsi1 in transgenic hot pepper plants enhances host resistance to viral, bacterial, and oomycete pathogens. *Molecular Plant-Microbe Interactions*, *15*, 983–989.

Shin, R., Park, C.-J., An, J.-M., & Paek, K.-H. (2003). A novel TMV-induced hot pepper cell wall protein gene (*CaTin2*) is associated with virus-specific hypersensitive response pathway. *Plant Molecular Biology*, *51*, 687–701.

Sijen, T., Wellink, J., Hendriks, J., Verver, J., & van Kammen, A. (1995). Replication of cowpea mosaic virus RNA1 and RNA2 is specifically blocked in transgenic *Nicotiana benthamiana* plants expressing the full-length replicase or movement protein genes. *Molecular Plant-Microbe Interactions*, *8*, 340–347.

Silva-Rosales, L., Lindbo, J. A., & Dougherty, W. G. (1994). Analysis of transgenic tobacco plants expressing a truncated form of a potyvirus coat protein nucleotide sequence. *Plant Molecular Biology, 24*, 929–939.
Simon, A. E., Roossinck, M. J., & Havelda, Z. (2004). Plant virus satellite and defective interfering RNAs: New paradigms for a new century. *Annual Review of Phytopathology, 42*, 415–437.
Singh, V. B., Haq, Q. M. R., & Malathi, V. G. (2013). Antisense RNA approach targeting Rep gene of *Mungbean yellow mosaic India virus* to develop resistance in soybean. *Archives of Phytopathology and Plant Protection, 46*, 2191–2207.
Sinisterra, X. H., Polston, J. E., Abouzid, A. M., & Hiebert, E. (1999). Tobacco plants transformed with a modified coat protein of tomato mottle begomovirus show resistance to virus infection. *Phytopathology, 89*, 701–706.
Sivamani, E., Brey, C. W., Dyer, W. E., Talbert, L. E., & Qu, R. (2000). Resistance to wheat streak mosaic virus in transgenic wheat expressing the viral replicase (NIb) gene. *Molecular Breeding, 6*, 469–477.
Sivamani, E., Brey, C. W., Talbert, L. E., Young, M. A., Dyer, W. E., Kaniewski, W. K., et al. (2002). Resistance to wheat streak mosaic virus in transgenic wheat expressing the viral coat protein gene. *Transgenic Research, 11*, 31–41.
Sivamani, E., Huet, H., Shen, P., Ong, C. A., de Kochko, A., Fauquet, C., et al. (1999). Rice plant (*Oryza sativa* L.) containing Rice tungo spherical virus (RTSV) coat protein transgenes are resistant to virus infection. *Molecular Breeding, 5*, 177–185.
Sivparsad, B. J., & Gubba, A. (2014). Development of transgenic sweet potato with multiple virus resistance in South Africa (SA). *Transgenic Research, 23*, 377–388.
Smith, H. A., Powers, H., Swaney, S., Brown, C., & Dougherty, W. G. (1995). Transgenic potato virus Y resistance in potato: Evidence for an RNA-mediated cellular response. *Phytopathology, 85*, 864–870.
Smith, N. A., Singh, S. P., Wang, M.-B., Stoutjesdijk, P. A., Green, A. G., & Waterhouse, P. M. (2000). Total silencing by intron-spliced hairpin RNAs. *Nature, 407*, 319–320.
Smith, H. A., Swaney, S., Parks, T. D., Wernsman, E. A., & Dougherty, W. G. (1994). Transgenic plant virus resistance mediated by untranslatable sense RNAs: Expression, regulation and fate of nonessential RNAs. *Plant Cell, 6*, 1441–1453.
Sohn, S., Kim, Y. H., Lee, H. S., Yi, B. Y., Hur, H. S., & Lee, J. Y. (2004). Resistance to heterologous *Potyvirus* in *Nicotiana tabacum* expressing the coat protein gene of *Soybean mosaic virus*. *Korean Journal of Breeding, 36*, 295–301.
Sokhandan-Bashir, N., Gillings, M., & Bowyer, J. W. (2012). A dual coat protein construct establishes resistance to passionfruit woodiness virus and cucumber mosaic virus. *Journal of Agricultural Science and Technology, 14*, 1105–1120.
Sokolova, M. A., Pugin, M. M., Shulga, O. A., & Skryabin, K. G. (1994). Construction of transgenic potato plants resistant to potato-virus-Y. *Molecular Biology, 28*, 646–649.
Soler, N., Plomer, M., Fagoaga, C., Moreno, P., Navarro, L., Flores, R., et al. (2012). Transformation of Mexican lime with an intron-hairpin construct expressing untranslatable versions of the genes coding for the three silencing suppressors of *Citrus tristeza virus* confers complete resistance to the virus. *Plant Biotechnology Journal, 10*, 597–608.
Sonada, S., & Nishiguchi, M. (1999). Virus resistance in gene-silenced transgenic plants with the coat protein gene of sweet potato feathery mottle potyvirus. *Annals of the Phytopathological Society of Japan, 65*, 297–300.
Song, E. K., Koh, H. K., Kim, J. K., & Lee, S. Y. (1999). Genetically engineered transgenic plants with the Domain 1 sequence of tobacco mosaic virus 126 kDa protein gene are completely resistant to viral infection. *Molecules and Cells, 9*, 569–575.

Song, Y.-R., Li, C., Hou, L.-L., Zhang, L.-Z., Ma, Q.-H., Peng, X.-X., et al. (1994). Construction of plant expression vectors and identification of transgenic potato plants. *Acta Botanica Sinica*, *36*, 842–848.

Song, G.-Q., Sink, K. C., Walworth, A. E., Cook, M. A., Allison, R. F., & Lang, G. A. (2013). Engineering cherry rootstock with resistance to Prunus necrotic ring spot virus through RNAi-mediated silencing. *Plant Biotechnology Journal*, *11*, 702–708.

Souza, M. T., Jr., Nickel, O., & Gonsalves, D. (2005). Development of virus resistant transgenic papayas expressing the coat protein gene from a Brazilian isolate of *Papaya ringspot virus*. *Fitopatologia Brasileira*, *30*, 357–365.

Spassova, M. I., Prins, T. W., Folkertsma, R. T., Klein-Lankhorst, R. M., Hille, J., Goldbach, R. W., et al. (2001). The tomato gene Sw5 is a member of the coiled coil, nucleotide binding, leucine-rich repeat class of plant resistance genes and confers resistance to TSWV in tobacco. *Molecular Breeding*, *7*, 151–161.

Spielmann, A., Krastanova, S., Douet-Orhant, V., & Gugerli, P. (2000). Analysis of transgenic grapevine (*Vitis rupestris*) and *Nicotiana benthamiana* plants expressing an *Arabis mosaic virus* coat protein gene. *Plant Science*, *156*, 235–244.

Spillane, C., Baulcombe, D. C., & Kavanagh, T. A. (1998). Genetic engineering of the potato cultivar Glenroe for increased resistance to potato virus X (PVX). *Irish Journal of Agricultural and Food Research*, *37*, 173–182.

Spillane, C., Verchot, J., Kavanagh, T. A., & Baulcombe, D. C. (1997). Concurrent suppression of virus replication and rescue of movement-defective virus in transgenic plants expressing the coat protein of potato virus X. *Virology*, *236*, 76–84.

Stanley, J., Frischmuth, T., & Ellwood, S. (1990). Defective viral DNA ameliorates symptoms of geminivirus infection in transgenic plants. *Proceedings of the National Academy of Sciences of the United States of America*, *87*, 6291–6295.

Stark, D. M., & Beachy, R. N. (1989). Protection against potyvirus infection in transgenic plants: Evidence for broad spectrum resistance. *Bio/Technology*, *7*, 1257–1262.

Stenger, D. C. (1994). Strain-specific mobilization and amplification of a transgenic defective-interfering DNA of the geminivirus beet curly top virus. *Virology*, *203*, 397–402.

Stommel, J. R., Tousignant, M. E., Wai, T., Pasini, R., & Kaper, J. M. (1998). Viral satellite RNA expression in transgenic tomato confers field tolerance to cucumber mosaic virus. *Plant Disease*, *82*, 391–396.

Sudarsono, Young, J. B., Woloshuk, S. L., Parry, D. C., Hellmann, G. M., Wernsman, E. A., et al. (1995). Transgenic Burley and flue-cured tobacco with resistance to four necrotic isolates of potato virus Y. *Phytopathology*, *85*, 1493–1499.

Sun, F., Xiang, Y., & Sanfaçon, H. (2001). Homology-dependent resistance to tomato ringspot neopvirus in plants transformed with the VPg-protease coding region. *Canadian Journal of Plant Pathology*, *23*, 292–299.

Sunitha, S., Shanmugapriya, G., Balamani, V., & Veluthambi, K. (2013). *Mungbean yellow mosaic virus* (MYMV) *AC4* suppresses post-transcriptional gene silencing and an *AC4* hairpin RNA reduces MYMV accumulation in transgenic tobacco. *Virus Genes*, *46*, 496–504.

Sunitha, S., Shivaprasad, P. V., Sujata, K., & Veluthambi, K. (2012). High frequency of T-DNA deletions in transgenic plants transformed with intron-containing hairpin RNA genes. *Plant Molecular Biology*, *30*, 158–167.

Suzuki, M., Masuta, C., Takanami, Y., & Kuwata, S. (1996). Resistance against cucumber mosaic virus in plants expressing the viral replicon. *FEBS Letters*, *379*, 26–30.

Swaney, S., Powers, H., Goodwin, J., Rosales, L. S., & Dougherty, W. G. (1995). RNA-mediated resistance with nonstructural genes from the tobacco etch virus genome. *Molecular Plant-Microbe Interactions*, *8*, 1004–1011.

Tacke, E., Salamini, F., & Rohde, W. (1996). Genetic engineering of potato for broad-spectrum protection against virus infection. *Nature Biotechnology*, *14*, 1597–1601.

Taliansky, M. E., Ryabov, E. V., & Robinson, D. J. (1998). Two distinct mechanisms of transgenic resistance mediated by groundnut rosette virus satellite RNA sequences. *Molecular Plant-Microbe Interactions*, *11*, 367–374.

Tan, X., Zhang, D., Wintgtens, C., Willingmann, P., Adam, G., & Heinze, C. (2012). A comparative testing of *Cucumber mosaic virus* (CMV)-based constructs to generate virus resistant plants. *American Journal of Plant Sciences*, *3*, 461–472.

Taschner, P. E. M., Van Marle, G., Brederode, F. T., Tumer, N. E., & Bol, J. F. (1994). Plants transformed with a mutant alfalfa mosaic virus coat protein gene are resistant to the mutant but not to wild-type virus. *Virology*, *203*, 269–276.

Tavert-Roudet, G., Ravelonandro, M., Bachelier, J.-C., & Dunez, J. (1998). Transgenic *Nicotiana benthamiana* plants containing the P1 gene of plum pox virus are resistant to virus challenge. *European Journal of Virology*, *104*, 103–107.

Tavladoraki, P., Benvenuto, E., Trinca, S., De Martinis, D., Cattaneo, A., & Galeffi, P. (1993). Transgenic plants expressing a functional single-chain Fv antibody are specifically protected from virus attack. *Nature*, *366*, 469–472.

Telem, R. S., Wani, S. H., Singh, N. B., Nandini, R., Sadhukhan, R., Bhattacharya, S., et al. (2013). Cisgenics—A sustainable approach to crop improvement. *Current Genomics*, *14*, 468–476.

Tenllado, F., García-Luque, I., Serra, M. T., & Díaz-Ruíz, J. R. (1995). *Nicotiana benthamiana* plants transformed with the 54-kDa region of the pepper mild mottle tobamovirus replicase gene exhibit two types of resistance responses against viral infection. *Virology*, *211*, 170–183.

Tennant, P. F., Gonsalves, C., Ling, K.-S., Fitch, M., Manshardt, R., Slightom, J. L., et al. (1994). Differential protection against papaya ringspot virus isolates in coat protein gene transgenic papaya and classically cross-protected papaya. *Phytopathology*, *84*, 1359–1366.

Thomas, P. E., Lawson, E. C., Zalewski, J. C., Reed, G. L., & Kaniewski, W. (2000). Extreme resistance to potato leafroll virus in potato cv. Russet Burbank mediated by the viral replicase gene. *Virus Research*, *71*, 49–62.

Thompson, J. R., & Tepfer, M. (2010). Assessment of the benefits and risks for engineered virus resistance. *Advances in Virus Research*, *76*, 33–56.

Tien, P., & Wu, G. S. (1991). Satellite RNA for the biocontrol of plant-disease. *Advances in Virus Research*, *39*, 321–339.

Timmerman-Vaughan, G. M., Pither-Joyce, M. D., Cooper, P. A., Russell, A. C., Goulden, D. S., Butler, R., et al. (2001). Partial resistance of transgenic peas to alfalfa mosaic virus under greenhouse and field conditions. *Crop Science*, *41*, 846–853.

Tomassoli, L., Ilardi, V., Barba, M., & Kaniewski, W. (1999). Resistance of transgenic tomato to cucumber mosaic cucumovirus under field conditions. *Molecular Breeding*, *5*, 121–130.

Tougou, M., Furutani, N., Yamagishi, N., Shizukawa, Y., Takahata, Y., & Hidaka, S. (2006). Development of resistant transgenic soybeans with inverted repeat-coat protein genes of soybean dwarf virus. *Plant Cell Reports*, *25*, 1213–1218.

Tougou, M., Yamagishi, N., Furutani, N., Shizukawa, Y., Takahata, Y., & Hidaka, S. (2007). *Soybean dwarf virus*-resistant transgenic soybeans with the sense coat protein gene. *Plant Cell Reports*, *26*, 1967–1975.

Tousch, D., Jacquemond, M., & Tepfer, M. (1994). Replication of cucumber mosaic virus satellite RNA from negative-sense transcripts produced either in vitro or in transgenic plants. *Journal of General Virology*, *75*, 1009–1014.

Trevisan, F., Mendes, B. M. J., Maciel, S. C., Vieira, M. L. C., Meletti, L. M. M., & Rezende, J. A. M. (2006). Resistance to *Passion fruit woodiness virus* in transgenic passionflower expressing the virus coat protein gene. *Plant Disease*, *90*, 1026–1030.

Tricoli, D. M., Carney, K. J., Russell, P. F., McMaster, J. R., Groff, D. W., Hadden, K. C., et al. (1995). Field evaluation of transgenic squash containing single or multiple virus coat protein gene constructs for resistance to cucumber mosaic virus, watermelon mosaic virus 2, and zucchini yellow mosaic virus. *Bio/Technology*, *13*, 1458–1465.

Trifonova, E. A., Romanova, A. V., Sangaev, S. S., Sapotsky, M. V., Malinkovsky, V. I., & Kochetov, A. V. (2012). Inducible expression of the gene of *Zinnia elegans* coding for extracellular ribonuclease in *Nicotiana tabacum* plants. *Biologia Plantarum, 56*, 571–574.
Trifonova, E. A., Sapotsky, M. V., Komarova, M. L., Scherban, A. B., Shummy, V. K., Polyakova, A. M., et al. (2007). Protection of transgenic tobacco expressing bovine pancreatic ribonuclease against tobacco mosaic virus. *Plant Cell Reports, 26*, 1121–1126.
Truve, E., Aaspollu, A., Honkanen, J., Puska, R., Mehto, M., Hassi, A., et al. (1993). Transgenic potato plants expressing mammalian 2′-5′ oligoadenylate synthetase are protected from potato virus-X infection under field conditions. *Bio/Technology, 11*, 1048–1052.
Tumer, N. E., Hwang, D.-J., & Bonness, M. (1997). C-terminal deletion mutant of pokeweed antiviral protein inhibits viral infection but does not depurinate host ribosomes. *Proceedings of the National Academy of Sciences of the United States of America, 94*, 3866–3871.
Tumer, N. E., Kaniewski, W., Haley, L., Gehrke, L., Lodge, J. K., & Sanders, P. (1991). The 2nd amino-acid of alfalfa mosaic-virus coat protein is critical for coat protein-mediated resistance. *Proceedings of the National Academy of Sciences of the United States of America, 88*, 2331–2335.
Tumer, N. E., O'Connell, K. M., Nelson, R. S., Sanders, P. R., Beachy, R. N., Fraley, R. T., et al. (1987). Expression of alfalfa mosaic virus coat protein gene confers cross-protection in transgenic tobacco and tomato plants. *The EMBO Journal, 6*, 1181–1188.
Tyagi, H., Rajasubramaniam, S., Rajam, M. V., & Dasgupta, I. (2008). RNA-interference in rice against Rice tungro bacilliform virus results in its decreased accumulation in inoculated rice plants. *Transgenic Research, 17*, 897–904.
Ultzen, T., Gielen, J., Venema, F., Westerbroek, A., de Haan, P., Tan, M.-L., et al. (1995). Resistance to tomato spotted wilt virus in transgenic tomato hybrids. *Euphytica, 85*, 159–168.
Vaira, A. M., Berio, T., Accotto, G. P., Vecchiati, M., & Allavena, A. (2000). Evaluation of resistance in *Osteospermum ecklonis* (DC.) Norl. plants transgenic for the *N* protein gene of tomato spotted wilt virus. *Plant Cell Reports, 19*, 983–988.
Vaira, A. M., Semeria, L., Crespi, S., Lisa, V., Allavena, A., & Accotto, G. P. (1995). Resistance to tospoviruses in *Nicotiana benthamiana* transformed with the N gene of tomato spotted wilt virus: Correlation between transgene expression and protection in primary transformants. *Molecular Plant-Microbe Interactions, 8*, 66–73.
van den Boogaart, T., Wen, F., Davies, J. W., & Lomonossoff, G. P. (2001). Replicase-mediated resistance against *Pea early browning virus* in *Nicotiana benthamiana* is an unstable resistance based upon posttranscriptional gene silencing. *Molecular Plant-Microbe Interactions, 14*, 196–203.
Van den Elzen, P. J. M., Huisman, M. J., Posthumus-Lutke Willink, D., Jongedijk, E., Hoekema, A., & Cornelissen, B. J. C. (1989). Engineering virus resistance in agricultural crops. *Plant Molecular Biology, 13*, 337–346.
Van den Heuvel, J. F. J. M., Van der Vlugt, R. A. A., Verbeek, M., de Haan, P. T., & Huttinga, H. (1994). Characteristics of a resistance-breaking isolate of potato virus Y causing potato tuber necrotic ringspot disease. *European Journal of Plant Pathology, 100*, 347–356.
Van der Vlugt, R. A. A., & Goldbach, R. W. (1993). Tobacco plants transformed with the potato virus Y^N coat protein gene are protected against different PVY isolates and against aphid-mediated infection. *Transgenic Research, 2*, 109–114.
Van der Vlugt, R. A. A., Ruiter, R. K., & Goldbach, R. (1992). Evidence for sense RNA-mediated protection to PVYN in tobacco plants transformed with the viral coat protein cistron. *Plant Molecular Biology, 20*, 631–639.
Van der Wilk, F., Posthumus-Lutke Willink, D., Huisman, M. J., Huttinga, H., & Goldbach, R. (1991). Expression of the potato leafroll virus coat protein gene in transgenic potato plants inhibits viral infection. *Plant Molecular Biology, 17*, 431–439.

Van Dun, C. M. P., & Bol, J. F. (1988). Transgenic tobacco plants accumulating tobacco rattle virus coat protein resist infection with tobacco rattle virus and pea early browning virus. *Virology*, *176*, 649–652.

Van Dun, C. M. P., Bol, J. F., & Van Vloten-Doting, L. (1987). Expression of alfalfa mosaic virus and tobacco rattle virus coat protein genes in transgenic tobacco plants. *Virology*, *159*, 299–305.

Van Dun, C. M. P., Overduin, B., Van Vloten-Doting, L., & Bol, J. F. (1988). Transgenic tobacco expressing tobacco streak virus or mutated alfalfa mosaic virus coat protein does not cross-protect against alfalfa mosaic virus infection. *Virology*, *164*, 383–389.

Vandenbussche, F., Peumans, W. J., Desmyter, S., Proost, P., Ciani, M., & Van Damme, E. J. M. (2004). The type-1 and type-2 ribosome-inactivating proteins from *Iris* confer transgenic tobacco plants local but not systemic protection against viruses. *Planta*, *220*, 211–221.

Vanderschuren, H., Alder, A., Zhang, P., & Gruissem, W. (2009). Dose-dependent RNAi-mediated geminivirus resistance in the tropical root crop cassava. *Plant Molecular Biology*, *70*, 265–272.

Vanderschuren, H., Moreno, I., Anjanappa, R. B., Zainuddin, I. M., & Gruissem, W. (2012). Exploiting the combination of natural and genetically engineered resistance to cassava mosaic and cassava brown streak viruses impacting cassava production in Africa. *PLoS One*, 7, e45277.

Vanderschuren, H., Stupak, M., Fütterer, J., Gruissem, W., & Zhang, P. (2007). Engineering resistance to geminiviruses—Review and perspectives. *Plant Biotechnology Journal*, *5*, 207–220.

Vannini, C., Iriti, M., Bracale, M., Locatelli, F., Faoro, F., Croce, P., et al. (2006). The ectopic expression of the rice Osmyb4 gene in Arabidopsis increases tolerance to abiotic, environmental and biotic stresses. *Physiological and Molecular Plant Pathology*, *69*, 26–42.

Vardi, E., Sela, I., Edelbaum, O., Livneh, O., Kuznetsova, L., & Stram, Y. (1993). Plants transformed with a citron of a potato virus Y protease (NIa) are resistant to virus infection. *Proceedings of the National Academy of Sciences of the United States of America*, *90*, 7513–7517.

Vaskin, M. F. S., Vidal, M. S., Alves, E. D., Farinelli, L., & de Oliveira, D. E. (2001). Co-suppression mediated virus resistance in transgenic tobacco plants harboring the 3′-untranslated region of Andean potato mottle virus. *Transgenic Research*, *10*, 489–499.

Vassilakos, N., Bem, F., Tzima, A., Barker, H., Reavy, B., Karanastasi, E., et al. (2008). Resistance of transgenic tobacco plants incorporating the putative 57-kDa polymerase read-through gene of *Tobacco rattle virus* against rub-inoculated and nematode-transmitted virus. *Transgenic Research*, *17*, 929–941.

Vasudevan, A., Oh, T.-K., Park, J. S., Lakshmi, S. V., Choi, B. K., Kim, S. H., et al. (2008). Characterization of resistance mechanism in transgenic *Nicotiana benthamiana* containing *Turnip crinkle virus* coat protein. *Plant Cell Reports*, *27*, 1731–1740.

Vazquez Rovere, C., Asurmendi, A., & Hopp, H. E. (2001). Transgenic resistance in potato plants expressing potato leaf roll virus (PLRV) replicase gene sequence is RNA-mediated and suggests the involvement of post-transcriptional gene silencing. *Archives of Virology*, *146*, 1337–1353.

Verma, V., Sharma, S., Devi, S. V., Rajasubramaniam, S., & Dasgupta, I. (2012). Delay in virus accumulation and low virus transmission from transgenic rice expressing *Rice tungro spherical virus* RNA. *Virus Genes*, *45*, 350–359.

Vidya, C. S. S., Manoharan, M., Kumar, C. T. R., Savithra, H. S., & Sita, G. L. (2000). *Agrobacterium*-mediated transformation of tomato (*Lycoperison esculentum* var. Pusa Ruby) with coat-protein gene of *Physalis* mottle tymovirus. *Journal of Plant Physiology*, *156*, 106–110.

Villani, M. E., Roggero, P., Bitti, O., Benvenuto, E., & Franconi, R. (2005). Immuno-modulation of cucumber mosaic virus infection by intrabodies selected in vitro from a stable single-framework phage display library. *Plant Molecular Biology*, *58*, 305–316.

Voloudakis, A. E., Aleman-Verdaguer, M.-E., Padgett, H. S., & Beachy, R. N. (2005). Characterization of resistance in transgenic *Nicotiana benthamiana* encoding N-terminal deletion and assembly mutants of the tobacco etch potyvirus coat protein. *Archives of Virology, 150*, 2567–2582.

Voss, A., Niersbach, M., Hain, R., Hirsch, H., Liao, Y., Kreuzaler, F., et al. (1995). Reduced virus infectivity in *N. tabacum* secreting a TMV-specific full size antibody. *Molecular Breeding, 1*, 39–50.

Vu, T. V., Choudhury, N. R., & Mukherjee, S. K. (2013). Transgenic tomato plants expressing artificial microRNAs for silencing the pre-coat and coat proteins of a begomovirus, Tomato leaf curl New Delhi virus, show tolerance to virus infection. *Virus Research, 172*, 35–45.

Wako, T., Terami, F., Hanada, K., & Tabei, Y. (2001). Resistance to *Zucchini yellow mosaic virus* (ZYMV) in transgenic cucumber plants (*Cucumis sativus* L.) harboring the coat protein gene of ZYMV. *Bulletin of the National Research Institute of Vegetables, Ornamental Plants and Tea, 16*, 175–186.

Wang, M.-B., Abbott, D. C., & Waterhouse, P. M. (2000). A single copy of a virus-derived transgene encoding hairpin RNA gives immunity to barley yellow dwarf virus. *Molecular Plant Pathology, 1*, 347–356.

Wang, X., Eggenberger, A. L., Nutter, F. W., Jr., & Hill, J. H. (2001). Pathogen-derived transgenic resistance to soybean mosaic virus in soybean. *Molecular Breeding, 8*, 119–127.

Wang, Y. Y., Gardner, R. C., & Pearson, M. N. (1997). Resistance of Vanilla necrosis potyvirus in transgenic *Nicotiana benthamiana* plants containing the coat protein gene. *Journal of Phytopathology, 145*, 7–15.

Wang, X., Kohalmi, S. E., Svircev, A., Wang, A., Sanfaçon, H., & Tian, L. (2013). Silencing of the host factor eIF(iso)4E gene confers plum pox virus resistance in plum. *PLoS One, 8*, e50627.

Wang, M.-B., Masuta, C., Smith, N. A., & Shimura, H. (2012). RNA silencing and plant viral diseases. *Molecular Plant-Microbe Interactions, 25*, 1275–1285.

Wang, P., & Tumer, N. E. (2000). Virus resistance mediated by ribosome inactivating proteins. *Advances in Virus Research, 55*, 325–355.

Wang, C.-X., Yang, M.-Z., Pan, N.-S., & Chen, Z.-L. (1993). Resistance to potato virus X infection in transgenic tobacco plants with coat protein gene of virus. *Acta Botanica Sinica, 35*, 819–824.

Wang, H.-Z., Zhao, P.-J., Xu, J.-C., Zhao, H., & Zhang, H.-S. (2003). Virus resistance in transgenic watermelon plants containing a WMV-2 coat protein gene. *Acta Genetica Sinica, 30*, 70–75.

Wang, H. Z., Zhao, P. J., & Zhou, X. Y. (2000). Regeneration of transgenic *Cucumis melo* and its resistance to virus diseases. *Acta Phytophylacica Sinica, 27*, 126–130.

Wang, P., Zoubenko, O., & Tumer, N. (1998). Reduced toxicity and broad spectrum resistance to viral and fungal infection in transgenic plants expressing pokeweed antiviral protein II. *Plant Molecular Biology, 38*, 957–964.

Watanabe, Y., Ogawa, T., Takahashi, H., Ishida, I., Takeuchi, Y., Yamamoto, M., et al. (1995). Resistance against multiple viruses in plants mediated by a double stranded-RNA specific ribonuclease. *FEBS Letters, 372*, 165–168.

Waterhouse, P. M., Graham, M. W., & Wang, M.-B. (1998). Virus resistance and gene silencing in plants can be induced by simultaneous expression of sense and antisense RNA. *Proceedings of the National Academy of Sciences of the United States of America, 95*, 13959–13964.

Wesley, S. V., Helliwell, C. A., Smith, N. A., Wang, M.-B., Rouse, D. T., Liu, Q., et al. (2001). Construct design for efficient, effective and high throughput gene silencing in plants. *Plant Journal, 27*, 581–590.

Whitham, S., McCormick, S., & Baker, B. (1996). The *N* gene of tobacco confers resistance to tobacco mosaic virus in transgenic tomato. *Proceedings of the National Academy of Sciences of the United States of America, 93*, 8776–8781.

Wintermantel, W. M., & Zaitlin, M. (2000). Transgene translatability increases effectiveness of replicase-mediated resistance to Cucumber mosaic virus. *Journal of Virology, 81*, 587–595.

Wittner, A., Palkovics, L., & Balázs, E. (1998). *Nicotiana benthamiana* plants transformed with the plum pox virus helicase gene are resistant to virus infection. *Virus Research, 53*, 97–103.

Wojtal, I., Pointek, P., Grzela, R., Jarmolowski, A., Zagórski, W., & Chroboczek, J. (2011). Analysis of potyviral terminal VPg-transgenic *Arabidopsis thaliana* plants. *Acta Biochimica Polonica, 58*, 349–353.

Wu, Y. F., Chen, Y., Liang, X. M., & Wang, X. Z. (2006). An experimental assessment of the factors influencing *Agrobacterium*-mediated transformation in tomato. *Russian Journal of Plant Physiology, 53*, 252–256.

Wu, H.-W., Yu, T.-A., Raja, J. A. J., Christofer, S. J., Wang, S.-L., & Yeh, S.-D. (2010). Double-virus resistance of transgenic oriental melon conferred by untranslatable chimeric construct carrying partial coat protein genes of two viruses. *Plant Disease, 94*, 1341–1347.

Wu, H.-W., Yu, T.-A., Raja, J. A. J., Wang, H.-C., & Yeh, S.-D. (2009). Generation of transgenic oriental melon resistant to *Zucchini yellow mosaic virus* by an improved cotyledon-cutting method. *Plant Cell Reports, 28*, 1053–1864.

Wu, H., Zhang, B., Gao, D., Xu, H., & Cheng, S. (2006). Disease resistance test of transgenic wheat lines with *WYMV-Nib8* gene and their application in breading. *Journal of Triticeae, 16*, 11–14.

Xiao, X. W., Chu, P. W. G., Frenkel, M. J., Tabe, L. M., Shukla, D. D., Hanna, P. J., et al. (2000). Antibody-mediated improved resistance to ClYVV and PVY infections in transgenic tobacco plants expressing a single-chain variable region antibody. *Molecular Breeding, 6*, 421–431.

Xu, D., Collins, G. B., Hunt, A. G., & Nielsen, M. T. (1997). Field resistance of transgenic burley tobacco lines and hybrids expressing the tobacco vein mottling virus coat protein gene. *Molecular Breeding, 3*, 319–330.

Xu, D. M., Collins, G. B., Hunt, A. G., & Nielsen, M. T. (1998). Resistance to alfalfa mosaic virus in transgenic burley tobaccos expressing the AMV coat protein gene. *Crop Science, 38*, 1661–1668.

Xu, H., Khalilian, H., Eweida, M., Squire, M., & Abouhaidar, M. (1995). Genetically engineered resistance to potato virus X in four commercial potato cultivars. *Plant Cell Reports, 15*, 91–96.

Xu, J., Schubert, J., & Altpeter, F. (2001). Dissection of RNA-mediated ryegrass mosaic virus resistance in fertile transgenic perennial ryegrass (*Lolium perenne* L.). *Plant Journal, 26*, 265–274.

Xu, B., Shang, H., & Wang, X. (2002). Cucumber mosaic virus and tobacco mosaic virus proliferation in transgenic chili peppers and its protoplast. *Virologica Sinica, 17*, 243–247.

Xu, B. L., Shi, G. Y., & Xue, Y. Y. (2005). Plant regeneration in transgenic Huanghemi (*Cucumis melo*) and identification of resistance to viral disease. *Journal of Fruit Science, 22*, 734–736.

Xu, L., Song, Y., Zhu, J., Guo, X., Zhu, C., & Wen, F. (2009). Conserved sequences of replicase gene-mediated resistance to *Potyvirus* through RNA silencing. *Journal of Plant Biology, 52*, 550–559.

Xue, B., Gonsalves, C., Provvidenti, R., Slightom, J. L., Fuchs, M., & Gonsalves, D. (1994). Development of transgenic tomato expressing high level of resistance to cucumber mosaic virus strains of subgroup-I and subgroup-II. *Plant Disease, 78*, 1038–1041.

Yadav, J. S., Ogwok, E., Wagaba, H., Patil, B. L., Bagewadi, B., Gaitin-Solis, E., et al. (2011). RNAi-mediated resistance to *Cassava brown streak Uganda virus* in transgenic cassava. *Molecular Plant Pathology, 12*, 677–687.

Yan, Y. T., Wang, J. F., Qiu, B. S., & Tien, P. (1997). Resistance to rice stripe virus conferred by expression of coat protein in transgenic indica rice plants regenerated from bombarded suspension culture. *Virologica Sinica, 12*, 260–269.

Yan, L., Xu, Z., Chen, K., Goldbach, R., & Prins, M. (2007). Resistance to Peanut stripe virus mediated by inverted repeat of the coat protein gene in transgenic tobacco plants. *Journal of Agricultural Biotechnology, 4*, 702–707.

Yan, F., Zhang, W., Xiao, H., Li, S., & Cheng, Z. (2007). Transgenic wheat expressing virus-derived hairpin RNA is resistant to *Barley yellow dwarf virus*. *Heriditas (Beijing), 29*, 97–102.

Yang, Y., Sherwood, T. A., Patte, C. P., Hieber, E., & Polston, J. E. (2004). Use of Tomato yellow leaf curl virus (TYLCV) Rep gene sequences to engineer TYLCV resistance in tomato. *Phytopathology, 94*, 490–496.

Yassaie, M., Afsharifar, A. R., Niazi, A., Salehzadeh, S., & Izadpanah, K. (2011). Induction of resistance to barley yellow dwarf virus (PAV) in bread wheat using post-transcriptional gene silencing (PTGS). *Iranian Journal of Plant Pathology, 47*, 15–18.

Yeh, S.-D., & Chu, F.-H. (1996). Production and evaluation of transgenic tobacco plants expressing the coat protein gene of passionfruit woodiness virus. *Botanical Bulletin of Academia Sinica, 37*, 181–190.

Yeh, S.-D., & Pon, J.-C. (1996). Coat protein-mediated resistance to tomato mosaic virus in systemic and local lesion hosts. *Plant Pathology Bulletin, 5*, 15–27.

Yepes, L. M., Fuchs, M., Slightom, J. L., & Gonsalves, D. (1996). Sense and antisense coat protein gene constructs confer high levels of resistance to tomato ringspot nepovirus in transgenic *Nicotiana* species. *Phytopathology, 86*, 417–424.

Yie, Y., Wu, Z. X., Wang, S. Y., Zhao, S. Z., Zhang, T. Q., Yao, G. Y., et al. (1995). Rapid production and field testing of homozygous transgenic tobacco lines with virus resistance conferred by expression of satellite RNA and coat protein of cucumber mosaic virus. *Transgenic Research, 4*, 256–263.

Yie, Y., Zhao, F., Zhao, S. Z., Liu, Y. Z., Liu, L., & Tien, P. (1992). High resistance to cucumber mosaic virus conferred by satellite RNA and coat protein in transgenic commercial tobacco cultivar G-140. *Molecular Plant-Microbe Interactions, 5*, 460–465.

Yoshikawa, N., Gotoh, S., Umezawa, M., Satoh, N., Satoh, H., Takahashi, T., et al. (2000). Transgenic *Nicotiana occidentalis* plants expressing the 50-kDa protein of *Apple chlorotic leaf spot virus* display increased susceptibility to homologous virus, but strong resistance to *Grapevine berry inner necrosis virus*. *Phytopathology, 90*, 311–316.

Yoshioka, K., Hanada, K., Harada, T., Minobe, Y., & Oosawa, K. (1993). Virus-resistance in transgenic melon plants that express the cucumber mosaic-virus coat protein gene and their progeny. *Japanese Journal of Breeding, 43*, 629–634.

Yu, T.-A., Chiang, C.-H., Wu, H.-W., Li, A.-M., Yang, C.-F., Chen, J.-H., et al. (2011). Generation of transgenic watermelon resistant to Zucchini yellow mosaic virus and Papaya ringspot virus type W. *Plant Cell Reports, 30*, 359–371.

Yu, Z., Zhao, S., & He, Q. (2007). High level resistance to *Turnip mosaic virus* in Chinese cabbage (*Brassica campestris* spp. *pekinensis* (Lour) Olsson) transformed with the antisense NIb gene using marker-free *Agrobacterium tumefaciens* infiltration. *Plant Science, 172*, 920–929.

Zaccomer, B., Cellier, F., Boyer, J.-C., Haenni, A.-L., & Tepfer, M. (1993). Transgenic plants that express genes including the 3′ untranslated region of the turnip yellow mosaic virus (TYMV) genome are partially protected against TYMV infection. *Gene, 136*, 87–94.

Zadeh, A. H., & Foster, G. D. (2004). Transgenic resistance to tobacco ringspot virus. *Acta Virologica, 48*, 145–152.

Zeng, Y., Wagner, E. J., & Cullen, B. R. (2002). Both natural and designed micro RNAs can inhibit the expression of cognate mRNAs when expressed in human cells. *Molecular Cell, 9*, 1327–1333.
Zeng, G., Ye, Y.-Y., Cao, M.-Q., Ma, R.-C., Tang, L.-C., & Yao, L. (2013). Research of high resistance to *Turnip mosaic virus* by coat protein gene segment. *Journal of Agricultural Science and Technology, 15*, 83–88.
Zhang, K., Biu, Y.-B., & Zhou, X.-P. (2005). Transgenic tobacco plants expressed dsRNA can prevent *Tobacco mosaic virus* infection. *Journal of Agricultural Biotechnology, 13*, 226–229.
Zhang, G., Chen, M., Li, L., Xu, Z., Chen, X., Guo, J., et al. (2009). Overexpression of the soybean GmERF3 gene, an AP2/ERF type transcription factor for increased tolerances to salt, drought, and diseases in transgenic tobacco. *Journal of Experimental Botany, 60*, 3781–3796.
Zhang, L., French, R., Langenberg, W. G., & Mitra, A. (2001). Accumulation of barley stripe mosaic virus is significantly reduced in transgenic wheat plants expressing a bacterial ribonuclease. *Transgenic Research, 10*, 13–19.
Zhang, Z.-Y., Fu, F.-L., Gou, L., Wang, H.-G., & Li, W.-C. (2010). RNA interference-based transgenic maize resistant to maize dwarf mosaic virus. *Journal of Plant Biology, 53*, 297–305.
Zhang, J.-Q., Li, H.-Y., Hu, X.-N., Shan, Z.-H., & Tang, G.-X. (2013). *Agrobacterium tumefaciens* mediated transformation of RNAi *CP* gene into soybean (*Glycine max* L.). *Acta Agronomica Sinica, 39*, 1594–1601.
Zhang, X., Li, H., Zhang, J., Zhang, C., Gong, P., Ziaf, K., et al. (2011). Expression of artificial microRNAs in tomato confers efficient and stable virus resistance in a cell-autonomous manner. *Transgenic Research, 20*, 569–581.
Zhang, H., Peng, X., Song, Y., & Li, T. (1997). Resistance to mixed PVY and PLRV infection in potato cultivars expressing dual PVY and PLRV coat protein genes. *Acta Botanica Sinica, 39*, 236–240.
Zhang, X., Sato, S., Ye, X., Dorrance, A. E., Morris, T. J., Clemente, T. E., et al. (2011). Robust RNAi-based resistance to mixed infection of three viruses in soybean plants expressing separate short hairpins from a single transgene. *Phytopathology, 101*, 1264–1269.
Zhang, C., Song, Y., Jiang, F., Li, G., Jiang, Y., Zhu, C., et al. (2012). Virus resistance in transgenic tobacco and rice by RNA interference using promoters with distinct activity. *Biologia Plantarum, 56*, 742–748.
Zhang, H., Song, Y., Pen, X., Li, T., Meng, Q., Cui, X., et al. (1996). Engineered resistance to mixed PVX and PVY infection in potato cultivars expressing both PVX and PVY coat protein genes. *Chinese Journal of Virology, 12*, 361–366.
Zhang, S. C., Tian, L., Svircev, A., Brown, D. C. W., Sibbald, S., Schneider, K. E., et al. (2006). Engineering resistance to *Plum pox virus* (PPV) through the expression of PPV-specific hairpin RNAs in transgenic plants. *Canadian Journal of Plant Pathology, 28*, 263–270.
Zhang, H., Tian, Y., Zhou, Y., Dang, B., Lan, H., Song, G., et al. (1999). Introduction of pokeweed antiviral protein cDNA into *Brassica napus* and acquisition of transgenic plants resistant to viruses. *Chinese Science Bulletin, 44*, 701–704.
Zhang, P., Vanderschuren, H., Fütterer, J., & Gruissem, W. (2005). Resistance to cassava mosaic disease in transgenic cassava expressing antisense RNAs targeting virus replication genes. *Plant Biotechnology Journal, 3*, 385–397.
Zhang, Z.-Y., Wang, Y.-G., Shen, X.-J., Li, L., Zhou, S.-F., Li, W.-C., et al. (2013). RNA interference-mediated resistance to maize dwarf mosaic virus. *Plant Cell, Tissue and Organ Culture, 113*, 571–578.
Zhang, L., Xie, X., Song, Y., Jiang, F., Zhu, C., & Wen, F. (2013). Viral resistance mediated by shRNA depends on the sequence similarity and mismatched sits between target sequence and siRNA. *Biologia Plantarum, 57*, 547–554.

Zhang, Z.-Y., Yang, L., Zhou, S.-F., Wang, H.-G., Li, W.-C., & Fu, F.-L. (2011). Improvement of resistance to maize dwarf mosaic virus mediated by transgenic RNA interference. *Journal of Biotechnology, 153*, 181–187.

Zhang, Z. J., Zhou, Z. X., Liu, Y. J., Jiang, Q. Y., You, M., Liu, G. M., et al. (1994). CMVcp gene transformation into pepper and expression in the offspring of the transgenic plants. *Acta Agriculturae Boreali-Sinica, 9*, 67–71.

Zhang, M.-Y., Zimmermann, S., Fischer, R., & Schillberg, S. (2008). Generation and evaluation of movement protein-specific single-chain antibodies for delaying symptoms of *Tomato spotted wilt virus* infection in tobacco. *Plant Pathology, 57*, 854–860.

Zhao, M.-A., An, S.-J., Lee, S.-C., Kim, D.-S., & Kang, B.-C. (2013). Overexpression of a single-chain variable fragment (*scFv*) antibody confers unstable resistance to TuMV in Chinese cabbage. *Plant Molecular Biology Reporter, 31*, 1203–1211.

Zhao, G. Q., & Mu, P. (2004). Genetic transformation of red clover using alfalfa mosaic virus coat protein gene and its virus resistance analysis. *Acta Botanica Boreali-Occidentalia Sinica, 24*, 1850–1855.

Zhao, G. Q., Mu, P., & Chu, P. (2005). Expression of alfalfa mosaic virus coat protein in white clover and its resistant against virus. *Journal of Agricultural Biotechnology, 13*, 230–234.

Zhou, Y., Yuan, Y., Yuan, F., Wang, M., Zhong, H., Gu, M., et al. (2012). RNAi-directed down-regulation of RSV results in increased resistance in rice (*Oryza sativa* L.). *Biotechnology Letters, 34*, 965–972.

Zhu, Y. J., McCafferty, H., Osterman, G., Lim, S., Agbayani, R., Lehrer, A., et al. (2011). Genetic transformation with untranslatable coat protein gene of *sugarcane yellow leaf virus* reduces virus titers in sugarcane. *Transgenic Research, 20*, 503–512.

Zhu, C., Song, Y., & Wen, F. (2008). Constructing transgenic tobacco for multiple-resistance against *Potato virus Y, Cucumber mosaic virus*, and *Tobacco mosaic virus*. *Scientia Agricultura Sinica, 41*, 1040–1047.

Zhu, C.-X., Song, Y.-Z., Yin, G.-H., & Wen, F.-J. (2009). Induction of RNA-mediated multiple virus resistance to *Potato virus Y, Tobacco mosaic virus* and *Cucumber mosaic virus*. *Journal of Phytopathology, 157*, 101–107.

Zhu, C., Song, Y., Zhang, S., Guo, X., & Wen, F. (2001). Production of transgenic Chinese cabbage by transformation with the CP gene of turnip mosaic virus. *Acta Phytopathologica Sinica, 31*, 257–264.

Zhu, X., Zhu, C., Song, Y., Wen, F., Liu, H., & Li, X. (2006). Resistance to Potato virus Y mediated by the 3′end segments of coat protein gene in the transgenic tobacco plants. *Scientia Agricultura Sinica, 39*, 1153–1158.

Zhu, J., Zhu, X., Wen, F., Bai, Q., Zhu, C., & Song, Y. (2004). Effect of cDNA fragments in different length derived from potato virus Y coat protein gene on the induction of RNA-mediated virus resistance. *Science in China. Series C, Life Sciences, 47*, 382–388.

Zhu, J., Zhu, C., Wen, F., & Song, Y. (2004). Comparison of resistance to *Potato virus Y* mediated by direct and inverted repeats of the coat protein gene segments in transgenic tobacco plants. *Acta Phytopathologica Sinica, 34*, 133–140.

Ziegler-Graff, V., Guilford, P. J., & Baulcombe, D. C. (1991). Tobacco rattle virus RNA-1 29K gene product potentiates viral movement and also affects symptom induction in tobacco. *Virology, 182*, 145–155.

Zimmermann, S., Schillberg, S., Liao, Y. C., & Fischer, R. (1998). Intracellular expression of TMV-specific single chain Fv fragments leads to improved virus resistance in *Nicotiana tabacum*. *Molecular Breeding, 4*, 369–379.

Zrachya, A., Kumar, P. P., Ramakrishnan, U., Levy, Y., Loyter, A., Arazi, T., et al. (2007). Production of siRNAs targeted against TYLCV coat protein transcripts leads to silencing of its expression and resistance to the virus. *Transgenic Research, 16*, 385–398.

CHAPTER THREE

Management of Whitefly-Transmitted Viruses in Open-Field Production Systems

Moshe Lapidot*[,1], James P. Legg[†], William M. Wintermantel[‡], Jane E. Polston[§]

*Institute of Plant Sciences, Volcani Center, ARO, Bet Dagan, Israel
[†]International Institute of Tropical Agriculture, Dar es Salaam, Tanzania
[‡]USDA-ARS, Salinas, California, USA
[§]Department of Plant Pathology, University of Florida, Gainesville, Florida, USA
[1]Corresponding author: e-mail address: lapidotm@volcani.agri.gov.il

Contents

Advances in Virus Research, Volume 90
ISSN 0065-3527
http://dx.doi.org/10.1016/B978-0-12-801246-8.00003-2

Abstract

Whiteflies are a key pest of crops in open-field production throughout the tropics and subtropics. This is due in large part to the long and diverse list of devastating plant viruses transmitted by these vectors. Open-field production provides many challenges to manage these viruses and in many cases adequate management has not been possible. Diseases caused by whitefly-transmitted viruses have become limiting factors in open-field production of a wide range of crops, i.e., bean golden mosaic disease in beans, tomato yellow leaf curl disease in tomato, cassava mosaic disease and cassava brown streak disease in cassava, and cotton leaf crumple disease in cotton. While host resistance has proven to be the most cost-effective management solution, few examples of host resistance have been developed to date. The main strategy to limit the incidence of virus-infected plants has been the application of insecticides to reduce vector populations aided to some extent by the use of selected cultural practices. However, due to concerns about the effect of insecticides on pollinators, consumer demand for reduced pesticide use, and the ability of the whitefly vectors to develop insecticide-resistance, there is a growing need to develop and deploy strategies that do not rely on insecticides. The reduction in pesticide use will greatly increase the need for genetic resistance to more viruses in more crop plants. Resistance combined with selected IPM strategies could become a viable means to increase yields in crops produced in open fields despite the presence of whitefly-transmitted viruses.

1. INTRODUCTION

Over the last 20 years, viruses transmitted by whiteflies have emerged as a global threat to crop production in a wide range of crops. This emergence is due in large part to the movement of plants and plant parts which distribute both vectors and viruses to new locations (Anderson et al., 2004). Whitefly and whitefly-transmitted viruses are primarily concerns in dicotyledonous crops. Many of the crops that are adversely affected by these viruses are economically significant and losses occur in both crops for export as well as those critical for subsistence. Many of these viruses are limiting factors in crop production. Some of the greatest losses occur in fiber crops such as cotton; vegetable crops such as cassava, cucurbits, tomato, pepper, common bean, various pulses; agronomic crops such as soybean; and biofuels such as Jatropha. Yield losses, which range from minimal to complete crop failure, depend upon the virus, the crop, the age of the crop at the time of infection, and the incidence of virus-infected plants. Crop resistance to most of these viruses or to feeding by the whitefly vectors has not been developed. In the absence of crop resistance, management of these viruses is usually very challenging, requiring the timely use of numerous management tactics with a heavy reliance on chemicals to limit the feeding, development, and movement of the vector(s).

2. WHITEFLIES AND THE VIRUSES THEY TRANSMIT

2.1. The whiteflies

Whiteflies (Order Hemiptera, Family Aleyrodidae) comprise more than 1500 species in approximately 126 genera (Martin, 2004). Of those many species, only five are known to transmit plant viruses: *Bemisia afer* (Priesner & Hosny), *Bemisia tabaci* species complex, *Parabemisia myricae* Kuwana (bayberry whitefly), *Trialeurodes abutilonea* Haldeman (banded wing whitefly), and *Trialeurodes vaporariorum* Westwood (greenhouse whitefly) (Gamarra et al., 2010; reviews by Hogenhout, Ammar, Whitfield, & Redinbaugh, 2008; Navas-Castillo, Fiallo-Olivé, & Sánchez-Campos, 2011; Ng & Falk, 2006). The diversity of whitefly vectors appears lower than it is in reality due to the current taxonomic organization of the *B. tabaci* species complex. This complex is actually composed of 34 distinguishable entities that are probable species based on both molecular and biological studies (reviewed by Polston, De Barro, & Boykin, 2014). When species names are eventually assigned to these entities, the number of whitefly vector species and the recognized diversity will increase significantly. Other whiteflies reported as vectors include *Trialeurodes ricini* (Misra), which was reported to transmit a begomovirus in Egypt, but this work has not been confirmed by any other reports (Idriss, Abdallah, Aref, Haridy, & Madkour, 1997). In addition, *B. tuberculata* was reported as a vector of viruses associated with frogskin disease of cassava (Angel, Pineda, Nolt, & Velasco, 1990) but phytoplasmas have also been shown to be associated with the disease (Alvarez et al., 2009) and its exact etiology remains uncertain.

Table 3.1 Summary of whitefly species and the virus genera they transmit

Virus family	Virus genus	No. of approved species[a]	Whitefly	Mode of transmission
Betaflexiviridae	*Carlavirus*	3[b]	*Bemisia tabaci* species complex	Nonpersistent/ Semi-persistent
Closteroviridae	*Crinivirus*	1	*Bemisia afer*	Semi-persistent
		4	*Bemisia tabaci* species complex	Semi-persistent
		4	*Trialeurodes abutilonea*	Semi-persistent
		4	*Trialeurodes vaporariorum*	Semi-persistent
Geminiviridae	*Begomovirus*	192	*Bemisia tabaci* species complex	Persistent, nonpropagative
		1	*Trialeurodes ricini*	Unknown
Potyviridae	*Ipomovirus*	4	*Bemisia tabaci* species complex	Semi-persistent
Secoviridae	*Torradovirus*	2[c]	*Bemisia tabaci* species complex	Undetermined
		1	*Trialeurodes vaporariorum*	Undetermined

[a]According to King, Lefkowitz, Adams, and Carstens (2011) except as noted.
[b]Includes *Cucumber vein-clearing virus* (Menzel, Abang, & Winter, 2011).
[c]Includes *Tomato necrotic dwarf virus* (Wintermantel & Hladky, 2013).

Table 3.1 briefly summarizes the families and genera of plant viruses, whitefly species known to transmit at least one species within those genera, and the mode of transmission. Four recent reviews cover these whitefly species and the viruses they transmit in more detail than will be presented here (Hogenhout et al., 2008; Navas-Castillo et al., 2011; Ng & Falk, 2006; Polston et al., 2014).

2.2. The viruses

The plant viruses transmitted by whiteflies have been reviewed extensively and readers are encouraged to consult these references for more detailed descriptions (Jones, 2003; Navas-Castillo et al., 2011; Tzanetakis, Martin, & Wintermantel, 2013; Legg et al., chapter 3).

2.2.1 *Betaflexiviridae, Carlavirus*

As of 2013, there were 52 virus species in the genus *Carlavirus* (www.ictvonline.org/virusTaxonomy.asp) of which only three, *Cowpea mild mottle virus* (CPMMV), *Cucumber vein-clearing virus* (CuVCV), and *Melon yellowing-associated virus* (MYaV), are known to be transmitted by whiteflies (Menzel et al., 2011; Nagata et al., 2005; Naidu et al., 1998). Carlaviruses are ssRNA viruses and are relatively easy to detect using standard types of assays—ELISA (commercially available for CPMMV) and reverse-transcription PCR. All three have been shown to be transmitted by *B. tabaci* MEAM1 (formerly known as the B biotype) although there is some confusion as to the manner of transmission. CuVCV was transmitted in a semi-persistent manner, while CPMMV has been reported to be transmitted in a nonpersistent manner by some and in a semi-persistent manner by others (Iwaki, Thongmeearkon, Prommin, Honda, & Hibi, 1982; Menzel et al., 2011; Muniyappa & Reddy, 1983; Rosario, Capobianco, Ng, Breitbart, & Polston, 2014). The manner of transmission of MYaV has not been reported. The confusion regarding CPMMV transmission may be due to the lack of identity of the whitefly species used in the studies, the variation among what are currently regarded as isolates of CPMMV but may be different species, as well as differences in methodologies used to determine the manner of transmission.

2.2.2 *Closteroviridae, Crinivirus*

The last 25 years have seen a rapid increase in the number of described species. These currently number 13 and all are transmitted by whiteflies (www.ictvonline.org/virusTaxonomy.asp). Criniviruses are difficult to recognize as their symptoms are not always apparent and are often readily mistaken for physiological or nutritional disorders or pesticide phytotoxicity. Detection is also difficult as the viruses occur in low concentrations and with less uniform distribution within the plant compared to other viruses. The genus now consists of three separate groups based largely on the sequences of their RNA-dependent RNA polymerase genes and vector transmission characteristics (Tzanetakis et al., 2013; Wintermantel, Hladky, Gulati-Sakhuja, et al., 2009). Criniviruses are transmitted in a semi-persistent manner and do not replicate within their whitefly vectors. Criniviruses can be transmitted by *B. tabaci*, *T. vaporariorum*, and *T. abutilonea* (Wintermantel, 2010). It is possible that additional vector species may exist but have yet to be identified. These three species are often common in areas in which criniviruses are found, and there is a clear relationship between prevalence of vector and of virus. Once a crinivirus is acquired, whiteflies remain viruliferous for 1–9 days depending on the vector and virus (Wintermantel, 2010; Wisler & Duffus, 2001; Wisler, Duffus, Liu, & Li, 1998). The viruses are not seed-borne, nor are they mechanically transmitted (Tzanetakis et al., 2013).

Like other members of the family *Closteroviridae*, criniviruses have long flexuous rod-shaped virions averaging between 650 and 1000 nm in length (Kreuze, Savenkov, & Valkonen, 2002; Liu, Wisler, & Duffus, 2000). Genomes are predominantly bipartite with RNA1 encoding functions predominantly associated with virus replication, and RNA2 (or RNAs 2 and 3 for *Potato yellow vein virus*) encoding as many as 10 proteins with a range of functions including virus encapsidation, cell-to-cell movement, and vector transmission (King et al., 2011). Crinivirus infections in some crops remain latent for nearly 3 weeks before symptoms appear and others remain latent until plants become coinfected with another virus (Tzanetakis et al., 2013). Such mixed infections can complicate identification of the primary virus causing disease because symptoms resulting from mixed infection with other viruses often induce different symptoms than those resulting from single infections. In other situations, the viruses involved in coinfections are obvious within different sections of infected plants exhibiting symptoms uniquely characteristic of each virus. Numerous studies have demonstrated that interactions between criniviruses and other coinfecting viruses have been known to influence the type and severity of symptoms observed on plants. In many cases, this leads to enhanced disease severity, as has been found with infection of *Sweet potato chlorotic stunt virus* (SPCSV) and members of the *Potyvirus* genus (Karyeija, Kreuze, Gibson, & Valkonen, 2000).

2.2.3 Geminiviridae, Begomovirus

The single-stranded circular DNA viruses in the genus *Begomovirus* and family *Geminiviridae* are the most studied of the whitefly-transmitted viruses. This is also the largest genus of whitefly-transmitted viruses with almost 200 recognized species (King et al., 2011). Transmission of these viruses by members of the *B. tabaci* species complex is persistent and nonpropagative (reviewed by Navas-Castillo et al., 2011). While there is no specificity of transmission of begomovirus species by members of the *B. tabaci* species complex, it has been shown that different members transmit the same virus with different efficiencies (Bedford, Briddon, Brown, Rosell, & Markham, 1994; Caciagli, Bosco, & Albitar, 1995; Idris, Smith, & Brown, 2001; Jiang, De Blas, Bedford, Nombela, & Muniz, 2004; Li, Hu, Xu, & Liu, 2010; McGrath & Harrison, 1995). For example, under the same conditions, whiteflies of the MEAM1 and Mediterranean clades transmitted *Tomato yellow leaf curl virus* (TYLCV) with equal efficiency, while whiteflies of the Asia II-1 clade transmitted the same virus about half as efficiently (Li et al., 2010). Differences in transmission efficiency have been shown to be due to the feeding habits and preferences of the vectors for the plant hosts used for acquisition and transmission. Efficiency is also affected by differences in the amount and distribution of begomoviruses among plant hosts (Azzam et al., 1994). The presence of selected endosymbionts (ex. *Hamiltonella spp.*) has also been shown to affect the transmission efficiency of begomoviruses (Gottlieb et al., 2010; Su et al., 2012). It has also been shown that at least one begomovirus can alter the settling, probing, and feeding behavior of the whitefly, thereby altering the transmission efficiency (Moreno-Delafuente, Garzo, Moreno, & Fereres, 2013).

2.2.4 *Potyviridae, Ipomovirus*

Species in the genus *Ipomovirus* are the only members of the family *Potyviridae* (single-stranded plus sense RNA genome) that are transmitted by whiteflies. Of the six approved species in this genus, four have been confirmed to be transmitted by members of the *B. tabaci* species complex at relatively low efficiencies (often below 50%) and in a semi-persistent manner. *Cassava brown streak virus* (CBSV) was shown to be transmitted by *B. tabaci* but not by *B. afer* (Maruthi et al., 2005). *Cucumber vein-yellowing virus* (CVYV) was shown to be transmitted in a semi-persistent manner by *B. tabaci* (Harpaz & Cohen, 1965; Mansour & Al-Musa, 1993). More recently, *Squash vein-yellowing virus* was demonstrated to be transmitted in a semi-persistent manner by *B. tabaci* MEAM1 (S. Webb, personal communication; Webb, Adkins, & Reitz, 2012). An eggplant-infecting strain of *Tomato mild mottling virus* (formerly Eggplant mild leaf mottle virus) was also recently shown to be transmitted in a semi-persistent manner by *B. tabaci* MEAM1 (Dombrovsky, Sapkota, Lachman, Pearlsman, & Antignus, 2013).

2.2.5 *Secoviridae, Torradovirus*

The *Torradovirus* genus (single-stranded plus sense RNA genome) is a recently established taxon with two approved species—*Tomato torrado virus* (ToTV) and *Tomato marchitez virus* (syn. Tomato apex necrosis virus), and four additional potential species, *Tomato chocolate virus*, *Tomato chocolate spot virus*, *Tomato necrotic dwarf virus* (ToNDV), and *Cassava torrado-like virus* (CsTLV) (Carvajal-Yepes et al., 2014; King et al., 2011; Larsen,

Duffus, & Liu, 1984; Verbeek, Dullemans, van den Heuvel, Maris, & van der Vlugt, 2007; Wintermantel & Hladky, 2013; www.ictvonline.org/virusTaxonomy.asp). Although little is currently known about the relationship of these recently emerged viruses with whiteflies, details are beginning to emerge. ToTV has been shown to be transmitted by members of the *B. tabaci* species complex and *T. vaporariorum* (Pospieszny et al., 2007; Amari et al., 2008). ToNDV was shown to be transmitted by *B. tabaci* New World1 during initial studies, and later by *B. tabaci* MEAM1 (Larsen et al., 1984; W. M. Wintermantel, pers. comm). Transmission of at least three torradovirus species by multiple whitefly species was recently shown to occur in a semi-persistent manner, requiring acquisistion and transmission periods of at least 2 hours (Verbeek, van Bekkum, Dullemans, & van der Vlugt, 2014).

3. MANAGEMENT OF WHITEFLY-TRANSMITTED VIRUSES USING PESTICIDES

Whiteflies are often managed primarily through multiple applications of a wide range of insecticides. In the absence of genetic resistance, insecticides play the dominant role in reducing whitefly populations and limiting the spread of viruses. This is true in the case of well-developed integrated management programs and even more so in cases where such recommendations do not exist. A wide range of insecticides is labeled for use against whiteflies (Table 3.2). These include pesticides with 11 different classifications of modes of action, as well as those that do not have an established mode of action. These insecticides vary in their target (whitefly eggs, immatures, and/or adults) and in their efficacy.

Optimal use of these pesticides requires several considerations. Selection of the insecticide and timing of application are best when applied using the results of field scouts who monitor whitefly populations. Insecticides should be used in a rotation where insecticides with different modes of action are applied so that development of resistance to any one pesticide by the whiteflies is prevented or delayed.

Of all the insecticides available the one class that has had the greatest impact on the management of whitefly-transmitted viruses is the neonicotinoids. These are generally applied as a soil drench or spray and due to their systemic and translaminar mobility within the plants are readily accessible to the feeding whiteflies. Their effect on whiteflies is rapid; they reduce whitefly populations very quickly and can impair the ability of whiteflies to transmit many plant viruses. The use of neonicotinoids expanded

Table 3.2 Summary of pesticides[a] recommended for use against whiteflies

Main group no. Primary site of action[b]	MoA code	Chemical subgroup or exemplifying active ingredient	Active ingredient	Notes
1 Acetylcholinesterase (AChE) inhibitors Nerve action	1A	Carbamates	Oxamyl	
	1B	Organophosphates	Methamidophos	Mix with a pyrethroid for whitefly control
			Acephate	Does not control silverleaf or sweet potato whiteflies
2 GABA-gated chloride channel antagonists Nerve action	2A	Cyclodiene, organochlorines	Endosulfan	
3 Sodium channel modulators Nerve action	3A	Pyrethroids, pyrethrins	Beta-cyfluthrin, bifenthrin, esfenvalerate, gamma-cyhalothrin, lambda-cyhalothrin, pyrethrins + piperonyl butoxide,[c] zeta-cypermethrin	
4 Nicotinic acetylcholine receptor (nAChR) agonists Nerve action	4A	Neonicotinoids	Acetamiprid, clothianidin, dinotefuran, imidacloprid, thiamethoxam	Foliar or soil drench

Continued

Table 3.2 Summary of pesticides recommended for use against whiteflies—cont'd

Main group no. Primary site of action	MoA code	Chemical subgroup or exemplifying active ingredient	Active ingredient	Notes
7 Juvenile hormone mimics Growth regulation	7C	Pyriproxyfen	Pyriproxyfen	Immatures of banded wing whitefly and silverleaf whitefly
9 Modulators of chordotonal organs Nerve action	9B	Pymetrozine	Pymetrozine	
15 Inhibitors of chitin biosynthesis, type 0 Growth regulation	15	Benzoylureas	Novaluron	
16 Inhibitors of chitin biosynthesis, type 1 Growth regulation	16	Buprofezin	Buprofezin	
21 Mitochondrial complex I electron transport inhibitors Energy metabolism	21A	METI acaricides and insecticides	Fenpyroxymate	
23 Inhibitors of acetyl CoA carboxylase Lipid synthesis, growth regulation	23	Tetronic and tetramic acid derivatives	Spiromesifen	Eggs and immatures
			Spirotetramat	
28 Ryanodine receptor modulators Nerve and muscle action	28	Diamides	Chlorantraniliprole	

Table 3.2 Summary of pesticides recommended for use against whiteflies—cont'd

Main group no. Primary site of action	MoA code	Chemical subgroup or exemplifying active ingredient	Active ingredient	Notes
Mixes of more than one active ingredient	3A, 28		Lambda-cyhalothrin, chlorantraniliprole	
	3A, 4A		Bifenthrin	
	3A, 4A		Bifenthrin, imidacloprid	
	3A, 4A		Lambda-cyhalothrin, thiamethoxan	
	3A, 6		Bifenthrin, Avermectin B1	
	3A, UN		Pyrethrins, azadirachtin	
	4A, 28		Thiamethoxam, chlorantraniliprole	
	16, 28		Buprofezin, flubendiamide	
UN Compounds of unknown or uncertain MoA	UN	Azadirachtin	Azadirachtin	
Not classified by IRAC	–	*n/a*	*Beauveria bassiana, Chromobacterium subtsugae* strain PRAA4-1, extract of *Chenopodium ambroisiodes,* extract of neem oil, insecticidal oil, insecticidal soap, *Isaria fumosorosea* Apopka strain 97	

[a]This is not an exhaustive list but does present a wide range of pesticides with reported activity against whiteflies. This list was prepared using the Vegetable Production Handbook for Florida 2013–2014 (www.omagdigital.com/publication/?i=175403).
[b]Uses the IRAC MoA Classification Scheme(Feb 2014, www.irac-online.org).
[c]Piperonyl butoxide is used as a synergist.

greatly over the last 20 years in terms of the number of crops to which they were applied as well as to the locations where they were routinely used. Until resistance to neonicotinoids develops in the whitefly population, this class of insecticides makes it possible to grow crops in many locations where whiteflies and the viruses they transmit exist.

Unfortunately, the neonicotinoids have been reported to have adverse effects on pollinators such as the European honeybee, bumble bees, birds, and they have been reported to contribute to Colony Collapse Disorder (CCD) (Blacquière, Smagghe, van Gestel, & Mommaerts, 2012; Desneux, Decourtye, & Delpuech, 2007; Mineau & Palmer, 2013). While the cause of CCD has not been resolved, the neonicotinoids have been implicated to play several roles that would be expected to contribute to a decline in pollinators. Honeybees exposed to neonicotinoids have been shown to have increased susceptibility to pathogens (Alaux et al., 2009; Di Prisco et al., 2013; Wu, Smart, Anelli, & Sheppard, 2012); impaired flight navigation, memory, and communication (Eiri & Nieh, 2012); reduced rates of foraging success and colony survival (Henry et al., 2012); and impaired olfactory learning and memory formation (Williamson & Wright, 2013). Studies with bumble bees have shown similar results (Whitehorn, O'Connor, Wackers, & Goulson, 2012) and neonicotinoids have been implicated as one of the causes of the dramatic decline in bumble bee diversity and populations in North America over the last 20 years (Cameron et al., 2011). However, recent studies have questioned these results suggesting that previous studies used abnormally high rates of exposure to neonicotinoids and that adverse effects were not seen at rates that pollinators would be expected to encounter in the field (Elston, Thompson, & Walters, 2013; Epstein et al., 2012). The most recent report from the USDA and EPA concludes that neonicotinoids are a less significant contributor than other factors in causing bee declines (Epstein et al., 2012).

The results of these studies have concerned both environmentalists and agriculturists. The lack of neonicotinoids will be likely to have a negative impact on the management of whitefly-transmitted viruses in the field. In the EU, three neonicotinoids (clothianidin, imidacloprid, and thiametoxam) have been banned from use for 2 years on flowering crops where bees actively forage but allowed on crops where bees are less active (http://www.nytimes.com/2013/04/30/business/global/30iht-eubees30.html?_r=2&; Gross, 2013). In the United States, the response has been more pragmatic and piecemeal. Although the US Environmental Protection Agency has been sued and petitioned to limit the use of neonicotinoids, no ban has as yet been imposed. However, adjustments on a piecemeal basis are taking place. For example, use of neonicotinoids for control of Citrus greening has been modified to continue its use but to minimize the exposure of honeybees (http://www.freshfromflorida.com/Divisions-Offices/Agricultural-Environmental-Services/Consumer-Services/Florida-Bee-Protection/Citrus-Greening).

4. MANAGEMENT OF WHITEFLY-TRANSMITTED VIRUSES USING CULTURAL PRACTICES

Whiteflies that transmit viruses are most commonly controlled through the use of host plant resistance or the application of insecticides. Cultural practices often play a secondary role in control regimes, but in many circumstances their use can be of critical importance in lessening the dependence on the two main tactics. Consequently, integrated control strategies for whitefly-transmitted viruses frequently include cultural practice components (Stansly & Natwick, 2010). Cultural practices cover a diverse array of activities that are all associated with the manner in which a crop is planted and managed throughout the course of the cropping cycle. Here, we discuss some of the most important and frequently applied cultural practices for the control of whiteflies and the viruses that they transmit.

4.1. Plastic soil mulches

Plastic (polyethylene) soil covers (mulch) are a popular strategy for protection of open-field production against whiteflies and the viruses they transmit (as well as against viruses transmitted by aphids and thrips) (Weintraub & Berlinger, 2004). There are two main approaches to plastic soil mulching—using colored (mainly yellow) plastic that attracts the whiteflies to the mulch instead of to the host, or silver or aluminum coated plastic mulch that strongly reflects light, which acts as a deterrent to the invading whiteflies. Both types of soil mulch interfere with the insect's ability to find the crop, and are most effective early in the season, before the developing plant canopy covers the mulch. Soil mulching is relatively easy to perform, relatively inexpensive, and has added benefits to the grower since mulching can change the plant microclimate; i.e., temperature, humidity, light, water, etc. Moreover, it has been demonstrated that dark mulches that block or reduce light penetration into the soil inhibit germination and growth of weeds, which cannot survive under the mulch (Lament, 1993; Ngouajio & Ernest, 2004). Indeed, in Israel the commercial yellow plastic mulch is actually yellow-on-brown (top side yellow, lower side brown) with a dual effect—protection against insects and inhibition of weeds

(Weintraub & Berlinger, 2004). By inducing favorable conditions for the plant, the mulch can positively affect plant growth and increase yield (Csizinszky, Schuster, & Kring, 1995). However, one of the problems of using plastic mulch is the difficulty of mulch disposal, as polyethylene is not easily degradable.

Yellow plastic mulch—The use of yellow plastic mulch to protect open-field tomato plants from the whitefly-borne TYLCV is a common practice in Israeli agriculture (Cohen & Lapidot, 2007; Polston & Lapidot, 2007). In 1962, Mound tested the attraction of whiteflies to different colors and demonstrated that yellow attracts whiteflies. It was suggested that yellow radiation is a component of the whitefly's host-selection mechanism (Mound, 1962). Testing the effect of yellow plastic mulch on tomato plants Cohen and Melamed-Madjar (1978) found that 28 days after germination (DAG) only 5% of the plants protected by yellow mulch had developed TYLCV symptoms, compared to over 20% of the nonmulched control plants. At 38 DAG, only 10% of the mulched plants exhibited TYLCV symptoms, compared to nearly 100% of the nonmulched control plants. At 48 DAG, 20% of the mulched plants showed TYLCV symptoms, and by 58 DAG incidence of symptomatic plants rose abruptly to 60% TYLCV infection, which was clearly better than the nonmulched control plants, but still unacceptable for the grower. It should be noted that only the yellow mulch, without any application of insecticides, protected the plants. Hence, it was concluded that the protection effect of the yellow mulch lasted about 3–5 weeks after transplanting, which is usually long enough to protect tomato plants during their critical period of susceptibility to begomovirus infection (Levy & Lapidot, 2008; Schuster, Stansly, & Polston, 1996). The effect of the yellow mulch decreases with time probably due to the increase over time of the ratio of plant canopy to mulch.

The controlling effect of yellow mulch is due to a combination of the whitefly's attraction to the yellow color of the mulch and its subsequent death due to dehydration induced by the high temperature of the mulch (Cohen, 1982; Cohen & Lapidot, 2007). It should be noted that the typical Israeli climate is semiarid—high temperature and low humidity. In the tomato-growing regions, soil temperatures exceeding 30 °C are quite common. It was demonstrated that at temperatures above 30 °C, in low-humidity conditions, whiteflies not feeding on a plant dehydrate within an hour (Cohen, 1982).

Another explanation for the effect of yellow mulch comes from the observation that many flying insects, including whiteflies, have higher landing rates on green-yellow surfaces. Landing induces probing in an attempt to feed, at which time the insect discriminates between "appropriate" and "inappropriate" hosts. If the host is inappropriate, as in the case of plastic mulch, the insect flies a short distance, lands, and probes again. After a number of such inappropriate landings, the insect is likely to fly away entirely (Finch & Collier, 2000; Hilje, Costa, & Stansly, 2001).

Reflective mulch—The successful use of reflective plastic mulch to delay the onset of whitefly infestations and infection by whitefly-transmitted viruses in open-field production is well established (Polston & Lapidot, 2007; Simmons, Kousik, & Levi, 2010; Smith, Koenig, McAuslane, & McSorley, 2000; Summers & Stapleton, 2002). The most effective reflective mulches are entirely or partially aluminized and reflect a lot of daylight. These are believed to reflect both visible and UV light which disorients whiteflies and decreases the landing rate of whiteflies on plants in the field. Like other mulches, the effectiveness decreases as the plant canopy increases and the mulch is covered. Reflective mulches are effective even when whitefly populations are expected to be high. Like the yellow mulch, this approach has the added benefit of interfering with other virus vectors (aphids and thrips), and also affects plant growth and increases yield, especially in cucurbits, which seem to grow better with the light reflected from the mulch (Greer & Dole, 2003). One negative aspect of reflective mulches is the discomfort that it can generate to humans working in the fields. The light can be nearly blinding, and the amount of heat reflected from the mulch makes working in the field nearly intolerable.

Interestingly, while reflective mulches were found to be highly effective in Florida in delaying whitefly infestation and reducing infection rate of tomato plants by the whitefly-transmitted *Tomato mottle virus*, yellow mulches were found to be less effective (Csizinszky, Schuster, & Kring, 1997; Csizinszky, Schuster, & Polston, 1999). The reason for this may be due to the very high level of humidity in Florida. Whiteflies, which were attracted to the yellow mulch probably were not dehydrated as quickly in Florida as they were in Israel, where relative humidity is much lower. Whiteflies attracted to the yellow mulch in Florida were still able to fly to a host plant and feed on it. In a climate with high relative humidity, the yellow mulch may actually attract whiteflies to the crop rather than protect it from whiteflies.

UV-reflective mulch has been used very successfully to reduce incidences of whiteflies and the whitefly-transmitted *Cucurbit leaf crumple virus* (CuLCrV) in zucchini squash (*Cucurbita pepo* L.) (Nyoike, Liburd, & Webb, 2008).

The reflective mulch was used with or without the systemic insecticide imidacloprid. It was found that the reflective mulch alone provided equal protection to squash plants against CuLCrV as did mulch combined with imidacloprid treatment. Hence, since no additional benefits were derived from combining reflective mulch with imidacloprid, it was suggested that the reflective mulch could be used on its own.

4.2. Virus-free seed/planting material

Planting seed or vegetatively propagated planting material that is free of viruses provides a crop with the optimal start to its growth cycle. Vegetatively propagated crops are particularly vulnerable to virus infection, as described in the part of this volume in which they are reviewed. Cassava and sweet potato are the two crops in this category that are most affected by whitefly-transmitted viruses. In sweet potato, which is more widely grown in developed and middle income countries, tissue culture is frequently used for the production of virus-free "seed," and tissue culture with virus indexing is now routinely used throughout growing areas of the United States in foundation seed programs (Clark et al., 2012). In Shangdong Province of China, the country which is the world's largest producer of sweet potato, there has been great success in farmer adoption of virus-free stocks of planting material. An impact assessment of this program demonstrated that more than 80% of the Province's growers were using virus-free seed by the end of a promotional program (Fuglie, Zhang, Salazar, & Walker, 1998). In cassava, tissue culture and virus indexing with thermotherapy has been routinely used for many years as a means of ensuring that germplasm exchanged between continents is free of virus (Frison, 1994). However, it is only recently that these approaches have been used at lower levels of country or region for the provision of virus-free tissue culture plantlets to large-scale "basic seed" producers. Pilot schemes are currently operating in several countries in Africa to develop formalized seed systems that incorporate the production of virus-free seed through tissue culture and certification standards for quality control at the various stages of planting material propagation (Yabeja, Mtunda, Shirima, Kanju, & Legg, 2013).

4.3. Crop placement—In space

Since whiteflies can fly over distances of several kilometers (Blackmer & Byrne, 1993; Cohen, Kern, Harpaz, & Ben Joseph, 1988), and many of the most important viruses that they transmit are persistent or semi-persistent (Duffus, 1987), new crops planted in proximity to older crops that are infested with whiteflies are vulnerable to being infected by viruses present in the neighboring older fields. Direct relationships between the levels of virus inoculum in surrounding fields and final incidences of virus disease recorded in test plots of the same crop have been demonstrated for both sweet potato (Aritua, Legg, Smit, & Gibson, 1999) and cassava (Legg et al., 1997). In order to minimize the risk of contamination of a new crop from external sources, it is necessary to locate the crop either at a site that is isolated from other infection sources, or that is upwind of the inoculum source (in an environment where there is a consistent prevailing wind). In the southwestern states of the United States, growers avoid planting cotton in close proximity to spring melons for this reason, and similarly, ensure that fall melons or vegetables are not planted too close to cotton (Ellsworth & Martinez-Carrillo, 2001). In cassava, the density of cultivation has been highlighted as a contributing factor to the rapid spread of cassava begomoviruses (Bock, 1994), as has the significance of prevailing wind direction on patterns of spread of these viruses into initially healthy crops (Fargette, Fauquet, Grenier, & Thresh, 1990). "Clean seed" programs for cassava in Tanzania make use of isolated, high elevation sites for pre-basic seed propagation in order to minimize the likelihood of infection by the whitefly-transmitted begomoviruses and CBSVs (Yabeja et al., 2013).

4.4. Crop placement—In time

Patterns of whitefly population increase and decline depend upon several environmental factors, the most important of which are the availability of hosts—determined by the date of planting—and the climatic variables of temperature and precipitation. In general, whiteflies are more abundant during periods of warm weather during which there is active crop growth (resulting from adequate soil moisture) as well as when crop host plants are young and rapidly growing. By careful manipulation of planting dates, therefore, it is often possible to reduce whitefly populations and the resulting incidence of the viruses that they transmit, although this should not be done in a way that makes growing conditions unfavorable for the crop. Mohamed (2012), working on cucurbits in Egypt, observed that using optimal combinations of variety, plant spacing and planting date could result in up to 20-fold reduction in *B. tabaci* populations. In the dry areas of northern Mexico, local regulations governing dates for planting and harvesting cotton, and enforcing host-free periods were successful in reducing pressure from whitefly populations, although these measures needed to be combined with the strategic but restrained use of insecticides (Ellsworth & Martinez-Carrillo, 2001). In cassava crops in West Africa, rapid increases in cassava mosaic disease (CMD) incidence were recorded in fields between November and June, while the rate of disease increase was much lower from July to October. Seventy percent of this variation could be attributed to whitefly vector numbers, temperature and radiation (Fargette, Jeger, Fauquet, & Fishpool, 1993). Similarly, in Uganda CMD spread was most rapid at one location in March/April, while at two others it was greatest in September/October (Legg, 1995). These patterns offer opportunities to reduce whitefly abundance and consequent virus spread through planting at times of the year when early vigorous stages of crop growth do not correspond with the period during which whiteflies are most abundant.

4.5. Trap crops

Traps crops have been shown to be effective in reducing populations of whiteflies, and therefore reducing the level of virus infection. In tomatoes in the south-eastern USA, squash planted around tomatoes acted as a trap crop, since whiteflies were more attracted to the squash plants than they were to the tomatoes (Schuster, 2004). Significant reductions in the abundance of *B. tabaci* whitefly adults in tomatoes surrounded by squash, compared with no-squash controls led to important reductions in the incidence of TYLCV in the "protected" tomato crop. There are numerous other examples, which while not relating directly to whitefly-transmitted viruses, nevertheless illustrate the potential benefit provided by trap crops for managing whitefly populations. *B. tabaci* has been shown to settle preferentially on cantaloupes in comparison with cotton (Castle, 2006). In leaf assays, there was a 67% preference for cantaloupes, which rose to 90% in full plant assays. When cantaloupes were planted around cotton crops, populations of eggs and early-stage nymphs on cantaloupe were more than 10 times those in the cotton for 9 out of 12 sampling dates through the season. However, the trap crop effect was not sufficient to prevent the action threshold from being reached. This fact highlights a general feature of cultural practices: they are seldom able to keep whitefly populations below threshold levels and similarly are rarely able to prevent virus transmission when applied in the absence of other "supporting" control measures (Castle, 2006). This is particularly true for whitefly-transmitted viruses, since a population of whiteflies that causes 100% virus infection may be significantly below a density that would cause physical damage. Tolerance levels for virus-transmitting whiteflies are much lower than those that only cause direct or indirect physical damage.

4.6. Intercropping

Mixtures of crops are not usually cultivated together in commercial agricultural environments. However, mixing crops together in a single field is a common practice in subsistence agricultural systems in developing countries. Breaking up a crop environment typically makes it less readily colonized by weak-flying sucking pests such as whiteflies, and this probably explains why many of the major whitefly outbreaks have been in large-scale commercial production situations. Several experimental examples illustrate the potential benefits of intercropping as a control measure for whiteflies. In Egypt, maize either intercropped or rotated with cucumber, tomato, or squash resulted in lower whitefly abundances in the vegetable crops and significantly lower incidences of CVYV in cucumber, *Squash leaf curl virus* (begomovirus) in squash, or TYLCV in tomato (Abd-Rabou & Simmons, 2012). In zucchini on Hawaii, both okra and sunnhemp planted as intercrops with zucchini resulted in significantly reduced *B. tabaci* whitefly populations, although these reductions did not result in significant yield differences when compared with monocrop controls (Manandhar, Hooks, & Wright, 2009). For cassava in Africa, intercropping experiments have been undertaken in West, Central and East Africa at various times in the last three decades. Mixing cassava with maize in Ivory Coast resulted in modest reductions in CMD incidence with some of the planting schemes tested (Fargette & Fauquet, 1988), and similar results were obtained with both maize and cowpea in Cameroon (Fondong, Thresh, & Zok, 2002). Since some intercrop planting arrangements "worked" while others—using the same intercrop—did not, it has proven difficult to disseminate intercropping extension messages that would be readily understood, and therefore adopted, by growers. This further highlights one of the difficulties in disseminating cultural control technologies—they are relatively knowledge intensive and require the input of substantial training efforts from public extension agencies.

4.7. Physical barriers

Crop colonization by whiteflies, and the virus transmission that may follow, can be hindered by placing physical barriers between flying whiteflies and the crop host plants that they seek. The usual ways of achieving this are through either temporary or permanent protection. Permanent protection, usually done by enclosing crop plants in permanent or semi-permanent insect-proof housing, is most appropriate for high value crops that are grown on a relatively small scale. For crops grown at a larger scale, where only temporary protection is sufficient, floating row covers or tunnels may be used. In permanent screenhouses, mesh can be used that excludes whiteflies and yet allows aphelinid parasitoids to enter, thereby promoting biological control. Screenhouses are used on a massive scale in the vegetable production zones of southern Europe, most famously including the "visible-from-space" expanse of plastic housing at Almeria, in southern Spain. In addition to providing good overall growing conditions for vegetables produced, the physical protection offered by this housing helps to reduce movement of whiteflies between crops and consequent virus spread. Integrated control strategies that combine protection of crops in screenhouses with biological control have resulted in drastic cuts in the levels of insecticide usage, while at the same time providing a more sustainable solution to the management of whitefly-transmitted viruses. In smaller scale, subsistence-oriented production systems, net tunnels constructed with locally available materials have been piloted for the protection of virus-free sweet potato planting material (Anon, 2012). Experiments conducted in Kenya have shown that over a period of 33 months, the use of these net tunnels led to increases in production and income of >100% (Anon, 2012). For an extended review on physical barriers, please see Chapter 1.

4.8. Physical traps

Whitefly populations can be reduced by physically removing these insects from the air space around crop plants in either protected- or open-field situations. The effect is clearly enhanced through the use of an attractant, which is typically the visual cue of the yellow color. Yellow sticky traps, that attract then kill whiteflies, have mainly been used for monitoring populations of the winged adults (Fishpool & Burban, 1994). In confined areas, such as screenhouses, where the air space is limited, it is possible for such traps to have a significant impact in reducing the overall whitefly population, thereby helping to control virus disease (Xi-Shu et al., 2008). In addition to demonstrating the large beneficial effect of the combined use of yellow sticky traps and biological control using the parasitoid *Eretmocerus* nr. *rajasthanicus*, Xi-Shu et al. (2008) also demonstrated that traps placed parallel to rows of tomato plants caught many more whiteflies than traps oriented perpendicular to the rows. While the benefits of sticky traps in protected environments have been shown (Lu, Bei, & Zhang, 2012), this same study demonstrated the absence of any significant beneficial effect in field-grown crops.

4.9. Conclusions

A detailed review of cultural control for *B. tabaci* whiteflies (Hilje et al., 2001) noted that there has been a disproportionately small amount of attention given to these approaches, while observing that this is likely a reflection of the difficulty of implementation of some of these measures. Cultural control tactics, such as managing planting dates, rotation systems or crop-free periods require a high degree of local coordination among growers, which is very often difficult to achieve. Other methods such as intercropping or the use of trap crops require substantial changes to the production system. Many successes have nevertheless been achieved, with the most significant coming from the physical protection of crops by using tunnels or screenhouses, isolation of fields, and from the use of virus-free seed. Even where the impacts of specific cultural control tactics are insufficient on their own to control whiteflies and whitefly-transmitted viruses, they may still play an important role in integrated control systems. Consequently, continued research to identify, enhance and implement these approaches within the context of integrated control programs is strongly merited.

5. GENETIC RESISTANCE

Genetic resistance in the host plant is considered highly effective in the defense against viral infection in the field. This is especially true for those viruses that have prolific vectors which can rapidly produce very high populations in the field and are hard to contain. Genetic resistance requires neither environmentally hazardous chemical application nor plant seclusion and can potentially be stable and long lasting. A disadvantage, however, is that genetic resistance requires the identification of resistance loci which are not always available, and in many cases are identified in wild species. Interspecific crossing programs for introgressing resistance from wild species into crop relatives can be long and laborious.

This section will cover issues essential for the development of resistance, such as the development of controlled procedures for inoculation by whiteflies, optimal plant age for resistance screening, breeding for virus resistance, the effect of virus-resistant genotypes on virus epidemiology and more;

elements that are related to all whitefly-transmitted viruses. However, due to the impact of the diseases induced by TYLCV, and to the large body of research and publications available for this virus, this section will emphasize genetic resistance to TYLCV which can be viewed as an example relevant to other whitefly-transmitted viruses.

5.1. Tomato yellow leaf curl virus

TYLCV, a monopartite begomovirus (family *Geminiviridae*) is one of the most devastating viruses in tomatoes in many tropical and subtropical regions worldwide (Lapidot & Friedmann, 2002; Moriones & Navas-Castillo, 2000; Navas-Castillo et al., 2011). Like all begomoviruses, TYLCV is transmitted by the whitefly *B. tabaci* in a circulative and persistent manner (Cohen & Harpaz, 1964; Rubinstein & Czosnek, 1997).

The viral circular ssDNA genome of nearly 2.8 kb contains six open reading frames (ORFs) that are organized directionally, two in the sense orientation and four in the complementary orientation (Gafni, 2003; Gronenborn, 2007; Lapidot & Polston, 2006). The bidirectional ORFs are separated by a ~250-bp intergenic region that contains elements for replication and bidirectional transcription (Gronenborn, 2007; Gutierrez, 1999; Hanley-Bowdoin, Settlage, Orozco, Nagar, & Robertson, 1999; Petty, Coutts, & Buck, 1988).

On the complementary strand, the *C1* gene encodes Rep (replication-associated protein) which is a multifunctional protein involved in viral replication and transcriptional regulation. This is the only viral protein absolutely required for viral replication (Gronenborn, 2007). The *C2* gene encodes TrAP(transcriptional activator protein), which enhances expression of the coat protein, and plays a role in the suppression of host defense responses as well as in viral systemic infection (Bisaro, 2006; Brough, Sunter, Gardiner, & Bisaro, 1992; Etessami, Saunders, Watts, & Stanley, 1991). The *C3* gene encodes the REn (replication enhancer protein) which acts by enhancing viral DNA accumulation in infected plants and interacts with Rep (Sunter, Hartitz, Hormuzdi, Brough, & Bisaro, 1990). The *C4* gene which is embedded within the *C1* gene, but in a different ORF, is implicated in viral pathogenicity and movement (Jupin, De Kouchkovsky, Jouanneau, & Gronenborn, 1994; Rigden, Krake, Rezaian, & Dry, 1994).

On the sense strand, the capsid protein encoded by *V1* is required for whitefly transmission, binds to viral ssDNA, may play a role in systemic movement, and acts as a nuclear shuttle protein that mediates movement of viral nucleic acid into the host-cell nucleus (Azzam et al., 1994; Briddon, Pinner, Stanley, & Markham, 1990; Kunik, Palanichelvam, Czosnek, Citovsky, & Gafni, 1998; Palanichelvam, Kunik, Citovsky, & Gafni, 1998; Rojas et al., 2001). The product of the *V2* ORF is involved in viral movement (Rojas et al., 2001; Wartig et al., 1997) and has been shown to act as a suppressor of RNA silencing (Zrachya et al., 2007).

TYLCV induces severe yield losses in tomato, which, depending on the age of the plant at the time of infection, can reach 100% (Lapidot et al., 1997; Levy & Lapidot, 2008). Two to three weeks after inoculation, the infected tomato plant displays pronounced disease symptoms that include upward cupping of the leaves, chlorosis of the leaf margins and severe stunting of the entire plant. In many tomato-growing areas, TYLCV has become the limiting factor for production of both open-field and protected cultivation systems (Lapidot & Friedmann, 2002).

TYLCV was first detected and identified in the northern part of Israel, following an outbreak of a new disease in tomatoes in 1959 (Cohen & Harpaz, 1964; Cohen & Lapidot, 2007). Similar disease symptoms associated with high populations of whiteflies were observed on tomatoes grown in the Jordan Valley in the late 1930s (Avidov, 1946). The outbreaks of tomato yellow leaf curl disease (TYLCD), which were sporadic in the 1960s, became a serious economic problem and by end of the 1970s, all tomato-growing regions in the eastern Mediterranean basin were affected by TYLCD (Hanssen & Lapidot, 2012). In the late 1980s, TYLCV particles were isolated and the virus was cloned and sequenced (GenBank accession no. X15656) and found to be a monopartite begomovirus (Navot, Pichersky, Zeidan, Zamir, & Czosnek, 1991). Shortly thereafter, another Mediterranean viral strain inducing TYLCD was cloned and sequenced—*Tomato yellow leaf curl Sardinia virus* (TYLCSV; GenBank accession no. X61153) (Kheyr-Pour et al., 1991). Over the years, especially with the advent of sequencing as a routine procedure, it became apparent that the name TYLCV had been given to a heterogeneous group of more than 10 virus species and their strains, all of which induce very similar disease symptoms in tomato (Moriones & Navas-Castillo, 2000; Navas-Castillo et al., 2011).

TYLCV, most probably emerged from the eastern Mediterranean, spread westward, and subsequently became recognized as a tomato pathogen throughout the Mediterranean basin (Cohen & Lapidot, 2007; Czosnek & Laterrot, 1997; Hanssen & Lapidot, 2012; Hanssen, Lapidot, & Thomma,

2010; Lefeuvre et al., 2010; Navas-Castillo et al., 2011). The disease continued to spread westward into the Caribbean, Central and North America, and eastward toward China, Japan, and Australia. Today, it is present in most tomato-growing areas worldwide (Lefeuvre et al., 2010; Navas-Castillo et al., 2011).

It should be noted that although TYLCV is primarily known as a pathogen of tomato, the virus can infect other agricultural plants. TYLCV induces severe symptoms in common bean (*Phaseolus vulgaris* L.) (Cohen & Antignus, 1994), the cut-flower lisianthus (*Eustoma grandiflorum*) (Cohen et al., 1995), while pepper (*Capsicum annuum*) was found to be a symptomless host of the virus (Morilla et al., 2005; Polston, Cohen, Sherwood, Ben-Joseph, & Lapidot, 2006).

5.2. Development of a controlled whitefly-mediated inoculation system

To succeed in a breeding program whose aim is to develop resistant cultivars to a virus, or any other pathogen for that matter, one must develop an accurate and reliable mass inoculation and selection system. Since many of the whitefly-transmitted viruses are only poorly if at all transmitted mechanically, it is essential to develop whitefly-mediated inoculation protocols, which will ensure high (preferably 100%) infection rates, and a standardized (as much as possible) inoculum pressure (for a review, see Lapidot, 2007).

Development of a whitefly-mediated inoculation system requires rearing of whiteflies. It requires a dedicated rearing facility suitable for rearing whiteflies on the one hand, but secluded so other insects do not penetrate and whiteflies do not escape. Polston and Capobianco (2013) present a detailed explanation of the conditions and considerations in rearing whiteflies for virus transmission. Maintenance of such a rearing facility is time consuming. Since whitefly populations reach very high numbers in the field, why not rely on spontaneous field inoculation? Surprisingly, spontaneous field infection has been shown to be largely inefficient, as many plants escape infection, even under heavy inoculation pressure (Vidavsky et al., 1998). Following planting of susceptible tomato plants in an area stricken with whiteflies and TYLCV, only 50% of the susceptible tomato plants were infected during the first month after planting. Despite high whitefly populations and available viral inoculum, 10% of the susceptible plants had escaped infection even 90 days after transplanting (Vidavsky et al., 1998). In another study, the percentage of viruliferous whiteflies in the general whitefly population in the field was found to be rather low

(Cohen et al., 1988). Depending on the TYLCV-susceptible host from which the whiteflies were collected, only 3–6% of the whiteflies collected in the field were able to transmit the virus (Cohen et al., 1988).

Spontaneous field inoculation has other disadvantages besides promoting inoculation escapees: field inoculation may lead to milder disease symptoms compared to controlled inoculation, probably due to late and unsynchronized infection. Pico, Diez, and Nuez (1998) assayed cultivated and wild tomato accessions for their resistance to TYLCV. They compared controlled whitefly inoculation with cage inoculation and with spontaneous field inoculation. It was concluded that the response of a resistant source to TYLCV may vary with the inoculation technique used and that controlled greenhouse inoculation corresponded to high inoculum levels, while spontaneous field inoculation corresponded to low inoculum levels. However, despite the low and delayed disease incidence following spontaneous field inoculation, it was possible to discard the most susceptible genotypes with field testing (Pico et al., 1998).

Another problem with spontaneous field inoculation is that there are other pathogens in the field, so a specific virus-resistant plant may become infected by an unrelated virus, or any other pathogen, and erroneously be considered susceptible. For evaluation of resistance to TYLCV in areas where diverse begomoviruses are present, this can be a serious concern. In field inoculation, the whitefly pressure, intensity of inoculation, level of viral inoculum, and plant age at time of inoculation are all unknown and variable. The elapsed time between whitefly acquisition and transmission of the virus is also unknown with regard to its impact in field inoculation. Whiteflies transmit begomoviruses in a persistent, circulative manner. However, it has been shown for a number of begomoviruses including TYLCV that although transmission may continue for the life span of the vector, transmission efficiency declines with time, which is clearly the case for the semi-persistent transmitted crini- and ipomoviruses (Caciagli et al., 1995; Cohen, Duffus, Larsen, Liu, & Flock, 1983; Cohen & Harpaz, 1964; Navas-Castillo et al., 2011; Rubinstein & Czosnek, 1997). Thus, the efficiency of spontaneous field inoculation is unknown and hence not reproducible.

5.3. When should we inoculate?

Another obstacle in the development of TYLCV (as well as other virus) resistance has been the lack of a standard method for the assessment of resistance (Lapidot, Ben Joseph, Cohen, Machbash, & Levy, 2006). Variability in assay conditions has led to contradictory results, where different resistance levels have been attributed to the same genetic sources (Pico et al., 1998; Vidavsky et al., 1998). The response of a plant to infection by a pathogen may be affected by test conditions such as temperature, light, growth conditions, inoculation pressure, and plant age (or developmental stage) at the time of infection. This latter phenomenon has been referred to as age-related or mature-plant resistance (Loebenstein, 1972). In some instances, it has been shown that mature plants resist or tolerate virus infection much better than plants infected at an early stage of development, leading to what appears (erroneously) to be increased viral resistance (Garcia-Ruiz & Murphy, 2001; Levy & Lapidot, 2008; Moriones, Aramburu, Riudavets, Arno, & Lavina, 1998).

To determine the effects of plant age on the expression of genetic resistance to TYLCV, tomato plants expressing different levels of resistance to TYLCV were inoculated at three different ages—14, 28, and 45 days after sowing (DAS). Resistance was assayed mainly by comparing yield components of inoculated plants to those of control, noninoculated plants of the same line or variety (Levy & Lapidot, 2008). It was found that plant age at inoculation had no effect on disease severity scores of the susceptible varieties, and little or no effect on those of the resistant varieties. In contrast, plant age at inoculation had a significant effect on the yield of all varieties tested. All the varieties suffered a significant yield reduction due to inoculation with TYLCV, but the older the plant was at time of inoculation, the TYLCV-induced yield reduction became smaller. Hence, it was concluded that there is an age-related (or mature-plant) resistance in tomato plants to TYLCV, regardless whether the tomato plants tested were susceptible or resistant to the virus.

The occurrence of age-related resistance raised another question—what is the optimal age for inoculation of the tomato plants when screening for TYLCV resistance? This may depend on the genetic material being screened: if segregating populations are being screened for individual resistant plants, then it is best to inoculate at the earliest possible stage, when the effect of the viral infection is most severe. This way the selected plants will indeed be those expressing the highest level of resistance. If, on the other hand, commercial hybrids are being tested for level of resistance, then inoculation at 28 DAS may be most suitable as most commercial tomato plants are sown in specialized and protected nurseries and transplanted to the field about 28 days later. Thus, from an agricultural point of view, 28 DAS may be the most appropriate stage for testing commercial hybrids as it mimics inoculation just following transplanting to the field.

5.4. Breeding tomatoes (*Solanum lycopersicum*) for resistance to TYLCV

There have been prolonged efforts to breed tomato cultivars resistant to TYLCV. Since all cultivated tomato accessions at the time were found to be extremely susceptible to the virus, wild tomato species were screened for their response to the virus in order to identify and introgress genes controlling resistance (Ji, Scott, Hanson, Graham, & Maxwell, 2007, reviewed in Lapidot & Friedmann, 2002; Lapidot & Polston, 2006; Vidavsky, 2007). Thus, breeding programs have been based on the introgression of resistance genes from accessions of wild origin into the cultivated tomato. Progress in breeding for TYLCV resistance has been slow, mainly due to the complex genetics of the resistance, the interspecific barriers between the wild and domesticated tomato species, and the need for a reliable screen for resistance to the virus (Ji, Scott, et al., 2007; Lapidot, 2007; Vidavsky, 2007). In spite of these challenges, TYLCV-resistant commercial tomato cultivars are available today from several seed companies.

Sources of resistance to TYLCV have been identified and introgressed from several wild tomato species, including: *Solanum pimpinellifolium*, *S. peruvianum*, *S. chilense*, and *S. habrochaites*. However, until now only five major resistance loci, termed *Ty-1* to *ty-5*, have been characterized and mapped to the tomato genome using molecular DNA markers (Ji, Scott, et al., 2007).

Resistance introgressed from *S. chilense* accession LA1969 was found to be controlled by a major partial dominant gene, termed *Ty-1*, and at least two additional modifier genes (Zamir et al., 1994). *Ty-1* was mapped to the top of chromosome 6, while the two modifiers were mapped to chromosomes 3 and 7 (Zamir et al., 1994). To the best of our knowledge, *Ty-1* is the most utilized TYLCV-resistance locus in tomato breeding programs worldwide, and most TYLCV-resistant commercial hybrids available today carry this locus.

Hanson et al. (2000) analyzed the resistant line H24, which contains resistance introgressed from accession B6013 of *S. habrochaites* (Kalloo & Banerjee, 1990). The authors screened resistant plants using what at the time they thought were three different isolates of TYLCV. It was however later found that those viral isolates were in fact three isolates of *Tomato leaf curl virus* and not TYLCV. The resistance that was found to be dominant was mapped to the bottom of chromosome 11, and was termed *Ty-2* (Hanson, Green, & Kuo, 2006).

A major partially dominant gene, which was introgressed from *S. chilense* accessions LA2779 and LA1932, was mapped to chromosome 6 and was termed *Ty-3* (Ji, Schuster, & Scott, 2007; Ji & Scott, 2006). The introgression derived from LA2779 was found also to contain *Ty-1*, suggesting a linkage between *Ty-1* and *Ty-3* (Ji, Schuster, et al., 2007). Indeed, in a recent study, it was shown that *Ty-1* and *Ty-3* are allelic, and that *Ty-1/Ty-3* code for an RNA-dependent RNA polymerase, suggesting that the resistance induced by these loci is via RNA silencing (Verlaan et al., 2011, 2013). The detailed mechanism of how TYLCV-resistance is mediated by *Ty-1/Ty-3* has yet to be elucidated.

Ty-4 was introgressed from *S. chilense* LA1932 and has been mapped to the long arm of chromosome 3. This locus is considered to be a minor one as it only accounted for 16% of the resistance when combined with *Ty-3* (Ji, Scott, Schuster, & Maxwell, 2009).

The TYLCV-resistant line TY172, carrying *Ty-5*, is thought to be derived from four different accessions formerly assigned as *S. peruvianum*: PI 126926, PI 126930, PI 390681, and LA0441 (Friedmann, Lapidot, Cohen, & Pilowsky, 1998). LA0441 was later subclassified as *S. arcanum* (Peralta, Knapp, & Spooner, 2005). TY172 is highly resistant to TYLCV: it produces minimal symptoms following infection and allows only low levels of viral DNA, and exhibited the highest level of resistance in a field trial which compared yield components of various resistant accessions following inoculation with TYLCV (Friedmann et al., 1998; Lapidot et al., 1997). Classical genetic studies have suggested that the resistance in TY172 is controlled by three genes exerting a partially dominant effect (Friedmann et al., 1998). Gene mapping showed that the resistance in TY172 was controlled by a previously unknown major recessive QTL, and four additional minor QTLs (Anbinder et al., 2009). The major QTL was mapped to chromosome 4 and was designated *Ty-5*.

Recently, the recessive resistance in the old commercial cultivar Tyking (Royal Sluis, The Netherlands) has been shown to colocalize with the resistance in TY172 (Hutton, Scott, & Schuster, 2012). The authors suggested that since one of the populations used by Anbinder et al. (2009) also showed recessive gene action that the locus in TY172 should therefore be renamed *ty-5*. Resistance derived from the cultivar Tyking has been used in many breeding programs. Interestingly, Bian et al. (2007) determined that resistance in the tomato line Fla. 653 was controlled by a recessive allele termed *tgr-1*. Fla. 653 has resistance derived from "Tyking" and is homozygous for *ty-5* (Hutton et al., 2012). In another study, Giordano, Silva-Lobo, Santana, Fonseca, and Boiteux (2005) also identified a recessive allele (termed *tcm-1*) derived from Tyking that was effective against bipartite begomoviruses. Hence, Hutton et al. (2012) hypothesized that both *tgr-1* and *tcm-1* describe the *ty-5* allele from Tyking, and speculated that this allele was introgressed from *S. peruvianum*.

5.5. Effect of TYLCV-resistant genotypes on virus epidemiology

As the use of TYLCV-resistant tomato cultivars in open-field cultivation becomes a common practice, the need to assess the potential effect of resistant varieties in TYLCV epidemiology becomes apparent. The potential of TYLCV-infected resistant genotypes to serve as virus reservoirs was studied in a greenhouse study (Lapidot, Friedmann, Pilowsky, Ben Joseph, & Cohen, 2001), and more recently in a field study (Srinivasan, Riley, Diffie, Sparks, & Adkins, 2012). In the first study, four different tomato genotypes exhibiting different levels of TYLCV resistance, ranging from fully susceptible to highly resistant, served as TYLCV-infected source plants. The survival and TYLCV acquisition and transmission rates for whiteflies having fed on the different infected tomato genotypes were examined.

Whitefly survival rates following feeding on the different source plants at 21 days postinoculation (DPI), shortly after the appearance of disease symptoms, were similar regardless of the plant genotype from which the virus was acquired. Significant differences in whitefly survival rates were found after whiteflies had fed on the infected source plants at 35 DPI, with the whitefly survival rate increasing with higher levels of resistance displayed by the source plant. This may have been due to the deleterious effect of TYLCV on the infected plant. At 35 DPI, the susceptible and moderately resistant genotypes exhibited pronounced disease symptoms, presumably making the plant less suitable for whitefly feeding. In contrast, the highly resistant genotypes hardly showed any disease symptoms, which would favor whitefly survival (Lapidot et al., 2001).

The TYLCV level in the whiteflies following feeding could be directly correlated to the virus level in the source plant: the higher the level in the source plant, the higher the TYLCV level in the whitefly. This correlation was the same, regardless of the time of feeding—21 or 35 DPI—and regardless of the state of the source plants. The severity of disease symptoms exhibited by the source plants did not seem to affect TYLCV acquisition by the whiteflies.

The transmission rate of whiteflies that had fed on infected source plants at 21 DPI was negatively correlated with the level of resistance displayed by the source plant. Therefore, the higher the resistance, the lower the transmission rate. However, at 35 DPI, transmission rates from the susceptible plants were lowest, presumably due to their poor condition. Transmission rates from source plants displaying a medium level of resistance were highest, with rates declining following feeding on source plants displaying higher levels of TYLCV resistance (Lapidot et al., 2001).

Based on these results, the authors postulated that a TYLCV-infected field of susceptible tomato plants might serve as a high-risk virus reservoir early after infection. However, as the plants deteriorate due to expression of disease symptoms, the potential of these plants to serve as a virus source declines. In contrast, a field of moderately resistant plants might serve as an effective virus reservoir throughout the season as plants do not deteriorate as badly as the susceptible genotypes. However, following infection in the field, highly resistant tomato genotypes pose the lowest risk to surrounding plants in terms of outbreaks of viral epidemics (Lapidot et al., 2001).

In the second study, Srinivasan et al. (2012) evaluated the effect of four different TYLCV-resistant and two susceptible commercial hybrids on whitefly population and TYLCV acquisition and transmission in the field, as well as in a greenhouse study. The different genotypes were evaluated in the field in two consecutive years. It was found that although whitefly populations in the field were not uniformly distributed, the different tomato genotypes exhibited minor differences in their ability to support whitefly populations. TYLCV-infection rate in the field was also the same among the different tomato genotypes, although the susceptible genotypes showed severe symptoms while most of the resistant genotypes showed no disease symptoms. TYLCV levels in whiteflies following acquisition from resistant genotypes were lower than from susceptible genotypes. These observations are consistent with the earlier study by Lapidot et al. (2001). However, in contrast to the earlier study, transmission rates following TYLCV acquisition from the different resistant and susceptible genotypes were the same—transmission ranged from 55% to 85% but the differences were not statistically significant (Srinivasan et al., 2012). There were a number of differences in the execution of the experiments between the two studies, but the main one was that while Lapidot et al. (2001) used a single whitefly per plant in the transmission experiments, Srinivasan et al. (2012) used 20 whiteflies per plant. This and other differences in experimental procedure could easily account for the differences in results. Nevertheless, Srinivasan et al. (2012) argued that their results demonstrate that under conditions of heavy inoculum and high vector pressure, similar to field conditions in many instances, TYLCV acquisition from resistant genotypes may result in efficient transmission to susceptible ones. Hence, they argue that tomato genotypes with a high level of TYLCV-resistance do not pose a lower risk to surrounding plants in the field (Srinivasan et al., 2012).

5.6. Bean (*P. vulgaris*) resistance to TYLCV

In 1997, an outbreak of an unknown disease in common bean (*P. vulgaris*) that caused severe losses was reported in Southern Spain (Navas-Castillo, Sanchez-Campos, & Diaz, 1999). The incidences of the unknown disease reached 80% in some fields. Symptoms consisted of downward curling, crumpling, thickening and elongation of leaves, and severe stunting of the plant. When infected early, plants showed dramatic stunting and abortion of new inflorescences, and production was entirely lost. It was soon discovered that causal agent of the disease was TYLCV. Moreover, since beans were used in Spain as an intercrop between tomato seasons, the bean plants served as a TYLCV reservoir and caused an increase in TYLCV epidemics in the tomatoes that were planted following the bean harvest (Sanchez Campos et al., 1999). It was concluded that there is a need for TYLCV-resistant bean cultivars (Navas-Castillo et al., 1999).

In Israel, it has been known for quite some time that common bean is susceptible to infection by TYLCV (Cohen & Antignus, 1994). Still, there have been no reports in Israel of TYLCV epidemics in beans, despite the fact that beans are grown in Israel and that TYLCV and its whitefly vector are present in all agricultural areas of the country. The major bean season in Israel is in early spring, thus, beans are planted and harvested before the build-up of large whitefly populations (see Section 4.4). However, this by itself does not seem to explain the lack of TYLCV epidemics in bean. It was postulated that the bean varieties being used by the growers were not susceptible to the virus. Hence, commercial varieties of common bean were screened for resistance to TYLCV (Lapidot, 2002). Out of the 42 varieties that were tested, 24 were found to be susceptible: the plants exhibited severe symptoms and accumulated high levels of viral DNA. Eighteen varieties were found to be resistant to the virus: one variety showed mild symptoms, while 17 showed no symptoms following inoculation. From the 17 symptomless varieties, plants of three varieties contained viral DNA while no viral DNA was detected in the plants of the other 14 varieties.

According to companies selling bean seeds, the two most popular bean varieties grown in Israel were found to be resistant to TYLCV (Lapidot, 2002). This may explain their popularity as well as the freedom from TYLCV epidemics in bean in Israel.

When the effect of bean plant age on viral inoculation by whiteflies was assayed, it was found that the success rate of TYLCV infection was highly dependent on bean plant age with the highest infection rates occurring in 14-day-old plants. Infection rates decreased when the inoculated plants were either younger or older than the optimum age of 14-day-old. A strong effect of plant age on infection success was also found when bean plants were inoculated with a different begomovirus, *Bean golden yellow mosaic virus* (BGYMV) (Morales & Niessen, 1988). Infection rates dropped from 100% infection in plants inoculated when they were 7-day-old plants, to 0% infection in plants inoculated at 12 days. Thus, the same phenomenon is observed in both studies—a distinct dependence of the rate of infection success with bean plant age.

In another study Monci, Garcia-Andres, Maldonado, and Moriones (2005) screened *P. vulgaris* breeding lines both in the field and in the greenhouse for resistance to TYLCV. High levels of resistance were found in the GG12 breeding line. There were no disease symptoms under field conditions as well as after controlled inoculation in the greenhouse. Following inoculation of segregating populations, the resistance was found to be controlled by a single dominant gene. Although the resistant plants did not show disease symptoms following inoculation with TYLCV, it was found that virus replication was not inhibited. Rather, it was found that viral systemic accumulation was strongly restricted in the resistant plants, suggesting that cell-to-cell or long-distance viral movement was impaired (Monci et al., 2005).

Recent reports indicate that TYLCV continues to be a problem for bean growers in southern Spain (Segundo et al., 2008), in Iran (Hedesh, Shams-Bakhsh, & Mozafari, 2011) and recently infection of common bean by TYLCV was also reported in China (Ji et al., 2012). However, it should be noted that in Latin America, where bean is a staple food, the most severe whitefly-transmitted diseases in bean cultivation are bean golden mosaic disease (BGMD) which is largely induced by the begomoviruses *Bean golden mosaic virus* and BGYMV. Indeed, major efforts are being made to breed common bean for resistance to BGMD (Morales & Jones, 2004).

5.7. Genetic resistance to the whitefly

As with development of viral resistance in the host, a potentially excellent control method against whiteflies and the viruses they transmit would be the development of whitefly resistance in the target host. There are several means by which resistance against insects can function. The presence of some plant secondary metabolites dissuades insects from settling on plants, in turn preventing the steady feeding that can lead to toxicity or virus transmission. Other metabolites may prevent oviposition, thereby reducing vector populations (Bleeker et al., 2011; Mutschler & Wintermantel, 2006; Nombela & Muniz, 2010). Resistance against whiteflies will complement resistance against the viruses they transmit and potentially reduce reliance on insecticides by reducing the frequency or the number (or both) of insecticide applications required to minimize insect populations. Host resistance to whiteflies can be combined with other methods of whitefly control. In a recent study, two whitefly-resistant genotypes of *Citrullus colocynthis* (L.), a wild relative of cultivated watermelon, were tested in the field in combination with the use of a reflective soil mulch, and were found to reduce whitefly populations (Simmons et al., 2010). The authors suggested that combining the use of reflective mulch and host plant resistance could additively suppress whitefly infestation. Although effective whitefly resistance has so far only been identified in a limited number of host plants (Nombela & Muniz, 2010; Simmons & Levi, 2002), and is mostly found in wild relatives of crop plants, the potential remains for such resistance in a number of crops that are affected by whiteflies and whitefly-transmitted viruses. Resistance to whitefly in some cotton genotypes reduced the number of insecticide applications required, thus lowering production costs while ensuring a marketable product (Chu et al., 1998). Host plant resistance to *B. tabaci* MEAM1 has been reported in two exotic melon accessions also identified as sources of resistance to *Cucurbit yellow stunting disorder virus* (CYSDV): PI 313970 and TGR-1551. Low-level resistance to *B. tabaci* MEAM1 was identified in PI 313970 in both greenhouse (Simmons & McCreight, 1996) and open-field (Boissot, Lafortune, Pavis, & Sauvion, 2003) studies. Similarly, TGR-1551 expressed low-level resistance to *B. tabaci* MED in a greenhouse study (Soria, López-Sesé, & Gómez-Guillamón, 1999). Other recent examples of the potential value of host plant resistance to whiteflies include the use of acylsugar-mediated resistance in tomato to reduce spread of TYLCV (Rodríguez-López et al., 2011). It was shown that in a no-choice experiment, 28 days after release of viruliferous whiteflies 60–80% (depending on the season) of the control tomato plants (*cv*. Moneymaker) were infected with TYLCV, while only 15–20% of the whitefly-resistant plants were infected. Similarly, preliminary studies using tomato breeding lines expressing acylsugars derived from a different wild *Solanum* species slowed the rate of infection by the crinivirus, *Tomato infectious chlorosis virus* in southern California fields (Mutschler & Wintermantel, 2006).

6. CASE STUDY 1: MANAGING BEGOMOVIRUSES AND IPOMOVIRUSES IN CASSAVA

Cassava is affected by three groups of whitefly-transmitted viruses: the cassava mosaic geminiviruses (CMGs), the CBSVs, and the CsTLVs. CMGs are present in both Africa (nine species) and South Asia (two species), CBSVs are present in coastal East Africa and the Great Lakes region of Central Africa (two species) and one newly described species, CsTLV—has been described from Colombia in Latin America (see Chapter XX which provides more detail on the cassava viruses). The CMGs and CBSVs are the biggest economic constraints to cassava production and have therefore been the subject of most of the effort to develop management programs. The CMGs cause CMD (Bock & Woods, 1983; Storey, 1936) which reduces plant growth, may lead to stunting and can cause yield losses of up to 90% in affected plants (Thresh, Fargette, & Otim-Nape, 1994). The CBSVs cause cassava brown streak disease (CBSD) (Hillocks, Raya, & Thresh, 1996; Storey, 1936; Winter et al., 2010). CBSD has a less obvious effect on cassava foliage than CMD, but causes a brown necrotic rot in the tuberous roots of affected plants. Yield losses of up to 70% have been reported (Hillocks, Raya, Mtunda, & Kiozia, 2001). Together, these two diseases cause more than US$ 1 billion of production losses annually in Africa (Legg, Owor, Sseruwagi, & Ndunguru, 2006; Thresh, Otim-Nape, Legg, & Fargette, 1997). CMD has been a target of control measures since the time of the earliest epidemics in the 1920s (Thresh & Cooter, 2005), while CBSD only began to "attract" more research attention since 2004, as it began to spread beyond its former confined distribution in coastal East Africa. Several major programs are currently being implemented to control both diseases in Africa.

6.1. Principal components of management strategies for cassava viruses

Cassava is a vegetatively propagated crop, and in almost all farming situations, new crops are planted using cuttings taken from mature stems.

Breeders use true botanical seed as part of their germplasm development work, and very rarely, tissue culture plantlets may be used to establish cassava plantings. The general use of vegetatively propagated cuttings for planting means that viruses may be readily carried from one crop to the next through planting material—leading to virus build-up or "degeneration"—unless preventive measures are taken. CMGs are transmitted persistently by *B. tabaci* (Dubern, 1994), while CBSVs are transmitted semi-persistently by the same insect (Jeremiah, 2012). These modes of transmission have important impacts on the epidemiological characteristics of CMD and CBSD (Legg et al., 2011) and therefore determine which control approaches are most suitable. In both cases, however, the suite of control tactics is similar and includes: host plant resistance, phytosanitation, other cultural control approaches, and vector control. Integrated management strategies combine several of these tactics to make the overall control impact stronger and more sustainable.

6.2. Host plant resistance to cassava viruses

Sources of resistance for both CMD and CBSD have been derived from introgressing resistance genes into cultivated cassava from wild relatives (Jennings, 1994). For CMD resistance, this has also been augmented through crosses with West African landraces which have contributed the single dominant gene—CMD2 (Akano, Dixon, Mba, Barrera, & Fregene, 2002). The speed and efficiency of cassava resistance breeding is currently being enhanced through the application of both molecular markers and next-generation sequencing approaches (Rabbi et al., 2014). Several of the most CMD-resistant varieties, bred using conventional approaches, are virtually immune to infection by CMGs. By contrast, conventional breeding has been relatively less successful in identifying and deploying high levels of resistance to CBSVs. For these ipomoviruses, transgenic approaches—based on RNAi technology—offer great potential for combining resistance to CBSVs with the farmer-preferred quality characteristics that are already present in conventionally bred CMD-resistant varieties. High levels of resistance have been demonstrated in cassava plants transformed with constructs derived from coat protein sequences of CBSVs (Yadav et al., 2011), and several transformed cassava varieties are currently being evaluated in confined field trials in Uganda and Kenya (Taylor et al., 2012).

6.3. Phytosanitation

Establishing new cassava plantings with healthy cuttings and maintaining the health of those plantings over the course of the growing season using phytosanitary practices have long been advocated as important approaches to managing cassava virus diseases (Calvert & Thresh, 2002). Plants derived from CMD-free cuttings yield significantly more than others established from infected cuttings, even if the initially healthy plants become infected during the course of the growing season (Fauquet & Fargette, 1990; Thresh, Fargette, et al., 1994). Furthermore, the later that plants become infected, the smaller the yield penalty (Thresh, Fargette, et al., 1994). Tissue culture (TC) is used by several strategically important laboratories and institutions to produce and exchange virus-indexed TC plantlets (Frison, 1994). Meristem tip culture combined with thermotherapy is effective in "cleaning up" cassava germplasm through virus elimination (Kartha & Gamborg, 1975). The primarily subsistence nature of the cassava crop in areas most affected by viruses, however, means that TC methods are not generally used to produce virus-free planting stock for larger scale field applications, as is done for other vegetatively propagated crops, such as sweet potato.

Roguing and selection of disease-free stems are widely advocated for the management of both CMD and CBSD. The differing patterns of spread of the viruses that cause these diseases, however, mean that these methods are not equally effective for the two diseases. Whitefly spread over medium to long distances is an important feature of the epidemiology of the CMGs, which means that roguing operations can be ineffective. The shorter distance spread of the semi-persistent CBSVs offers greater potential for managing CBSD through area-wide phytosanitation incorporating roguing. The cryptic symptoms of CBSD mean that effective training on accurate symptom recognition needs to be incorporated into phytosanitation programs.

6.4. Other cultural practices

Several other cultural practices have been used in attempts to control CMD and CBSD. Examples of these are presented in Section 4.2 and include managing the placement of the crop in space and/or time, and intercropping. The greatest benefits can be achieved by planting cassava crops in such a way that the degree of infestation by whitefly vectors is reduced. This can be achieved most effectively by planting during a season where the young crop growth stages occur at a time when whiteflies are not abundant, or by selecting a location for planting that is unfavorable for whiteflies. Intercropping may provide marginal benefits in reducing infection by CMGs (Fondong et al., 2002), but the relative success of control outcomes depends on the intercrop design, and the benefits offered are probably outweighed by the difficulties entailed in modifying the farming system.

6.5. Vector control

Super-abundant populations of *B. tabaci* whiteflies have driven the expansion of the pandemics of severe CMD and CBSD through large parts of East and Central Africa (Legg et al., 2011, 2014). However, surprisingly little attention has been directed toward the development and application of control tactics for this insect vector. Currently, there is virtually no field-level management of whiteflies being undertaken in cassava plantings in either Africa or South Asia, where whitefly-borne viruses are the major constraint.

Since cassava is grown primarily as a subsistence crop, there is little current use of inputs—either fertilizer or pesticides—in its cultivation in Africa. There has been some use of imidacloprid, or neonicotinoid equivalents, but typically only in experimental situations where whitefly exclusion is required in order to make comparisons between virus-infected and uninfected plants or between plots infected with different virus species (Owor, Legg, Okao-Okuja, Obonyo, & Ogenga-Latigo, 2004). As the commercialization of the crop progresses, however, it is likely that insecticides will see more widespread use in production.

Attempts have been made to determine whether sources of whitefly resistance (to *Aleurotrachelus socialis* Bondar) carried by some Latin American cassava genotypes (e.g., MEcu72) are also effective against African cassava *B. tabaci*. Although abundances of *B. tabaci* on these genotypes were significantly lower than the average for African cassava genotypes, the degree of resistance was much less than that recorded for *A. socialis* (Omongo et al., 2012). There is considerable current interest, however, in investigating the potential for the use of RNAi technologies involving the transformation of cassava plants with gene constructs that have the potential to provide much higher levels of whitefly resistance than are currently available. Some of the first practical examples of this for *Bemisia* control have recently been demonstrated (Upadhyay et al., 2011).

A diverse set of natural enemies have been reported from *B. tabaci* in several African countries, of which the most widely occurring are the aphelinid parasitoid wasps, *Encarsia* spp. and *Eretmocerus* spp. (Legg & James, 2005). Significant levels of parasitism have been reported, and there may be opportunities for these to be enhanced through manipulation of the cropping environment. However, the local parasitoids may only make an effective contribution to whitefly control if part of a multicomponent integrated management approach (Asiimwe et al., 2007). Although classical biological control has proven highly effective against pests introduced to Africa, such as the cassava mealybug [*Phenacoccus manihoti* Mat. -Ferr.] and the cassava green mite [*Mononychellus tanajoa* (Bondar)], the prospects of similar success for *Bemisia* seem less promising, since this genus of whiteflies is considered to be indigenous to Africa (Campbell, Steffen-Campbell, & Gill, 1996). Introductions of *Eretmocerus hayati* Rose & Zolnerowich to Australia for the control of *B. tabaci* MEAM1 have been highly successful, however, as this exotic parasitoid species has greatly augmented the existing activity of the 11 local species of *Eretmocerus* and *Encarsia* reported to attack *B. tabaci* (De Barro & Coombs, 2009). This success has encouraged recent initiatives to introduce *E. hayati* to East Africa to evaluate its potential effectiveness against African cassava-colonizing *B. tabaci*. Biological control is unlikely to be effective in a single-component control strategy and offers greatest potential as a component within an integrated whitefly management strategy also incorporating host plant resistance and cultural methods.

6.6. Integrated control strategies

The term integrated control has most commonly been used to refer to the combination of pesticide application with other management tactics, very often with a goal of reducing overall levels of pesticide usage (Naranjo & Ellsworth, 2009). Since pesticides are not usually used by cassava growers in Africa and South Asia, this term has rarely been used to describe multicomponent control strategies. Such approaches have been widely used for cassava virus disease management, however, and have primarily involved the combination of host plant virus resistance and phytosanitary measures to protect the health of cassava plants before and after planting. As cassava virus disease pandemics have spread through large parts of East and Central Africa, governments, NGOs and international partners have responded through the mass deployment of resistant varieties (Walsh, 2012). The initial target of these mitigation efforts was the pandemic of severe CMD, whose most rapid period of spread occurred during the 1990s, and the varieties multiplied had high levels of CMD resistance (Dixon et al., 2003; Legg, Kapinga, Teri, & Whyte, 1999; Thresh, Otim-Nape, & Jennings, 1994). Since the mid-2000s, these large-scale germplasm "roll-out" programs have been coupled with the application of quality management protocols (QMP), which comprise procedures for the assurance of specified minimum virus disease levels in multiplied and disseminated crops of improved cassava varieties (Jennings, 1994). In order to strengthen standards and avoid inadvertent spread of cryptic CBSD infection, a large-scale virus testing program was undertaken during the Great Lakes Cassava Initiative—a six-country regional cassava virus management program (Smith, 2014). By combining virus testing at a small number of high-level germplasm multiplication sites, with the large-scale application of QMP measures at primary (regional), secondary (district), and tertiary (community) multiplication sites, it was possible to deliver high-quality planting material with little or no cassava virus disease to the ultimate targets of the program—small-scale growers. Although the initial stages of this effort were hindered by the lack of resistance to CBSD in varieties being multiplied, improvements were achieved through the course of the initiative as CBSD-tolerant varieties were identified and incorporated.

On-going and future targets for cassava virus disease control work will be to improve resistance to CBSD using both conventional and transgenic methods, identify sources of whitefly resistance, and combine the use of improved germplasm with other cultural and biological controls to strengthen the effectiveness of integrated control. Looking further ahead, the significant risks of spread of cassava viruses, both within and between the continents in which cassava is grown (Africa, Asia, and Latin America), demand that great attention is given to strengthening surveillance and quarantine procedures.

7. CASE STUDY 2: MANAGEMENT OF CRINIVIRUSES

There is a tremendous amount of research on management of begomoviruses, but only a limited amount of information on management of other types of whitefly-transmitted viruses. This section will address management of criniviruses (reviewed by Tzanetakis et al., 2013), which share a number of similarities with begomoviruses, but differ greatly in mode of transmission by whitefly vectors and therefore management.

Over the past 25 years, the number of members in this genus has expanded through identification and characterization of new viruses affecting a wide range of crop and weed hosts throughout tropical and subtropical areas of the world where whitefly vectors are prevalent, as well as in greenhouse production facilities. In the 1980s, *Lettuce infectious yellows virus* and *Beet pseudo-yellows virus* (BPYV) were the only well-known whitefly-transmitted viruses with virions composed of flexuous rods, and it was not until many years later that BPYV was added to the genus after genome analysis confirmed that it too was bipartite like other members. Throughout the 1990s and beyond, a wide array of criniviruses have been characterized (Celix, Lopez-Sese, Almarza, Gomez-Guillamon, & Rodriguez-Cerezo, 1996; Duffus, Larsen, & Liu, 1986; Duffus, Liu, & Wisler, 1996; Duffus, Liu, Wisler, & Li, 1996; Liu, Li, Wisler, & Duffus, 1997; Martin, 2004; Martın, Velasco, Segundo, Cuadrado, & Janssen, 2008; Okuda, Okazaki, Yamasaki, Okuda, & Sugiyama, 2010; Salazar, Muller, Querci, Zapata, & Owens, 2000; Tzanetakis et al., 2004; Winter et al., 1992) and the genus now consists of three separate groups based largely on genetic relationships and vector transmission characteristics (Tzanetakis et al., 2013; Wintermantel, Hladky, Gulati-Sakhuja, et al., 2009).

Symptoms of criniviruses are not always as apparent as those of other plant viruses. Whereas begomoviruses often produce bright yellow symptoms on leaves, along with distortion and curling, criniviruses often cause symptoms that are readily mistaken for physiological or nutritional disorders or pesticide phytotoxicity. Depending on the host plant affected, these symptoms include interveinal yellowing of leaves, an associated loss of photosynthetic capability, leaf brittleness, reduced plant vigor, yield reductions, and early senescence (Tzanetakis et al., 2013; Wintermantel, 2010). Many crinivirus infections remain latent for nearly 3 weeks before symptoms appear, and as a result these viruses can be moved on transplants without knowledge the plants are infected. Symptoms are usually most apparent on the middle and older parts of plants, with new growth appearing normal and symptoms progressing toward newer growth over time. In a few crops, including strawberry and sweet potato, crinivirus infection can remain latent until plants become coinfected with another virus resulting in symptom development due to synergism between the crinivirus and the coinfecting virus.

Such mixed infections can complicate identification of the primary virus causing disease because symptoms resulting from mixed infection with other viruses often induce different symptoms than those resulting from single infections. In other situations, the viruses involved in coinfections are obvious with different sections of infected plants exhibiting symptoms uniquely characteristic of each virus. An example would be CYSDV infection of melon with coinfection by either a begomovirus or potyvirus. All three types of virus produce unique symptoms on melon. CYSDV produces interveinal yellowing beginning near the crown and progressing outward down the vines, whereas mosaic symptoms resulting from infection by *Watermelon mosaic virus* or leaf curl symptoms resulting from CuLCrV are usually found near the ends of vines.

Numerous studies have demonstrated that interactions between criniviruses and other coinfecting viruses have been known to influence the type and severity of symptoms observed on plants. In most documented cases, this leads to enhanced disease severity, as has been found with infection of SPCSV and members of the Potyvirus genus (Karyeija et al., 2000). Similar effects were later found with coinfection between SPCSV and viruses of other genera and families (Cuellar, De Souza, Barrantes, Fuentes, & Kreuze, 2011; Untiveros, Fuentes, & Salazar, 2007). In some hosts, such as strawberry, coinfection involving a crinivirus and another virus produces severe symptoms, whereas plants infected by either virus alone remained asymptomatic (Tzanetakis, Wintermantel, et al., 2006; Tzanetakis et al., 2004). Much remains to be determined regarding the interactions between the coinfecting viruses and their hosts and how this leads to increased severity; however, such interactions complicate management of virus diseases in the field.

7.1. CYSDV: Managing crinivirus infection in the field

CYSDV was widely studied following its establishment in the Imperial Valley of California. In the fall of 2006, CYSDV was identified affecting cucurbit production throughout the southwestern US (California and Arizona), as well as nearby Sonora in Mexico, resulting in widespread infection of the fall melon crop (Brown, Guerreo, Matheron, Olsen, & Idris, 2007; Kuo, Rojas, Gilbertson, & Wintermantel, 2007). Although at that time it was believed the host range of CYSDV was restricted to cucurbit crops (Celix et al., 1996), the virus survived the largely cucurbit-free winter months to infect a limited number of plants the following spring, and again infect nearly the entire fall crop in 2007. This pattern is now well established in the region. Populations of *B. tabaci* accumulate gradually during the spring melon season, with infection developing late in the season with only limited impact on yield during the spring season (Chu et al., 2007). In contrast, the exceptionally high populations of *B. tabaci* during the fall melon season in the desert region of southwestern US results in rapid and efficient transmission of CYSDV to melon, with infection occurring in seedling plants. Following the establishment of CYSDV in the region, research demonstrated that the virus was able to infect a broad range of common weed and crop plants prevalent in the desert production region (Wintermantel, Hladky, Cortez, et al.,

2009; Wintermantel, Hladky, Gulati-Sakhuja, et al., 2009). The establishment of CYSDV in this important region where most US winter melons are produced prompted an aggressive research effort toward development of effective management strategies for control of the virus and to mitigate CYSDV-induced losses.

7.2. Identification and management of crop and weed reservoir hosts

Studies demonstrated that the presence of several weed or alternate crop hosts in the region, most of which were symptomless when infected (Wintermantel, Hladky, Cortez, et al., 2009). Subsequent work focused on determining which of these newly identified hosts were of epidemiological significance. Although CYSDV was able to infect several weed hosts as well as lettuce (*Lactuca sativa*) and snap bean (*P. vulgaris*), titers were much lower in non-cucurbit hosts than in melon and other cucurbits (Wintermantel, Gilbertson, & Natwick, 2014). In some cases, the CYSDV titer in the host plant was directly related to the efficiency of virus transmission to melon; however, in several host plants transmission rates did not correspond to virus titer, indicating a complex relationship influencing the ability of different host plants to serve as efficient virus reservoirs for transmission to melon. Importantly, one of the most widely planted crops in the region, alfalfa (*Medicago sativa*), although a host of CYSDV, is very inefficient as a source for transmission of the virus to cucurbits (Wintermantel et al., 2014). Knowledge of reservoir hosts is important toward reducing sources of virus in the field through reduction of source plants when possible. By targeting weed management against virus reservoir hosts and plants on which the whitefly vector feeds or reproduces, spread of the virus can be limited when combined with other practices.

7.3. Genetic resistance to the virus

Traditionally, management of whitefly-transmitted viruses in the desert production regions of the southwestern US has been predominantly through control of the whitefly vector using insecticidal control. Several neonicotinoid formulations are currently used in conjunction with other chemistries, for control of *B. tabaci* populations. Still, even aggressive insecticidal methods are ineffective at reducing populations of *B. tabaci* MEAM1 (the whitefly vector common in the southwestern US) sufficiently to mitigate virus spread. CYSDV has established itself in the wild hosts and cultivated crops of the region, and insecticidal control alone has not been sufficient to obtain marketable melons when plants become infected early as occurs during the fall production season. Therefore, an aggressive resistance breeding program was developed. A number of laboratories are focused on development of CYSDV resistance, particularly in cucumber and cantaloupe melon (Aguilar, Abad, & Aranda, 2006; Eid, Abou-Jawdah, El-Mohtar, Sobh, & Havey, 2006; López-Sesé & Gómez-Guillamón, 2000; Marco, Aguilar, Abad, Gomez-Guillamon, & Aranda, 2003; McCreight & Wintermantel, 2011). Interestingly, the first few years of research on host plant resistance to CYSDV in California's Imperial Valley were conducted without any measures to control *B. tabaci* MEAM1, but it became evident that control of this insect was essential for resistance to CYSDV to be more fully expressed. Recent studies have shown that combining host plant resistance with effective vector control has been very effective at maximizing the effectiveness of resistance (McCreight J. D. & Wintermantel W. M., unpublished) and offers potential for improved management in production fields once resistance is introgressed into cultivated melon.

In cantaloupe melon, each resistance source has the potential to reduce severity, but breeding studies have demonstrated that combining resistance sources results in stronger resistance than can be achieved with individual genes (McCreight & Wintermantel, 2011). There are currently two independent sources of resistance to CYSDV in melon germplasm: TGR-1551 from Zimbabwe (López-Sesé & Gómez-Guillamón, 2000) and PI 313970 from India (McCreight & Wintermantel, 2008). Resistance in TGR-1551 was initially reported to be dominant (López-Sesé & Gómez-Guillamón, 2000); however, it is possible the resistance may be codominant and complex (Sinclair, 2003). Alternatively, CYSDV resistance in TGR-1551 may be affected by environmental variation (Rubio, Abou-Jawdah, Lin, & Falk, 2001). Most importantly, although the resistance in TGR-1551 can be effective, its performance has been variable depending on the conditions and locations where it has been evaluated. The single recessive gene for resistance in PI 313970 (McCreight & Wintermantel, 2011) can also reduce disease severity in the field, but the exceptionally high populations of viruliferous whiteflies that occur in the southwestern US during the summer and fall can be too much for this resistance source alone. Results of a cross of PI 313970 × TGR-1551 in 2009 and 2010 suggest the possibility of higher and more uniform levels of resistance when their genes are combined (McCreight & Wintermantel, 2011). However, introgression of these sources of resistance from exotic melons into commercially favored sweet cantaloupe melon is a formidable task.

7.4. Crinivirus management in transplanted crops

Many field crops begin in nurseries with propagation by seed or through cuttings or runners. In some crops, grafting is increasing in popularity as a means of introducing vigorous or highly resistant root systems that will benefit the plant once it is transplanted to the field. Any movement or manipulation of plant material inherently introduces the risk of virus infection. It is critical that nursery operations routinely monitor grafting stock for the most critical viruses that could affect the crop once it is in the field. This is particularly important when nursery facilities are located in areas known to harbor viruses of concern for the nursery crop or their insect vectors. Although such measures are important for preventing infection of nursery stock with all viruses, it is an especially significant concern with regard to criniviruses. As noted, criniviruses have a lengthy latent period in most host plants ranging from slightly under 3–4 weeks depending on the plant and virus. Due to the extended latent period, a crinivirus introduced in the nursery can easily remain symptomless until it is transplanted in the field, resulting in introduction of the virus to the initial field and potentially to an entire production region. Although criniviruses are not as easily graft-transmitted as many other plant viruses, these viruses can be introduced to healthy plant material through graft unions. They are also maintained in rooted cuttings and can be difficult to monitor. *Strawberry pallidosis-associated virus* is known to increase in titer during the winter months, but titers can decrease to nearly undetectable levels during the summer (Tzanetakis et al., 2004). Such cycling of virus titers may occur with other crinivirus infections as well, but such studies have not been conducted. Consequently, effective monitoring should be performed on nursery stock throughout the year, not only as plants are prepared for movement to the field.

7.5. Summary

Successful management of criniviruses in field production systems is best achieved through integrated pest management (IPM). Resources available increased tremendously over the past decade as the library of epidemiological knowledge of this important and emerging genus has grown. Host range information is largely established for most members of the genus and sources of host plant resistance are being developed for some. Effective management of weed hosts and carefully managed crop rotation and proximity will reduce spread among fields. It is anticipated that new sources of host plant resistance against both criniviruses and their whitefly vectors will continue to be identified, adding to the arsenal of protective measures available for crop production. Currently there are a number of options for insecticidal control of whitefly vectors. Effective management to prolong functionality of insecticides, coupled with virus and/or insect resistance will also be valuable for management of criniviruses in field production systems. Admittedly, identification of resistance sources will require time for some of the more recently characterized members of the genus. However, through strategic use of host plant and/or insect resistance when available, monitoring and internal management of nursery stock, and efficient use of pesticides, crinivirus infection of field crops can be minimized.

8. CONCLUDING REMARKS

Whiteflies have become a key pest in modern agriculture, mainly in open-field production. This is largely due to the long list of devastating plant viruses transmitted by the whiteflies. Some of the diseases caused by whitefly-transmitted viruses have become a limiting factor in open-field production, and important examples include: BGMD in beans, TYLCD in tomato, CMD and CBSD in cassava, and Cotton leaf curl disease (CLCuD) in cotton. The main strategy to stop whitefly infestation in the field is the application of insecticides. However, due to growing consumer demand for cleaner produce, and the whitefly's ability to develop insecticide-resistance, there is a growing demand to reduce insecticide application. This will increase the need for effective IPM strategies to reduce infestation by whiteflies and the viruses they transmit. Genetic resistance in the host—to the whitefly as well as to the whitefly-transmitted viruses, combined with other IPM strategies could become a viable solution to open-field production in the days of reduced application of insecticides.

Management of whitefly-transmitted viruses is challenging. While host resistance to the virus is the best approach, the number and diversity of the viruses and crops affected mean that there are few examples of diseases being managed by this approach. The importance of resistance for management of these viruses cannot be overstated, and as such there is a tremendous need to develop resistance to whitefly-transmitted viruses in many crops. In most cases, the focus of management is the reduction in whitefly populations and the reduction of inoculum sources.

The most effective management schemes that reduce whitefly populations and limit virus spread are those that use multiple approaches simultaneously. This also involves professional crop scouts who can help tailor the timing and type of insecticide application to keep costs as low as possible and maximize effectiveness. These schemes can be expensive and demanding of resources and education and work best when there is a support structure to provide data essential for development and use of appropriate management tools (Adkins et al., 2011). Knowledge of the alternative hosts of the virus, the identity and feeding preferences of the vector, information on expected changes in whitefly populations throughout the year, and knowledge of alternative hosts of the whitefly all contribute to the design of effective management recommendations. Growers that lack such resources cannot manage these viruses very effectively and often must abandon the crop for alternative crops that are not affected. Further investments in the development of host resistance as well as in the development of effective integrated management tactics will allow us to produce both food and biofuel crops despite the presence of these viruses and their vector.

REFERENCES

Abd-Rabou, S., & Simmons, A. M. (2012). Some cultural strategies to help manage *Bemisiatabaci* (Hemiptera: Aleyrodidae) and whitefly-transmitted viruses in vegetable crops. *African Entomology*, *20*, 371–379.

Adkins, S., Webster, C. G., Kousik, C. S., Webb, S. E., Roberts, P. D., Stansly, P. A., et al. (2011). Ecology and management of whitefly-transmitted viruses of vegetable crops in Florida. *Virus Research*, *159*, 110–114.

Aguilar, J. M., Abad, J., & Aranda, M. A. (2006). Resistance to Cucurbit yellow stunting disorder virus in cucumber. *Plant Disease*, *90*, 583–586.

Akano, A., Dixon, A., Mba, C., Barrera, E., & Fregene, M. (2002). Genetic mapping of a dominant gene conferring resistance to cassava mosaic disease. *Theoretical and Applied Genetics*, *105*, 521–525.

Alaux, C., Brunet, J., Dussaubat, C., Mondet, F., Tchamitchan, S., Cousin, M., et al. (2009). Interactions between Nosema microspores and a neonicotinoid weaken honeybees (*Apis mellifera*). *Environmental Microbiology*, *12*, 774–782.

Alvarez, E., Mejía, J. F., Llano, G. A., Loke, J. B., Calari, A., Duduk, B., et al. (2009). Characterization of a phytoplasma associated with frogskin disease in cassava. *Plant Disease*, *93*, 1139–1145.

Amari, K., Gonzalez-Ibeas, D., Gomez, P., Sempere, R. N., Sanchez-Pina, M. A., Aranda, M. A., et al. (2008). Tomato torrado virus is transmitted by *Bemisia tabaci* and infects pepper and eggplant in addition to tomato. *Plant Disease*, *92*, 1139.

Anbinder, I., Reuveni, M., Azari, R., Paran, I., Nahon, S., Shlomo, H., et al. (2009). Molecular dissection of *Tomato leaf curl virus* resistance in tomato line TY172 derived from *Solanum peruvianum*. *Theoretical and Applied Genetics*, *119*, 519–530.

Anderson, P. K., Cunningham, A. A., Patel, N. G., Morales, F. J., Epstein, P. R., & Daszak, P. (2004). Emerging infectious diseases of plants: Pathogen pollution, climate change and agrotechnology drivers. *Trends in Ecology & Evolution, 19*, 535–544.

Angel, J. C., Pineda, B. L., Nolt, B., & Velasco, A. C. (1990). Mosca blanca (Homoptera: Aleyrodidae) asociadas a transmisi´on de virus en yuca. *Fitopatologı´a Colombiana, 13*, 65–71 (in spanish).

Anon (2012). *Net tunnels to protect sweetpotato planting material from disease: A guide to construct and maintain tunnels*. Lima, Peru: CIP. http://sweetpotatoknowledge.org/seedsystem/Brochure%20Net%20Tunnel.pdf/view.

Aritua, V., Legg, J. P., Smit, N. E. J. M., & Gibson, R. W. (1999). Effect of local inoculum on the spread of sweet potato virus disease: Widespread cultivation of a resistant sweet potato cultivar limits infection of susceptible cultivars. *Plant Pathology, 48*, 655–661.

Asiimwe, P., Ecaat, J. S., Otim, M., Gerling, D., Guershon, M., Kyamanywa, S., et al. (2007). Life table analysis of mortality factors affecting populations of *Bemisia tabaci* on cassava in Uganda. *Entomologia Experimentalis et Applicata, 122*, 37–44.

Avidov, H. Z. (1946). Tobacco whitefly in Israel. *Hassadeh*, 1–33 (in Hebrew).

Azzam, O., Frazer, J., De La Rosa, D., Beaver, J. S., Ahlquist, P., & Maxwell, D. P. (1994). Whitefly transmission and efficient ssDNA accumulation of bean golden mosaic geminivirus require functional coat protein. *Virology, 204*, 289–296.

Bedford, I. D., Briddon, R. W., Brown, J. K., Rosell, R. C., & Markham, P. G. (1994). Geminivirus transmission and biological characterization of whitefly *Bemisia tabaci* biotypes from different geographic regions. *The Annals of Applied Biology, 125*, 311–325.

Bian, X. Y., Thomas, M. R., Rasheed, M. S., Saeed, M., Hanson, P., De Barro, P. J., et al. (2007). A recessive allele (tgr-1) conditioning tomato resistance to geminivirus infection is associated with impaired viral movement. *Phytopathology, 97*, 930–937.

Bisaro, D. M. (2006). Silencing suppression by geminivirus proteins. *Virology, 344*, 158–168.

Blackmer, J. L., & Byrne, D. N. (1993). Flight behaviour of *Bemisia tabaci* in a vertical flight chamber: Effect of time of day, sex, age and host quality. *Physiological Entomology, 18*, 223–232.

Blacquière, T., Smagghe, G., van Gestel, C. A. M., & Mommaerts, V. (2012). Neonicotinoids in bees: A review on concentrations, side-effects and risk assessment. *Ecotoxicology, 21*, 973–992.

Bleeker, P. M., Diergaarde, P. J., Ament, K., Schutz, S., Johne, B., Dijkink, J., et al. (2011). Tomato-produced 7-epizingiberene and R-curcumene act as repellents to whiteflies. *Phytochemistry, 72*, 68–73.

Bock, K. R. (1994). The spread of African cassava mosaic geminivirus in coastal and western Kenya. *Tropical Science, 34*, 92–101.

Bock, K. R., & Woods, R. D. (1983). The etiology of African cassava mosaic disease. *Plant Disease, 67*, 994–995.

Boissot, N., Lafortune, D., Pavis, C., & Sauvion, N. (2003). Field resistance to *Bemisia tabaci* in *Cucumis melo*. *HortScience, 38*, 77–80.

Briddon, R. W., Pinner, M. S., Stanley, J., & Markham, P. G. (1990). Geminivirus coat protein gene replacement alters insect specificity. *Virology, 177*, 85–94.

Brough, C. L., Sunter, G., Gardiner, W. E., & Bisaro, D. M. (1992). Kinetics of tomato golden mosaic virus DNA replication and coat protein promoter activity in *Nicotiana tabacum* protoplasts. *Virology, 187*, 1–9.

Brown, J. K., Guerreo, J. C., Matheron, M., Olsen, M., & Idris, A. M. (2007). Widespread outbreak of *Cucurbit yellow stunting disorder virus* (CYSDV) in the Sonoran plateau region of the Western U.S. and Pacific coast of Mexico. *Plant Disease, 91*, 773.

Caciagli, P., Bosco, D., & Albitar, L. (1995). Relationships of the sardinian isolate of *tomato yellow leaf curl geminivirus* with its whitefly vector *Bemisia-Tabaci* Gen. *European Journal of Plant Pathology, 101*, 163–170.

Calvert, L. A., & Thresh, J. M. (2002). The viruses and virus diseases of cassava. In A. C. Bellotti, R. J. Hillocks, & J. M. Thresh (Eds.), *Cassava: Biology, production and utilization* (pp. 237–260). Wallingford, UK: CABI.

Cameron, S. A., Lozier, J. D., Strange, J. P., Koch, J. B., Cordes, N., Solter, L. F., et al. (2011). Patterns of widespread decline in North American bumble bees. *PNAS*, *108*, 662–667.

Campbell, B. C., Steffen-Campbell, J. D., & Gill, R. (1996). Origin and radiation of whiteflies: An initial molecular phylogenetic assessment. In D. Gerling, & R. T. Mayer (Eds.), *Taxonomy, Biology, Damage, Control and Management*Andover, UK: Intercept.

Carvajal-Yepes, M., Olaya, C., Lozano, I., Cuervo, M., Castaño, M., & Cuellar, W. J. (2014). Unraveling complex viral infections in cassava (Manihot esculenta Crantz.) from Colombia. *Virus Research*, *186*, 76–86.

Castle, S. J. (2006). Concentration and management of *Bemisia tabaci* in cantaloupe as a trap crop for cotton. *Crop Protection*, *25*, 574–584.

Celix, A., Lopez-Sese, A., Almarza, N., Gomez-Guillamon, M. L., & Rodriguez-Cerezo, E. (1996). Characterization of *Cucurbit yellow stunting disorder virus*, a *Bemisia tabaci*-transmitted closterovirus. *Phytopathology*, *86*, 1370–1376.

Chu, C. C., Barnes, E., Natwick, E. T., Chen, Y., Ritter, D., & Henneberry, T. J. (2007). Trap catches of the sweetpotato whitefly (*Homoptera: Aleyrodidae*) in the Imperial Valley, California, from 1996 to 2002. *Insect Science*, *14*, 165–170.

Chu, C. C., Natwick, E. T., Perkins, H. H., Brushwood, D. E., Henneberry, T. J., Castle, S. J., et al. (1998). Upland cotton susceptibility to *Bemisia argentifolii* (Homoptera: Aleyrodidae) infestations. *Journal of Cotton Science*, *2*, 1–9.

Clark, C. A., Davis, J. A., Abed, J. A., Cuellar, W. J., Fuentes, S., Kreuze, J. F., et al. (2012). Sweetpotato viruses: 15 years of progress on understanding and managing complex diseases. *Plant Disease*, *96*, 168–185.

Cohen, S. (1982). Control of whitefly vectors of viruses by color mulches. In K. F. Harris, & K. Maramorosch (Eds.), *Pathogens, vectors and plant diseases, approaches to control* (pp. 46–56). New York: Academic Press.

Cohen, S., & Antignus, Y. (1994). Tomato yellow leaf curl virus, a whitefly-borne geminivirus of tomatoes. *Advances in Disease Vector Research*, *10*, 259–288.

Cohen, S., Duffus, J. E., Larsen, R. C., Liu, H. Y., & Flock, R. A. (1983). Purification, serology, and vector relationships of squash leaf curl virus, a whitefly-transmitted geminivirus. *Phytopathology*, *73*, 1669–1673.

Cohen, J., Gera, A., Ecker, R., Ben Joseph, R., Perlsman, M., Gokkes, M., et al. (1995). Lisianthus leaf curl—A new disease of lisianthus caused by tomato yellow leaf curl virus. *Plant Disease*, *79*, 416–420.

Cohen, S., & Harpaz, I. (1964). Periodic rather than continual acquisition of a new tomato virus by its vector, the tobacco whitefly (*Bemisia tabaci* Gennadius). *Entomologia Experimentalis et Applicata*, *7*, 155–166.

Cohen, S., Kern, J., Harpaz, I., & Ben Joseph, R. (1988). Epidemiological studies of the *Tomato yellow leaf curl virus* (TYLCV) in the Jordan Valley, Israel. *Phytoparasitica*, *16*, 259–270.

Cohen, S., & Lapidot, M. (2007). Appearance and expansion of TYLCV: A historical point of view. In H. Czosnek (Ed.), *Tomato yellow leaf curl virus disease* (pp. 3–12). The Netherlands: Springer.

Cohen, S., & Melamed-Madjar, V. (1978). Prevention by soil mulching of the spread of *Tomato yellow leaf curl virus* transmitted by *Bemisia tabaci* (Gennadius) (Hemiptera: Aleyrodidae) in Israel. *Bulletin of Entomological Research*, *68*, 465–470.

Csizinszky, A. A., Schuster, D. J., & Kring, J. B. (1995). Color mulches influence yield and insect pest populations in tomatoes. *Journal of the American Society for Horticultural Science*, *120*, 778–784.

Csizinszky, A. A., Schuster, D. J., & Kring, J. B. (1997). Evaluation of color mulches and oil sprays for yield and for the control of silverleaf whitefly, *Bemisia argentifolii* (Bellows and Perring) on tomatoes. *Crop Protection, 16*, 475–481.

Csizinszky, A. A., Schuster, D. J., & Polston, J. E. (1999). Effect of UV-reflective mulches on tomato yields and on the silverleaf whitefly. *HortScience, 34*, 911–914.

Cuellar, W. J., De Souza, J., Barrantes, I., Fuentes, S., & Kreuze, J. F. (2011). Distinct cavemoviruses interact synergistically with *Sweet potato chlorotic stunt virus* (genus *Crinivirus*) in cultivated sweet potato. *The Journal of General Virology, 92*, 1233–1243.

Czosnek, H., & Laterrot, H. (1997). A worldwide survey of tomato yellow leaf curlviruses. *Archives of Virology, 142*, 1391–1406.

De Barro, P. J., & Coombs, M. T. (2009). Post-release evaluation of *Eretmocerus hayati* Zolnerowich and Rose in Australia. *Bulletin of Entomological Research, 99*, 193–206.

Desneux, N., Decourtye, A., & Delpuech, J. M. (2007). The sublethal effects of pesticides on beneficial arthropods. *Annual Review of Entomology, 52*, 81–106.

Di Prisco, G., Cavaliere, V., Annoscia, D., Varricchio, P., Caprio, E., Nazzi, F., et al. (2013). Neonicotinoid clothianidin adversely affects insect immunity and promotes replication of a viralpathogen in honey bees. *Proceedings of the National Academy of Sciences of the United States of America, 110*, 18466–18471.

Dixon, A. G. O., Bandyopadhyay, R., Coyne, D., Ferguson, M., Ferris, R. S. B., Hanna, R., et al. (2003). Cassava: From a poor farmer's crop to a pacesetter of African rural development. *Chronica Horticulturae, 43*, 8–14.

Dombrovsky, A., Sapkota, R., Lachman, O., Pearlsman, M., & Antignus, Y. (2013). A new aubergine disease caused by a whitefly-borne strain of *Tomato mild mottle virus* (TomMMoV). *Plant Pathology, 62*, 750–759.

Dubern, J. (1994). Transmission of African cassava mosaic geminivirus by the whitefly (*Bemisia tabaci*). *Tropical Science, 34*, 82–91.

Duffus, J. E. (1987). Whitefly-transmitted plant viruses. *Current Topics in Vector Research, 4*, 73–91.

Duffus, J. E., Larsen, R. C., & Liu, H.-Y. (1986). *Lettuce infectious yellows virus*-a new type of whitefly-transmitted virus. *Phytopathology, 76*, 97–100.

Duffus, J. E., Liu, H.-Y., & Wisler, G. C. (1996). Tomato infectious chlorosis virus-A new clostero-like virus transmitted by Trialeurodes vaporariorum. *European Journal of Plant Pathology, 102*, 219–226.

Duffus, J. E., Liu, H. Y., Wisler, G. C., & Li, R. (1996). Lettuce chlorosis virus—A new whitefly-transmitted closterovirus. *European Journal of Plant Pathology, 102*, 591–596.

Eid, S., Abou-Jawdah, Y., El-Mohtar, C., Sobh, H., & Havey, M. (2006). Tolerance in Cucumber to Cucurbit yellow stunting disorder virus. *Plant Disease, 90*, 645–649.

Eiri, D., & Nieh, J. C. (2012). A nicotinic acetylcholine receptor agonist affects honey bee sucrose responsiveness and decreases waggle dancing. *The Journal of Experimental Biology, 215*, 2022–2029.

Ellsworth, P. C., & Martinez-Carrillo, J. L. (2001). IPM for *Bemisia tabaci*: A case study from North America. *Crop Protection, 20*, 853–869.

Elston, C., Thompson, H. M., & Walters, K. F. A. (2013). Sub-lethal effects of thiamethoxam, a neonicotinoid pesticide, and propiconazole, a DMI fungicide, on colony initiation in bumblebee (*Bombus terrestris*) micro-colonies. *Apidologie, 44*, 563–574.

Epstein, D., Frazier, J. L., Purcell-Miramontes, M., Hackett, K., Rose, R., Erickson, T., et al. (2012). Report on the National Stakeholders Conference on honey bee health. In *National Honey Bee Health Stakeholder Conference Steering Committee.* file://ad.ufl.edu/ifas/PLP/Users/jep/Profile/My%20Documents/MANUSCRIPTS%20IN%20PREPARATION/MANAGEMENT%20OF%20WF%20TRANS%20VIRUSES%202014/Honeybee%20Problem/ReportHoneyBeeHealth%20USDA%20and%20EPA.pdf.

Etessami, P., Saunders, K., Watts, J., & Stanley, J. (1991). Mutational analysis of complementary-sense genes of African cassava mosaic virus DNA A. *The Journal of General Virology, 72,* 1005–1012.

Fargette, D., & Fauquet, C. (1988). A preliminary study of intercropping maize and cassava on the spread of African cassava mosaic virus by whiteflies. *Aspects of Applied Biology, 17,* 195–202.

Fargette, D., Fauquet, C., Grenier, E., & Thresh, J. M. (1990). The spread of African cassava mosaic virus into and within cassava fields. *Journal of Phytopathology, 130,* 289–302.

Fargette, D., Jeger, M., Fauquet, C., & Fishpool, L. D. C. (1993). Analysis of temporal disease progress of *African cassava mosaic virus*. *Phytopathology, 84,* 91–98.

Fauquet, C., & Fargette, D. (1990). *African cassava mosaic virus*: Etiology, epidemiology and control. *Plant Disease, 74,* 404–411.

Finch, S., & Collier, R. H. (2000). Host-plant selection by insects—A theory based on 'appropriate/inappropriate landings' by pest insects of cruciferous plants. *Entomologia Experimentalis et Applicata, 96,* 91–102.

Fishpool, L. D. C., & Burban, C. (1994). *Bemisia tabaci*: The whitefly vector of African cassava mosaic geminivirus. *Tropical Science, 34,* 55–72.

Fondong, V., Thresh, J. M., & Zok, S. (2002). Spatial and temporal spread of cassava mosaic virus disease in cassava grown alone and when intercropped with maize and/or cowpea. *Journal of Phytopathology, 150,* 365–374.

Friedmann, M., Lapidot, M., Cohen, S., & Pilowsky, M. (1998). A novel source of resistance to *Tomato yellow leaf curl virus* exhibiting a symptomless reaction to viral infection. *Journal of the American Society for Horticultural Science, 123,* 1004–1007.

Frison, E. (1994). Sanitation techniques for cassava. *Tropical Science, 34,* 146–153.

Fuglie, K. O., Zhang, L., Salazar, L. F., & Walker, T. (1998). *Economic impact of virus-free sweet potato seed in Shandong Province, China.* http://www.eseap.cipotato.org/MF-ESEAP/Fl-Library/Eco-Imp-SP.pdf (accessed March 6, 2014).

Gafni, Y. (2003). *Tomato yellow leaf curl virus*, the intracellular dynamics of a plant DNA virus. *Molecular Plant Pathology, 4,* 9–15.

Gamarra, H. A., Fuentes, S., Morales, F. J., Glover, R., Malumphy, C., & Barker, I. (2010). *Bemisia afer* sensu lato, a vector of *Sweet potato chlorotic stunt virus*. *Plant Disease, 94,* 510–514.

Garcia-Ruiz, H., & Murphy, J. F. (2001). Age-related resistance in bell pepper to *Cucumber mosaic virus*. *The Annals of Applied Biology, 139,* 307–317.

Giordano, L. B., Silva-Lobo, V. L., Santana, F. M., Fonseca, M. E. N., & Boiteux, L. S. (2005). Inheritance of resistance to the bipartite Tomato chlorotic mottle begomovirus derived from Lycopersicon esculentum cv. 'Tyking'. *Euphytica, 143,* 27–33.

Gottlieb, Y., Zchori-Fein, E., Mozes-Daube, N., Kontsedalov, S., Skaljac, M., Brumin, M., et al. (2010). The transmission efficiency of *Tomato yellow leaf curl virus* by the whitefly *Bemisia tabaci* is correlated with the presence of a specific symbiotic bacterium species. *Journal of Virology, 84,* 9310–9317.

Greer, L., & Dole, J. M. (2003). Aluminum foil, aluminum-painted, plastic and degradable mulches increase yields and decrease insect-vectored viral diseases of vegetables. *HortTechnology, 13,* 276–284.

Gronenborn, B. (2007). The *Tomato yellow leaf curl virus* genome and function of its proteins. In H. Czosnek (Ed.), *Tomato yellow leaf curl virus disease* (pp. 67–84). The Netherlands: Springer.

Gross, M. (2013). EU ban puts spotlight on complex effects of neonicotinoids. *Current Biology, 23,* R462–R464.

Gutierrez, C. (1999). Geminivirus DNA replication. *Cellular and Molecular Life Sciences, 56,* 313–329.

Hanley-Bowdoin, L., Settlage, S. B., Orozco, B. M., Nagar, S., & Robertson, D. (1999). Geminiviruses: Models for plant DNA replication, transcription, and cell cycle regulation. *Critical Reviews in Plant Sciences, 18*, 71–106.

Hanson, P. M., Bernacchi, D., Green, S., Tanksley, S. D., Muniyappa, V., Padmaja, S., et al. (2000). Mapping a wild tomato introgression associated with *Tomato yellow leaf curl virus* resistance in a cultivated tomato line. *Journal of the American Society for Horticultural Science, 125*, 15–20.

Hanson, P., Green, S. K., & Kuo, G. (2006). Ty-2, a gene on chromosome 11 conditioning geminivirus resistance in tomato. *Tomato Genetic Cooperative Report, 56*, 17–18.

Hanssen, M. I., & Lapidot, M. (2012). Major tomato viruses in the Mediterranean basin. In G. Loebenstein, & H. Lecoq (Eds.), *Viruses and virus diseases of vegetables in the Mediterranean Basin in the series "Advances in Virus Research": Vol. 84* (pp. 31–66). UK: Academic Press.

Hanssen, I. M., Lapidot, M., & Thomma, P. H. J. B. (2010). Emerging viral diseases of tomato crops. *Molecular Plant-Microbe Interactions, 23*, 539–548.

Harpaz, I., & Cohen, S. (1965). Semipersistent relationship between *Cucumber vein yellowing virus* (CVYV) and its vector the tobacco whitefly *Bemisia tabaci*. *Phytopathologische Zeitschrift, 54*, 240–248.

Hedesh, R. M., Shams-Bakhsh, M., & Mozafari, J. (2011). Evaluation of common bean lines for their reaction to *Tomato yellow leaf curl virus*-Ir2. *Crop Protection, 30*, 163–167.

Henry, M., Béguin, M., Requier, F., Rollin, O., Odoux, J.-F., Aupinel, P., et al. (2012). A common pesticide decreases foraging success and survival in honey bees. *Science, 336*, 348–350.

Hilje, L., Costa, H. S., & Stansly, P. A. (2001). Cultural practices for managing *Bemisia tabaci* and associated viral diseases. *Crop Protection, 20*, 801–812.

Hillocks, R. J., Raya, M., & Thresh, J. M. (1996). The association between root necrosis and above ground symptoms of brown streak virus infection of cassava in Southern Tanzania. *International Journal of Pest management, 42*, 285–289.

Hillocks, R. J., Raya, M., Mtunda, K., & Kiozia, H. (2001). Effects of brown streak virus disease on yield and quality of cassava in Tanzania. *Journal of Phytopathology, 149*, 1–6.

Hogenhout, S. A., Ammar, E. D., Whitfield, A. E., & Redinbaugh, M. G. (2008). Insect-vector interactions with persistently transmitted viruses. *Annual Review of Phytopathology, 46*, 327–359.

Hutton, S. F., Scott, J. W., & Schuster, D. J. (2012). Recessive resistance to *Tomato yellow leaf curl virus* from the tomato cultivar Tyking is located in the same region as Ty-5 on chromosome 4. *HortScience, 47*, 324–327.

Idris, A. M., Smith, S. E., & Brown, J. K. (2001). Ingestion, transmission, and persistence of *Chino del tomate virus* (CdTV), a new world begomovirus, by old and new world biotypes of the whitefly vector *Bemisia tabaci*. *Annals of Applied Biology, 139*, 45–154.

Idriss, M., Abdallah, N., Aref, N., Haridy, G., & Madkour, M. (1997). Biotypes of the castor bean whitefly *Trialeurodesricini* (Misra) (Hom., Aleyrodidae) in Egypt: Biochemical characterization and efficiency of geminivirus transmission. *Journal of Applied Entomology, 121*, 501–509.

Iwaki, M., Thongmeearkon, P., Prommin, M., Honda, Y., & Hibi, J. (1982). Whitefly transmission and some properties of *Cowpea mild mottle virus* on soybean in Thailand. *Plant Disease, 66*, 265–268.

Jennings, D. L. (1994). Breeding for resistance to *African cassava mosaic virus* in East Africa. *Tropical Science, 34*, 110–122.

Jeremiah, S. (2012). *The role of whitefly (Bemisia tabaci) in the spread and transmission of cassava brown streak disease in the field*. PhD thesis (p. 229). Tanzania: University of Dar es Salaam.

Ji, Y. H., Cai, Z. D., Zhou, X. W., Liu, Y. M., Xiong, R. Y., Zhao, T. M., et al. (2012). First report of *Tomato yellow leaf curl virus* infecting common bean in China. *Plant Disease, 96*, 1229.

Ji, Y., Schuster, D. J., & Scott, J. W. (2007). Ty-3, a begomovirus resistance locus near the Tomato yellow leaf curl virus resistance locus Ty-1 on chromosome 6 of tomato. *Molecular Breeding, 20*, 271–284.

Ji, Y., & Scott, J. W. (2006). *Ty-3*, a begomovirus resistance locus linked to *Ty-1* on chromosome 6 of tomato. *Tomato Genetic Cooperative Report, 56*, 22–25.

Ji, Y., Scott, J. W., Hanson, P., Graham, E., & Maxwell, D. P. (2007). Sources of resistance, inheritance, and location of genetic loci conferring resistance to members of the tomato-infecting begomoviruses. In H. Czosnek (Ed.), *Tomato yellow leaf curl virus disease* (pp. 343–362). The Netherlands: Springer.

Ji, Y., Scott, J. W., Schuster, D. J., & Maxwell, D. P. (2009). Molecular mapping of Ty-4, a new Tomato yellow leaf curl virus resistance locus on chromosome 3 of Tomato. *Journal of the American Society for Horticultural Science, 134*, 281–288.

Jiang, Y. X., De Blas, C., Bedford, I. D., Nombela, G., & Muniz, M. (2004). Effect of *Bemisia tabaci* biotype in the transmission of *Tomato yellow leaf curl Sardinia virus* (TYLCSV-ES) between tomato and common weeds. *Spanish Journal of Agricultural Research, 2*, 115–119.

Jones, D. R. (2003). Plant viruses transmitted by whiteflies. *European Journal of Plant Pathology, 109*, 195–219.

Jupin, I., De Kouchkovsky, F., Jouanneau, F., & Gronenborn, B. (1994). Movement of tomato yellow leaf curl geminivirus (TYLCV): Involvement of the protein encoded by ORF C4. *Virology, 204*, 82–90.

Kalloo, G., & Banerjee, M. K. (1990). Transfer of tomato leaf curl virus resistance from *Lycopersicon hirsutum f. glabratum* to *L. esculentum*. *Plant Breeding, 105*, 156–159.

Kartha, K. K., & Gamborg, O. L. (1975). Elimination of cassava mosaic disease by meristem culture. *Phytopathology, 65*, 826–828.

Karyeija, R. F., Kreuze, J. F., Gibson, R. W., & Valkonen, J. P. T. (2000). Synergistic interactions of a potyvirus and a phloem-limited crinivirus in sweet potato plants. *Virology, 269*, 26–36.

Kheyr-Pour, A., Bendahmane, M., Matzeit, M., Accotto, G. P., Crespi, S., & Gronenborn, B. (1991). *Tomato yellow leaf curl virus* from Sardinia is a whitefly-transmitted monopartite geminivirus. *Nucleic Acids Research, 19*, 6763–6769.

King, A. M., Lefkowitz, Q. E., Adams, M. J., & Carstens, E. B. (Eds.). (2011). *Virus taxonomy: Ninth report of the International Committee on taxonomy of viruses*. San Diego: Academic Press.

Kreuze, J. F., Savenkov, E. I., & Valkonen, J. P. T. (2002). Complete genome sequence and analyses of the subgenomic RNAs of *Sweet potato chlorotic stunt virus* several new features for the genus *Crinivirus*. *Journal of Virology, 76*, 9260–9270.

Kunik, T., Palanichelvam, K., Czosnek, H., Citovsky, V., & Gafni, Y. (1998). Nuclear import of the capsid protein of Tomato yellow leaf curl virus (TYLCV) in plant and insect cells. *The Plant Journal, 13*, 393–399.

Kuo, Y.-W., Rojas, M. R., Gilbertson, R. L., & Wintermantel, W. M. (2007). First report of *Cucurbit yellow stunting disorder virus* in California and Arizona, in association with *Cucurbit leaf crumple virus* and *Squash leaf curl virus*. *Plant Disease, 91*, 330.

Lament, W. J. (1993). Plastic mulches for the production of vegetable crops. *HortTechnology, 3*, 35–39.

Lapidot, M. (2002). Screening common bean (*Phaseolus vulgaris*) for resistance to *Tomato yellow leaf curl virus*. *Plant Disease, 86*, 429–432.

Lapidot, M. (2007). Screening for TYLCV-resistant plants using whitefly-mediated inoculation. In H. Czosnek (Ed.), *Tomato yellow leaf curl virus disease* (pp. 329–342). The Netherlands: Springer.

Lapidot, M., Ben Joseph, R., Cohen, L., Machbash, Z., & Levy, D. (2006). Development of a scale for evaluation of *Tomato yellow leaf curl virus*-resistance level in tomato plants. *Phytopathology*, *96*, 1404–1408.

Lapidot, M., & Friedmann, M. (2002). Breeding for resistance to whitefly-transmitted geminiviruses. *The Annals of Applied Biology*, *140*, 109–127.

Lapidot, M., Friedmann, M., Lachman, O., Antignus, Y., Nahon, S., Cohen, S., et al. (1997). Comparison of resistance level to *Tomato yellow leaf curl virus* among commercial cultivars and breeding lines. *Plant Disease*, *81*, 1425–1428.

Lapidot, M., Friedmann, M., Pilowsky, M., Ben Joseph, R., & Cohen, S. (2001). Effect of host plant resistance to *Tomato yellow leaf curl virus* (TYLCV) on virus acquisition and transmission by its whitefly vector. *Phytopathology*, *91*, 1209–1213.

Lapidot, M., & Polston, J. E. (2006). Resistance to *Tomato yellow leaf curl virus* in tomato. In G. Loebenstein, & J. P. Carr (Eds.), *Natural resistance mechanisms of plants to viruses* (pp. 503–520). The Netherlands: Springer.

Larsen, R. C., Duffus, J. E., & Liu, H.-Y. (1984). Tomato necrotic dwarf virus—A new type of whitefly-transmitted virus. *Phytopathology*, *74*, 795.

Lefeuvre, P., Martin, D. P., Harkins, G., Lemey, P., Gray, A. J. A., Meredith, S., et al. (2010). The spread of *Tomato yellow leaf curl virus* from the Middle East to the world. *PLoS Pathogen*, *6*, e1001164.

Legg, J. P. (1995). *The ecology of Bemisia tabaci (Gennadius) (Homoptera), vector of African cassava mosaic geminivirus in Uganda.* Doctoral thesis, UK: University of Reading, 183 p.

Legg, J., & James, B. (2005). Whiteflies as vectors of plant viruses in cassava and sweetpotato in Africa: Conclusions and recommendations. In P. K. Anderson, & F. Morales (Eds.), *Whiteflies and whitefly-borne viruses in the tropics: Building a knowledge base for global action* (pp. 98–111). Cali, Colombia: Centro Internacional de Agricultura Tropical.

Legg, J., James, B., Cudjoe, A., Saizonou, S., Gbaguidi, B., Ogbe, F., et al. (1997). A regional collaborative approach to the study of ACMD epidemiology in sub-Saharan Africa. In E. Adipala, J. S. Tenywa, & M. W. Ogenga-Latigo (Eds.), *Proceedings of the African Crop Science Conference, Pretoria, 13–17 January, 1997* (pp. 1021–1033). Kampala, Uganda: Makerere University.

Legg, J. P., Jeremiah, S. C., Obiero, H. M., Maruthi, M. N., Ndyetabula, I., Okao-Okuja, G., et al. (2011). Comparing the regional epidemiology of the cassava mosaic and cassava brown streak pandemics in Africa. *Virus Research*, *159*, 161–170.

Legg, J. P., Kapinga, R., Teri, J., & Whyte, J. B. A. (1999). The pandemic of *Cassava mosaic virus* disease in East Africa: Control strategies and regional partnerships. *Roots*, *6*, 10–19.

Legg, J. P., Owor, B., Sseruwagi, P., & Ndunguru, J. (2006). Cassava mosaic virus disease in East and Central Africa: Epidemiology and management of a regional pandemic. *Advances in Virus Research*, *67*, 355–418.

Legg, J. P., Sseruwagi, P., Boniface, S., Okao-Okuja, G., Shirima, R., Bigirimana, S., et al. (2014). Spatio-temporal patterns of genetic change amongst populations of cassava *Bemisia tabaci* whiteflies driving virus pandemics in East and Central Africa. *Virus Research*, *186*, 61–75.

Levy, D., & Lapidot, M. (2008). Effect of plant age at inoculation on expression of genetic resistance to tomato yellow leaf curl virus. *Archives of Virology*, *153*, 171–179.

Li, M., Hu, J., Xu, F.-C., & Liu, S. S. (2010). Transmission of *Tomato yellow leaf curl virus* by two invasive biotypes and a Chinese indigenous biotype of the whitefly *Bemisia tabaci*. *International Journal of Pest Management*, *56*, 275–280.

Liu, H.-Y., Li, R. H., Wisler, G. C., & Duffus, J. E. (1997). Characterization of *Abutilon yellows virus*-a new clostero-like virus transmitted by banded-wing whitefly (*Trialeurodes abutilonea*). *Phytopathology*, *87*, S58–S59.

Liu, H.-Y., Wisler, G. C., & Duffus, J. E. (2000). Particle length of whitefly-transmitted criniviruses. *Plant Disease*, *84*, 803–805.

Loebenstein, G. (1972). Localization and induced resistance in virus-infected plants. *Annual Review of Phytopathology*, *10*, 177–206.

López-Sesé, A. I., & Gómez-Guillamón, M. L. (2000). Resistance to *Cucurbit yellowing stunting disorder virus* (CYSDV) in *Cucumis melo* L. *HortScience*, *35*, 110–113.

Lu, Y., Bei, Y., & Zhang, J. (2012). Are yellow sticky traps an effective method for control of sweetpotato whitefly, *Bemisia tabaci* in the greenhouse or field? *Journal of Insect Science*, *12*, 1–12.

Manandhar, R., Hooks, C. R. R., & Wright, M. G. (2009). Influence of cover crop and intercrop systems on *Bemisia argentifolli* (Hemiptera: Aleyrodidae) infestation and associated squash silverleaf disorder in zucchini. *Environmental Entomology*, *38*, 442–449.

Mansour, A., & Al-Musa, A. (1993). *Cucumber vein yellowing virus*; host range and virus vector relationships. *Journal of Phytopathology*, *137*, 73–79.

Marco, C. F., Aguilar, J. M., Abad, J., Gomez-Guillamon, M. L., & Aranda, M. A. (2003). Melon resistance to *Cucurbit yellow stunting disorder virus* is characterized by reduced virus accumulation. *Phytopathology*, *93*, 844–852.

Martin, J. H. (2004). Whiteflies of Belize (Hemiptera: Aleyrodidae). Part 1: Introduction and account of the subfamily Aleurodicinae Quaintance & Baker. *Zootaxa*, *681*, 1–119.

Martın, G., Velasco, L., Segundo, E., Cuadrado, I. M., & Janssen, D. (2008). The complete nucleotide sequence and genome organization of *Bean yellow disorder virus*, a new member of the genus Crinivirus. *Archives of Virology*, *153*, 999–1001.

Maruthi, M. N., Hillocks, R. J., Mtunda, K., Raya, M. D., Muhanna, M., Kiozia, H., et al. (2005). Transmission of Cassava brown streak virus by *Bemisia tabaci* (Gennadius). *Journal of Phytopathology*, *153*, 307–312.

McCreight, J. D., & Wintermantel, W. M. (2008). Potential new sources of genetic resistance in melon to *Cucurbit yellow stunting disorder virus*. In M. Pitrat (Ed.), *Cucurbitaceae 2008. Proceedings of the IXth EUCARPIA meeting on genetics and breeding of Cucurbitaceae, May 21–24th, 2008*, (pp. 173–179). Avignon (France): INRA.

McCreight, J. D., & Wintermantel, W. M. (2011). Genetic resistance in melon PI 313970 to *Cucurbit yellow stunting disorder virus*. *HortScience*, *46*, 1582–1587.

McGrath, P. F., & Harrison, B. D. (1995). Transmission of tomato leaf curl geminiviruses by *Bemisia tabaci*: Effects of virus isolate and vector biotype. *Annals of Applied Biology*, *126*, 307–316.

Menzel, W., Abang, M. M., & Winter, S. (2011). Characterization of *Cucumber vein-clearing virus*, a whitefly (*Bemisia tabaci* G.)-transmitted carlavirus. *Archives of Virology*, *156*, 2309–2311.

Mineau, P., & Palmer, C. (2013). *The impact of the nation's most widely used insecticides on birds*. http://www.abcbirds.org/abcprograms/policy/toxins/Neonic_FINAL.pdf.

Mohamed, M. A. (2012). Impact of planting dates, spaces and varieties on infestation of cucumber plants with whitefly, *Bemisia tabaci* (Genn.). *The Journal of Basic & Applied Zoology*, *65*, 17–20.

Monci, F., Garcia-Andres, S., Maldonado, J. A., & Moriones, E. (2005). Resistance to monopartite begomoviruses associated with the bean leaf crumple disease in *Phaseolus vulgaris* controlled by a single dominant gene. *Phytopathology*, *95*, 819–826.

Morales, F. J., & Jones, P. G. (2004). The ecology and epidemiology of whitefly-transmitted viruses in Latin America. *Virus Research*, *100*, 57–65.

Morales, F. J., & Niessen, A. I. (1988). Comparative responses of selected *Phaseolus vulgaris* germplasm inoculated artificially and naturally with *Bean golden mosaic virus*. *Plant Disease*, *72*, 1020–1023.

Moreno-Delafuente, A., Garzo, E., Moreno, A., & Fereres, A. (2013). A plant virus manipulates the behavior of its whitefly vector to enhance its transmission efficiency and spread. *PLoS One*, *8*, e61543. http://dx.doi.org/10.1371/journal.pone.0061543.

Morilla, G., Janssen, D., Garcia-Andres, S., Moriones, E., Cuadrado, I. M., & Bejarano, E. R. (2005). Pepper (*Capsicum annuum*) is a dead-end host for *Tomato yellow leaf curl virus*. *Phytopathology*, *95*, 1089–1097.

Moriones, E., Aramburu, J., Riudavets, J., Arno, J., & Lavina, A. (1998). Effect of plant age at time of infection by tomato spotted wilt tospovirus on the yield of field-grown tomato. *European Journal of Plant Pathology*, *104*, 295–300.

Moriones, E., & Navas-Castillo, J. (2000). *Tomato yellow leaf curl virus*, an emerging virus complex causing epidemics worldwide. *Virus Research*, *71*, 123–134.

Mound, L. A. (1962). Studies on the olfaction and colour sensitivity of *Bemisia tabaci* (GENN.) (Homoptera, Aleurodidae). *Entomologia Experimentalis et Applicata*, *5*, 99–104.

Muniyappa, Y., & Reddy, D. V. R. (1983). Transmission of cowpea mild mottle virus by Bemisia tabaci in a nonpersistent manner. *Plant Disease*, *67*, 391–393.

Mutschler, M. A., & Wintermantel, W. M. (2006). Reducing virus associated crop loss through resistance to insect vectors. In G. Loebenstein, & J. P. Carr (Eds.), *Natural resistance mechanisms of plants to viruses* (pp. 241–260). New York: Springer.

Nagata, T., Alves, D. M. T., Inoue-Nagata, A. K., Tian, T. Y., Kitajima, E. W., Cardoso, J. E., et al. (2005). A novel melon flexivirus transmitted by whitefly. *Archives of Virology*, *150*, 379–387.

Naidu, R. A., Gowda, S., Satyanarayana, T., Boyko, V., Reddy, A. S., Dawson, W. O., et al. (1998). Evidence that whitefly-transmitted *Cowpea mild mottle virus* belongs to the genus Carlavirus. *Archives of Virology*, *143*, 769–780.

Naranjo, S. E., & Ellsworth, P. C. (2009). 50 years of the integrated control concept: Moving the model and implementation forward in Arizona. *Pest Management Science*, *65*, 1267–1286.

Navas-Castillo, J., Fiallo-Olivé, E., & Sánchez-Campos, S. (2011). Emerging virus diseases transmitted by whiteflies. *Annual Review of Phytopathology*, *49*, 219–248.

Navas-Castillo, J., Sanchez-Campos, S., & Diaz, J. A. (1999). *Tomato yellow leaf virus*-Is causes a novel disease of common bean and severe epidemics in tomato in Spain. *Plant Disease*, *83*, 29–32.

Navot, N., Pichersky, E., Zeidan, M., Zamir, D., & Czosnek, H. (1991). *Tomato yellow leaf curl virus*: A whitefly-transmitted geminivirus with a single genomic component. *Virology*, *185*, 151–161.

Ng, J. C. K., & Falk, B. W. (2006). Virus-vector interactions mediating nonpersistent and semipersistent transmission of plant viruses. *Annual Review of Phytopathology*, *44*, 183–212.

Ngouajio, M., & Ernest, J. (2004). Light transmission through colored polyethylene mulches affects weed populations. *HortScience*, *39*, 1302–1304.

Nombela, G., & Muniz, M. (2010). Host plant resistance for the management of *Bemisia tabaci*: A multi-crop survey with emphasis on tomato. In P. Stansly, & S. E. Naranjo (Eds.), *Bemisia: Bionomics and management of a global pest* (pp. 357–384). New York, USA: Springer.

Nyoike, T. W., Liburd, O. E., & Webb, S. E. (2008). Suppression of whiteflies, *Bemisia tabaci* (Hemiptera : Aleyrodidae) and incidence of *Cucurbit leaf crumple virus*, a whitefly-transmitted virus of zucchini squash new to Florida, with mulches and imidacloprid. *Florida Entomologist*, *91*, 460–465.

Okuda, M., Okazaki, S., Yamasaki, S., Okuda, S., & Sugiyama, M. (2010). Host range and complete genome sequence of *Cucurbit chlorotic yellows virus*, a new member of the genus *Crinivirus*. *Phytopathology*, *100*, 560–566.

Omongo, C. A., Kawuki, R., Bellotti, A. C., Alicai, T., Baguma, Y., Maruthi, M. N., et al. (2012). African cassava whitefly, Bemisia tabaci, resistance in African and South American cassava genotypes. *Journal of Integrative Agriculture*, *11*, 327–336.

Owor, B., Legg, J. P., Okao-Okuja, G., Obonyo, R., & Ogenga-Latigo, M. W. (2004). The effect of cassava mosaic geminiviruses on symptom severity, growth and root yield of a cassava mosaic virus disease-susceptible cultivar in Uganda. *Annals of Applied Biology, 145*, 331–337.

Palanichelvam, K., Kunik, T., Citovsky, V., & Gafni, Y. (1998). The capsid protein of tomato yellow leaf curl virus binds cooperatively to single-stranded DNA. *The Journal of General Virology, 79*, 2829–2833.

Peralta, I. E., Knapp, S., & Spooner, D. M. (2005). New species of wild tomatoes (*Solanum* Section *Lycopersicon*: Solanaceae) from Northern Peru. *Systematic Botany, 30*, 424–434.

Petty, I. T. D., Coutts, R. H. A., & Buck, K. W. (1988). Transcriptional mapping of the coat protein gene of *Tomato golden mosaic virus*. *Journal of General Virology, 69*, 1359–1365.

Pico, B., Diez, M., & Nuez, F. (1998). Evaluation of whitefly-mediated inoculation techniques to screen *Lycopersicon esculentum* and wild relatives for resistance to *Tomato yellow leaf curl virus*. *Euphytica, 101*, 259–271.

Polston, J. E., & Capobianco, H. (2013). Transmitting plant viruses using whiteflies. 2013. *Journal of Visualized Experiments, 81*, e4332. http://dx.doi.org/10.3791/4332.

Polston, J. E., Cohen, L., Sherwood, T. A., Ben-Joseph, R., & Lapidot, M. (2006). Capsicum Species: Symptomless Hosts and Reservoirs of *Tomato yellow leaf curl virus*. *Phytopathology, 96*, 447–452.

Polston, J. E., De Barro, P., & Boykin, L. M. (2014). Transmission specificities of plant viruses with the newly identified species of the *Bemisia tabaci* species complex. *Pest Management Science*. http://dx.doi.org/10.1002/ps.3738.

Polston, J. E., & Lapidot, M. (2007). Management of *Tomato yellow leaf curl virus*: US and Israel perspectives. In H. Czosnek (Ed.), *Tomato yellow leaf curl virus disease* (pp. 251–262). The Netherlands: Springer.

Pospieszny, H., Borodynko, N., Obrepalska-Steplowska, A., & Hasiow, B. (2007). The first report of *Tomato torrado virus* in Poland. *Plant Disease, 91*, 1364.

Rabbi, I. Y., Hamblin, M. T., Kumar, P. L., Gedil, M. A., Ikpan, A. S., Jannink, J. L., et al. (2014). High-resolution mapping of resistance to cassava mosaic geminiviruses in cassava using genotyping-by-sequencing and its implications for breeding. *Virus Research, 186*, 87–96.

Rigden, J. E., Krake, L. R., Rezaian, M. A., & Dry, I. B. (1994). ORF C4 of tomato leaf curl geminivirus is a determinant of symptom severity. *Virology, 204*, 847–850.

Rodríguez-López, M. J., Garzo, E., Bonani, J. P., Fereres, A., Fernández-Muñoz, R., & Moriones, E. (2011). Whitefly resistance traits derived from the wild tomato *Solanum pimpinellifollium* affect the preference and feeding behavior of *Bemisia tabaci* and reduce the spread of tomato yellow leaf curl virus. *Phytopathology, 101*, 1191–1201.

Rojas, M. R., Jiang, H., Salati, R., Xoconostle-Cázares, B., Sudarshana, M. R., Lucas, W. J., et al. (2001). Functional analysis of proteins involved in movement of the monopartite begomovirus, *Tomato yellow leaf curl virus*. *Virology, 291*, 110–125.

Rosario, K., Capobianco, H., Ng, T. F. F., Breitbart, M., & Polston, J. E. (2014). Metagenomic analysis of DNA and RNA viruses in whiteflies leads to the discovery and characterization of *Cowpea mild mottle virus* in Florida. *PLoS One, 9*, e86748. http://dx.doi.org/10.1371/journal.pone.0086748.

Rubinstein, G., & Czosnek, H. (1997). Long-Term association of *Tomato Yellow Leaf Curl Virus* with its whitefly vector *Bemisia tabaci*—Effect on the insect transmission capacity, longevity and fecundity. *Journal of General Virology, 78*, 2683–2689.

Rubio, L., Abou-Jawdah, Y., Lin, H.-X., & Falk, B. W. (2001). Geographically distant isolates of the crinivirus *Cucurbit yellow stunting disorder virus* show very low genetic diversity in the coat protein gene. *The Journal of General Virology, 82*, 929–933.

Salazar, L. F., Muller, G., Querci, M., Zapata, J. L., & Owens, R. A. (2000). Potato yellow vein virus: Its host range, distribution in South America and identification as a crinivirus transmitted by *Trialeurodes vaporariorum*. *The Annals of Applied Biology, 137*, 7–19.

Sanchez Campos, S., Navas Castillo, J., Camero, R., Soria, C., Diaz, J. A., & Moriones, E. (1999). Displacement of *Tomato yellow leaf curl virus* (TYLCV)-Sr by TYLCV-Is in tomato epidemics in Spain. *Phytopathology, 89*, 1038–1043.

Schuster, D. J. (2004). Squash as a trap crop to protect tomato from whitefly-vectored tomato yellow leaf curl. *International Journal of Pest Management, 50*, 281–284.

Schuster, D. J., Stansly, P. A., & Polston, J. E. (1996). Expressions of plant damage of Bemisia. In D. Gerling, & R. T. Mayer (Eds.), *Bemisia1995: Taxonomy, biology, damage, control and management* (pp. 153–156). UK: Intercept.

Segundo, E., Carmona, M. P., Saez, E., Velasco, L., Martin, G., Ruiz, L., et al. (2008). Occurrence and incidence of viruses infecting green beans in south-eastern Spain. *European Journal of Plant Pathology, 122*, 579–591.

Simmons, A. M., Kousik, C. S., & Levi, A. (2010). Combining reflective mulch and host plant resistance for sweet potato whitefly (Hemiptera: Aleyrodidae) management in watermelon. *Crop Protection, 29*, 898–902.

Simmons, A. M., & Levi, A. (2002). Sources of whitefly (Homoptera: Aleyrodidae) resistance in *Cirrullus* for the improvement of cultivated watermelon. *HortScience, 37*, 581–584.

Simmons, A. M., & McCreight, J. D. (1996). Evaluation of melon for resistance to *Bemisia argentifolii* (Homoptera: Aleyrodidae). *Journal of Economic Entomology, 89*, 1663–1668.

Sinclair, J. W. (2003). *Screening for resistance to cucurbit yellow stunting disorder virus, gummy stem blight, and monosporascus root rot and detection of RAPD markers associated with QTL for soluble solids, sugars, and vitamin C in melon (Cucumis melo L.)*. PhD Diss., Dept. Hort. Sci., Texas A&M Univ., College Station.

Smith, J. (2014). *The Great Lakes Cassava Initiative: Case study—GLCI distribution of cassava planting material and consequences of testing for cassava brown streak disease*. http://www.crsprogramquality.org/storage/pubs/agenv/glci-case-study-cassava-distribution.pdf, (accessed March 12, 2014).

Smith, H. A., Koenig, R. L., McAuslane, H. J., & McSorley, R. (2000). Effect of silver reflective mulch and a summer squash trap crop on densities of immature *Bemisia argentifolii* (Homoptera: Aleyrodidae) on organic bean. *Journal of Economic Entomology, 93*, 726–731.

Soria, C., López-Sesé, A. I., & Gómez-Guillamón, M. L. (1999). Resistance of *Cucumis melo* against *Bemisia tabaci* (Homoptera: Aleyrodidae). *Environmental Entomology, 28*, 831–835.

Srinivasan, R., Riley, D., Diffie, S., Sparks, A., & Adkins, S. (2012). Whitefly population dynamics and evaluation of whitefly-transmitted *Tomato yellow leaf curl virus* (TYLCV)-resistant tomato genotypes as whitefly and TYLCV reservoirs. *Journal of Economic Entomology, 105*, 1447–1456.

Stansly, P. A., & Natwick, E. T. (2010). Integrated systems for managing *Bemisia tabaci* in protected and open field agriculture. In P. A. Stansly, & S. E. Naranjo (Eds.), *Bemisia: Bionomics and management of a global pest* (pp. 467–497). Dordrecht, Heidelberg, London, New York: Springer.

Storey, H. H. (1936). Virus diseases on East African plants. VI. A progress report on the studies of the diseases of cassava. *East African Agricultural Journal, 2*, 34–39.

Su, Q., Pan, H., Liu, B., Chu, D., Xie, W., Wu, Q., et al. (2012). Insect symbiont facilitates vector acquisition, retention, and transmission of plant virus. *Scientific Reports, 3*, 1–6. http://dx.doi.org/10.1038/srep01367.

Summers, C. G., & Stapleton, J. J. (2002). Use of UV reflective mulch to delay the colonization and reduce the severity of *Bemisia argentifolii* (Homoptera: Aleyrodidae) infestations in cucurbits. *Crop Protection, 21*, 921–928.

Sunter, G., Hartitz, M. D., Hormuzdi, S. G., Brough, C. L., & Bisaro, D. M. (1990). Genetic analysis of tomato golden mosaic virus: ORF AL2 is required for coat protein accumulation while ORF AL3 is necessary for efficient DNA replication. *Virology, 179*, 69–77.

Taylor, N. J., Halsey, M., Gaitán-Solís, E., Anderson, P., Gichuki, S., Miano, D., et al. (2012). The VIRCA Project: Virus resistant cassava for Africa. *GM Crops & Food*, *3*, 93–103.

Thresh, J. M., & Cooter, R. J. (2005). Strategies for controlling cassava mosaic virus disease in Africa. *Plant Pathology*, *54*, 587–614.

Thresh, J. M., Fargette, D., & Otim-Nape, G. W. (1994). Effects of African cassava mosaic geminivirus on the yield of cassava. *Tropical Science*, *34*, 26–42.

Thresh, J. M., Otim-Nape, G. W., & Jennings, D. L. (1994). Exploiting resistance to *African cassava mosaic virus*. *Aspects of Applied Biology*, *39*, 51–60.

Thresh, J. M., Otim-Nape, G. W., Legg, J. P., & Fargette, D. (1997). African cassava mosaic virus disease: The magnitude of the problem. *African Journal of Root and Tuber Crops*, *2*, 13–18.

Tzanetakis, I. E., Halgren, A. B., Keller, K. E., Hokanson, S. C., Maas, J. L., McCarthy, P. L., et al. (2004). Identification and detection of a virus associated with strawberry pallidosis disease. *Plant Disease*, *88*, 383–390.

Tzanetakis, I. E., Martin, R. R., & Wintermantel, W. M. (2013). Epidemiology of criniviruses: An emerging problem in world agriculture. *Frontiers in Microbiology*, *119*, 1–15. http://dx.doi.org/10.3389/fmicb.2013.00119.

Tzanetakis, I. E., Wintermantel, W. M., Cortez, A. A., Barnes, J. E., Barrett, S. M., Bolda, M. P., et al. (2006). Epidemiology of *Strawberry pallidosis associated virus* and occurrence of pallidosis disease in North America. *Plant Disease*, *90*, 1343–1346.

Untiveros, M., Fuentes, S., & Salazar, L. F. (2007). Synergistic interaction of *Sweet potato chlorotic stunt virus* (*Crinivirus*) with carla-, cucumo-, ipomo-, and potyviruses infecting sweet potato. *Plant Disease*, *91*, 669–676.

Upadhyay, S. K., Chandrashekar, K., Thakur, N., Verma, P. C., Borgio, J. F., Singh, P. K., et al. (2011). RNA interference for the control of whiteflies (*Bemisia tabaci*) by oral route. *Journal of Biosciences*, *36*, 153–161.

Verbeek, M., Dullemans, A. M., van den Heuvel, J. F., Maris, P. C., & van der Vlugt, R. A. (2007). Identification and characterisation of *Tomato torrado virus*, a new plant picorna-like virus from tomato. *Archives of Virology*, *152*, 881–890.

Verbeek, M., van Bekkum, P. J., Dullemans, A. M., & van der Vlugt, R. A. (2014). Torradoviruses are transmitted in a semi-persistent and stylet-borne manner by three whitefly vectors. *Virus Research*, *186*, 55–60.

Verlaan, M. G., Hutton, S. F., Ibrahem, R. M., Kormelink, R., Visser, R. G., Scott, J. W., et al. (2013). The *Tomato yellow leaf curl virus* resistance genes Ty-1 and Ty-3 are allelic and code for DFDGD-class RNA-dependent RNA polymerases. *PLoS Genetics*, *9*, e1003399.

Verlaan, M. G., Szinay, D., Hutton, S. F., de Jong, H., Kormelink, R., Visser, R. G., et al. (2011). Chromosomal rearrangements between tomato and *Solanum chilense* hamper mapping and breeding of the TYLCV resistance gene *Ty-1*. *The Plant Journal*, *68*, 1093–1103.

Vidavsky, F. S. (2007). Exploitation of resistance genes found in wild tomato species to produce resistant cultivars; pile up of resistant genes. In H. Czosnek (Ed.), *Tomato yellow leaf curl virus disease* (pp. 363–372). The Netherlands: Springer.

Vidavsky, F., Leviatov, S., Milo, J., Rabinowitch, H. D., Kedar, N., & Czosnek, H. (1998). Response of tolerant breeding lines of tomato, *Lycopersicon esculentum*, originating from three different sources (*L. peruvianum, L. pimpinellifolium and L. chilense*) to early controlled inoculation by *tomato yellow leaf curl virus* (TYLCV). *Plant Breeding*, *117*, 165–169.

Walsh, S. (2012). *Seed system innovations in the Great Lakes Cassava Initiative*. http://www.crsprogramquality.org/storage/pubs/agenv/glci-case-study-seed-system-innovations.pdf (accessed March 11, 2014).

Wartig, L., Kheyr-Pour, A., Noris, E., De Kouchkovsky, F., Jouanneau, F., Gronenborn, B., et al. (1997). Genetic analysis of the monopartite tomato yellow leaf curl geminivirus: Roles of V1, V2, and C2 ORFs in viral pathogenesis. *Virology, 228*, 132–140.

Webb, S. E., Adkins, S., & Reitz, S. R. (2012). Semipersistent whitefly transmission of *Squash vein yellowing virus*, causal agent of viral watermelon vine decline. *Plant Disease, 96*, 839–844.

Weintraub, P. G., & Berlinger, M. J. (2004). Physical control in greenhouse and field crops. In A. R. Horowitz, & I. Ishaaya (Eds.), *Insect pest management* (pp. 302–318). Berlin: Springer.

Whitehorn, P. R., O'Connor, S., Wackers, F. L., & Goulson, D. (2012). Neonicotinoid pesticide reduces bumble bee colony growth and queen production. *Science, 336*, 351–352.

Williamson, S. M., & Wright, G. A. (2013). Exposure to multiple cholinergic pesticides impairs olfactory learning and memory in honeybees. *The Journal of Experimental Biology, 216*, 1799–1807.

Winter, S., Koerbler, M., Stein, B., Pietruszka, A., Paape, M., & Butgereitt, A. (2010). Analysis of cassava brown streak viruses reveals the presence of distinct virus species causing cassava brown streak disease in East Africa. *Journal of General Virology, 91*, 1365–1372.

Winter, S., Purac, A., Leggett, F., Frison, E. A., Rossel, H. W., & Hamilton, R. I. (1992). Partial characterization and molecular cloning of a closterovirus from sweet potato infected with sweet potato virus disease complex from Nigeria. *Phytopathology, 82*, 869–875.

Wintermantel, W. M. (2010). Transmission efficiency and epidemiology of criniviruses. In P. Stansly, & S. E. Naranjo (Eds.), *Bemisia: Bionomics and Management of a Global Pest* (pp. 319–331). New York, USA: Springer.

Wintermantel, W. M., Gilbertson, R. L., & Natwick, E. T. (2014). Epidemiology of emergent *Cucurbit yellow stunting disorder virus* in Imperial Valley California and evaluation of potential reservoir hosts. *Phytopathology*, (submitted).

Wintermantel, W. M., & Hladky, L. L. (2013). Genome characterization of Tomato necrotic dwarf virus, a Torradovirus from southern California. *Phytopathology, 103*, S160.

Wintermantel, W. M., Hladky, L. L., Cortez, A. A., & Natwick, E. T. (2009). A new expanded host range of *Cucurbit yellow stunting disorder virus* includes three agricultural crops. *Plant Disease, 93*, 685–690.

Wintermantel, W. M., Hladky, L. L., Gulati-Sakhuja, A., Li, R., Liu, H.-Y., & Tzanetakis, I. E. (2009). The complete nucleotide sequence and genome organization of *Tomato infectious chlorosis virus*: A distinct crinivirus most closely related to lettuce infectious yellows virus. *Archives of Virology, 154*, 1335–1341.

Wisler, G. C., & Duffus, J. E. (2001). Transmission properties of whitefly-borne criniviruses and their impact on virus epidemiology. In K. F. Harris, O. P. Smith, & J. E. Duffus (Eds.), *Virus-insect-plant interactions* (pp. 293–308). San Diego, CA, USA: Academic Press.

Wisler, G. C., Duffus, J. E., Liu, H.-Y., & Li, R. H. (1998). Ecology and epidemiology of whitefly-transmitted closteroviruses. *Plant Disease, 82*, 270–280.

Wu, J. Y., Smart, M. D., Anelli, C. A., & Sheppard, W. S. (2012). Honey bees (*Apis mellifera*) reared in brood combs containing high levels of pesticide residues exhibit increased susceptibility to Nosema (Microsporidia) infection. *Journal of Invertebrate Pathology, 109*, 326–329.

Xi-Shu, G., Wen-Jun, B., Wei-Hong, X., Yi-Chuan, B., Bai-Ming, L., & Tong-Xian, L. (2008). Population suppression of *Bemisia tabaci* (Hemiptera: Aleyrodidae) using yellow sticky traps and *Eretmocerus* nr. *rajasthanicus* (Hymenoptera: Aphelinidae) on tomato plants in greenhouses. *Insect Science, 15*, 263–270.

Yabeja, J. W., Mtunda, K., Shirima, R., Kanju, E., & Legg, J. P. (2013). Clean seed systems: An intervention for the management of cassava virus diseases in sub-Saharan Africa. In *12th symposium of the International Society of Tropical Root Crops—Africa Branch, 30th September—5th October, 2013, Accra, Ghana.*

Yadav, J. S., Ogwok, E., Wagaba, H., Patil, B. L., Bagewadi, B., Alicai, T., et al. (2011). RNAi-mediated resistance to *Cassava brown streak Uganda virus* in transgenic cassava. *Molecular Plant Pathology*, *12*, 677–687.

Zamir, D., Ekstein-Michelson, I., Zakay, Y., Navot, N., Zeidan, M., Sarfatti, M., et al. (1994). Mapping and introgression of a *Tomato yellow leaf curl virus* tolerance gene, TY-1. *Theoretical and Applied Genetics*, *88*, 141–146.

Zrachya, A., Glick, E., Levy, Y., Arazi, T., Citovsky, V., & Gafni, Y. (2007). Suppressor of RNA silencing encoded by *Tomato yellow leaf curl virus*-Israel. *Virology*, *358*, 159–165.

CHAPTER FOUR

Control of Plant Virus Diseases in Cool-Season Grain Legume Crops

Khaled M. Makkouk*, Safaa G. Kumari†, Joop A.G. van Leur‡, Roger A.C. Jones§,¶,1

*National Council for Scientific Research, Beirut, Lebanon
†International Centre for Agricultural Research in the Dry Areas (ICARDA), Tunis, Tunisia
‡NSW Department of Primary Industries, Tamworth Agricultural Institute, New South Wales, Australia
§School of Plant Biology and Institute of Agriculture, University of Western Australia, Nedlands, Western Australia, Australia
¶Department of Agriculture and Food, South Perth, Western Australia, Australia
[1]Corresponding author: e-mail address: roger.jones@agric.wa.gov.au

Contents

Advances in Virus Research, Volume 90
ISSN 0065-3527
http://dx.doi.org/10.1016/B978-0-12-801246-8.00004-4

Abstract

Cool-season grain legume crops become infected with a wide range of viruses, many of which cause serious diseases and major yield losses. This review starts by discussing which viruses are important in the principal cool-season grain legume crops in different parts of the world, the losses they cause and their economic impacts in relation to control. It then describes the main types of control measures available: host resistance, phytosanitary measures, cultural measures, chemical control, and biological control. Examples are provided of successful deployment of the different types of measures to control virus epidemics in cool-season grain legume crops. Next it emphasizes the need for integrated approaches to control because single control measures used alone rarely suffice to adequately reduce virus-induced yield losses in these crops. Development of effective integrated disease management (IDM) strategies depends on an interdisciplinary team approach to (i) understand the ecological and climatic factors which lead to damaging virus epidemics and (ii) evaluate the effectiveness of individual control measures. In addition to using virus-resistant cultivars, other IDM components include sowing virus-tested seed stocks, selecting cultivars with low seed transmission rates, using diverse phytosanitary or cultural practices that minimize the virus source or reduce its spread, and using selective pesticides in an environmentally responsible way. The review finishes by briefly discussing the implications of climate change in increasing problems associated with control and the opportunities to control virus diseases more effectively through new technologies.

1. INTRODUCTION

Lentil (*Lens culinaris* Med.), chickpea (*Cicer arietinum* L.), faba bean (*Vicia faba* L.), pea (*Pisum sativum* L.), and lupin (*Lupinus* spp.) are infected naturally by more than 70 different viruses worldwide, and the number of viruses involved continues to increase (Boiteux, de Avila, Giordano, Lima, & Kitajima, 1995; Bos, Hampton, & Makkouk, 1988; Hampton, 1984; Jones & McLean, 1989; Kumar, Kumari, & Waliyar, 2008; Kumari, Larsen, Makkouk, & Bashir, 2009; Kumari & Makkouk, 2007; Makkouk, 1994; Makkouk, Kumari, Hughes, Muniyappa, & Kulkarni, 2003; Makkouk, Pappu, & Kumari, 2012; Schwinghamer, Johnstone, & Johnston-Lord, 1999; Schwinghamer, Thomas, Parry, Schilg, & Dann, 2007; Thomas, Parry, Schwinghamer, & Dann, 2010; Thomas et al., 2004). However, at any one specific location, normally, only a few of these viruses are of major economic concern. Yield losses resulting from virus attack vary greatly, from little or none with *Vicia cryptic virus* (VCV, genus *Alphacryptovirus*, family *Partitiviridae*) to complete crop failure when conditions permit widespread virus infection at the vulnerable early plant growth stage, as with *Faba bean necrotic yellows virus* (FBNYV, genus *Nanovirus*, family *Nanoviridae*) and some members of the family *Luteoviridae* (Makkouk et al., 1994). It is not intended here to review all viruses that infect the full range of cool-season grain legume crops, as other comprehensive publications already cover this (e.g., Boswell & Gibbs, 1983; Bos et al., 1988; Jones & McLean, 1989; Loebenstein & Thottappilli, 2003; Makkouk & Kumari, 2009; Makkouk, Kumari, et al., 2003; Makkouk et al., 2012). A summary of the main features of viruses reported to infect lentil, chickpea, faba bean, pea, and lupin naturally in three major world regions is presented in Table 4.1. In this review, the focus will be on management of viruses affecting lentil, chickpea, faba bean, pea, and lupin with examples mainly from Mediterranean and temperate climatic regions. However, management of virus diseases in these cool-season grain legumes growing under cool conditions, e.g., at higher altitude, in regions of the world with subtropical or tropical climates, follows similar principles.

2. SURVEYS, IMPORTANCE, LOSSES, ECONOMICS IN RELATION TO VIRUS CONTROL AND CONTROL METHODS

Information on virus incidence and the seed yield and quality losses that virus infection causes is necessary to develop an understanding of their economic importance in production of cool-season grain legume crops in different regions of the world. Obtaining such information requires large-scale surveys to establish virus incidence and field experiments to quantify seed yield and quality losses. The understanding obtained is crucial when decisions over deploying control measures are taken, as it is used in determining the "economic threshold" above which the expense and effort of deploying control measures is justified against a particular virus disease (Jones, 2004).

Large-scale surveys done over the last two decades in Central and West Asian and North African (CWANA) countries showed that the most widespread viruses in lentil, chickpea, and faba bean are FBNYV, *Chickpea chlorotic stunt virus* (CpCSV, genus *Polerovirus*, family *Luteoviridae*), *Bean leafroll virus* (BLRV, genus *Luteovirus*, family *Luteoviridae*), *Beet western yellows virus* (BWYV, genus *Polerovirus*, family *Luteoviridae*), *Soybean dwarf virus* (SbDV, genus *Luteovirus*, family *Luteoviridae*), *Chickpea chlorotic dwarf virus* (CpCDV,

Table 4.1 The main features of viruses reported to naturally infect lentil, chickpea, faba bean, pea, and lupin in Asia (As), Africa (Af), and Australia (Au)

Virus	Crop infected	Manner of transmission	Virus particle shape and size (nm) diameter/length	Occurrence	References
Alfalfa mosaic virus (AMV, genus *Alfamovirus*, family *Bromoviridae*)	FB, Cp, Le, P, Lu	Sa, Se,[a] Aphids NP	Ba, 30–57	As, Af, Au	Abraham and Makkouk (2002), Bekele et al. (2005), El-Attar, Ghabrial, and Nour Eldin (1971), El-Muadhidi et al. (2001), Fortass and Bos (1991), Ismail and Hassan (1995), Jaspars and Bos (1980), Jones and Coutts (1996), Latham and Jones (2001b), Makkouk et al. (1994), Makkouk, Bahamish, Kumari, and Lotf (1998), Makkouk, Bashir, Jones, and Kumari (2001), Makkouk, Kumari, Shahraeen, et al. (2003), Nour and Nour (1962), Kaiser et al. (1968), Salama and El-Behadli (1979), Tadesse et al. (1999)
Bean leafroll virus (BLRV, genus *Luteovirus*, family *Luteoviridae*)	FB, P, Le, Cp	Aphids P	Is, 27	As, Af, Au	Ait Yahia et al. (1999), Ashby (1984), Bekele et al. (2005), El-Muadhidi et al. (2001), Fortass and Bos (1991), Horn, Makkouk, Kumari, van den Heuvel, and Reddy (1995), Makkouk et al. (1994), Makkouk, Bos, Azzam, Kumari, and Rizkallah (1988), Makkouk, Kumari, Shahraeen, et al. (2003), Mustafayev, Kumari, Attar, and Zeynal (2011), Najar et al. (2000), Najar, Kumari, Attar, and Lababidi (2011)

Bean yellow mosaic virus (BYMV, genus *Potyvirus*, family *Potyviridae*)	FB, P, Cp, Le, Lu	Sa, Se, Aphids NP	Fi, 750	As, Af, Au	Abraham et al. (2000), Aftab, Mughal, and Ghafoor (1989), Ait Yahia, Aitouada, Illoul, and Tair (1997), Al-Musa, Al-Haj, Mansour, and Janakat (1987), Bayaa et al. (1998), Bekele et al. (2005), Bos (1970), Cheng and Jones (1999), El-Muadhidi et al. (2001), Fortass and Bos (1991), Gamal-Eldin, El-Amrety, Mazyad, and Rizkallah (1982), Jones and McLean (1989), Latham and Jones (2001a), Makkouk et al. (1994), Makkouk, Kumari, and Al-Daoud (1992), Makkouk, Lesemann, and Haddad (1982), Makkouk, Bahamish, et al. (1998), Makkouk, Bos, Azzam, et al. (1988), Makkouk, Kumari, Shahraeen, et al. (2003), McKirdy, Jones, Latham, and Coutts (2000), Najar et al. (2000), Russo, Kishtah, and Tolba (1981), Shiying et al. (2007), Tadesse et al. (1999), Younis, Shagrun, and Khalil (1992)
Beet western yellows virus (BWYV, genus *Polerovirus*, family *Luteoviridae*)	FB, Cp, Le	Aphids P	Is, 26	As, Af, Au	Bekele et al. (2005), Coutts, Hawkes, and Jones (2006), Duffus (1972), El-Muadhidi et al. (2001), Fortass, van der Wilk, van den Heuvel, and Goldbach (1997), Horn et al. (1995), Johnstone and Duffus (1984), Latham and Jones (2001a), Mustafayev et al. (2011), Kumari et al. (2008), Najar et al. (2000), Makkouk, Bashir, Jones, et al. (2001), Makkouk, Kumari, Shahraeen, et al. (2003), Shiying et al. (2007), Tadesse et al. (1999)

Continued

Table 4.1 The main features of viruses reported to naturally infect lentil, chickpea, faba bean, pea, and lupin in Asia (As), Africa (Af), and Australia (Au)—cont'd

Virus	Crop infected	Manner of transmission	Virus particle shape and size (nm) diameter/length	Occurrence	References
Broad bean mottle virus (BBMV, genus *Bromovirus*, family *Bromoviridae*)	FB, Le, Cp	Sa, Se, Beetles	Is, 26	As, Af	Al-Ani and El-Azzawi (1987), Bekele et al. (2005), Fortass and Diallo (1993), Fortass and Bos (1991), Gibbs (1972), Makkouk, Bos, Azzam, et al. (1988), Makkouk, Bos, Rizkallah, Assam, Katul (1988), Makkouk, Hamed, Hussein, and Kumari (2003), Murant, Abu Salih, and Goold (1974), Najar et al. (2000), Zagh and Ferault (1980)
Broad bean stain virus (BBSV, genus *Comovirus*, family *Comoviridae*)	FB, Le, P	Sa, Se, Beetles	Is, 25	As, Af	Abraham et al. (2000), Allam, Gamal Eldin, and Riskallah (1979), Bayaa et al. (1998), Fortass and Bos (1991), Gibbs and Smith (1979), Makkouk, Bos, Azzam, Katul, and Rizkallah (1987), Makkouk, Bos, Azzam, et al. (1988), Makkouk et al. (1992), Makkouk, Hamed, Hussein, et al. (2003), Makkouk, Kumari, Shahraeen, et al. (2003), Najar et al. (2000)
Broad bean true mosaic virus (BBTMV, genus *Comovirus*, family *Comoviridae*)	FB	Sa, Se, Beetles	Is, 25	As, Af	Abraham et al. (2000), Fortass and Bos (1991), Gibbs and Paul (1970), Makkouk, Bos, Azzam, et al. (1988), Mazyad, El-Hammady, and Tolba (1975), Nienhaus and Saad (1967)

Broad bean wilt virus (BBWV, genus *Fabavirus* family *Comoviridae*)	FB, Le	Sa, Se, Aphids NP	Is, 25	As, Af, Au	Al-Musa, Al-Haj, and Monayer (1986), Buchen-Osmond, Crabtree, Gibbs, and McLean, (1988), El-Muadhidi et al. (2001), Makkouk, Bos, Azzam, et al. (1988), Makkouk, Kumari, and Bos (1990), Makkouk et al. (1992), Makkouk et al. (1994), Parvin and Izadpanah (1978), Tadesse et al. (1999)
Chickpea chlorotic dwarf virus (CpCDV, genus *Mastrevirus*, family *Geminiviridae*)	FB, Cp, Le	Leafhoppers	Ge, 15 × 25	As, Af	Abraham et al. (2000), El-Muadhidi et al. (2001), Hamed and Makkouk (2002), Horn, Reddy, Roberts, and Reddy (1993), Kumari, Makkouk, Attar, Ghulam, and Lesemann (2004), Makkouk, Dafalla, Hussein, and Kumari (1995), Makkouk, Bahamish, et al. (1998), Makkouk, Bashir, Jones, et al. (2001), Makkouk, Kumari, Shahraeen, et al. (2003), Makkouk, Rizkallah, Kumari, Zaki, and Abul Enein (2003)
Chickpea chlorotic stunt virus (CpCSV, genus *Polerovirus*, family *Luteoviridae*)	FB, Cp, Le, P	Aphids P	Is, 28	As, Af	Abraham, Menzel, Lesemann, Varrelmann, and Vetten (2006), Abraham, Menzel, Varrelmann, and Vetten (2009), Asaad et al. (2009), Kumari, Makkouk, Asaad, Attar, and Hlaing Loh (2007), Kumari et al. (2008), Mustafayev et al. (2011), Najar et al. (2011)

Continued

Table 4.1 The main features of viruses reported to naturally infect lentil, chickpea, faba bean, pea, and lupin in Asia (As), Africa (Af), and Australia (Au)—cont'd

Virus	Crop infected	Manner of transmission	Virus particle shape and size (nm) diameter/length	Occurrence	References
Cucumber mosaic virus (CMV, genus *Cucumovirus*, family *Bromoviridae*)	FB, P, Le, Cp, Lu	Sa, Se, Aphids NP	Is, 28	As, Af, Au	Abraham and Makkouk (2002), Bashir and Malik (1993), Dhingra, Chenulu, and Verma (1979), El-Maataoui and El-Hassani (1984), El-Muadhidi et al. (2001), Francki, Mossop, and Hatta (1979), Ismail and Hassan (1995), Jones and Coutts (1996), Jones and McLean (1989), Kaiser, Danesh, Okhovat, and Mossahebi (1968), Latham, Jones, and McKirdy (2001), Makkouk, Bos, Azzam, et al. (1988), Makkouk, Bashir, Jones, et al. (2001), Milles and Ahmed (1984), Mouhanna, Makkouk, and Ismail (1994), Najar et al. (2000), Rangaraju and Chenulu (1981), Shiying et al. (2007), Tadesse et al. (1999)
Faba bean necrotic yellows virus (FBNYV, genus *Nanovirus*, family *Nanoviridae*)	FB, Cp, Le, P	Aphids P	Is, 18	As, Af	Fadel, Khalil, and Shagrun (2005), Al-Nsour, Mansour, Al-Musa, and Salem (1998), Bekele et al. (2005), El-Amri (1999), El-Muadhidi et al. (2001), Katul et al. (1993), Kumari et al. (2008), Mustafayev et al. (2011), Makkouk et al. (1994), Makkouk, Bahamish, et al. (1998), Makkouk, Bashir, Jones, et al. (2001), Makkouk, Hamed, Hussein, et al. (2003), Makkouk, Kumari, Shahraeen, et al. (2003), Najar et al. (2000, 2011), Tadesse et al. (1999), Zidan, Khalil, and Shagrun (2002)

Milk vetch dwarf virus (MDV, genus *Nanovirus*, family *Nanoviridae*)	FB, P	Aphids P	Is, 26	As	Inouye, Inouye, and Mitsuhata (1968), Kumari et al. (2010), Sano, Wada, Hashimoto, Matsumoto, and Kojima (1998)
Pea early browning virus (PEBV, genus *Tobravirus*, family unassigned)	FB	Sa, Se, Nematodes	Short rod 105–215	Af	Bos, Mahir, and Makkouk (1993), Harrison (1973), Lockhart and Fischer (1976), Mahir, Fortass, and Bos (1992), Makkouk and Kumari (1998)
Pea enation mosaic virus-1 (PEMV-1, genus *Enamovirus*, family *Luteoviridae*)	FB, P, Le, Cp	Sa, Se, Aphids P	Is, 28	As, Af	Farzadfar and Izadpanah (2001), Fortass and Bos (1991), Makkouk, Bos, Azzam, et al. (1988), Makkouk, Kumari, and Bayaa (1999), Makkouk, Kumari, Sarker, and Erskine (2001), Makkouk, Hamed, Hussein, et al. (2003), Makkouk, Kumari, Shahraeen, et al. (2003), Peters (1982), Tadesse et al. (1999)
Pea seed-borne mosaic virus (PSbMV, genus *Potyvirus*, family *Potyviridae*)	FB, P, Cp, Le	Sa, Se, Aphids NP	Fi, 770	As, Af, Au	Abraham et al. (2000), Allam et al. (1979), Bayaa et al. (1998), Fortass and Bos (1991), Hampton and Mink (1975), Kaiser et al. (1968), Kumari and Makkouk (2007), Latham and Jones (2001b), Makkouk, Bos, Azzam, et al. (1988), Makkouk, Bos, Horn, and Srinivasa Rao (1993), Makkouk, Bashir, Jones, et al. (2001), Mustafayev et al. (2011), Shiying et al. (2007), Tadesse et al. (1999)

Continued

Table 4.1 The main features of viruses reported to naturally infect lentil, chickpea, faba bean, pea, and lupin in Asia (As), Africa (Af), and Australia (Au)—cont'd

Virus	Crop infected	Manner of transmission	Virus particle shape and size (nm) diameter/length	Occurrence	References
Soybean dwarf virus (SbDV, genus *Luteovirus*, family *Luteoviridae*)	FB, Le, Cp	Aphids P	Is, 25	As, Af, Au	Bekele et al. (2005), El-Muadhidi et al. (2001), Johnstone, Liu, and Duffus (1984), Johnstone and Guy (1986), Makkouk et al. (1997), Makkouk, Kumari, and van Leur (2002), Makkouk, Kumari, Shahraeen, et al. (2003), Najar, Kumari, Makkouk, and Daaloul (2003), Tadesse et al. (1999)
Subterranean clover stunt virus (SCSV, genus *Nanovirus*, family *Nanoviridae*)	FB, P	Aphids P	Is, 17–19	Au	Chu and Helms (1988)

[a]Seed transmission of the virus.

FB, faba bean; Le, lentil; Cp, chickpea; Lu, lupin; P, pea; Sa, Sap; Se, seeds; Ba, bacilliform; Fi, filamentous; Ge, gemini; Is, isometric; NP, nonpersistent transmission; P, persistent transmission; SP, semipersistent transmission.

genus *Mastrevirus*, family *Geminiviridae*), *Bean yellow mosaic virus* (BYMV, genus *Potyvirus*, family *Potyviridae*), *Broad bean mottle virus* (BBMV, genus *Bromovirus*, family *Bromoviridae*), *Pea enation mosaic virus-1* (PEMV-1, genus *Enamovirus*, family *Luteoviridae*), and *Pea seed-borne mosaic virus* (PSbMV, genus *Potyvirus*, family *Potyviridae*). Similarly, recent detailed surveys in Australia found the most widespread viruses in lentil, pea, chickpea, lupin, and faba bean to be BLRV, SbDV, BWYV, BYMV, PSbMV, *Subterranean clover stunt virus* (genus *Nanovirus*, family *Nanoviridae*), *Alfalfa mosaic virus* (AMV, genus *Alfamovirus*, family *Bromoviridae*), and *Cucumber mosaic* virus (CMV, genus *Cucumovirus*, family *Bromoviridae*). In the north–west of the Indian subcontinent, the most prevalent viruses in lentil and chickpea were CMV, PSbMV, CpCDV, FBNYV, *Broad bean wilt virus* (BBWV, genus *Fabavirus*, family *Comoviridae*), *Broad bean true mosaic virus* (genus *Comovirus*, family *Comoviridae*), *Broad bean stain virus* (BBSV, genus *Comovirus*, family *Comoviridae*), *Pea early browning virus* (PEBV, genus *Tobravirus*, family unassigned), *Milk vetch dwarf virus* (MDV, genus *Nanovirus*, family *Nanoviridae*), and several other viruses are widespread in some of these crops in specific regions within specific countries (e.g., Abraham et al., 2006; Ait Yahia, Aitouada, Hadj Arab, Belfendess, & Sarni, 1997; Ait Yahia, Aitouada, Illoul, et al., 1997; Babin, Ortiz, Castro, & Romero, 2000; Bayaa et al., 1998; Bos et al., 1993; Cheng & Jones, 1999; Fortass & Bos, 1991; Freeman & Aftab, 2001; Freeman et al., 2013; Fresno et al., 1997; Horn et al., 1995; Jones & Coutts, 1996; Katul et al., 1993; Kumari et al., 2007, 2008; Larsen & Webster, 1999; Latham & Jones, 2001a; Makkouk, Bashir, Jones, et al., 2001; Makkouk, Bos, Azzam, et al., 1988; Makkouk et al., 1995, 1999, 1994; Mustafayev et al., 2011; Najar et al., 2000; Shiying et al., 2007; Tadesse et al., 1999; van Leur, Aftab, Manning, Bowring, & Riley, 2013; van Leur, Makkouk, Freeman, & Schilg, 2003). The most widespread and important viruses of pea internationally include PEMV, PEBV, FBNYV, PSbMV, BLRV, BYMV, and BWYV (Bos et al., 1988; Larsen, 2000; Latham & Jones, 2001a; Salam et al., 2011). Figure 4.1 provides examples of the types of symptoms and damage viruses induce in cool-season grain legumes globally.

Yield loss studies are best done in the field where infection spreads naturally rather than by comparing yields from single virus-infected and healthy plants, as the latter reflects a "worse-case scenario" where there is no compensatory growth by healthy plants (e.g., Jones, 2006; Latham, Jones, & Coutts, 2004). The growth stage when plants become virus-infected and the proportion of plants infected in the crop are critical factors in determining the extent of seed yield losses. Losses are generally greatest when plants become infected at vulnerable early growth stages and incidence approaches

Figure 4.1 (A) Symptoms caused by *Cucumber mosaic virus* (CMV) infection in chickpea: leaf chlorosis in kabuli chickpea (left), leaf reddening in desi chickpea (right). (B) Symptoms of leaf chlorosis and plant stunting (smaller shoots on right) caused by infection with *Chickpea chlorotic dwarf virus* (right), healthy shoot (single shoot on left). (C) Symptoms of leaf chlorosis and bunching of young leaves caused by CMV infection in a narrow-leafed lupin plant. (D) Marked plant stunting with chlorosis and bunching of young leaves caused by luteovirus infection in a lentil plant (left), a healthy plant (right). (E) Chlorosis and bunching of young leaves and marked plant stunting caused by *Bean yellow mosaic virus* (BYMV) infection in lentil plant (left), less severely affected plants (right). (F) Leaf symptoms of mild mosaic and chlorotic spotting caused by *Pea seed-borne mosaic virus* infection in a faba bean plant (right), healthy plant (left). (G) Leaf symptoms of stiffness, interveinal chlorosis, and marginal necrosis caused by infection with *Bean leafroll virus* in a plant of faba bean (left), healthy plant (right). (H) Chlorotic marginal leaf mottle caused by infection with *Broad bean stain virus* in a leaf of faba bean. (I) Chlorotic leaf spotting symptoms caused by *Broad bean mottle virus* infection in leaves of faba bean. (See the color plate.)

100%. Examples of quantitative estimates of seed yield losses from field experiments with natural spread include ones with CMV and BYMV in lupin where yield losses of up to 66% occurred (Bwye, Jones, & Proudlove, 1994; Jones, 2001; Jones, Coutts, & Cheng, 2003), CMV in chickpea where yield losses were up to 80% (Jones, Coutts, Latham, & McKirdy, 2008), and PSbMV in pea where yield losses of up to 25% occurred (Coutts, Prince, & Jones, 2009). Yield losses from field experiments with other combinations of virus and cool-season grain legume crops are given by Alkhalaf,

Kumari, Haj Kasem, Makkouk, and Al-Chaabi (2010), Boswell and Gibbs (1983), Bos et al. (1988), Cockbain (1983), Franz, Makkouk, and Vetten (1997), Kraft and Hampton (1980), Kumari and Makkouk (1995), Makkouk, Bos, Azzam, et al. (1988), Makkouk, Kumari, and Bos (1993), and Makkouk et al. (1990).

Several viruses impair the quality of crop legume seeds, thereby rendering them less attractive to purchasers. For example, PSbMV induces necrotic rings and line patterns, cracking, and malformation in seed coats of pea, faba bean, lentil, and chickpea (Coutts, Prince, & Jones, 2008; Latham & Jones, 2001b; Makkouk, Bos, et al., 1993; Makkouk, Kumari, et al., 1993). Likewise, BBSV infection leads to undesirable staining of faba bean seed coats, which renders the seeds less valuable for canning (Bos et al., 1988). Figure 4.2 provides examples of the types of symptoms and damage viruses induce in seeds of cool-season grain legumes globally.

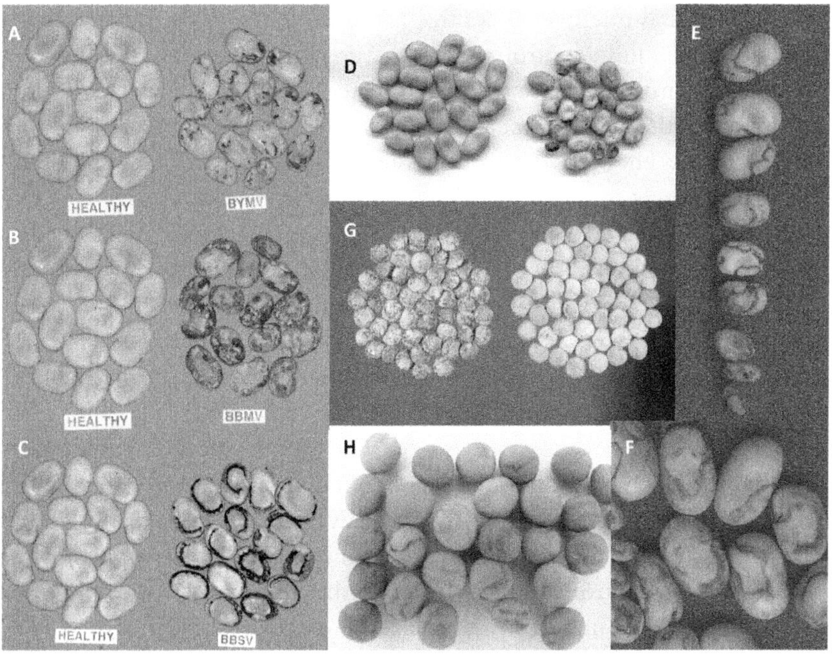

Figure 4.2 Symptoms of necrosis, reduction in size, and malformation in seeds of faba bean, virus-infected (right) and healthy (left): causal viruses are *Bean yellow mosaic virus* (A), *Broad bean mottle virus* (B), and *Broad bean stain virus* (BBSV) (C). Symptoms of malformation, reduction in size and necrotic rings caused by *Pea seed-borne mosaic virus* (PSbMV) in faba bean seeds (D), virus-infected (right), healthy (left). Necrotic rings caused by PSbMV in broad bean seeds (E, F). Symptoms of necrosis in lentil seeds caused by BBSV (G), virus-infected (left), healthy (right). Necrotic rings caused by PSbMV in pea seed (H). (See the color plate.)

In addition to requiring an understanding of the economic importance of a particular virus to a crop, decisions over which control measures to use require detailed knowledge from ecological and epidemiological studies. They also require information on the effectiveness of each control measure from field experiments. Control measures are aimed either at decreasing the virus source or at preventing virus spread within the legume crop, usually by a vector. The different types of control measures available can be classified as host resistance, phytosanitary, cultural, chemical, and biological. Of these, the cultural and phytosanitary types are very diverse. Biological control measures are rarely suitable for use with cool-season grain legumes, while chemical measures, if selected, need always to be deployed in an environmentally responsible way.

Control is optimized through integrated disease management (IDM) approaches, which combine all possible measures that operate in different ways such that they complement each other and can be applied together in farmers' fields as one overall control package (Jones, 2004, 2006; Thresh, 2003, 2006). Thus, control measures can be classified as (i) those that control the virus, (ii) those that are directed toward avoidance of vectors or reducing their occurrence, and (iii) those that integrate more than one method. They can also be classified as "selective," i.e., those that specifically target a single virus, virus strain, or vector species and may not be transferable between pathosystems, or "nonselective," i.e., those that are effective against a wide range of virus pathogens or their vectors. In general, phytosanitary and agronomic measures are nonselective, whereas host resistance and most biological control measures are selective. Chemical control measures have low selectivity when a general pesticide that kills several vector species is applied but high selectivity when a specific pesticide kills only one vector species is employed. However, a drawback to selective measures, including the phytosanitary measure of roguing, is that deploying them widely may result in inadvertent selection of variants of virus or vector that are more difficult to control. Host resistance, chemical, and biological control measures decrease the rate of virus spread but are inactive against the initial virus source, whereas phytosanitary control measures and many cultural measures address both source and spread. When combining control measures within IDM strategies, success is optimized by including measures that are selective and others that are nonselective, that address both kinds of sources (external and internal) and phases (early and late) of virus spread, and that operate in as many different ways as possible (Jones, 2006).

3. HOST RESISTANCE

3.1. Selection and breeding for virus resistance

Breeding for virus resistance can be greatly simplified if the resistance is based on a single gene or a few genes that code for a qualitative response (high level of resistance or immunity). On the other hand, resistance that is based on a quantitative (partial resistance) response or based on a larger number of genes is far more difficult to identify and to utilize in a practical breeding program. Contrary to the experience when breeding for resistance to fungal diseases, single genes for virus resistance have often proved to be durable (Garcia-Arenal & McDonald, 2003; Harrison, 2002; Parlevliet, 2002). Virus resistance can be based on dominant or recessive genes. Recessive genes generally express themselves only in homozygous genotypes, complicating their detection, especially in out-crossing host plant species like faba bean. A summary of available cool-season grain legume genotypes and identified genes for virus resistance is shown in Tables 4.2 and 4.3.

3.2. Resistance to potyviruses

Numerous studies have been carried out to determine the genetic basis of resistance to potyviruses. Provvidenti and Hampton (1992) reviewed the resistance to 56 potyviruses in 334 plant species and found simple inheritance, that was either dominant (60 genes) or recessive (39 genes).

The PSbMV–pea pathosystem is one of the best researched virus pathosystems worldwide. After the first description of the virus (Musil, 1966), three clearly distinct pathotypes (P1, P2, and P4) were identified (Alconero, Provvidenti, & Gonsalves, 1986). Differential reactions of pea germplasm lines to these pathotypes lead to four recessive resistance genes being postulated (Provvidenti & Hampton, 1992). More recently, large advances in the understanding of the genetic basis of this pathosystem have been made through the use of sophisticated molecular techniques. For example, Johansen, Lund, Hjulsager, and Laursen (2001) demonstrated that PSbMV pathotypes can be explained by the properties of two viral cistrons and predicted the existence of a fourth pathotype, P3. This pathotype was subsequently identified in a faba bean germplasm accession originating from Nepal (Hjulsager, Lund, & Johansen, 2002). Research by Gao et al. (2004) indicated the existence of only two recessive resistance genes (*sbm1* and *sbm2*): The *sbm1* gene confers resistance to all four PSbMV pathotypes,

Table 4.2 Inheritance of resistance to virus diseases in lentil, chickpea, faba bean, pea, and lupin

Crop	Genes reported	Resistance to	Mode of inheritance	References
Lentil	*sbv*	PSbMV	Recessive	Haddad, Meuhlbauer, and Hampton (1978)
Faba bean	*bym-1*, *bym-2*	BYMV	Recessive	Rohloff and Stupnagel (1984), Schmidt, Rollowitz, Schimanski, and Kegler (1985)
Pea	*mo*	BYMV	Recessive	Yen and Fry (1956)
	lr	BLRV	Recessive	Drijfhout (1968)
	sbm1, *sbm1*^{1}, *sbm2*	PSbMV	Recessive	Hagedorn and Gritton (1973), Provvidenti and Alconero (1988), Gao et al. (2004)
	En	PEMV	Dominant	Schroeder and Barton (1958)
	cyv,*cyv-2*	ClYVV	Recessive	Provvidenti (1987)
Lupin	*Ncm-1*	CMV	Dominant	Jones and Latham (1996), Jones and Burchell (2004)
	Nbm-1	BYMV	Dominant	Cheng and Jones (2000), Cheng, Jones, and Thackray (2002), Thackray, Smith, Cheng, Perry, and Jones (2002), Jones and Smith (2005)

Note: Additional useful references that cover resistances to several viruses in faba bean are those of Schmidt, Geissler, Karl, and Schmidt, 1986; Schmidt, Carl, and Meyer, 1988; Schmidt, Meyer, Haack, and Karl, 1989.

while a different allele of the *sbm1* gene, *sbm1*^{1}, confers resistance only to the P1 and P2 pathotypes. The *sbm2* gene only confers resistance to the P2 and P3 pathotypes. Both genes also control, or are strongly linked to, separate genes that control resistance to a number of other potyviruses (Bruun-Rasmussen et al., 2007). For example, *sbm2* is linked to, or the same as, the *mo* BYMV resistance gene. In addition to the complete resistance coded by the *sbm* genes, a wide range of partial PSbMV resistance can be observed among different pea genotypes (Latham & Jones, 2001a, 2001b; van Leur, Kumari, Aftab, Leonforte, & Moore, 2013). Resistance to seed transmission in particular would be highly effective in controlling the virus, as the magnitude of field infection and virus-induced losses depend to a large extent on the initial inoculum provided by PSbMV-infected seed (Coutts et al., 2009). However, the infection of pea seed embryos by PSbMV is influenced by a large number of factors (Roberts, Wang, Thomas, & Maule, 2003).

Table 4.3 Faba bean, lentil, chickpea, pea, and lupin genotypes resistant to virus diseases

Crop/genotype	Resistant to	References
Faba bean		
2N138	BYMV	Ghad and Bernier (1984)
BPL 5247 through to BPL 5255	BYMV	Makkouk and Kumari (1995)
BPL 5271 through to BPL 5285	BLRV	Makkouk, Fazlali, Kumari, and Farzadfar (2002)
Several	BYMV, CMV, AMV, PSbMV	McKirdy et al. (2000), Latham et al. (2001), Latham and Jones (2001b)
Pea		
B442-15, B445-66	PSbMV	Baggett and Hampton (1977)
VR 74-410-2, VR 74-1	PSbMV	Kraft and Giles (1978)
Vervil, Fioletovo, Krasnodarkski 87, Jewel	BYMV	Ovchinmikova et al. (1987)
PI 140295	PEMV	Schroeder and Barton (1958)
Patea, Puke, Piri, Pania	BLRV	Kraft and Kaiser (1993)
Several	AMV, BYMV, CMV, PSbMV	McKirdy et al. (2000), Latham et al. (2001), Latham and Jones (2001b)
Lupin		
P26849, P26853, P26858, P26849, P26815, WTD1191, Motiv 368, Motiv 369, Teo, Popiel, P26859	CMV	Jones and Latham (1996)
All narrow-leafed lupin cultivars	BYMV	Cheng and Jones (2000), Jones and Smith (2005)
Several	AMV, CMV	Jones and Cowling (1995), Jones (2001), McKirdy et al. (2000), Jones, Pearce, Prince, and Coutts (2008)

Continued

Table 4.3 Faba bean, lentil, chickpea, pea, and lupin genotypes resistant to virus diseases—cont'd

Crop/genotype	Resistant to	References
Lentil		
PI 212610, PI 251786, PI 297745, PI 368645	PSbMV	Haddad et al. (1978)
Red Chief, Palouse	PSbMV	Kumari and Makkouk (1995)
PI 472547, PI 472609	PEMV	Aydin, Meuhlbauer, and Kaiser (1987)
ILL 7163	BYMV, CMV	McKirdy et al. (2000)
ILL 5480	AMV	Latham and Jones (2001b)
ILL 75	BLRV, FBNYV, SbDV	Makkouk, Bashir, Jones, et al. (2001), Makkouk, Kumari, and Lesemann (2001), Makkouk, Kumari, Sarker, et al. (2001)
ILL 74, ILL 85, ILL 213, ILL 214, ILL 6816	FBNYV, BLRV	Makkouk, Kumari, Sarker, et al. (2001)
Red Chief	BBSV	Makkouk and Kumari (1990)
Several	AMV, BYMV, CMV, PSbMV	McKirdy et al. (2000), Latham et al. (2001), Latham and Jones (2001b)
Chickpea		
ICC 1781, ICC 8203	CMV	Chalam, Reddy, Subbayya, Nene, and Beniwal (1986)
ICC 607, ICC 6999, ICC 1468, ICC 2162, ICC 2342, ICC 3440, ICC 3598, ICC 4045, ICC 11550	BYMV	Chalam et al. (1986)
PI 315826, PI 450843, PI 450975, PI 450693, PI 439829, PI 450906, PI 451594, PI 450867, PI 450763	PEMV	Larsen and Porter (2010)
ICC 10466	Chickpea stunt	Nene (1988)
Several	BYMV, CMV, PSbMV	McKirdy et al. (2000), Latham et al. (2001); Latham and Jones (2001b), Coutts et al. (2008)

Differences among pea genotypes in PSbMV seed-to-plant transmission were found to vary greatly between growing seasons, so selection for resistance to seed transmission might not prove a practical objective (van Leur, Kumari, et al., 2013).

Less attention has been given to PSbMV resistance in lentils, even though PSbMV causes a major problem in traditional lentil growing regions in the Mediterranean region, Indian subcontinent, and Ethiopia. Haddad et al. (1978) found four lentil germplasm accessions from Afghanistan, the former Soviet Union, Greece, and Yugoslavia (PI 212610, PI 251786, PI 297745, and PI 368648) to be immune to PSbMV with this resistance controlled by a single recessive gene, sbv. However, Hampton (1982) reported differential reactions when testing these lines with different PSbMV pathotypes, so likely more than a single gene is operating within the lentil-PSbMV pathosystem. Kumari and Makkouk (1995) identified two US pea cultivars (Red Chief and Palouse) with partial resistance to PSbMV. This resistance was characterized by very low grain yield losses due to infection and low seed transmission rate. In another study, Makkouk and Kumari (1990) found that Redchief had the lowest seed transmission rate (0.2%) and yield loss (14%) among the 19 lentil genotypes they evaluated.

Although BYMV is not considered to be among the major viruses in pea, the single recessive *mo/sbm2* gene, which controls complete BYMV resistance, is found at a high frequency in US pea cultivars (Kasimor, Bagget, & Hampton, 1997). It was also found at a surprising high incidence in an evaluation for virus resistance in Australian field pea cultivars (van Leur, Kumari, et al., 2013), possibly resulting from the use of US cultivars as parents in the Australian pea breeding programs.

Contrary to the situation in pea, BYMV causes a widespread and economically important disease of faba bean. Ghad and Bernier (1984) identified an inbred faba bean line (2N138) to be immune to "mosaic" and "necrotic" BYMV strains under field and greenhouse conditions. Makkouk and Kumari (1995) screened a large collection of faba bean germplasm from the ICARDA Gene Bank against local Syrian BYMV strains and found nine BYMV-resistant pure lines after several cycles of inbreeding and selection of BYMV-free plants. Resistance to BYMV in faba bean was found to be controlled by two recessive complementary genes, *bym-1* and *bym-2* (Rohloff & Stupnagel, 1984; Schmidt et al., 1985).

Strain-specific, systemic hypersensitive resistance to BYMV is present in all narrow-leafed lupin cultivars. It is controlled by dominant gene *Nbm-1*, and probably another gene, and confers useful "epidemic rate limiting" resistance in the field (Cheng et al., 2002; Jones, 2005; Jones et al., 2003; Jones & Smith, 2005; Thackray et al., 2002). The systemic hypersensitive response is typical of early infection in the field, but when late infection occurs at the growth stage when pods are first forming, BYMV induces black pod syndrome (necrosis of pods) but does not kill the whole plant. This syndrome apparently represents expression of *Nbm-1* in old plants (Kehoe, Coutts, Buirchell, & Jones, 2014). Some narrow-leafed, white and yellow lupin genotypes have useful "infection resistance" against natural aphid inoculation with BYMV (Jones, 2001; Jones & McLean, 1989).

3.3. Resistance to luteo-, polero-, and nanoviruses

Among the persistently transmitted viruses, BLRV is the most widespread and has received most research attention. Variability in susceptibility, resistance, and/or tolerance to BLRV among pea cultivars has been reported (Hubbeling, 1956), but only a limited number of studies have been published on the genetic basis of BLRV resistance in peas. Both Drijfhout (1968) in the Netherlands and Baggett and Hampton (1991) in the USA concluded that the resistance was based on a single recessive gene, but commented on the difficulties in distinguishing between BLRV-resistant and susceptible segregants. Baggett and Hampton (1991) did not exclude the possibility of additional, additive, genes for BLRV resistance. van Leur, Kumari, et al. (2013) quantified BLRV resistance in Australian pea cultivars, but did not find immunity. Crampton and Watts (1968) in New Zealand concluded that resistance to pea leaf-roll/top-yellows (likely a combination of BWYV and SbDV) was based on additive and dominant genes.

Using repeated cycles of selection in inbred populations under artificial inoculation, resistance to BLRV was found in 15 faba bean genotypes coded BPL 5271 through to BPL 5285 (Makkouk, Kumari, et al., 2002). Six lentil genotypes were identified with combined resistance (Makkouk, Kumari, Sarker, et al., 2001): ILL 75 had resistance to BLRV, FBNYV, and SbDV; and ILL 74, ILL 85, ILL 213, ILL 214, and ILL 6816 were resistant to FBNYV and BLRV. Severe epidemics caused by a combination of persistently transmitted viruses (BWYV, BLRV, SbDV, and unidentified luteoviruses) are reported in chickpeas in the northern Australian grain belt. The chickpea variety "Gully" was developed through mass selection from an Iranian germplasm accession (NEC-1971) under high virus pressure in 1992, but does not have complete resistance (Anonymous, 1999).

After severe FBNYV epidemics in the Mediterranean region during the 1990s, ICARDA embarked on a selection program to identify resistance in its mandate crops. Resistance was readily identified in lentils (Makkouk, Kumari, Sarker, et al., 2001), but for faba bean repeated selection cycles were needed to develop genotypes with adequate resistance (S. G. Kumari, unpublished).

PEMV is considered to be among the most important viruses in all legume growing regions, except Australia. Schroeder and Barton (1958) found only 1 accession (PI 140295 originating from Iran) of 171 field pea germplasm accessions that had surviving plants (4 of 301) under severe PEMV inoculum pressure. The progenies of the surviving plants, designated G168 (Geneva Selection 168), were used extensively as parents in the US pea breeding programs. The PEMV resistance in G168, based on a single dominant gene (*Le*), is complete and proved to be durable. Aydin, Meuhlbauer, and Kaiser (1987) evaluated 29 lentil lines and identified two germplasm accessions, PI 472547 (originating from India) and PI 472636 (from Iran) with partial resistance when inoculated with PEMV by aphids. Larsen and Porter (2010) tested 499 USDA chickpea gene bank accessions (desi and kabuli germplasm) from a wide range of origins and identified nine accessions (eight desi and one kabuli) originating from either Iran or India with a high level of resistance (not immunity). Their results as well as the earlier reports of resistant pea and lentil germplasm prompted them to suggest that Iran and India may constitute a center for PEMV-resistant pea germplasm.

3.4. Resistance to alphamo- and cucumoviruses

Ford and Baggett (1965) screened field pea lines from the USDA Plant Introduction Collection for virus resistance by mechanically inoculating plants with a white clover-derived AMV isolate and identified 31 accessions that remained symptomless. However, none of these lines proved resistant in a similar study with another AMV isolate (Hagedorn, 1968). Timmerman-Vaughan et al. (2001) concluded that no effective host-derived AMV resistance is available in pea and therefore investigated pathogen-derived resistance, but only found partial resistance in transgenic plants. Latham and Jones (2001a, 2001b) detected different levels of partial AMV resistance among Australian pea cultivars and breeding lines in field testing, indicating possibilities for breeding programs to improve resistance levels. However, all chickpea cultivars proved highly susceptible to AMV. Chalam et al. (1986) reported several chickpea accessions with high levels of CMV (and BYMV) resistance. Screening of chickpea and lentil germplasm accessions from a wide range of geographical origins by mechanical inoculation with Syrian AMV and CMV strains failed to identify immune genotypes, but different levels of symptom severity were found (S. G. Kumari, unpublished).

CMV is of particular importance in lupins, mainly because of high seed transmission rates and the key role of sowing infected seed stocks in introducing sources of the virus to lupin crops. Hypersensitive resistance to CMV was found in several yellow lupin (*Lupinus luteus*) cultivars and was controlled by gene *Ncm-1*. *Lupinus hispanicus* Boiss. & Reuter genotypes P26849, P26853, and P26859 also had hypersensitive resistance to CMV, and pearl lupin (*Lupinus mutabilis*) P26956 had extreme CMV resistance. White lupin (*Lupinus albus*) is not a host of CMV (Jones & Burchell, 2004; Jones & Latham, 1996). Resistance to seed transmission, a trait that determines low seed transmission rates, is present in some genotypes of narrow-leafed (*Lupinus angustifolius*) and yellow lupins (Jones & Cowling, 1995; Jones & Latham, 1996). The Australian lupin breeding program screens all advanced breeding lines to select ones with low levels of CMV seed transmission, which is a quantitatively inherited trait (Jones & Cowling, 1995). Use of cultivars with low inherent seed transmission rates has contributed greatly to the diminished importance of CMV epidemics in narrow-leafed lupin CMV-prone areas of southwest Australia (Jones, 2001, 2006).

3.5. Breeding for vector resistance

Host resistance to the insect vector is not expected to help reduce spread of nonpersistently aphid-transmitted viruses, such as CMV, BYMV, BCMV, PSbMV, and AMV, in legume crops. With such viruses, when the legume host is not preferred for aphid colonization, the end result tends to be shorter probes and more movement and consequently more virus spread. For example, CMV spreads readily in chickpea fields despite the fact that chickpea is highly resistant to aphids (Jones & Coutts, 1996; Jones et al., 2008).

3.6. Transgenic virus resistance

An advantage of the transgenic approach is that virus resistance can be introduced to improve the performance of widely grown virus-susceptible crop legume cultivars that are popular because of the good commercial attributes they posses. However, there is considerable opposition to the release of genetically modified plants and their widespread use for controlling virus diseases of grain legumes seems unlikely for many years to come.

4. PHYTOSANITARY MEASURES

4.1. Healthy seeds

Use of healthy seed stocks deserves special attention in legumes. Around 50% of viruses affecting leguminous crops are seed-borne (Bos et al., 1988), whereas in small-grain cereals, only two are recorded so far (Diekmann & Putter, 1995; Jones, Coutts, Mackie, & Dwyer, 2005). Where seed transmission rates are high, as with PSbMV in pea, use of healthy seed is an effective control measure (Hampton, 1984). Production of healthy pea seeds can be achieved in areas of low virus pressure and by monitoring mother plants and seed lots obtained from them using sensitive virus detection methods. Sowing virus-infected legume seed stocks results in infection foci scattered at random throughout a crop, and this provides a much more potent virus inoculum source than when the source is external to the crop. Spread resulting in considerable yield losses normally occurs when the level of infection in the seed sown is high and the environmental conditions are conducive to early and considerable vector activity (e.g., Coutts et al., 2009; Jones, 2001; Thackray, Diggle, Berlandier, & Jones, 2004). To establish what different levels of seed infection mean in terms of yield losses, field experimentation is needed in which seed lots with different levels of infection are sown at different locations under different climatic conditions (e.g., Bwye et al., 1994; Coutts et al., 2009). Such information can be used to establish defined "threshold" values for seed infection rates that identify acceptable levels of risk of economic loss. With lupin, the commercial testing service in Australia is based on representative samples of 1000 seeds from farmers' seed stocks. When sowing lupin crops in lower risk areas, a "threshold level" of <0.5% CMV seed infection is sufficiently conservative to avoid serious yield loses but for high-risk areas a more stringent threshold of <0.1% seed infection applies (Jones, 2000). Similar recommendations apply to PSbMV infection of pea seed stocks (Coutts et al., 2009). Seed transmission of BBSV was reduced to zero when seeds were exposed to 70 °C for 28 days; however, this treatment caused an unacceptable reduction (57%) in seed germination (Kumari & Makkouk, 1996). Such an approach could be useful to eliminate seed infection in germplasm accessions deposited in gene banks, but is not economical for commercial production.

4.2. Roguing

The removal of symptomatic plants, known as roguing, is a phytosanitary control measure widely used to remove sources of virus infection from within crops. It is only effective when symptoms are conspicuous and when the symptomatic plants are removed early, before vectors have visited them, and is more effective where vector numbers are low and when a virus is being transmitted persistently by insect vectors (e.g., Thresh, 1982, 2003, 2006). It is feasible in small-scale subsistence crops where labor is plentiful but not when legume crops are large, such as the >100 ha crops common in Australia and North America. When practiced two to three times early during the growing season, roguing of FBNYV-infected crops was effective in minimizing the incidence of primary infection foci inside small faba bean fields in Egypt (Makkouk, Vetten, Katul, Franz, & Madkour, 1998).

4.3. Removal of wild hosts, volunteers, and crop residues

A standard phytosanitary control measure is to remove sources of virus infection or vector hosts that are not part of the crop so as to minimize the initial virus infection source and number of vectors (Jones, 2004, 2006; Thresh, 1982, 2003, 2006). Weed and volunteer legume crop hosts within or near crops constitute key virus sources as do nearby harvested or finished legume crops. Removing such sources greatly diminishes the rate at which virus epidemics develop within legume crops and is often recommended as part of IDM approaches, e.g., with seed-borne viruses in cool-season grain legume breeding and selection plots (Jones, 2001; Jones, Latham, & Coutts, 2004). Prior to advocating such an approach for FBNYV control in faba bean, more information on the relative importance of wild legumes as sources of FBNYV infection is needed (Makkouk, Vetten, et al., 1998).

5. CULTURAL PRACTICES

5.1. Isolation, avoiding successive side-by-side plantings, and crop rotation

A standard cultural virus control measure when sowing a crop is to ensure safe planting distances and isolation from potentially infected crops or other potential virus sources, such as nearby legume pastures. Crop rotation helps avoid infection sources consisting of volunteer crop legumes that may have become infected through seed or have survived from previous crops. On small farms or in the plots of breeding programs, avoiding successive side-by-side plantings is a similar control measure that helps to minimize exposure to initial infection sources (Jones, 2004, 2006; Thresh, 2003, 2006). Such measures are important components of IDM approaches to controlling virus spread in cool-season grain legumes (Jones, 2001; Jones et al., 2004).

5.2. Early canopy development and high plant density

Promotion of early canopy development is a cultural control measure achieved best by sowing seed early at high seeding rates with close row spacing. It is particularly important with seed-borne viruses when insect vector activity starts late. Such an approach is effective because, before insect vectors arrive, there is then time for less vigorous, seed-infected plants to become shaded over and killed as foliage covers them, and consequently removes them as potential infection sources. Early canopy development can also be important in removing early current-season (secondarily) infected plants such that they are no longer available as virus sources. In addition, high seeding rate resulting in high plant density is important on its own as a control measure because it dilutes the proportion of infected plants and favors compensatory growth of healthy plants. For example, early canopy development and high plant density are both important components of an IDM strategy for controlling CMV in lupin (Jones, 2001).

5.3. Sowing date

A close relationship exists between sowing date and the extent of virus spread in grain legume crops. Greatest virus spread occurs when seed is sown such that crops emerge, and plants are therefore at their most vulnerable early growth stage, just before or during the peak period for insect vector activity. Predictive models and decision support systems often depend on forecasting when vectors will arrive in crops, with early vector arrival associated with rapid spread, high subsequent yield losses, and a need to deploy control measures early, e.g., with CMV and BYMV in lupin (Jones, Salam, Maling, Diggle, & Thackray, 2010; Maling, Diggle, Thackray, Siddique, & Jones, 2008; Thackray et al., 2004). Manipulation of planting date to avoid exposing young plants to peak vector populations at their most vulnerable early phase of growth is a standard virus control measure that is widely recommended for use with crop legumes (e.g., A'Brook, 1964; Jones, 1991, 2001, 2004, 2006; Thresh, 1982, 2003, 2006). In Syria and Egypt, faba bean crops planted early are often severely attacked by FBNYV, leading to 100% infection (Makkouk, Vetten, et al., 1998). In such circumstances, farmers plough the crop under and replant with another faba bean crop. Delaying sowing until later results in lower infection and consequently less crop losses due to fewer viruliferous aphid vectors arriving from neighboring virus sources.

5.4. Narrow row spacing and high seeding rate

Combining narrow row spacing and high seeding rates to generate early canopy formation before the time when most virus spread occurs is widely used not only to shade out virus-infected source plants within the crop (see 5.2 above) but also to decrease the landing rates of incoming viruliferous aphid vectors and by doing so diminish spread of aphid-borne viruses in legume crops. This landing rate effect is because bare soil around plants attracts migrants of most aphid species while groundcover, as provided by a plant canopy meeting across rows, repels them (Jones, 2001, 2004, 2006; Thresh, 1982, 2003, 2006). An example from Australia is where wide row spacing and low plant density promoted greater aphid landings and increased BYMV spread from external sources into lupin crops (Jones, 1993). Promoting early canopy formation by sowing at high seeding rates with close row spacing is therefore a recommended control measure for BYMV in lupin (Jones, 2001). However, although high seeding rates are beneficial with CMV in lupin, the benefit that close row spacing provides in diminishing aphid landing rates needs to be balanced against the benefit provided by rapid removal of seed-infected source plants that occurs when wide row sowing is used at high seeding rates. The latter occurs because having far more plants present within each row greatly increases the within-row shading effect on infected seedlings by the numerous, vigorous healthy seedlings (Bwye, Jones, & Proudlove, 1999).

5.5. Groundcover

Minimum tillage, which minimizes stubble burial, and stubble retention at sowing or adding stubble mulches that help cover the ground, are standard cultural virus control measures that help to diminish the landing rates of aphid vectors and so decrease virus spread. Like early plant canopy cover, they act by minimizing exposure of bare ground between plants and so repel incoming aphids. They are important control measures in the early phase of crop growth before canopy closure covers the soil surface between rows, and are most beneficial when stands are thin and row spacing wide (Jones, 2001, 2004). For example, when Jones (1994) examined the effect of straw spread on the soil surface before germination on the subsequent incidence of BYMV that was spreading into plots of lupin from nearby infected clover pasture, straw mulch greatly decreased the rate and amount of virus spread, leading to increased seed yields. The effect of straw was attributed to decreased landing rates of incoming viruliferous vector alates. The recommendation that resulted was that stubble should be retained on the soil surface at seeding to assist in management of BYMV in lupin crops, especially where wide row spacing is practiced. Retaining stubble in minimum tillage also has an effect on microclimate, including reducing wind speed (Cutforth, McConkey, Ulrich, Miller, & Angadi, 2002), which might alter aphid landing rates. In other studies, when CMV and BYMV spread into single row plots of lupin, reflective mulch placed around the rows repelled aphids and so decreased spread of both viruses (Jones, 1991). Deploying reflective mulch was therefore part of an IDM package of measures recommended for control of BYMV and CMV in small plots within lupin breeding programs (Jones, 2001). Assad, Kumari, Haj-Kassem, Al-Chaabi, and Malhotra (2012) found that the incidence of luteoviruses in chickpea crops was decreased by 40% when zero tillage was applied, while chickpea yield was increased by 62% in comparison with traditional tillage. However, more testing of this approach was needed to develop recommendations for use by farmers.

5.6. Early maturing cultivars

Because virus epidemics take time to build up within crops and physiologically old plants are generally more resistant to virus infection than physiologically young plants ("mature plant resistance"), epidemics tend to do more damage to cultivars that grow for long periods than to ones that mature rapidly, causing greater yield losses in them. Also, it is possible to time the sowing of short duration cultivars so they approach maturity and express mature plant resistance to virus infection during the period of greatest insect vector activity. Sowing early maturing cultivars is therefore a recommendation for virus control in grain legume crops (e.g., Jones, 2001, 2004).

5.7. Nonhost barrier, tall nonhost cover crop, and mixed cropping with a nonhost

Deploying nonhost barriers around the perimeters of crops to separate them from external virus sources is a useful cultural control measure sometimes employed with insect-vectored viruses. The nonhost acts not just a physical barrier but also a "virus cleansing" barrier. With nonpersistently aphid-borne viruses, the explanation is that when incoming migrant aphids probe nonhost plants while in search of their preferred hosts, they lose a non-persistently aphid-borne virus from their mouthparts, thereby diminishing the amount of virus introduced once the aphids arrive at the crop. Similarly, with most persistently transmitted viruses, if the viruliferous insect vector stays and colonizes the nonhost, by the time its progeny move on to the susceptible crop, they are no longer viruliferous. Tall nonhost cover crops and mixed cropping or interplanting with a nonhost crop help to diminish virus incidence in similar ways (Fereres, 2000; Jones, 2001, 2004; Thresh, 2003). Nonhost barrier crops of oats are deployed as "cleansing barriers" to decrease spread of BYMV from nearby pasture into lupin crops. However, although mixture of oats with lupin decreased BYMV spread, it was not a practical proposition because of adverse effects on lupin growth (Jones, 1993, 2005). Use of nonhost buffers in between small plots is a component of IDM strategies to minimize infection with seed-borne viruses in lupin, field pea, lentil, chickpea, and faba bean breeding and selection programs (Jones, 1991; Jones et al., 2004).

5.8. Planting upwind or using windbreaks

Planting upwind of potential virus sources and using windbreaks to redirect wind carrying insect vectors away from or around a small susceptible crop are standard cultural virus control measures (Jones, 2004, 2006; Thresh, 1982, 2006). Both are recommended as components of IDM strategies against plant viruses, but such recommendations are often generic rather than based on actual data for the specific combination of virus and crop concerned.

6. CHEMICAL CONTROL

Application of insecticides helps to decrease the spread of some legume viruses vectored by insects (Broadbent, 1957). However, it is often ineffective because success with it depends on factors such as the mechanism of transmission of the virus and the mode of action of the pesticide selected. Moreover, insecticides should always be applied sparingly as they become ineffective when vectors develop resistance to them, and their overuse results in unwanted side effects with environmental and economic consequences, such as buildup of toxic residues, loss of beneficial natural enemies of vectors, and unforeseen accumulation of other pests or pathogens. There are several general reviews on this subject (e.g., Irwin & Nault, 1996;

Perring, Gruenhagen, & Farrar, 1999; Raccah, 1986), as well as more specific reviews regarding chemical control of aphid-transmitted viruses (Schepers, 1989) and leafhopper-transmitted viruses (Heinrichs, 1979). In general, success in decreasing virus spread by chemical control of vectors is considerably greater with persistently than with nonpersistently transmitted viruses. This is mostly because incoming viruliferous vectors carrying nonpersistently transmitted viruses tend not to be killed fast enough to prevent probing and consequently virus inoculation to sprayed healthy plants.

6.1. Nonpersistent viruses

As mentioned above, in general, most types of insecticides are relatively ineffective at controlling nonpersistently aphid-borne viruses. Oil sprays can be used instead (Simons, 1982), but are rarely cost effective because of the repeated applications required. The most effective types of insecticide for control of nonpersistently aphid-borne viruses are the newer generation synthetic pyrethroids (Loebenstein & Raccah, 1980). This is because of their rapid knockdown and greater antifeedant activity. However, although it significantly decreased CMV spread in lupin under conditions when imidacloprid, methamidophos, and triazamate did not, one to two applications of the newer generation pyrethroid alpha-cypermethrin did not control the virus sufficiently well to provide reproducible yield increases (Thackray, Jones, Bwye, & Coutts, 2000). It could not therefore be recommended generally for CMV control in lupin crops, although multiple sprays of alpha-cypermethrin are worthwhile using with high value lupin seed crops and breeders plots (Jones, 2001).

6.2. Persistent viruses

Some success was obtained with chemical control of luteoviruses, such as BLRV and related viruses, in cool-season grain legumes. For example, field experiments at ICARDA showed that a seed treatment with the systemic neonicotinoid insecticide imidacloprid at a rate of 0.5–2.8 g a.i./kg seed gave significant protection of lentil and faba bean plots against FBNYV infection, which lasted for 2 months after sowing. Incidence of FBNYV in faba bean was reduced from 28% in untreated plots to 2% and 1% in plots treated with 1.4 and 2.8 g a.i./kg of seed, respectively, and the yield loss from 37% in untreated plots to 0% in plots treated with either 1.4 or 2.8 g a.i./kg of seed. Moreover, although it had no effect on the yield of virus-resistant genotypes, this seed treatment significantly improved the yields of moderately virus-resistant and susceptible lentil genotypes (Makkouk & Kumari, 2001). Such treatment could prove useful in areas where infection with FBNYV is likely to occur early in the growing season. Because such chemicals are used at very low concentration (100–200 g/ha), this makes them more environmentally acceptable than many of the older generation of systemic insecticides normally applied as foliar sprays. However, due to the environmental impact, chemical control should still only be considered when other control approaches are insufficient to achieve economic yields in infected crops.

In a field experiments in Ghab region, Syria, the infection of spring-sown chickpea cultivars infected with luteoviruses was significantly reduced by application of imidacloprid seed dressing in comparison with thiomethoxan insecticide: the rate of reduction in virus incidence was 43%, while yield was increased by 20%. However, when imidacloprid was applied as a seed treatment for winter sowings, it had no significant effects on these parameters. By contrast, reduction in the rate of virus infection was 27% less following previous application of thiomethoxan insecticide to spring-sown chickpea, when seed yield was decreased by only 8%. The increased performance of thiomethoxan on winter-sown chickpea was a 51% reduction in virus incidence associated with a yield increase of 22%, which was significantly better than when imidacloprid was applied to winter-sown plants. The best results were achieved by applying both insecticides together for both winter- and spring-sown chickpeas, the rates of reduction in virus infection being 66% and 65%, respectively, while yield was increased by 31% and 32%, respectively in comparison with the nontreated control (Assad et al., 2012).

7. BIOLOGICAL CONTROL

Cool-season grain legume crops are rarely grown under protected cropping where the use of biological control agents, such as insect parasitoids or predators, is useful to control virus spread by insect vectors in crops (e.g., Jones, 2004). In the field, such biological approaches have proved to be largely ineffective.

8. INTEGRATED APPROACHES

Individual control measures used alone often bring only small benefits and may become ineffective in the long term. When control measures that act in diverse ways are combined into IDM tactics, their effects are complementary resulting in far more effective overall control. Such IDM approaches for virus diseases of grain legumes combine available host resistance, phytosanitary, cultural and chemical control measures (Jones, 2001, 2004, 2006; Makkouk, 1996). Selecting the best mix of measures for each virus–crop combination and production situation requires knowledge of the epidemiology of the causal virus and the mode of action and effectiveness of each individual control measure. Each strategy needs to be affordable by farmers and fulfill the requirements of being environmentally and socially responsible. It must also be compatible with control measures already in use against other pests and pathogens (Jones, 2004). However, it is not anticipated that, when cost is an overriding factor, farmers with limited resources will necessarily use all of the different measures within an individual IDM strategy. In such situations, each strategy is best viewed as a package of measures from which the farmer can select the ones to use that are most appropriate to his or her particular circumstances. In contrast, with high value legume crops or when the virus epidemic is drastically depleting yield and so is of paramount importance, it may be worthwhile to deploy them all.

An example of an IDM approach being widely used for practical virus control in a grain legume crop is one for FBNYV in faba bean. This strategy combines planting late in the growing season, use of high seeding rate, application of one or two systemic insecticide sprays, that are well timed during the early stages of the crop development, and roguing infected plants early in the growing season. Its use has led to significant reduction of FBNYV infection and more profitable faba bean production in Egypt (Makkouk, Kumari, Hughes, et al., 2003).

Field experiments were conducted by Al-Jallad, Kumari, and Ismail (2007) in coastal Syria to investigate the effects of a number of management components (planting date, plant density, seed dressing with imidacloprid, cereal border (wheat), and foliar spray with insecticide and mineral oil) in reducing the spread of aphid-transmitted faba bean viruses. Results indicated that the combination of late planting, imidacloprid seed treatment, and a plant density of 33 plants/m^2 was an effective management option to reduce virus disease incidence.

An example for chickpea comes from northern Sudan, where trials showed that using a combination of partial virus resistance, delayed sowing date, and shorter intervals between irrigation led to reduced incidence of CpCDV and significantly increased chickpea grain yield (Hamed & Makkouk, 2002). An example for lupin comes from southwestern

Australia, where the crop is grown on a large scale with low inputs. Here, lupin yields in high-risk areas for BYMV and CMV are optimized by sowing virus-tested lupin seed stocks with minimal virus contents and sowing cultivars with inherently low seed transmission rates (CMV only); using perimeter nonhost oat barriers and avoiding fields with large perimeter: area ratios (BYMV only); isolation from neighboring legume crops or pastures; promoting early canopy development, generating high plant densities, adjusting row spacing, direct drilling into retained stubble, sowing early maturing cultivars, maximizing weed control, and crop rotation (both viruses). The differences between the measures recommended for the different viruses reflect the internal seed-borne source with CMV versus the external clover pasture source with BYMV (Jones, 2001). Another example from the same region is the strategy being deployed to minimize infection with CMV in the lupin breeding program described by Jones (2001). Here, protection of the small plots from infection is paramount rather than the cost of the control measures or environmental issues, so a very comprehensive range of individual measures can be used, including regular insecticide applications.

When information on the effectiveness of individual control measures is limited or lacking for a specific virus disease, crop, and growing situation suffering from damaging virus epidemics, the farmers' need for advice on how to control the virus is still immediate. However, obtaining information from field experiments on the effectiveness of individual control measures is expensive and may take several years. This is too long for the farmer to wait. In these circumstances, an "interim IDM strategy" is required which is devised using available epidemiological knowledge and "generic" information on control measures known to work well with other similar combinations of virus and legume crop (Jones et al., 2004). However, such interim approaches should be followed up by field experimentation on control measures and collection of epidemiological information from field situations where the strategy is, or is not, being used so as to help validate and fine-tune it.

An example of an "interim IDM strategy" which is very comprehensive is the one devised by Jones et al. (2004) to control seed and aphid-borne CMV, AMV, BYMV, and PSbMV in field pea, chickpea, faba bean, and lentil plots at breeding, selection, and seed increase sites. An "interim IDM strategy" that uses some of these measures and is affordable to farmers is available for use against the same four seed-borne viruses in commercial crops of these pulses in Australia (R. A. C. Jones, unpublished).

9. IMPLICATIONS OF CLIMATE CHANGE AND NEW TECHNOLOGIES

The rapid pace at which the climate is currently changing is directly related to by the rapid increase in the human population. Climate change is likely to diminish the effectiveness of certain control measures, such as some cultural control measures and temperature-sensitive single gene resistance. Viral epidemics are projected to become less predictable, causing increasing difficulties in suppressing them successfully using current management technologies. Understanding what alterations climate change is likely to cause to the occurrence and importance of virus disease epidemics in cool-season grain legumes is therefore important. Gaining such an understanding is made more difficult by the effects of climate alterations on behavior of different kinds of invertebrate and fungal vectors and the increasing impact of human activities (Jones, 2009, 2014a; Jones & Barbetti, 2012).

Rapidly increasing technological innovation is providing an increasingly important means to help address the threat plant virus disease epidemics pose to future food security. This innovation is making available new techniques that provide considerable opportunities to magnify and accelerate the achievements of worldwide research on plant virus control measures. New technologies can help ensure much more efficient and intelligent choices that ensure correct decisions are made over which control measures to employ. For example, recent innovations in remote sensing and precision agriculture provide valuable information about virus epidemics occurring at continental, regional, or district scales (via satellites) and within individual crops (mostly via lightweight unmanned aerial vehicles), and exactly where to target control measures. Moreover, improvements in information systems and innovations in modeling improve (i) understanding of virus epidemics and ability to predict them and (ii) delivery to end-users of advice on control measure deployment (e.g., Jones, 2014b).

10. CONCLUSIONS

Plant breeding programs aiming to breed for virus resistance into cool-season grain legumes often concentrate on incorporation of major genes that provide of extreme resistance (immunity) or single gene hypersensitive resistance (which is often strain specific) in genotypes already adapted to the target environment. However, such resistances are not always available, while a wide range of partial resistance categories can be found for most virus diseases. In many cases, eliminating susceptible genotypes by screening breeding material for virus resistance in sites with a high frequency of natural virus infection can be sufficient to avoid severe virus-induced losses. Field screening of large collections of established cultivars and germplasm accessions can be particularly useful in quantifying levels of partial resistance to a range of viruses. For example, differences in susceptibility in lentil genotypes are reported for BYMV, CMV, and AMV (Latham & Jones, 2001b; Latham, Jones, & McKirdy, 2001; McKirdy et al., 2000), and in chickpea for CMV (Latham et al., 2001), BYMV and PSbMV (Latham & Jones, 2001b; McKirdy et al., 2000). However, even though breeding virus-resistant cool-season grain legume cultivars is a critical control measure, often only a few virus-resistant cultivars are widely used by farmers. One of the reasons is the fact that in some regions around the world the virus situation is unstable. In one season, a specific virus against which there are virus-resistant cultivars can dominate the scene, whereas in the following season, it could become a minor causal agent and other viruses the cultivars are not resistant to may come to the fore. Such complications happen because virus vectors are greatly affected by weather conditions. Understanding the macro agroecological (landscape) system is therefore critical for devising sustainable solutions.

The most effective approach for virus disease control in legume crops is through IDM, that is, by crop management or ecosystem management. Experience over several decades has shown that legume viruses can rarely be controlled adequately just by applying chemicals to the virus-infected crop. The complicated ecology of many of the viruses that affect legume crops, and for that matter, crops in general, calls for strategies which are long term, sustainable, friendly to the environment, and economically feasible to (1) resource-poor subsistence farmers in developing countries or (2) large-scale, low input, or small-scale high input farmers in developed or developing countries. The soundest approach possible is to exploit all available control measures and recommend their use in an integrated manner. Farmers can then decide which combination is most appropriate to deploy in their particular circumstances.

It is evident from the above review that many approaches are available to control virus diseases of legume crops, and also that, in both developing and developed countries, a number of economically important virus diseases are being controlled by comprehensive IDM approaches based on existing knowledge. Successful implementation of IDM relies on sound epidemiological knowledge of the virus and—especially—the virus vector, and reliable information on the modes of action and effectiveness of potential control measures. However, when such information is insufficient for a specific combination of virus and legume crop in a specific region, an "interim" approach toward devising an IDM strategy is possible by employing available "generic" information on control measures from related systems. Such an approach proved successful in devising IDM tactics for grain legume programs in Australia (Jones et al., 2004). However, "interim IDM strategies" should always be validated subsequently and the tactics included need to be fine-tuned as needed.

An important obstacle for dissemination of reliable IDM practices is ensuring their adoption on a sufficiently large scale by farmers. This applies in both developing and developed countries. In some cases, researchers have devised effective IDM strategies that function successfully in minimizing the losses caused by a particular virus disease. However, the information was not then disseminated to farmers because of serious limitations in extension services, which, unfortunately, have often declined rather than improved in recent years in many countries. This is an area that requires special attention and is beyond the scope of this review. But it is important to emphasize that extension methodology needs to be improved and more resources allocated to help effectively enhance dissemination of viral IDM approaches. In addition, wherever possible, cool-season grain legume virus researchers themselves should be encouraged to become more involved in technology transfer (Thresh, 2003, 2006).

REFERENCES

A'Brook, J. (1964). The effect of planting date and spacing on the incidence of groundnut rosette disease and of the vector *Aphis craccivora* Koch, at Mokawa, Northern Nigeria. *Annals of Applied Biology*, *54*, 199–208.

Abraham, A., & Makkouk, K. M. (2002). The incidence and distribution of seed transmitted viruses in pea and lentil seed lots in Ethiopia. *Seed Science and Technology*, *30*, 567–574.

Abraham, A., Makkouk, K. M., Gorfu, D., Lencho, A. G., Ali, K., Tadessi, N., et al. (2000). Survey of faba bean (*Vicia faba* L.) virus diseases in Ethiopia. *Phytopathologia Mediterranea*, *39*, 277–282.

Abraham, A. D., Menzel, W., Lesemann, D. E., Varrelmann, M., & Vetten, H. J. (2006). *Chickpea chlorotic stunt virus*: A new *Polerovirus* infecting cool-season food legumes in Ethiopia. *Phytopathology*, *96*, 437–446.

Abraham, A. D., Menzel, W., Varrelmann, M., & Vetten, H. J. (2009). Molecular, serological and biological variation among chickpea chlorotic stunt virus isolates from five countries of North Africa and West Asia. *Archives of Virology*, *154*, 791–799.

Aftab, M., Mughal, S. M., & Ghafoor, A. (1989). Occurrence and identification of bean yellow mosaic virus from faba bean in Pakistan. *Indian Journal of Virology*, *5*, 88–93.

Ait Yahia, A., Aitouada, M., Hadj Arab, K., Belfendess, R., & Sarni, K. (1997). Identification of chickpea stunt viruses in Algeria. *OEPP/EPPO Bulletin, 27*, 265–268.

Ait Yahia, A., Aitouada, M., Hadj Arab, K., Belfendess, R., Sarni, K., & Ouadi, K. (1999). Identification and characterization of bean leaf roll luteoviruses (BLRV), a major component of chickpea stunt disease in Algeria. In *Proceedings of the 2nd regional symposium for cereal and legume diseases, Nabeul, Tunisia, 10–12 November* (pp. 289–293).

Ait Yahia, A., Aitouada, M., Illoul, H., & Tair, M. I. (1997). First occurrence of bean yellow mosaic potyvirus on chickpea in Algeria. *OEPP/EPPO Bulletin, 27*, 261–263.

Al-Ani, R. A., & El-Azzawi, Q. K. (1987). Effect of infection with broad bean mottle and bean yellow mosaic viruses on nitrogen fixation in faba bean. *Journal of Agricultural Sciences (Iraq), 18*, 199–212.

Alconero, R., Provvidenti, R., & Gonsalves, D. (1986). Three pea seedborne mosaic virus pathotypes from pea and lentil germ plasm. *Plant Disease, 70*, 783–786.

Al-Jallad, R., Kumari, S. G., & Ismail, I. D. (2007). Integrated management of aphid-transmitted faba bean viruses in the Coastal Area of Syria. *Arab Journal of Plant Protection, 25*, 175–180.

Alkhalaf, M., Kumari, S. G., Haj Kasem, A., Makkouk, K. M., & Al-Chaabi, S. (2010). *Bean yellow mosaic virus* on cool-season food legumes and weeds: Distribution and its effect on faba bean yield and control in Syria. *Arab Journal of Plant Protection, 28*, 38–47.

Allam, E. K., Gamal Eldin, A. S., & Riskallah, L. R. (1979). Some viruses affecting broad bean in Egypt. *Egyptian Journal of Phytopathology, 11*, 67–77.

Al-Musa, A., Al-Haj, H., Mansour, A., & Janakat, S. (1987). Properties of bean yellow mosaic virus occurring on broad bean in the Jordan valley. *Dirasat, 14*, 135–140.

Al-Musa, A. M., Al-Haj, H. A., & Monayer, L. O. (1986). Light and electron microscopy of the Jordanian isolate of broad bean wilt virus. *Dirasat, 8*, 57–62.

Al-Nsour, A. H., Mansour, A., Al-Musa, A., & Salem, N. (1998). Distribution and incidence of faba bean necrotic yellows virus in Jordan. *Plant Pathology, 47*, 510–515.

Anonymous (1999). Chickpea variety 'Gully'. *Plant Varieties Journal, 12*, 21–23.

Asaad, N. Y., Kumari, S. G., Hasj-Kassem, A. A., Shalaby, A., Al-Chaabi, S., & Malhotra, R. S. (2009). Detection and characterization of Chickpea chlorotic stunt virus in Syria. *Journal of Phytopathology, 157*, 756–761.

Ashby, J. W. (1984). Bean leaf roll virus. In *CMI/AAB descriptions of plant viruses no. 286*. Wellesbourne, Warwick, UK: Association of Applied Biologists.

Assad, N., Kumari, S. G., Haj-Kassem, A. A., Al-Chaabi, S., & Malhotra, R. (2012). Effect of some insecticides, agricultural measures on control of viruses causing yellowing on chickpea in Syria. *Arab Journal of Plant Protection, 30*, 86–94.

Aydin, H., Meuhlbauer, F. J., & Kaiser, W. J. (1987). Pea enation mosaic virus resistance in lentil (*Lens culinaris*). *Plant Disease, 71*, 635–638.

Babin, M., Ortiz, V., Castro, S., & Romero, J. (2000). First detection of faba bean necrotic yellows virus in Spain. *Plant Disease, 84*, 707.

Baggett, J. R., & Hampton, R. O. (1977). Oregon B442-15 and B445-16 pea seed-borne mosaic virus resistant lines. *HortScience, 12*, 635–638.

Baggett, J. R., & Hampton, R. O. (1991). Inheritance of viral bean leaf roll tolerance in peas. *Journal of the American Society for Horticultural Science, 116*, 728–731.

Bashir, M., & Malik, B. A. (1993). Natural occurrence of cucumber mosaic virus in chickpea in Pakistan. *International Chickpea Newsletter, 29*, 10.

Bayaa, B., Kumari, S. G., Akkaya, A., Erskine, W., Makkouk, K. M., Turk, Z., et al. (1998). Survey of major biotic stresses of lentil in South-East Anatolia, Turkey. *Phytopathologia Mediterranea, 37*, 88–95.

Bekele, B., Kumari, S. G., Ali, K., Yusuf, A., Makkouk, K. M., Aslake, M., et al. (2005). Survey for viruses affecting legume crops in Amhara and Oromia Regions in Ethiopia. *Phytopathologia Mediterranea, 44*, 235–246.

Boiteux, L. S., de Avila, A. C., Giordano, L. B., Lima, M. I., & Kitajima, E. W. (1995). Apical chlorosis disease of chickpea (*Cicer arietinum*) caused by tomato spotted wilt virus in Brazil. *Journal of Phytopathology, 143*, 629–631.
Bos, L. (1970). Bean yellow mosaic virus. In *CMI/AAB descriptions of plant viruses no. 40.* Wellesbourne, Warwick, UK: Association of Applied Biologists.
Bos, L., Hampton, R. O., & Makkouk, K. M. (1988). Viruses and virus diseases of pea, lentil, faba bean and chickpea. In R. J. Summerfield (Ed.), *World crops: Cool season food legumes* (pp. 591–615). Dordrecht, The Netherlands: Kluwer Academic Publishers.
Bos, L., Mahir, M. A. M., & Makkouk, K. M. (1993). Some properties of pea early-browning tobravirus from faba bean (*Vicia faba* L.) in Libya. *Phytopathologia Mediterranea, 32*, 7–13.
Boswell, K. F., & Gibbs, A. J. (1983). Viruses of legumes 1983. In *Descriptions and keys from VIDE.*Canberra, Australia: The Australian National University, Research School of Biological Sciences 139 pp.
Broadbent, L. (1957). Insecticidal control of the spread of plant viruses. *Annual Review of Entomology, 2*, 339–354.
Bruun-Rasmussen, M., Moller, I. S., Tulinius, G., Hansen, J. K. R., Lund, O. S., & Johansen, I. E. (2007). The same allele of translation initiation factor 4E mediates resistance against two *Potyvirus* spp. in *Pisum sativum. Molecular Plant-Microbe Interactions, 20*, 1075–1182.
Buchen-Osmond, C., Crabtree, K., Gibbs, A., & McLean, G. (1988). *Viruses of plants in Australia.* Canberra, ACT: Australian National University Printing Service.
Bwye, A. M., Jones, R. A. C., & Proudlove, W. (1994). Effects of sowing seed with different levels of infection, plant density and the growth stage at which plants first develop symptoms of cucumber mosaic virus infection in narrow-leafed lupins (*Lupinus angustifolius*). *Australian Journal of Agricultural Research, 45*, 1395–1412.
Bwye, A. M., Jones, R. A. C., & Proudlove, W. (1999). Effects of different cultural practices on spread of cucumber mosaic virus in narrow-leafed lupins (*Lupinus angustifolius*). *Australian Journal of Agricultural Research, 50*, 985–996.
Chalam, T. V., Reddy, J., Subbayya, J., Nene, N. L., & Beniwal, S. P. S. (1986). Screening of chickpea for resistance to cucumber mosaic and bean yellow mosaic viruses. *International Chickpea Newsletter, 14*, 25–26.
Cheng, Y., & Jones, R. A. C. (1999). Distribution and incidence of the necrotic and non-necrotic strains of bean yellow mosaic virus in wild and crop lupins. *Australian Journal of Agricultural Research, 50*, 589–599.
Cheng, Y., & Jones, R. A. C. (2000). Biological properties of necrotic and non-necrotic strains of bean yellow mosaic virus in cool season grain legumes. *Annals of Applied Biology, 136*, 215–227.
Cheng, Y., Jones, R. A. C., & Thackray, D. J. (2002). Deploying strain specific hypersensitive resistance to diminish temporal virus spread. *Annals of Applied Biology, 140*, 69–79.
Chu, P. W. G., & Helms, K. (1988). Novel virus-like particles containing circular single-stranded DNA's associated with subterranean clover stunt disease. *Virology, 167*, 38–49.
Cockbain, A. J. (1983). Viruses and virus-like diseases of *Vicia faba* L. In P. D. Hebblethwaite (Ed.), *The faba bean (*Vicia faba *L.): A basis for improvement* (pp. 421–462). London: Butterworths.
Coutts, B. A., Hawkes, J. R., & Jones, R. A. C. (2006). Occurrence of *Beet western yellows virus* and its aphid vectors in over-summering broad-leafed weeds and volunteer crop plants in the grain belt region of south-western Australia. *Australian Journal of Agricultural Research, 57*, 975–982.
Coutts, B. A., Prince, R. T., & Jones, R. A. C. (2008). Further studies on *Pea seed-borne mosaic virus* in cool-season crop legumes: Responses to infection and seed quality defects. *Australian Journal of Agricultural Research, 59*, 1130–1145.

Coutts, B. A., Prince, R. T., & Jones, R. A. C. (2009). Quantifying effects of seedborne inoculum on virus spread, yield losses, and seed infection in the *Pea seed-borne mosaic virus* field pea pathosystem. *Phytopathology, 99*, 1156–1167.

Crampton, M. J., & Watts, L. E. (1968). Genetic studies of pea leaf-roll (top-yellows) virus resistance in *Pisum sativum*. *New Zealand Journal of Agricultural Research, 11*, 771–783.

Cutforth, H. W., McConkey, B. G., Ulrich, D., Miller, P. R., & Angadi, S. V. (2002). Yield and water use efficiency of pulses seeded directly into standing stubble in the semiarid Canadian prairie. *Canadian Journal of Plant Science, 82*, 681–686.

Dhingra, K. L., Chenulu, V. V., & Verma, A. (1979). A leaf reduction disease of *Cicer arietinum* in India, caused by a cucumovirus. *Current Science, 48*, 486–488.

Diekmann, M., & Putter, C. A. J. (1995). Small grain temperate cereals. *FAO/IPGRI technical guidelines for the safe movement of germplasm no. 14* (67 pp).

Drijfhout, E. (1968). Testing for pea leaf roll virus and inheritance of resistance in peas. *Euphytica, 17*, 224–235.

Duffus, J. E. (1972). Beet western yellows virus. In *CMI/AAB descriptions of plant viruses no. 89* (4 p). Wellesbourne, Warwick, UK: Association of Applied Biologists.

El-Amri, A. (1999). Identification and repartition of faba bean necrotic yellows virus (FBNYV) in Morocco. *Al Awamia, 99*, 19–26.

El-Attar, S., Ghabrial, S. A., & Nour Eldin, F. (1971). A strain of alfalfa mosaic virus on broad bean in the Arab Republic of Egypt. *Agricultural Research Review (Egypt), 49*, 277–284.

El-Maataoui, M., & El-Hassani, A. (1984). Cucumber mosaic virus of chickpea in Morocco. *International Chickpea Newsletter, 10*, 14–15.

El-Muadhidi, M. A., Makkouk, K. M., Kumari, S. G., Jerjess, M., Murad, S. S., Mustafa, R. R., et al. (2001). Survey for legume and cereal viruses in Iraq. *Phytopathologia Mediterranea, 40*, 224–233.

Fadel, S., Khalil, J., & Shagrun, M. (2005). First record of Faba bean necrotic yellows virus and a luteovirus in the faba bean crop (*Vicia faba* L.) in Libya. *Arab Journal of Plant Protection, 23*, 132.

Farzadfar, Sh., & Izadpanah, K. (2001). Sources and properties of the Iranian isolate of *Pea enation mosaic virus*. *Iranian Journal of Plant Pathology, 37*, 77.

Fereres, A. (2000). Barrier crops as a cultural control measure of non-persistently transmitted aphid-borne viruses. *Virus Research, 71*, 221–223.

Ford, R. E., & Baggett, J. R. (1965). Reactions of plant introduction lines of *Pisum sativum* to alfalfa mosaic, clover yellow mosaic and pea streak viruses, and to powdery mildew. *Plant Disease Reporter, 49*, 787–789.

Fortass, M., & Bos, L. (1991). Survey of faba bean (*Vicia faba* L.) for viruses in Morocco. *Netherlands Journal of Plant Pathology, 97*, 369–380.

Fortass, M., & Diallo, S. (1993). Broad bean mottle bromovirus in Morocco; curculionid vectors, and natural occurrence in food legumes other than faba bean (*Vicia faba* L.). *Netherlands Journal of Plant Pathology, 99*, 219–226.

Fortass, M., van der Wilk, F., van den Heuvel, J. E. J., & Goldbach, R. W. (1997). Molecular evidence for the occurrence of beet western yellows virus on chickpea in Morocco. *European Journal of Plant Pathology, 103*, 481–484.

Francki, R. I. B., Mossop, D. W., & Hatta, T. (1979). Cucumber mosaic virus. In *CMI/AAB descriptions of plant viruses no. 213* (6 p). Wellesbourne, Warwick, UK: Association of Applied Biologists.

Franz, A., Makkouk, K. M., & Vetten, H. J. (1997). Host range of faba bean necrotic yellows virus and potential yield loss in infected faba bean. *Phytopathologia Mediterranea, 36*, 94–103.

Freeman, A. J., & Aftab, M. (2001). Surveying for and mapping of viruses in pulse crops in south-eastern Australia. In *Proceedings of the Australasian plant pathology society 13th biennial conference, 24–27 September 2001, Cairns, Queensland* (p. 149).

Freeman, A. J., Spackman, M., Aftab, M., McQueen, V., King, S., van Leur, J. A. G., et al. (2013). Comparison of tissue blot immunoassay and high throughput PCR for virus-testing samples from a south eastern Australian pulse virus survey. *Australasian Plant Pathology, 42*, 675–683. http://dx.doi.org/10.1007/s13313-013-0252-9.

Fresno, J., Castro, S., Babin, M., Carazo, G., Molina, A., De Blas, C., et al. (1997). Virus disease of broad bean in Spain. *Plant Disease, 81*, 112.

Gamal-Eldin, A. S., El-Amrety, A. A., Mazyad, H. M., & Rizkallah, L. R. (1982). Effect of bean yellow mosaic and broad bean wilt viruses on broad bean yield. *Agriculture Research Review (Egypt), 60*, 195–204.

Gao, Z., Johansen, I. E., Eyers, S., Thomas, C. L., Ellis, T. H. N., & Maule, A. J. (2004). The potyvirus recessive resistance gene, *sbm1*, identifies a novel role for translation initiation factor eIF4E in cell-to-cell trafficking. *The Plant Journal, 40*, 376–385.

Garcia-Arenal, F., & McDonald, B. A. (2003). An analysis of the durability of resistance to plant viruses. *Phytopathology, 93*, 941–952.

Ghad, I. P. S., & Bernier, C. C. (1984). Resistance in faba bean (*Vicia faba*) to bean yellow mosaic virus. *Plant Disease, 68*, 109–111.

Gibbs, A. J. (1972). Broad bean mottle virus. In *CMI/AAB descriptions of plant viruses no. 101* (4 p). Wellesbourne, Warwick, UK: Association of Applied Biologists.

Gibbs, A. J., & Paul, H. L. (1970). Echtes ackerbohnemosaik-virus. In *CMI/AAB descriptions of plant viruses no. 20* (4 p). Wellesbourne, Warwick, UK: Association of Applied Biologists.

Gibbs, A. J., & Smith, H. G. (1979). Broad bean stain virus. In *CMI/AAB descriptions of plant viruses no. 29* (3 p). Wellesbourne, Warwick, UK: Association of Applied Biologists.

Haddad, N. J., Meuhlbauer, F. J., & Hampton, R. O. (1978). Inheritance of resistance to pea seed-borne mosaic virus in lentils. *Crop Science, 18*, 613–615.

Hagedorn, D. J. (1968). Disease reaction of *Pisum sativum* plant introductions to three legume viruses. *Plant Disease Reporter, 52*, 160–162.

Hagedorn, D. J., & Gritton, E. T. (1973). Inheritance of resistance to the pea seed-borne mosaic virus. *Phytopathology, 63*, 1130–1133.

Hamed, A. A., & Makkouk, K. M. (2002). Occurrence and management of *Chickpea chlorotic dwarf virus* in chickpea fields in northern Sudan. *Phytopathologia Mediterranea, 41*, 193–198.

Hampton, R. O. (1982). Incidence of the lentil strain of pea seedborne mosaic virus as a contaminant of *Lens culinaris* germ plasm. *Phytopathology, 72*, 695–698.

Hampton, R. O. (1984). Diseases caused by viruses. In D. J. Hagedorn (Ed.), *Compendium of pea diseases and pests* (pp. 31–37). St. Paul, MN: American Phytopathological Society.

Hampton, R. O., & Mink, G. I. (1975). Pea seed-borne mosaic virus. In *CMI/AAB descriptions of plant viruses no. 146* (4 p). Wellesbourne, Warwick, UK: Association of Applied Biologists.

Harrison, B. D. (1973). Pea early-browning virus. In *CMI/AAB descriptions of plant viruses no. 120* (4 p). Wellesbourne, Warwick, UK: Association of Applied Biologists.

Harrison, B. D. (2002). Virus variation in relation to resistance-breaking in plants. *Euphytica, 124*, 181–189.

Heinrichs, E. A. (1979). Control of leafhopper and plant hopper vectors of rice viruses. In K. Maramorosch, & K. F. Harris (Eds.), *Leafhopper vectors and plant disease agents* (pp. 529–560). New York: Academic Press.

Hjulsager, C. K., Lund, O. S., & Johansen, I. E. (2002). A new pathotype of Pea seedborne mosaic virus explained by properties on the P3-6k1- and viral genome-linked protein (VPg)-coding regions. *Molecular Plant-Microbe Interactions, 15*, 169–171.

Horn, N. M., Makkouk, K. M., Kumari, S. G., van den Heuvel, H. F., & Reddy, D. V. R. (1995). Survey of chickpea (*Cicer arientinum* L.) for chickpea stunt disease and associated viruses in Syria, Turkey and Lebanon. *Phytopathologia Mediterranea, 34*, 192–198.

Horn, N. M., Reddy, S. V., Roberts, I. M., & Reddy, D. V. R. (1993). Chickpea chlorotic dwarf virus, a new leafhopper-transmitted geminivirus of chickpea in India. *Annals of Applied Biology, 122*, 467–479.

Hubbeling, N. (1956). Resistance to top yellows and *Fusarium* wilt in peas. *Euphytica, 5*, 71–86.

Inouye, T., Inouye, N., & Mitsuhata, K. (1968). Yellow dwarf of pea and broad-bean caused by milk vetch dwarf virus. *Annals of the Phytopathological Society of Japan, 34*, 28–35.

Irwin, M. E., & Nault, L. R. (1996). Virus/vector control. In G. S. Persley (Ed.), *Biotechnology and integrated pest management* (pp. 304–322). London: CAB International.

Ismail, I. D., & Hassan, M. H. M. (1995). Survey of seed-borne viruses of faba bean in Sebha region south of Libya. *Journal University of Sebha, 2*, 95–109.

Jaspars, E. M. J., & Bos, L. (1980). Alfalfa mosaic virus. In *CMI/AAB descriptions of plant viruses no. 229* (4 p). Wellesbourne, Warwick, UK: Association of Applied Biologists.

Johansen, I. E., Lund, O. S., Hjulsager, C. K., & Laursen, J. (2001). Recessive resistance in *Pisum sativum* and potyvirus pathotype resolved in a gene-for-cistron correspondence between host and virus. *Journal of Virology, 75*, 6609–6614.

Johnstone, G. R., & Duffus, J. E. (1984). Some luteovirus diseases in Tasmania caused by beet western yellows and subterranean clover red leaf viruses. *Australian Journal of Agricultural Research, 35*, 821–830.

Johnstone, G. R., & Guy, P. L. (1986). Epidemiology of viruses persistently transmitted by aphids. In *Proceedings of the workshop on epidemiology of plant virus diseases IX/1–IX/7, international society of plant pathology, Orlando, FL, USA* (pp. 6–8).

Johnstone, G. R., Liu, H.-Y., & Duffus, J. E. (1984). First report of a subterranean clover red leaf-like virus in the Western Hemisphere (Abstr.). *Phytopathology, 74*, 795.

Jones, R. A. C. (1991). Reflective mulch decreases the spread of two non-persistently aphid transmitted viruses to narrow-leafed lupin (*Lupinus angustifolius*). *Annals of Applied Biology, 118*, 79–85.

Jones, R. A. C. (1993). Effects of cereal borders, admixture with cereals and plant density on the spread of bean yellow mosaic potyvirus into narrow-leafed lupins (*Lupinus angustifolius*). *Annals of Applied Biology, 122*, 501–518.

Jones, R. A. C. (1994). Effect of mulching with cereal straw and row spacing on spread of bean yellow mosaic potyvirus into narrow-leafed lupins (*Lupinus angustifolius*). *Annals of Applied Biology, 124*, 45–58.

Jones, R. A. C. (2000). Determining threshold levels for seed-borne virus infection in seed stocks. *Virus Research, 71*, 171–183.

Jones, R. A. C. (2001). Developing integrated disease management strategies against non-persistently aphid-borne viruses: A model programme. *Integrated Pest Management Reviews, 6*, 15–46.

Jones, R. A. C. (2004). Using epidemiological information to develop effective integrated virus disease management strategies. *Virus Research, 100*, 5–30.

Jones, R. A. C. (2005). Patterns of spread of two non-persistently aphid-borne viruses in lupin stands under four different infection scenarios. *Annals of Applied Biology, 146*, 337–350.

Jones, R. A. C. (2006). Control of plant virus diseases. *Advances in Virus Research, 67*, 205–244.

Jones, R. A. C. (2009). Plant virus emergence and evolution: Origins, new encounter scenarios, factors driving emergence, effects of changing world conditions, and prospects for control. *Virus Research, 141*, 113–130.

Jones, R. A. C. (2014a). Plant virus ecology and epidemiology: Historical perspectives, recent progress and future prospects. *Annals of Applied Biology, 164*, 320–347.

Jones, R. A. C. (2014b). Trends in plant virus epidemiology: Opportunities from new or improved technologies. *Virus Research, 186*, 3–19.

Jones, R. A. C., & Barbetti, M. J. (2012). Influence of climate change on plant disease infections and epidemics caused by viruses and bacteria. *CAB Reviews*, 7(22), 1–32. Retrieved from, http://www.cabi.org/cabreviews.

Jones, R. A. C., & Burchell, G. M. (2004). Resistance to *Cucumber mosaic virus* in *Lupinus mutabilis* (Pearl lupin). *Australasian Plant Pathology*, *33*, 591–593.

Jones, R. A. C., & Coutts, B. A. (1996). Alfalfa mosaic and cucumber mosaic virus infection in chickpea and lentil: incidence and seed transmission. *Annals of Applied Biology*, *129*, 491–506.

Jones, R. A. C., Coutts, B. A., & Cheng, Y. (2003). Yield limiting potential of necrotic and non-necrotic strains of bean yellow mosaic virus in narrow-leafed lupin (*Lupinus angustifolius*). *Australian Journal of Agricultural Research*, *54*, 849–859.

Jones, R. A. C., Coutts, B. A., Latham, L. J., & McKirdy, S. J. (2008). Cucumber mosaic virus infection of chickpea stands: Temporal and spatial patterns of spread and yield-limiting potential. *Plant Pathology*, *57*, 842–853.

Jones, R. A. C., Coutts, B. A., Mackie, A. E., & Dwyer, G. I. (2005). Seed transmission of *Wheat streak mosaic virus* shown unequivocally in wheat. *Plant Disease*, *89*, 1048–1050.

Jones, R. A. C., & Cowling, W. A. (1995). Resistance to seed transmission of cucumber mosaic virus in narrow-leafed lupins (*Lupinus angustifolius*). *Australian Journal of Agricultural Research*, *46*, 1339–1352.

Jones, R. A. C., & Latham, L. J. (1996). Natural resistance to cucumber mosaic virus in lupin species. *Annals of Applied Biology*, *129*, 523–542.

Jones, R. A. C., Latham, L. J., & Coutts, B. A. (2004). Devising integrated disease management tactics against plant viruses from 'generic' information on control measures. *Agricultural Science, Australia*, *17*, 10–18.

Jones, R. A. C., & McLean, G. D. (1989). Virus diseases of lupins. *Annals of Applied Biology*, *114*, 609–637.

Jones, R. A. C., Pearce, R. M., Prince, R. T., & Coutts, B. A. (2008). Natural resistance to *Alfalfa mosaic virus* in different lupin species. *Australasian Plant Pathology*, *37*, 112–116.

Jones, R. A. C., Salam, M. U., Maling, T., Diggle, A. J., & Thackray, D. J. (2010). Principles of predicting epidemics of plant virus disease. *Annual Review of Phytopathology*, *48*, 179–203.

Jones, R. A. C., & Smith, L. J. (2005). Inheritance of hypersensitive resistance to *Bean yellow mosaic virus* in narrow-leafed lupin (*Lupinus angustifolius*). *Annals of Applied Biology*, *146*, 539–543.

Kaiser, W. J., Danesh, D., Okhovat, M., & Mossahebi, H. (1968). Disease of pulse crops (edible legumes) in Iran. *Plant Disease Reporter*, *52*, 687–691.

Kasimor, K., Bagget, J. R., & Hampton, R. O. (1997). Pea cultivar susceptibility and inheritance of resistance to the lentil strain (pathotype P2) of pea seedborne mosaic virus. *Journal of the American Society for Horticultural Science*, *122*, 325–328.

Katul, L., Vetten, H. J., Maiss, E., Makkouk, K. M., Lesemann, D. E., & Casper, R. (1993). Characterization and serology of virus-like particles associated with faba bean necrotic yellows. *Annals of Applied Biology*, *123*, 629–647.

Kehoe, M. A., Coutts, B. A., Buirchell, B., & Jones, R. A. C. (2014). Black pod syndrome of *Lupinus angustifolius* is caused by late infection with *Bean yellow mosaic virus*. *Plant Disease*, *98*, 739–745.

Kraft, J. M., & Giles, R. A. (1978). Registration of VR 74-410-2 and VR 1492-1 pea germplasm (Reg. Nos. GP 19 to 20). *Crop Science*, *18*, 1099.

Kraft, J. M., & Hampton, R. O. (1980). Crop losses from pea seed-borne mosaic virus in six processing pea cultivars. *Plant Disease*, *64*, 922–924.

Kraft, J. M., & Kaiser, W. J. (1993). Screening for disease resistance in pea. In K. B. Singh, & M. C. Saxena (Eds.), *Breeding for stress tolerance in cool-season food legumes* (pp. 123–144). Chichester, UK: John Wiley & Sons.

Kumar, P. L., Kumari, S. G., & Waliyar, F. (2008). Virus diseases of chickpea. In G. P. Rao, P. L. Kumar, & R. J. H. Penna (Eds.), *Characterization, diagnosis and management of plant viruses: Vol 3. Vegetable and pulse crops* (pp. 213–234). Texas, USA: Studium Press LLC.

Kumari, S. G., Larsen, R., Makkouk, K. M., & Bashir, M. (2009). Virus diseases of lentil and their control. In W. Erskine, F. J. Muehlbauer, A. Sarker, & B. Sharma (Eds.), *The lentil: Botany, production and uses* (pp. 306–325). UK: CABI.

Kumari, S. G., & Makkouk, K. M. (1995). Variability among twenty lentil genotypes in seed transmission rates and yield loss induced by pea seed-borne mosaic potyvirus infection. *Phytopathologia Mediterranea, 34*, 129–132.

Kumari, S. G., & Makkouk, K. M. (1996). Inactivation of broad bean stain comovirus in lentil seeds by dry heat treatment. *Phytopathologia Mediterranea, 35*, 124–126.

Kumari, S. G., & Makkouk, K. M. (2007). Virus diseases of faba bean (*Vicia faba* L.) in Asia and Africa. *Plant Viruses, 1*, 93–105.

Kumari, S. G., Makkouk, K. M., Asaad, N., Attar, N., & Hlaing Loh, M. (2007). *Chickpea chlorotic stunt virus* affecting cool-season food legumes in West Asia and North Africa. In *Abstract book of 10th international plant virus epidemiology symposium, on the theme "Controlling Epidemics of Emerging and Established Plant Virus Diseases—The Way Forward", 15–19 October 2007, Hyderabad, India* (p. 157).

Kumari, S. G., Makkouk, K. M., Attar, N., Ghulam, W., & Lesemann, D.-E. (2004). First report of *Chickpea chlorotic dwarf virus* infecting spring chickpea in Syria. *Plant Disease, 88*, 424.

Kumari, S. G., Makkouk, K. M., Loh, M., Negassi, K., Tsegay, S., Kidane, R., et al. (2008). Viral diseases affecting chickpea crop in Eritrea. *Phytopathologia Mediterranea, 47*, 42–49.

Kumari, S. G., Rodoni, B., Vetten, H.-J., Hlaing Loh, M., Freeman, A., van Leur, J. A. G., et al. (2010). Detection and partial characterization of *Milk vetch dwarf virus* isolates from faba bean (*Vicia faba* L.) in Yunnan Province, China. *Journal of Phytopathology, 158*, 35–39.

Larsen, R. C. (2000). Foliar diseases caused by viruses. In J. M. Kraft, & F. L. Pfleger (Eds.), *Compendium of pea diseases and pests* (pp. 32–39) (2nd ed.). St. Paul, MN: American Phytopathological Society Press.

Larsen, R. C., & Porter, L. D. (2010). Identification of novel sources of resistance to *Pea enation mosaic virus* in chickpea germplasm. *Plant Pathology, 59*, 42–47.

Larsen, R. C., & Webster, D. M. (1999). First report of bean leafroll luteovirus infecting pea in Italy. *Plant Disease, 83*, 399.

Latham, L. J., & Jones, R. A. C. (2001a). Distribution and incidence of virus infection in experimental plots, commercial crops and seed stocks of cool season crop legumes. *Australian Journal of Agricultural Research, 52*, 397–413.

Latham, L. J., & Jones, R. A. C. (2001b). Alfalfa mosaic and pea seed-borne mosaic viruses in cool season crop, annual pasture and forage legumes: Susceptibility, sensitivity and seed transmission. *Australian Journal of Agricultural Research, 52*, 710–790.

Latham, L. J., Jones, R. A. C., & McKirdy, S. J. (2001). Cucumber mosaic cucumovirus infection of cool season crop, annual pasture and forage legumes: Susceptibility, sensitivity and seed transmission. *Australian Journal of Agricultural Research, 52*, 683–689.

Latham, L. J., Jones, R. A. C., & Coutts, B. A. (2004). Yield losses caused by virus infection in four combinations of non-persistently aphid-borne virus and cool-season crop legume. *Australian Journal of Agricultural Research, 44*, 57–63.

Lockhart, B. E. L., & Fischer, H. U. (1976). Some properties of an isolate of pea early browning virus occurring in Morocco. *Phytopathology, 66*, 1391–1394.

Loebenstein, G., & Raccah, B. (1980). Control of non-persistently transmitted aphid-borne viruses. *Phytoparasitica, 8*, 221–235.

Loebenstein, G., & Thottappilli, G. (Eds.). (2003). *Virus and virus-like diseases of major crops in developing countries*. Dordrecht, The Netherlands: Kluwer Academic Publishers.

Mahir, M. A.-M., Fortass, M., & Bos, L. (1992). Identification and properties of a deviant isolate of the broad bean yellow band serotype of pea early-browning virus from faba bean (*Vicia faba*) in Algeria. *Netherlands Journal of Plant Pathology, 98*, 237–252.
Makkouk, K. M. (1994). Viruses and virus diseases of cool season food legumes in West Asia and North Africa. *IPA Journal of Agricultural Research (Iraq), 4*, 98–115.
Makkouk, K. M. (1996). Integrated management of virus disease affecting cereal and legume crops. In B. Ezzahiri, A. Lyamani, A. Farih, & M. El-Yamani (Eds.), *Proceedings du symposium régional sur les maladies des céréales et des legumineuses alimentaires, INRA, Rabat, Morocco* (pp. 319–322).
Makkouk, K. M., Bahamish, H. S., Kumari, S. G., & Lotf, A. (1998). Major viruses affecting faba bean (*Vicia faba* L.) in Yemen. *Arab Journal of Plant Protection, 16*, 98–101.
Makkouk, K. M., Bashir, M., Jones, R. A. C., & Kumari, S. G. (2001). Survey for viruses in lentil and chickpea crops in Pakistan. *Journal of Plant Diseases and Protection, 108*, 258–268.
Makkouk, K. M., Bos, L., Azzam, O. I., Katul, L., & Rizkallah, A. (1987). Broad bean stain virus: Identification detectability in faba bean leaves and seeds, occurrence in West Asia and North Africa and possible wild hosts. *Netherlands Journal of Plant Pathology, 93*, 97–106.
Makkouk, K. M., Bos, L., Azzam, O. I., Kumari, S., & Rizkallah, A. (1988). Survey of viruses affecting faba bean in six Arab countries. *Arab Journal of Plant Protection, 6*, 53–61.
Makkouk, K. M., Bos, L., Horn, N. M., & Srinivasa Rao, B. (1993). Screening for virus resistance in cool-season legumes. In K. B. Singh, & M. C. Saxena (Eds.), *Breeding for stress tolerance in cool-season food legumes* (pp. 179–192). UK: Wiley-Sayce Co-Publication.
Makkouk, K. M., Bos, L., Rizkallah, A., Azzam, O. I., & Katul, L. (1988). Broad bean mottle virus: Identification, serology, host range and occurrence on faba bean (*Vicia faba*) in West Asia and North Africa. *Netherlands Journal of Plant Pathology, 94*, 195–212.
Makkouk, K. M., Dafalla, G., Hussein, M., & Kumari, S. G. (1995). The natural occurrence of chickpea chlorotic dwarf geminivirus in chickpea and faba bean in the Sudan. *Journal of Phytopathology, 143*, 465–466.
Makkouk, K. M., Damsteegt, V., Johnstone, G. R., Katul, L., Lesemann, D.-E., & Kumari, S. G. (1997). Identification and some properties of soybean dwarf luteovirus affecting lentil in Syria. *Phytopathologia Mediterranea, 36*, 135–144.
Makkouk, K. M., Fazlali, Y., Kumari, S. G., & Farzadfar, S. (2002). First record of *Beet western yellows, Chickpea chlorotic dwarf, Faba bean necrotic yellows* and *Soybean dwarf viruses* affecting chickpea and lentil crops in Iran. *Plant Pathology, 51*, 387.
Makkouk, K. M., Hamed, A. A., Hussein, M., & Kumari, S. G. (2003). First report of *Faba bean necrotic yellows virus* (FBNYV) infecting chickpea (*Cicer arietinum*) and faba bean (*Vicia faba*) crops in Sudan. *Plant Pathology, 52*, 412.
Makkouk, K. M., & Kumari, S. G. (1990). Variability among 19 lentil genotypes in seed transmission rates and yield loss induced by broad bean stain virus infection. *LENS Newsletter, 17*, 31–33.
Makkouk, K. M., & Kumari, S. G. (1995). Screening and selection of faba bean (*Vicia faba* L.) germplasm for resistance to bean yellow mosaic potyvirus. *Journal of Plant Diseases and Protection, 102*, 461–466.
Makkouk, K. M., & Kumari, S. G. (1998). Further serological characterization of two tobravirus isolates from Algeria and Libya. *Pakistan Journal of Biological Sciences, 1*, 303–306.
Makkouk, K. M., & Kumari, S. G. (2001). Reduction of spread of three persistently aphid-transmitted viruses affecting legume crops by seed-treatment with Imidacloprid (Gaucho®). *Crop Protection, 20*, 433–437.
Makkouk, K. M., & Kumari, S. G. (2009). Epidemiology and integrated management of persistently transmitted aphid-borne viruses of legume and cereal crops in West Asia and North Africa. *Virus Research, 141*, 209–218.

Makkouk, K. M., Kumari, S. G., & Al-Daoud, R. (1992). Survey of viruses affecting lentil (*Lens culinaris* Med.) in Syria. *Phytopathologia Mediterranea, 31*, 188–190.
Makkouk, K. M., Kumari, S. G., & Bayaa, B. (1999). First report of pea enation mosaic virus affecting lentil (*Lens culinaris* Medik.) in Syria. *Plant Disease, 83*, 303.
Makkouk, K. M., Kumari, S. G., & Bos, L. (1990). Broad bean wilt virus: Host range, purification, serology, transmission characteristics, and occurrence in faba bean in West Asia and North Africa. *Netherlands Journal of Plant Pathology, 96*, 291–300.
Makkouk, K. M., Kumari, S. G., & Bos, L. (1993). Pea seed-borne mosaic virus: Occurrence in faba bean (*Vicia faba* L.) and lentil (*Lens culinaris* Med.) in West Asia and North Africa, and further information on host range, purification, serology and transmission characteristics. *Netherlands Journal of Plant Pathology, 99*, 115–124.
Makkouk, K. M., Kumari, S. G., Hughes, J. d'A., Muniyappa, V., & Kulkarni, N. K. (2003). Other legumes: Faba bean, chickpea, lentil, pigeonpea, mungbean, blackgram, lima bean, horegram, bambara groundnut and winged bean. In G. Loebenstein, & G. Thottappilly (Eds.), *Virus and virus-like diseases of major crops in developing countries* (pp. 447–476). Dordrecht, The Netherlands: Kluwer Academic Publishers.
Makkouk, K. M., Kumari, S. G., & Lesemann, D.-E. (2001). First record of *Pea enation mosaic virus* naturally infecting chickpea and grasspea crops in Syria. *Plant Disease, 85*, 1032.
Makkouk, K. M., Kumari, S. G., Sarker, A., & Erskine, W. (2001). Registration of six lentil germplasm lines with combined resistance to viruses. *Crop Science, 41*, 931–932.
Makkouk, K. M., Kumari, S. G., Shahraeen, N., Fazlali, Y., Farzadfar, Sh., Ghotbi, T., et al. (2003). Identification and seasonal variation of viral diseases of chickpea and lentil in Iran. *Journal of Plant Diseases and Protection, 110*, 157–169.
Makkouk, K. M., Kumari, S. G., & van Leur, J. A. G. (2002). Screening and selection of faba bean (*Vicia faba* L.) germplasm resistant to *Bean leafroll virus*. *Australian Journal of Agricultural Research, 53*, 1077–1082.
Makkouk, K. M., Lesemann, D. E., & Haddad, N. A. (1982). Bean yellow mosaic virus from broad bean in Lebanon: Incidence, host range, purification, and serological properties. *Journal of Plant Diseases and Protection, 89*, 59–66.
Makkouk, K. M., Pappu, H. R., & Kumari, S. G. (2012). Virus diseases of peas, beans and faba bean in the Mediterranean region. *Advances in Virus Research, 84*, 367–402.
Makkouk, K. M., Rizkallah, L., Kumari, S. G., Zaki, M., & Abul Enein, R. (2003). First record of *Chickpea chlorotic dwarf virus* (CpCDV) affecting faba bean (*Vicia faba*) crops in Egypt. *Plant Pathology, 52*, 413.
Makkouk, K. M., Rizkallah, L., Madkour, M., El-Sherbeiny, M., Kumari, S. G., Amriti, A. W., et al. (1994). Survey of faba bean (*Vicia faba* L.) for viruses in Egypt. *Phytopathologia Mediterranea, 33*, 207–211.
Makkouk, K. M., Vetten, H. J., Katul, L., Franz, A., & Madkour, M. A. (1998). Epidemiology and control of faba bean necrotic yellows virus. In A. Hadidi, R. K. Khetarpal, & H. Koganezawa (Eds.), *Plant virus disease control* (pp. 534–540). St. Paul, MN: APS Press, The American Phytopathological Society.
Maling, T., Diggle, A. J., Thackray, D. J., Siddique, K. H. M., & Jones, R. A. C. (2008). An epidemiological model for externally sourced vector-borne viruses applied to *Bean yellow mosaic virus* in lupin crops in a Mediterranean-type environment. *Phytopathology, 98*, 1280–1290.
Mazyad, H., El-Hammady, M., & Tolba, M. A. (1975). The broad bean true mosaic disease in Egypt. *Annals of Agricultural Science, Moshtohor, 4*, 87–94.
McKirdy, S. J., Jones, R. A. C., Latham, L. J., & Coutts, B. A. (2000). Bean yellow mosaic potyvirus infection of alternative annual pasture, forage and cool season crop legumes: Susceptibility, sensitivity and seed transmission. *Australian Journal of Agricultural Research, 51*, 325–345.

Milles, P. R., & Ahmed, A. H. (1984). Host range and properties of cucumber mosaic virus (CMV-Su) infecting *Vicia faba* in Sudan. *FABIS Newsletter, 9*, 31–33.
Mouhanna, A. M., Makkouk, K. M., & Ismail, I. D. (1994). Survey of virus disease of wild and cultivated legumes in the coastal region of Syria. *Arab Journal of Plant Protection, 12*, 12–19.
Murant, A. F., Abu Salih, H. S., & Goold, R. A. (1974). Viruses from broad bean in the Sudan. *Annual Report Scottish Horticultural Research Institute for 1973*, 67.
Musil, M. (1966). Über das Vorkommen des Virus des Blatrollens der Erbse in der Slowakei (Vorläufige Mitteilung). *Biologia (Bratislava), 21*, 133–138.
Mustafayev, E., Kumari, S. G., Attar, N., & Zeynal, A. (2011). Viruses infecting chickpea and lentil crops in Azerbaijan. *Australasian Plant Pathology, 40*, 612–620.
Najar, A., Kumari, S. G., Attar, N., & Lababidi, S. (2011). Present status of some virus diseases affecting legume crops in Tunisia and partial characterization of *Chickpea chlorotic stunt virus*. *Phytopathologia Mediterranea, 50*, 310–315.
Najar, A., Kumari, S. G., Makkouk, K. M., & Daaloul, A. (2003). A survey of viruses affecting faba bean (*Vicia faba*) in Tunisia includes first record of *Soybean dwarf virus*. *Plant Disease, 87*, 1151.
Najar, A., Makkouk, K. M., Boudhir, H., Kumari, S. G., Zarouk, R., Bessai, R., et al. (2000). Viral diseases of cultivated legume and cereal crops in Tunisia. *Phytopathologia Mediterranea, 39*, 423–432.
Nene, N. Y. (1988). Multiple-disease resistance in grain legumes. *Annual Review of Phytopathology, 26*, 203–217.
Nienhaus, F., & Saad, A. T. (1967). First report on plant virus diseases in Lebanon, Jordan and Syria. *Zeitschrift fur Pflanzenkrankheiten und Pflanzenschutz, 74*, 459–471.
Nour, M. A., & Nour, J. J. (1962). A mosaic disease of *Dolichos lablab* and diseases of other crops caused by alfalfa mosaic virus in the Sudan. *Phytopathology, 52*, 427–432.
Ovchinmikova, A. M., Zelenov, A. N., Plovanova, T. A., Larionova, L. I., Azarova, E. F., & Androkhima, R. M. (1987). Problems of immunity in pulse crops. *Zashchita Rastenii (Moscow), 9*, 23–24.
Parlevliet, J. E. (2002). Durability of resistance against fungal, bacterial and viral pathogens; present situation. *Euphytica, 124*, 147–156.
Parvin, S., & Izadpanah, K. (1978). Broad bean wilt virus-identification, host range and distribution in the Fars province of Iran. *Iranian Journal of Agricultural Research, 6*, 81–90.
Perring, T. M., Gruenhagen, N. M., & Farrar, C. A. (1999). Management of plant viral diseases through chemical control of insect vectors. *Annual Review of Entomology, 44*, 457–481.
Peters, D. (1982). Pea enation mosaic virus. In *CMI/AAB descriptions of plant viruses no. 257* (4 p). Wellesbourne, Warwick, UK: Association of Applied Biologists.
Provvidenti, R. (1987). Inheritance of resistance to clover yellow vein virus in *Pisum sativum*. *Journal of Heredity, 70*, 126–128.
Provvidenti, R., & Alconero, R. (1988). Inheritance of resistance to a lentil strain of pea seed-borne mosaic virus in *Pisum sativum*. *Journal of Heredity, 79*, 45–47.
Provvidenti, R., & Hampton, R. O. (1992). Sources of resistance to viruses in the *Potyviridae*. *Archives of Virology, 5*, 189–211.
Raccah, B. (1986). Nonpersistent viruses: Epidemiology and control. In K. Maramorosch, F. A. Murphy, & A. J. Shatkin (Eds.), *Advances in virus research* (pp. 387–429). Toronto: Academic Press.
Rangaraju, R., & Chenulu, V. V. (1981). Occurrence of interveinal chlorosis of lentil in India. *Current Science, 50*, 191–192.
Roberts, I. M., Wang, D., Thomas, C. L., & Maule, A. J. (2003). *Pea seed-borne mosaic virus* seed transmission exploits novel symplastic pathways to infect the pea embryo and is, in part, dependent upon chance. *Protoplasma, 222*, 31–43.

Rohloff, H., & Stupnagel, R. (1984). Resistance to bean yellow mosaic virus in *Vicia faba. FABIS Newsletter, 10*, 29.

Russo, M., Kishtah, A. A., & Tolba, M. A. (1981). A disease of lentil caused by bean yellow mosaic virus in Egypt. *Plant Disease, 65*, 611–612.

Salam, M. U., Davidson, J. A., Thomas, G. J., Ford, R., Jones, R. A. C., Lindbeck, K. D., et al. (2011). Advances in winter pulse pathology research in Australia. *Australasian Plant Pathology, 40*, 549–567.

Salama, E. S., & El-Behadli, A. H. (1979). Strain of alfalfa mosaic virus on broad bean in Iraq. *Bulletin of the Natural History Research Centre, 7*, 101–112.

Sano, Y., Wada, M., Hashimoto, Y., Matsumoto, T., & Kojima, M. (1998). Sequences of ten circular ssDNA components associated with the milk vetch dwarf virus genome. *Journal of General Virology, 79*, 3111–3118.

Schepers, A. (1989). Chemical control. In A. K. Minks, & R. Harrewijn (Eds.), *Aphids, their biology, natural enemies and control* (pp. 89–112). Amsterdam, The Netherlands: Elsevier.

Schmidt, H. E., Carl, F., & Meyer, U. (1988). Resistance of field bean, *Vicia faba* L. ssp. minor (Peterm. Em Harz) Rothm. to pea enation mosaic virus. *Archiv für Phytopathologie und Pflanzenschutz, 24*, 77–79.

Schmidt, H. E., Geissler, K., Karl, E., & Schmidt, H. B. (1986). A line of field bean (*Vicia faba* L.) with combined resistance to BYMV and ClYVV and *Aphis faba* Scop. *Archiv für Phytopathologie und Pflanzenschutz, 22*, 87–99.

Schmidt, H. E., Meyer, U., Haack, I., & Karl, E. (1989). Detection, accumulation and characterization of multiple resistance in field bean, *Vicia faba* L. ssp. minor (Peterm. Em Harz) Rothm. to bean yellow mosaic virus and pea enation mosaic viruses. *Archiv für Zuchtungsforschung, 19*, 193–196.

Schmidt, H. E., Rollowitz, W., Schimanski, H. H., & Kegler, H. (1985). Detection of resistance genes against bean yellow mosaic virus in *Vicia faba* L. *Archiv für Phytopathologie und Pflanzenschutz, 21*, 83–85.

Schroeder, W. T., & Barton, D. W. (1958). The nature and inheritance of resistance to the pea enation mosaic virus in garden pea, *Pisum sativum* L. *Phytopathology, 48*, 628–632.

Schwinghamer, M. W., Johnstone, G. R., & Johnston-Lord, C. F. (1999). First records of bean leafroll luteovirus in Australia. *Australasian Plant Pathology, 28*, 260.

Schwinghamer, M. W., Thomas, J. E., Parry, J. N., Schilg, M. A., & Dann, E. K. (2007). First record of natural infection of chickpea by *Turnip mosaic virus. Australasian Plant Disease Notes, 2*, 41–43.

Shiying, B., Xiaoming, W., Zhendong, Z., Xuxiao, Z., Kumari, S. G., Freeman, A., et al. (2007). Survey for faba bean and field pea viruses in Yunnan Province, China. *Australasian Plant Pathology, 36*, 347–353.

Simons, J. N. (1982). Use of soil sprays and reflective surfaces for control of insect-transmitted plant viruses. In K. F. Harris, & K. Maramorosch (Eds.), *Pathogens, vectors and plant diseases: Approaches to control* (pp. 71–93). New York: Academic Press.

Tadesse, N., Ali, K., Gorfu, D., Abraham, A., Lencho, A., Ayalew, M., et al. (1999). Survey for chickpea and lentil virus diseases in Ethiopia. *Phytopathologia Mediterranea, 38*, 149–158.

Thackray, D. J., Diggle, A. J., Berlandier, F. A., & Jones, R. A. C. (2004). Forecasting aphid outbreaks and *Cucumber mosaic virus* epidemics in lupin crops in a Mediterranean-type environment. *Virus Research, 100*, 67–82.

Thackray, D. J., Jones, R. A. C., Bwye, A. M., & Coutts, B. A. (2000). Further studies on the effects of insecticides on aphid vector numbers and spread of cucumber mosaic virus in narrow-leafed lupins (*Lupinus angustifolius*). *Crop Protection, 19*, 121–139.

Thackray, D. J., Smith, L. J., Cheng, Y., Perry, J. N., & Jones, R. A. C. (2002). Effect of strain-specific hypersensitive resistance on spatial patterns of virus spread. *Annals of Applied Biology, 141*, 45–59.

Thomas, J. E., Parry, J. N., Schwinghamer, M. W., & Dann, E. K. (2010). Two novel mastreviruses from chickpea (*Cicer arietinum*). *Archives of Virology*, *155*, 1777–1788.
Thomas, J. E., Schwinghamer, M. W., Parry, J. N., Sharman, M., Schilg, M. A., & Dann, E. K. (2004). First report of *Tomato spotted wilt virus* in chickpea (*Cicer arietinum*) in Australia. *Australasian Plant Pathology*, *33*, 597–599.
Thresh, J. M. (1982). Cropping practices and virus spread. *Annual Review of Phytopathology*, *20*, 193–218.
Thresh, J. M. (2003). Control of plant virus diseases in Sub-Saharan Africa: The possibility and feasibility of an integrated approach. *African Crops Science Journal*, *11*, 199–223.
Thresh, J. M. (2006). Control of tropical plant virus diseases. *Advances in Virus Research*, *67*, 245–295.
Timmerman-Vaughan, G. M., Pither-Joyce, M. D., Cooper, P. A., Russell, A. C., Goulden, D. S., Butler, R., et al. (2001). Partial resistance of transgenic peas to *Alfalfa mosaic virus* under greenhouse and field conditions. *Crop Science*, *41*, 846–853.
van Leur, J. A. G., Aftab, M., Manning, W., Bowring, A., & Riley, M. J. (2013). A severe outbreak of chickpea viruses in northern New South Wales, Australia, during 2012. *Australasian Plant Disease Notes*, *8*, 49–53. http://dx.doi.org/10.1007/s13314-013-0093-y.
van Leur, J. A. G., Kumari, S., Aftab, M., Leonforte, A., & Moore, S. (2013). Virus resistance of Australian pea (*Pisum sativum*) varieties. *New Zealand Journal of Crop & Horticultural Science*, *41*, 86–101. http://dx.doi.org/10.1080/01140671.2013.781039.
van Leur, J. A. G., Makkouk, K. M., Freeman, A., & Schilg, M. A. (2003). Occurrence of viruses in faba bean on the Liverpool Plains, northern New South Wales. In *Proceedings of the 8th international congress of plant pathology, 2–7 February 2003* (p. 265).New Zealand: Christchurch. (Abstr.).
Yen, D. E., & Fry, P. R. (1956). The inheritance of immunity to pea mosaic virus. *Australian Journal of Agricultural Research*, 7, 272–280.
Younis, H. A., Shagrun, M., & Khalil, J. (1992). Isolation of bean yellow mosaic virus from broad bean plants in Libya. *Libyan Journal of Agriculture*, *13*, 165–170.
Zagh, S., & Ferault, A. C. (1980). A broad bean virus diseases occurring in Algeria. *Annals of Phytopathology*, *12*, 153–159.
Zidan, F., Khalil, J., & Shagrun, M. (2002). Survey and identification of pea viruses in Libya. *Arab Journal of Plant Protection*, *20*, 154–156.

CHAPTER FIVE

Control of Cucurbit Viruses

Hervé Lecoq*[,1], **Nikolaos Katis**[†]
*INRA, UR407, Station de Pathologie Végétale, Montfavet Cedex, France
[†]Faculty of Agriculture, Forestry and Natural Environment, School of Agriculture, Plant Pathology Lab, Aristotle University of Thessaloniki, Thessaloniki, Greece
[1]Corresponding author: e-mail address: herve.lecoq@avignon.inra.fr

Contents

Abstract

More than 70 well-characterized virus species transmitted by a diversity of vectors may infect cucurbit crops worldwide. Twenty of those cause severe epidemics in major production areas, occasionally leading to complete crop failures. Cucurbit viruses' control is based on three major axes: (i) planting healthy seeds or seedlings in a clean environment, (ii) interfering with vectors activity, and (iii) using resistant cultivars. Seed disinfection and seed or seedling quality controls guarantee growers on the sanitary status of their planting material. Removal of virus or vector sources in the crop environment can significantly delay the onset of viral epidemics. Insecticide or oil application may reduce virus spread in some situations. Diverse cultural practices interfere with or prevent vector reaching the crop. Resistance can be obtained by grafting for soil-borne viruses, by cross-protection, or generally by conventional breeding or genetic engineering. The diversity of the actions that may be taken to limit virus spread in cucurbit crops and their limits will be discussed. The ultimate goal is to provide farmers with technical packages that combine these methods within an integrated disease management program and are adapted to different countries and cropping systems.

Advances in Virus Research, Volume 90
ISSN 0065-3527
http://dx.doi.org/10.1016/B978-0-12-801246-8.00005-6

1. INTRODUCTION

Cucurbit crops are affected by an ever-growing number of viral diseases; more than 70 virus species belonging to the major plant virus genera are now described as infecting cultivated cucurbits in nature (Lecoq, 2003; Lecoq & Desbiez, 2012; Lovisolo, 1980; Provvidenti, 1996). The diversity of the viruses may be related to the ecological and genetic diversity of their hosts (Lecoq, 2003). Cucurbits are grown throughout the world in a great variety of agro-ecosystems ranging from the highly sophisticated soil-less cucumber production in heated glasshouses of Northern Europe to the traditional rain-fed cultivation of watermelon in the Sudano-Sahelian region. These variable environments constitute more or less favorable ecological niches for specific viruses and their vectors. Four major cucurbit species are cultivated worldwide: cucumber (*Cucumis sativus*), melon (*Cucumis melo*), squash and pumpkin (*Cucurbita pepo*, C. *moschata*, and C. *maxima*), and watermelon (*Citrullus lanatus*) and, for each of these species, there are many local landraces, cultivars, breeding lines, and hybrids (Robinson & Decker-Walters, 1997).

These four major cultivated cucurbits are among the main vegetable crops cultivated in the world. They are consumed as mature or immature fruits, fresh or cooked, and are an important ingredient of many local dishes (Robinson & Decker-Walters, 1997). Cucurbits were cultivated in 2012 on over 8.7 million ha from temperate to tropical climatic regions and produced overall more than 227 million tons of fruit (FAOSTAT, 2014). Developing and emerging countries are the main cucurbit producers and China is far ahead from other countries. China produced over 48 million tons of cucumber on 1.15 million ha, followed by Turkey (1.7 million tons, 63.000 ha), Iran, the Russian Federation, and Ukraine. China is the first producer of melon with 17.5 million tons on 605.000 ha, followed by Iran (1.5 million tons, 82.000 ha), Turkey, Egypt, and India. China also produced 7 million tons of squash on 383.000 ha, followed by India (4.9 million tons, 510.000 ha), the Russian Federation, Iran, and the United States. Finally, China produced 70 million tons of watermelon on 1.83 million ha, followed by Iran (3.8 million tons, 145.000 ha), Turkey, Brazil, and Egypt (data for 2012, FAOSTAT, 2014). Although accurate figures are lacking, if we take conservative estimates of 1% of the cucurbits grown worldwide being virus infected, and that the mean virus content is around 500 ng/kg of fresh fruit, these production data may infer that over 1000 tons of viruses are produced yearly in cultivated cucurbit fruits.

The increase in cucurbit virus incidences reported in the world is probably due to the improvement of plant virus identification methods, but is also a consequence of changes in cultural practices (i.e., greenhouses or low tunnel crops) and of global warming that may have drastic impact on vector populations. Globalization increased exchanges of seed and plant materials around the world and contributed substantially to virus dissemination at the planet scale. This is illustrated by the increase of cucurbit virus number in recent years: 23, 55, and 70 different viruses were reported in cucurbit crops in 1980, 2003, and 2014, respectively (Lecoq, 2003; Lovisolo, 1980, this chapter). Some of these viruses are widespread and cause major yield losses, while others remain restricted to limited geographical regions or to specific cropping systems where they are of minor economic importance.

Among the new virus species, some are typical "emerging" viruses that become suddenly widely spread in a short period of time and this often follows the rapid increase and dissemination of their natural vectors. Viruses transmitted by the whitefly *Bemisia tabaci* are a typical example of this situation. Within the last 15 years, 10 new begomoviruses, ipomoviruses, and criniviruses have been reported infecting cucurbits, mainly in Asia and America, and new occurrences are continuously reported in various countries (Navas-Castillo, Fiallo-Olive, & Sanchez-Campos, 2011; Romay, Lecoq, & Desbiez, 2014). Control measures have to be developed or adapted for each newly described virus; some are generic and can be applied with a similar efficacy to all viruses having the same vectors (Table 5.1), while others such as resistance generally require specific breeding programs.

Virus infections in cucurbits may have major consequences on marketable yields: gross yield reductions, as a consequence of poor fruit setting and fruit quality depreciation due to various virus symptoms: discoloration, more or less severe deformations, necrotic symptoms, and eventually imperfect maturation (Blancard, Lecoq, & Pitrat, 1994). Often, individual plants are infected by more than one virus, leading to a combination of symptoms on fruits. It is difficult to have a real indication of the global economic impact of virus infection in cucurbit crops at a world scale, due to the diversity of market requirements in different countries. However, nearly complete yield losses caused by early infections by severe viruses such as *Cucumber mosaic virus* (CMV), *Zucchini yellow mosaic virus* (ZYMV), or begomoviruses have been recorded in many cases. For example, losses caused to the watermelon industry in Florida by the emerging whitefly-borne *Squash vein yellowing virus* (SqVYV) have been estimated to be over 60 million U.S. $ in 2004–2005 (Kousik et al., 2012).

Table 5.1 Major cucurbit viruses classified by their vectors and transmission modes

Vector transmission	Genus	Species	Distribution	First report
Aphid-borne nonpersistent	*Cucumovirus*	*Cucumber mosaic virus* (CMV)	Worldwide	1916
	Potyvirus	*Moroccan watermelon mosaic virus* (MWMV)	Africa, Europe	1974
		Papaya ringspot virus (PRSV)	Worldwide tropical and subtropical	1949
		Watermelon mosaic virus (WMV)	Worldwide temperate and Mediterranean	1954
		Zucchini yellow mosaic virus (ZYMV)	Worldwide	1981
Aphid-borne persistent	*Polerovirus*	*Cucurbit aphid-borne yellows virus* (CABYV)	Worldwide	1992
Whitefly-borne semipersistent	*Crinivirus*	*Cucurbit yellow stunting disorder virus* (CYSDV)	Europe, Middle East, United States	1982
	Ipomovirus	*Cucumber vein yellowing virus* (CVYV)	Middle East, Europe	1960
		Squash vein yellowing virus (SqVYV)	United States	2007
Whitefly-borne persistent	*Begomovirus*	*Cucurbit leaf curl virus* (CuLCuV)	North America	2000
		Squash leaf curl virus (SLCV)	North America, Middle East	1983
		Tomato leaf curl New Delhi virus (ToLCNDV)	Asia, Europe	1995
		Watermelon chlorotic stunt virus (WmCSV)	East Africa, Arabic Peninsula, Middle East	1987
Beetle-borne semipersistent	*Comovirus*	*Squash mosaic virus* (SqMV)	Worldwide	1941
Thrips-borne persistent	*Tospovirus*	*Melon yellow spot virus* (MYSV)	Japan	1992
		Watermelon silver mottle virus (WSMoV)	Asia	1982

Table 5.1 Major cucurbit viruses classified by their vectors and transmission modes—cont'd

Vector transmission	Genus	Species	Distribution	First report
Fungus borne	*Carmovirus*	*Melon necrotic spot virus* (MNSV)	Worldwide	1966
No vector	*Tobamovirus*	*Cucumber green mottle mosaic virus* (CGMMV)	Worldwide	1935

At present, there is no direct treatment that can be applied on infected plants directly in the field and cure them. Therefore, control methods rely only on preventing plants from becoming infected by viruses for as long as possible within the crop cycle. A thorough knowledge of viral biology is a prerequisite for making innovative advances in virus control measures (Jones, 2004).

Methods of virus control in cucurbit crops comprise of three major components (Atkins et al., 2011; Fereres & Moreno, 2011; Jones, 2004; Quiot, Labonne, & Marrou, 1982; Zitter & Simons, 1980):

- Use of healthy seeds in a healthy environment. This requires the implementation of seed production schemes that guarantee the farmers with access to virus-free seeds or seedlings and the elimination as much as possible of virus or vector sources around the fields.
- Disturbance of the activity and efficiency of vectors. This most often involves adaptation of cultural or phytosanitary practices depending on the biology of viruses and their vectors and on the cultivation types.
- Production of plants more resistant to viral infection. This can be achieved by several means, including grafting on resistant rootstocks, cross-protection, and of course more generally by selecting resistant varieties either by conventional breeding or by genetic engineering.

Implementation of prophylactic measures, such as using healthy seeds or preventing/delaying virus spread by vectors, results only in partial virus control. However, delayed infection is sometimes sufficient to prevent significant economic losses. For viruses with air-borne insect vectors, control measures include protection of nurseries from the vectors, careful weeding to eliminate virus or vector reservoirs before planting, plastic mulching, and covering the plants with insect-proof nets or floating covers. For viruses with soil-borne vectors, control measures include the use of grafting, soil sterilization (when technically and legally possible), or fungicide addition to nutrient solutions. Disinfection of pruning or harvesting tools prevents dissemination of mechanically transmitted viruses. For some viruses such as ZYMV, cross-protection using a mild-virus strain has been used successfully to control severe strains (Lecoq, Lemaire, & Wipf-Scheibel, 1991).

The use of virus-resistant cultivars, when commercially available, is probably the easiest and cheapest way to control viral diseases at the farm's level. Breeding for resistance still relies mainly upon germplasm evaluation and introgression of the resistance gene(s) into commercially acceptable cultivars (Lecoq, Moury, Desbiez, Palloix, & Pitrat, 2004; Moury, Fereres, Garcia-Arenal, & Lecoq, 2011). In many countries, systematic germplasm evaluations have been conducted both by public institutions and by private breeding companies, often in collaboration, which has led to the description of a number of virus-resistance genes in the major cucurbit crops (Call & Wehner, 2011; Dogimont, 2011; Guner & Wehner, 2003; Paris & Brown, 2004). Whenever an emerging virus is described in cucurbits, experimental protocols have to be developed in order to look again for resistance in germplasm resources.

The development of biotechnology has provided new possibilities for creating virus-resistant cultivars. It is worth noting that squash cultivars resistant to ZYMV and *Watermelon mosaic virus* (WMV) were the first virus-resistant transgenic plants that were commercially available in the United States, as early as 1994 (Medley, 1994; Tricoli et al., 1995).

2. GROWING HEALTHY SEEDS IN A HEALTHY ENVIRONMENT

One of the main factors conditioning the earliness of virus epidemics in cucurbits is the availability of virus sources within the crop or in its close environment. For seed-borne viruses, primary virus infection can occur as early as the seedling stage. In recent years, most of the farmers in developed countries buy their seedlings from reliable nursery units and this has substantially improved the quality of the starting planting material. Virus sources in the environment can be very diverse and strongly depend on virus biology, stability, and host range. For very stable viruses such as tobamoviruses or carmoviruses, sources can be infested soil or greenhouse structures contaminated by infected plant debris and/or seed, while, for others, it can be alternative hosts (i.e., weeds, ornamentals, etc.) or infected senescent crops growing nearby the field.

2.1. Use of healthy seeds and seedlings

Only a small number of cucurbit viruses are known to be seed borne (Table 5.2), although this number could be higher since this property is not documented for a number of recently described virus species (i.e., new tobamovirus or fabavirus species). Use of infected seeds does not only affect individual farmers but also poses a problem for international seed trade and a risk of introduction of a new disease into a country. For instance, seed transmission has been often evoked as an explanatory factor of the rapid

Table 5.2 Cucurbit viruses reported to be transmitted by seeds

Genus	Species	Mode of transmission	Crop
Aureusvirus	*Cucumber leaf spot virus*	Not determined	Cucumber
Carmovirus	*Melon necrotic spot virus*	Seed-coat contamination, vector-assisted transmission	Melon
Comovirus	*Squash mosaic virus*	Embryo infection	Melon, squash, watermelon
Cucumovirus	*Cucumber mosaic virus*	Not determined	Squash
Nepovirus	*Tobacco ringspot virus*	Embryo infection	Cucumber, melon, squash
Potyvirus	*Telfairia mosaic virus*	Not determined	Telfairia
	Zucchini yellow mosaic virus	Not determined	Squash
Sobemovirus	*Snake melon asteroid mosaic virus*	Embryo infection	Melon
Tobamovirus	*Cucumber green mottle mosaic virus*	Seed-coat contamination	Cucumber, melon, watermelon
	Kyuri green mottle mosaic virus	Seed-coat contamination	Cucumber
	Zucchini green mottle mosaic virus	Seed-coat contamination	Squash
Tymovirus	*Melon rugose mosaic virus*	Embryo infection	Melon

spread of ZYMV throughout the world (Lecoq & Desbiez, 2008). Also, infected seeds of *Lagenaria* sp. and other watermelon rootstocks have introduced *Cucumber green mottle mosaic virus* (CGMMV) in many countries causing serious yield losses (Boubourakas, Hatziloukas, Antignus, & Katis, 2004). This emphasizes the need to check the sanitary status of both the rootstock and the cultivar/hybrid seeds. Seed transmission is also an important issue for plant breeders when they exchange germplasm accessions originating from various regions in the world.

2.1.1 Modes of seed transmission

Most viruses may infect mother tissue of the seed (including seed coats) but in cucurbits only a few, such as *Squash mosaic virus* (SqMV) or *Tobacco ringspot virus* (TRSV), can reach the embryo (Nolan & Campbell, 1984). These viruses multiply in the developing embryo during germination leading to a rapid systemic infection of the plantlet. Virus may become inactivated during storage and transmission of SqMV in different seed lots dropped from 11–31% to 0% after 2 years of storage (Powell & Schlegel, 1970).

Generally, plant viruses rapidly lose their infectivity in seed coats during the process of seed drying. Very stable viruses, such as CGMMV, may remain infectious in dried seed coats and eventually infect the plantlet by wounds caused to the rootlets when they come out of the infected seed coat during germination and/or during transplanting (Choi, 2001).

An original mechanism of seed transmission is the vector-assisted seed transmission (VAST) described for *Melon necrotic spot virus* (MNSV) in melon (Campbell, Wipf-Scheibel, & Lecoq, 1996). Infectious MNSV virus particles occur in the internal and external teguments of the seeds, and they are released into the soil as the seeds become soaked with water and germinate. MNSV may be then transmitted to seedling roots if zoospores of the vector, *Olpidium bornovanus*, are present and acquire the virus in the soil. No seed transmission is observed if the same seeds are sown in sterile soil (Campbell et al., 1996).

Accurate estimation of seed-transmission rates is better obtained by grow-out tests rather than by DAS-ELISA tests conducted on individual seeds or on groups of seeds. Indeed, DAS-ELISA may detect noninfectious virus particles contaminating seed coats, greatly overestimating the number of infected plantlets. For instance, although CGMMV and SqMV were detected by DAS-ELISA on individual seeds at rates of 84% and 52%, respectively, only 2% and 6% of the seedlings were infected, respectively (Choi, 2001; Lecoq, Piquemal, Michel, & Blancard, 1988).

2.1.2 How to prevent seed transmission

An initial way to prevent virus-seed transmission is to produce seeds in countries or regions where seed-borne viruses are not prevalent or to grow mother plants under conditions where they cannot be reached by viruliferous vectors (see Section 2).

For viruses, such as CGMMV which is carried externally on the seed, disinfection by chemicals (trisodium phosphate, sodium hypochlorite, hydrochloric acid, pectinase, etc.) or dry heat treatment (at 72 °C for 24–72 h) can completely eliminate seed transmission without affecting germination rates (Choi, 2001; Sang-Min, Sang-Hyun, Jung-Myung, Kyu-Ock, & Kook-Hyung, 2003). Similarly, MNSV has been eliminated from melon seeds by heat treatment for 144 h at 70 °C. However, in the same study, seeds immersed in either 10% trisodium phosphate solution for 3 h or 0.1 N HCl solution for 30 min reduced but did not effectively eliminate MNSV from seeds. However, a reliable test to evaluate both seed-disinfection efficiency and MNSV seed-transmission rates is still lacking due to the difficulty to implement experimental conditions for demonstrating VAST (Herrera-Vasquez, Cordoba-Selles, Cebrian, Alfaro-Fernandez, & Jorda, 2009).

Such disinfection treatments are routinely applied by the major seed production companies prior to marketing. In contrast, these treatments are seldom applied in regions where seed multiplication is conducted on a smaller scale or when farmers produce their own seeds.

For viruses infecting the embryo, quality control tests using biological, serological, or molecular methods must be implemented. Adapted sampling and statistical methods may allow an estimation of seed-transmission rate for specific seed lots (Masmoudi et al., 1994). Subsequently, according to local regulations and individual seed company quality policy, infected seed lots may be either destroyed or marketed if the estimated seed-transmission rate is under an acceptable threshold to be determined according to the virus and its means of dissemination in the field.

2.1.3 Seedling quality

A major part of cucurbit crops, particularly in intensive production areas, is planted with grafted or regular seedlings rather than sown directly with seeds. It is therefore important for all viruses in general to start the crop with virus- and vector-free seedlings. This measure is nowadays more easily achieved as the majority of the farmers does not produce the seedlings themselves but instead buy them from well-equipped nurseries, which have the facilities and the infrastructure to produce high-quality seedlings. Also, it is recommended to avoid using transplants that have been produced in areas where a particular virus is considered a serious problem. This practice has been recommended for the control of whitefly-transmitted viruses in Spain and elsewhere (Janssen et al., 2003).

When farmers are still producing their own seedlings, they should produce them in disinfected potting soil and under insect-proof structures (see Section 3) to prevent early infections by viruliferous vectors.

2.2. Limiting virus sources near cucurbit crops

Limiting virus sources in the neighboring to the crop environment requires a thorough knowledge of virus biology, including host range and means of dissemination (Jones, 2004). For very stable viruses, such as tobamoviruses or carmoviruses, that are easily mechanically transmitted, virus sources may be greenhouse structures, nutrient solutions, soils, or rockwool blocks contaminated by infected plant debris. In this case, simple disinfection procedures could be enough to limit potential sources of infection (Astier, Albouy, Maury, Robaglia, & Lecoq, 2007).

For other viruses, sources could be either arable weeds, infected ornamentals, or crops growing at close vicinity.

2.2.1 Weeds: Important virus sources

The importance of weeds and more generally of the wild flora as sources of important viruses has been established since the beginning of plant virology. As early as 1925, Doolittle and Walker demonstrated the role of various weeds as reservoirs of CMV for cucumber crops and proposed weed elimination as a control method. Since then, much more CMV reservoirs have been identified, and the importance of weeds in the epidemiology of major plant viruses has been abundantly illustrated (Duffus, 1971; Thresh, 1981). ZYMV and WMV are two potyviruses having similar vectors but contrasting epidemiological profiles in cucurbits in France; ZYMV epidemics are irregular and generally late while WMV epidemics are regular and early. Differences in weed contamination rates (0.2% and 4.6% for ZYMV and WMV, respectively) appeared as a major factor explaining the differences in the epidemiological patterns of these two viruses (Lecoq, Wipf-Scheibel, Nozeran, Millot, & Desbiez, 2014).

Often a plant species can be an alternative host for several cucurbit viruses; *Lamium amplexicaule* may be naturally infected by *Cucurbit aphid-borne yellows virus* (CABYV), CMV, WMV, and ZYMV whereas *Ecballium elaterium* is a host of CABYV and *Cucurbit yellow-stunting disorder virus* (CYSDV; Lecoq, 1992, unpublished data; Orphanidou & Katis, unpublished data). The total destruction of alternative hosts is not feasible for obvious practical and environmental reasons. Careful weeding within and around fields is nevertheless a generally recommended measure, although its efficiency and feasibility may vary greatly from one ecosystem to another (Quiot et al., 1982). Control of cucurbit weeds adjacent to watermelon fields has been recommended for the control of whitefly-transmitted viruses such as SqVYV (Atkins et al., 2011). However, since weeds are also often hosting insect vectors which will disseminate viruses when the weeds will dry off, it is recommended to spray the weeds with an insecticide before pulling them out or applying herbicide. It should be taken into consideration that weeds are also hosts of predators and parasitoids of insect vectors and therefore may have a positive effect on vector control (Fereres & Moreno, 2011).

2.2.2 Other potential virus sources

Other underestimated possible virus sources are infected cucurbit fruits (Lecoq, Desbiez, Wipf-Scheibel, & Girard, 2003) and cucurbits or ornamentals growing in private gardens. Cucurbits grown in small gardens in residential areas are important sources of ZYMV and WMV affecting melon and watermelon in the Imperial Valley (California; Perring, Farrar, Mayberry, & Blua, 1992). More recently, some ornamental species such as begonia, larkspur, or horned violet were identified as possible ZYMV sources in southeastern France (Lecoq et al., 2014).

In specific cases such as protected crops, eradication, i.e., the elimination of an infected plant as soon as the first symptoms are observed can be used to stop or further limit the virus spread. However, it should be considered that these plants can possibly act as efficient virus sources before the first symptoms appear.

2.2.3 Avoiding overlapping crops

Cucurbit crops have a short-vegetative cycle, and very often farmers will have several successive plantings during the year, starting by early spring crops protected by plastic (tunnels) followed by one or several cycles of field crops. Due to time constraints, farmers may abandon the early crops after they finish harvesting them instead of destroying the crops and as a result they may become infested with various insect vectors (aphids, thrips, and whiteflies) and infected with a number of viruses. Such crops can be efficient sources of both viruses and insect vectors. In the meantime, if a new crop is planted nearby, it will get infected very quickly as viruliferous insects move from the adjacent cucurbit crops. To avoid that, it is recommended for the farmers to destroy the affected crop soon after the harvest and this practice has been adopted by many farmers nowadays (Kousik et al., 2012).

For example, to prevent the spread of viruses and their insect vectors into neighboring cucurbit crops, farmers in Florida destroy or "burn down" the cucurbit fields after harvest using fast-killing herbicides in order to eliminate them both as virus and/or vector sources for adjacent newly established cucurbit crops (Atkins et al., 2011). To avoid movement of viruliferous insects into neighboring crops as the plants collapse, the farmers add insecticide and/or oil to the herbicide mixture.

Such situations could also be avoided whenever possible by a proper spatial distribution of plantings in the farm. For instance, it is recommended to plant late crops upwind older crops than downwind (Coutts, Kehoe, & Jones, 2011).

Ideally, it would be efficient to create a break in the infection cycle by having crop-free periods. This is practically difficult to implement in cucurbit crops due to the number of cucurbit types cultivated and the diversity of growing conditions. However, a crop-free period from mid-June to mid-July considerably reduced incidence of aphid- and whitefly-transmitted viruses in the Arava Valley in Israel (Ucko, Cohen, & Ben-Joseph, 1998). A similar one-month cucurbit-free period was suggested to limit whitefly-borne viruses in Southern Spain (Janssen et al., 2003). In regions with a more diversified agriculture, changing planting dates may be sufficient to avoid peaks in vector populations (Zitter & Simons, 1980). However, to apply this, information on vector population dynamics is needed. It is also important to avoid interplanting. In the last few years, in order to gain time in greenhouse cucurbit production, farmers may interplant young seedlings among the old and often infected ones. As a result, the newly established young seedlings are getting infected soon by the viruliferous insect vectors present on the old plants.

3. ALTERING THE ACTIVITY OF VECTORS

One of the major components of the prophylactic measures to control cucurbit viruses is based on the prevention of vectors to reach the cultivated plant. Different actions can be taken according to the means of virus dissemination (Table 5.1). Interestingly, some cultural practices are effective against all viruses transmitted by the same vector types on the same modes. This may simplify the implementation of adapted cultural practices.

3.1. Actions against viruses transmitted by manual operations

Several cucurbit viruses are transmitted mechanically or by contact including tobamoviruses, MNSV, SqMV, and even less stable viruses such as ZYMV (Choi, 2001; Coutts, Kehoe, & Jones, 2013). Mechanical transmission can occur both during the preparation of seedlings in the nurseries and later on in the field, during pruning operation, plant manipulation or when harvesting fruits. A major way to prevent virus spread is by careful hand cleaning and disinfection of grafting and pruning tools with solutions such as 3% trisodium phosphate or 20% skimmed milk. In protected crops, where plants are subjected to recurrent manual operations, when a primary infection is observed in a site it is advised to work first in the units without infection and to finish operations on the infected ones.

3.2. Actions against soil-borne viruses

Control of fungal vectors of cucurbit viruses is rather difficult. Chemical disinfection of the soil with methyl bromide has been used worldwide and for many years it was the most effective treatment that prevented spread of MNSV. However, application of this chemical has been banned since 2005, according to the Montreal protocol because of environmental concerns. Soil disinfection can still be done by soil steaming or solarization, but the effect of these treatments on MNSV transmission/spread is not yet well documented. In hydroponic cultivation, the surfactant Agral 20® was shown to efficiently eliminate *O. bornovanus* zoospores in nutrient solutions thus preventing MNSV spread in soil-less cucumber cultivation (Tomlinson & Thomas, 1986). Composting of infected melon plants after a crop cycle for 30 or 60 days proved to be efficient to destroy MNSV and *O. bornovanus* infectivity (Aguilar, Guirado, Melero-Vara, & Gomez, 2010).

3.3. Actions against insect-borne viruses

These measures are mainly intended to reduce vector populations and to prevent or limit the contact of viruliferous insects with cultivated plants. They are not specific for a particular virus and are generally efficient for all insect-borne viruses. However, they might differ slightly for protected crops and for field crops.

3.3.1 Protected crops

Protected cucurbit crops have been steadily increasing during the last 30 years as a means to gain in earliness, to favor plant growth by controlling the environment and to extend the cucurbit production periods. However, if protected crops offer better growing conditions to cucurbits, they also provide favorable conditions for whitefly or aphid rapid multiplication. Managing vector populations has long been considered one of the primary ways to control insect-borne viruses particularly in the closed environment of protected crops (Perring, Gruenhagen, & Farrar, 1999). A number of strategies have been adopted for controlling insect vectors with the use of insecticides being the most popular (Perring et al., 1999; Zitter & Simons, 1980). Various insecticides have been successfully used for the control of aphids and whiteflies although resistance to the insecticides has been reported worldwide and therefore insects are becoming more difficult to control/eradicate (Roditakis et al., 2009; Schuster, Stanley, Polston, Gilreath, & McAvoy, 2007). Other negative impacts of pesticide use on the human health and the environment have prompted continuous efforts to enhance biological control of aphids and whiteflies (Gerling, Alomar, & Arno, 2001; Hilje, Costa, & Stansly, 2001). Nowadays, different approaches have been adopted for biological control of whiteflies and aphids and they play an important role in managing these pests in greenhouses worldwide. For example, the parasite *Encarsia formosa* and the fungus *Verticillium lecanii* have been successfully used as biological agents against *B. tabaci*, but their use did not reduce virus dissemination.

Effective control of whitefly populations and whitefly-transmitted viruses has been achieved by installing insect-proof nets on glasshouse or plastic tunnels' openings including doors and openings for ventilation. In the Almeria region (Spain), an efficient control of *B. tabacci* and the viruses it transmits such as CYSDV and *Cucumber vein yellowing virus* (CVYV) was achieved by creating an indoor structure of continuous insect-proof nets with double doors (Janssen et al., 2003).

An interesting new development in the control of insect-borne viruses in protected crops has been the use of plastic which reduces substantially the UV irradiation which enters the plastic tunnels and results in reduced numbers of insect vectors such as whiteflies (Antignus, Nestel, Cohen, & Lapidot, 2001) and aphids (Chyzik, Dobrinin, & Antignus, 2003). This happens because insects need the UV light in order to see and orientate and therefore when UV is blocked, their behavior is affected (Raviv & Antignus, 2004). As a result, UV-absorbing plastic reduces insect infestation of the covered crop and the incidence and spread of the viruses they transmit become limited (Antignus, Lapidot, Hadar, Messika, & Cohen, 1998). This strategy has been effective against whiteflies and whitefly-transmitted viruses (Antignus et al., 2001; see also chapter one by Antignus et al., this book) and also against aphids and aphid-borne viruses (Legarrea, Weintraub, Plaza, Vinuela, & Fereres, 2012).

3.3.2 Field crops

The actions that can limit vector activities in field crops are basically of three types: the direct application of chemicals to the plant (insecticides, oils), the modification of the crop environment (barrier crops, intercropping, etc.), and the use of plastic material repelling or preventing vectors (i.e., aphids and whiteflies) to reach the plant (mulches, row covers, etc.) (Fereres & Moreno, 2011; Jones, 2004; Loebenstein & Raccah, 1980; Quiot et al., 1982).

The effect of chemical control of insect vectors on limiting the transmission and spread of viruses depends on the transmission manner (Perring et al., 1999). Insecticides' applications have been generally found inefficient in limiting the spread of nonpersistently aphid-transmitted viruses (Perring et al., 1999) and generally are more effective in controlling semi- and persistently transmitted viruses. Indeed, large numbers of winged aphids are visiting cucurbit crops and insecticides generally do not act quickly enough to prevent the short-assay probes during which they transmit nonpersistent viruses. For example, endosulfan applications did not prevent potyvirus spread in watermelon crops in Florida (Webb & Linda, 1993). On the other hand, field studies with the semipersistently transmitted SqVYV showed that significantly fewer SqVYV-symptomatic fruits were produced by plants treated with insecticides to control whiteflies (Kousik, Webster, Turechek, Stanley, & Roberts, 2010). In this case, reducing whitefly population was effective in limiting the virus-induced yield reduction, but the emergence of insecticide-resistant whiteflies can rapidly compromise this protection (Atkins et al., 2011). Generally, although insecticides are thought to be more beneficial against persistently transmitted viruses (Perring et al., 1999) this was not the case with CABYV, a virus transmitted persistently by aphids (Lecoq, 1999) where insecticide applications did not limit its spread.

Oil applications (ultra-fine oil, JMS stylet oil, etc.) can delay primary infections of nonpersistenly transmitted viruses when inoculum pressure is moderate, but become inefficient when field infection rates are over 10–20% (Webb & Linda, 1993). However, it should be taken into consideration that the cost of oil application is quite high (application every 4–7 days to cover new vegetation) and that phytotoxicity may be observed (Loebenstein & Raccah, 1980). In watermelon, regular oil sprays slightly delayed virus epidemics, but a decrease in yield was observed in treated plots, probably due to phytotoxicity (Webb & Linda, 1993).

Border crops can also prevent or delay viruliferous aphids alighting on the crop. Border crops may have two effects on vectors; if they are tall enough, they can act as a physical barrier to aphids or whiteflies movement, in a similar way as classical windbreaks. For nonpersistently transmitted viruses, the barriers can also act as natural "sinks" were aphids carrying non-persistent viruses can "cleanse" their mouth parts. Obviously, border crops should be nonhost (immune) for both the vector and the virus (Fereres, 2000; Hooks & Fereres, 2006). To be efficient, barrier crops should be tall enough to really interfere with vector movement. Millet or sorghum were efficient in protecting cucurbit crops from potyviruses while barrier crops with a lower canopy (peanut, soybean, or lablab) have been ineffective (Coutts et al., 2011; Damicone, Edelson, Sherwood, Myers, & Motes, 2007).

Intercropping of pumpkin with sorghum substantially reduced the incidence of the nonpersistently transmitted aphid-borne viruses PRSV and WMV from 96% to 43% (Damicone et al., 2007). Trap cropping may also be a supplement to chemical pesticides in which a preferred plant species (immune to the virus) is used to attract vectors and keep them away from a less-preferred main crop (Shelton & Badenes-Perez, 2006). Although this technique has been used, it is uncertain if it is a viable management tool for whiteflies such as *B. tabaci* or aphids as the results are rather confusing (Lee, Nyrop, & Sanderson, 2009).

Other actions include the use of mulches and row covers (Fig. 5.1A and B). It is very well established that migrating insects such as aphids and whiteflies are attracted to surfaces that reflect green-yellow color (Doring & Chiitka, 2007) and repelled by highly reflective surfaces (Simmons, Kousik, & Levi, 2010; Summers, Mitchell, & Stapleton, 2004). In addition, these insects fly away after probing on plastic sheets, a response that was termed as "rejection flight" by Kring (1972). Therefore, reflective mulches of different colors (transparent, black, brown, silver, etc.) have been used by farmers in tropical and temperate regions to prevent aphids and whiteflies landing on susceptible cucurbit crops to reduce the incidence of the viruses they do transmit (Greer & Dole, 2003; Orozco-Santos, Perez-Zamora, & Lopez-Arriaga, 1995). However, mulches confer

Figure 5.1 Different methods used to control cucurbit viruses. (A) Plastic mulches repel aphid and whitefly vectors and delay virus spread. (B) Row covers prevent insect vectors from reaching the plants, but they should be partially removed at flowering stage to allow pollination. (C) Cucumber plant grafted on *Cucurbita* rootstock (indicated by arrow). (D) Mild-strain protection against ZYMV in zucchini squash: left protected plants and right nonprotected plants. (E) Machine developed to inoculate seedlings in nurseries with ZYMV mild strain. (F) An example of *Cucurbita* genetic resources in which virus resistance can be looked for. (G) Susceptible squash cultivar inoculated by a severe ZYMV strain. (H) Same cultivar carrying a transgenic resistance to ZYMV, inoculated the same day with the same strain as the susceptible plant in (G). (See the color plate.)

protection to the early stages of the crop because their efficiency decreases when the plant canopy covers the mulching material. An abundant literature reports the delay in aphid- or whitefly-borne virus spread observed with the use of polyethylene mulches either transparent (Lecoq, 1992; Orozco-Santos et al., 1995; Fig. 5.1A), white (Abd-Rabou & Simmons, 2012), yellow (Antignus, 2000), or silver (Abou-Jawdah et al., 2000; Stapleton & Summers, 2002). Increase in yields compared to bare soil can be considerable; for instance, the yield of export quality melon was increased by up to 450% in melon mulched with transparent polyethylene (Orozco-Santos et al., 1995). Despite the benefits of reflective mulches, which also include weed management, soil warming, and water preservation, the cost and difficulty of disposing used material have limited their use in some regions, while in others it became a common practice. Biodegradable synthetic latex spray mulches have been developed, but they did not provided significant protection against aphid-borne viruses (Stapleton & Summers, 2002).

Row covers of different types (unwoven, perforated plastics, etc.) can also be used (Fig. 5.1B); they physically prevent winged aphids from reaching the plants, but they must be removed at the flowering stage to allow insect pollination necessary for cucurbit fruit production (Lecoq, 1992; Walters, 2003). UV-absorbing plastics can also be used to cover field crops and limit virus incidence (Legarrea, Betancourt, et al., 2012).

4. MAKING CUCURBITS RESISTANT TO VIRUSES

The use of cultivars resistant to viruses is probably the easiest way to control viral diseases at the farmer's level. Rendering a plant resistant to viruses can be achieved by several means. Grafting the desired susceptible cultivar onto a virus-resistant rootstock has proved to be very efficient for protecting cucumber, melon, and watermelon from soil-borne viruses (i.e., carmoviruses, necroviruses, tobamoviruses, etc.). Mild-strain protection has been developed successfully for some of the most damaging mosaic-inducing cucurbit viruses. But the most practical way remains the use of cultivars that are genetically resistant to viruses. A number of commercially available cultivars now possess efficient virus-resistance genes introduced either through conventional breeding or through genetic engineering. However, it should be pointed out that the later can be used only in countries where growing genetically modified organisms (GMOs) is permitted by law.

4.1. Grafting on resistant rootstock

Grafting has been practiced in cucumber crops in Japan for almost a century to improve plant development in adverse biotic or abiotic conditions (Sakata, Ohara, & Sugiyama, 2008). Grafting has been progressively extended to other crops (melon and watermelon) to favor plant growth in poor soil conditions (low temperature, salinity, wet conditions, etc.) and to control soil-borne diseases (Fusarium wilt, nematodes, viruses, etc.). Grafting is now a common practice for protected and field cucumber, melon, and especially for watermelon cultivation (Fig. 5.1C). For example, China produces more than half of the world's watermelon and cucumber production and about 20% of these are grafted (Davis et al., 2008). The recent development of cucurbit grafting is mainly due to the selection of new rootstock genotypes with better grafting compatibility, the use of robots reducing the cost of grafting and the ban on methyl bromide, the major soil disinfectant used until now (Choi, 2001). Grafting on resistant rootstock (*Lagenaria siceraria*, various *Cucurbita* sp. including interspecific crosses between *C. moschata* and *C. maxima* and resistant melons) are now commonly used to prevent soil-borne infection by MNSV or CGMMV (Boubourakas et al., 2004; Choi, 2001; Cohen, Burger, Herev, Koren, & Edelstein, 2007; Huitron-Ramirez, Ricardez-Salinas, & Camacho-Ferre, 2009).

4.2. Mild-strain cross-protection

Mild-strain cross-protection is the art of using viruses against viruses. It is based on a dogma valid for most plant viruses; when a plant is infected by a virus strain, it cannot be infected by another strain of the same virus. The rationale of mild-strain cross-protection consists in inoculating a virus strain causing only mild symptoms and having no or limited impact on yield into a plant in order to protect it against severe strains of the same virus (Fuchs, Ferreira, & Gonsalves, 1997; Lecoq, 1998). Although suspected to be related to plant RNA silencing, the molecular mechanisms involved in cross-protection have not yet been fully determined (Ratcliff, MacFarlane, & Baulcombe, 1999). Nevertheless, cross-protection has been applied at a commercial scale against several severe viruses of horticultural crops (i.e., cucurbits, tomato, papaya, citrus, etc.). Local regulations and registration steps should be strictly respected before considering mild-strain field tests or commercial use.

4.2.1 Cross-protection against ZYMV

ZYMV is the first of a series of emerging viruses that threatened the cucurbit industry in major production areas during the last decades. Soon after its first description, a mild variant that occurred naturally in a greenhouse experiment has been isolated (Lecoq et al., 1991). This mild strain (ZYMV-WK) cumulated several favorable properties; it produced mild symptoms in major cultivated cucurbits although multiplying in infected plants at levels similar to those of severe strains, it was not aphid transmissible and it protected efficiently against the predominant ZYMV severe strains (Fig. 5.1D). Although ZYMV-WK differed from its original severe strain by a single mutation in the HC-Pro (Gal-On & Raccah, 2000), ZYMV-WK appeared genetically stable and reversion to severe forms were only very rarely observed. It was easy to multiply and to keep either as dried, frozen, or lyophilized material. ZYMV-WK has been used successfully in Europe, Israel, and Hawaii for several years in commercial scale (Fuchs, Ferreira, et al., 1997; Lecoq, 1998; Yarden et al., 2000). This success was in great part achieved through an efficient inoculum production scheme and the development of an inoculation machine that allowed inoculating in nurseries hundreds of cucurbit seedlings within a short time (Yarden et al., 2000; Fig. 5.1E). The use of cross-protection against ZYMV declined in recent years for two major reasons: the development of ZYMV resistant or tolerant cucumber and zucchini squash commercial cultivars and the emergence in the Middle East of very devastating cucurbit begomoviruses against which ZYMV-WK conferred of course no protection.

4.2.2 Other mild strains of cucurbit viruses

Mild strains may be obtained either as natural occurring variants (as ZYMV-WK) or after various treatments including high or low temperature, nitrous acid, or UV irradiation treatments (Fuchs, Ferreira, et al., 1997; Lecoq, 1998; Nischiguchi & Kobayashi, 2011). After a combination of these treatments, mild strains of CMV, CGMMV, PRSV, WMV, and ZYMV have been obtained and tested under field conditions. Multiple inoculations with three attenuated viruses have even been tested successfully to protect cucumber from viral diseases (Kosaka & Fukunishi, 1997). More recently, a mild strain (AG) has been obtained by site-directed mutagenesis on a ZYMV cDNA infectious clone (Gal-On & Raccah, 2000). Determinants of mild symptoms' expression are generally single mutations. Interestingly, these mutations occur in viral proteins known to be involved in gene-silencing suppression by different viruses (Nischiguchi & Kobayashi, 2011).

4.2.3 *Limitations of cross-protection*

As any biological control method, cross-protection has some limitations. First, mild strains are not available for all major cucurbit viruses and it cannot be considered for viruses, which are not easily mechanically inoculated. Mild-strain phenotypes are generally caused by single mutations and the possibility that a mild isolate reverse toward the severe form cannot be ruled out. Examples of synergisms between mild strains and nontarget viruses have been reported in noncucurbit hosts. Although not yet documented, there are possibilities of recombination between a mild strain and another virus, leading to a new viral entity. But the major limitation is specificity. Indeed, cross-protection has been proved to be efficient mainly against strains with high percentages of nucleotide identity. With more divergent isolates, cross-protection may be only partial or even may not operate at all, leading to protection breakdowns. For instance, ZYMV-WK was inefficient in protecting cucurbits against the ZYMV isolates from Réunion Island (Lecoq et al., 1991).

Despite these limitations, the versatility of cross-protection makes it easy to implement in a timely manner as new problems arise (Fuchs, Ferreira, et al., 1997; Lecoq, 1998). It should be used only for combating viruses presenting a real threat for a crop and for which no other alternative control methods are currently available. Cross-protection can also be rapidly applied to a range of cultivars including local landraces thus preserving cucurbit crops biodiversity.

4.3. Conventional breeding for resistance

Use of resistant hybrids/cultivars is the easiest, most efficient, and environmentally friendly strategy of combating serious cucurbit viral diseases at the farm's level (Lecoq, Wisler, & Pitrat, 1998). Therefore, searching for genes/traits for resistance to viruses in the biodiversity of cultivated species and in their wild relatives has been investigated since the beginning of the 20th century. In many countries, systematic germplasm evaluations have been conducted both by public institutions and by private breeding companies, often in collaboration, to search for virus resistances (Fig. 5.1F). This has lead to the description of a number of virus-resistance factors in the major cucurbit crop species (Table 5.3; Call & Wehner, 2011; Dogimont, 2011; Guner & Wehner, 2003; Paris & Brown, 2004). Although considerable efforts have been devoted to identify virus-resistant genes in cucurbits only in a few cases that they were successfully transferred to cultivated hybrids/cultivars

(Sugiyama, 2013; Tables 5.3 and 5.4). There are indeed a number of commercial hybrids/cultivars highly resistant or tolerant to severe viruses, but there is still a need in the market for new virus-resistant cucurbit species to combat old and emerging viral diseases all over the world. Resistance is based on a single or a few genes, which may be dominant or recessive. The ease to transfer resistance traits to advanced breeding lines will depend on its dominance, its inheritance, and the availability of simple methods differentiating resistant from susceptible plants. Combining several resistances to viral and fungal pathogens in a single cultivar or hybrid requires a great effort and many years by conventional breeding; although the recent development of marker-assisted selection significantly facilitates this process.

4.3.1 Diversity of virus-resistance mechanisms in cucurbits

A virus-resistance factor is any property that can disturb the virus life cycle at the cell, organ, plant, or population levels (Lecoq et al., 2004; Moury et al., 2011). The resistance phenotype is best described by comparison with susceptible cultivars. In addition to phenotype description, in a few cases the molecular mechanisms involved in resistance have been deciphered. A few examples will show the various levels where resistance in cucurbits can interfere with virus-infection cycle.

Resistance may occur as early as the virus inoculation step. A resistance to virus transmission by *Aphis gossypii*, the major aphid species-colonizing melon, has been described in melon. *A. gossypii* is an efficient vector of aphid-borne viruses including CMV, potyviruses, and CABYV. This resistance is complete with the nonpersistently transmitted CMV, WMV, ZYMV, and PRSV and partial with CABYV. It is specific for *A. gossypii* and is not effective for other efficient vectors such as *Myzus persicae* or *Aphis fabae* (Lecoq, Cohen, Pitrat, & Labonne, 1979; Lecoq, Labonne, & Pitrat, 1980). The resistance is controlled by a single dominant gene (*Vat* for *Virus aphid transmission*) which also controls antixenosis and antibiosis resistance to *A. gossypii* (Pitrat & Lecoq, 1980). *Vat* encodes a coiled-coil, nucleotide binding site, leucine-rich repeat protein (Pauquet et al., 2004). Therefore, it is likely that the recognition between the Vat protein and an aphid gene product triggers plant responses preventing aphid feeding, aphid fecundity and survival, and nonspecific antiviral responses (Pauquet et al., 2004).

Several mechanisms may lead to virus sequestration in a cell or in a few cells close to the inoculation site. Sometimes replication does not occur (immunity) or occurs but is limited to a few neighboring cells. This resistance to cell-to-cell movement may be or may not be accompanied

Table 5.3 Conventional virus resistances available in cucurbit genetic resources and genetic determinants when known

Genus	Species	Cucumber	Melon	Squash	Watermelon
Cucumovirus	*Cucumber mosaic virus*	*Cmv*[a] and other genes	Oligogenic recessive	*Cmv*	
Potyvirus	*Moroccan watermelon mosaic virus*	*mwm*[b]	ND		
	Papaya ringspot virus	*prsv* or *Prsv-2*	*Prv1*	*prv*	*prv*
	Watermelon mosaic virus	*Wmv* or oligogenic recessive	*Wmr* or *wmr*	*Wmv*	Oligogenic recessive
	Zucchini yellow mosaic virus	*zym*	*Zym-1* and two complementary genes	*Zym* and additional genes	*zym*
Polerovirus	*Cucurbit aphid-borne yellows virus*	ND[c]	*cab1* and *cab2*	ND	
Crinivirus	*Cucurbit yellow stunting disorder virus*		*Cys*		
Ipomovirus	*Cucumber vein yellowing virus*	ND	*Cvy-1*		
	Squash vein yellowing virus				ND
Begomovirus	*Squash leaf curl virus*			*Slc*	
	Watermelon chlorotic stunt virus		ND		ND
Tospovirus	*Melon yellow spot virus*	ND			

Continued

Table 5.3 Conventional virus resistances available in cucurbit genetic resources and genetic determinants when known—cont'd

Genus	Species	Cucumber	Melon	Squash	Watermelon
Carmovirus	*Melon necrotic spot virus*		*nsv* or *Mnr-1* and *Mnr-2*		
Tobamovirus	*Cucumber green mottle mosaic virus*		*cgmmv-1* and *cgmmv2*		

[a]Dominant genes have a first capital letter.
[b]Recessive genes have a first lower case letter.
[c]Inheritance not yet established.

Table 5.4 Virus resistances (conventional or transgenic) and commercial cultivars (conventional or transgenic) available in melon, cucumber, watermelon, and squash.

	Melon	Cucumber	Watermelon	*Cucurbita* sp.
CMV	+, C, T	++, C, T	T	++, C, TC
WMV	+, C, T	++, C, T	+, T	+/++, C, TC
ZYMV	+, C, T	++, C, T	++, T	+/++, C, TC
PRSV	++, C, T	+, C, T	+, T	+, T
CGMMV	+, T		T	
CFMMV		T		
CABYV	++	++		++, C
MNSV	++, C			
SqMV	+, T			+, T
CVYV		+, C, T		
A. gossypii	++, C			

+, Intermediate or ++, high level of conventional virus resistance available in genetic resources; C, commercial conventional virus-resistant cultivars have been released; T, experimental transgenic virus-resistant material has been obtained; and TC, commercial transgenic virus-resistant cultivars have been released.

by a necrotic reaction (hypersensitivity). Immunity to MNSV has been associated to one recessive gene (*nsv*) in melon (Coudriet, Kishaba, & Bohn, 1981). Analyses of protoplasts of susceptible and resistant melons inoculated with MNSV have shown that the resistance trait is expressed at the single cell level (Diaz, Nieto, Moriones, Truniger, & Aranda, 2004).

Nieto et al. (2006) identified the *nsv* locus as coding for the melon eukaryotic translation initiation factor 4E (*Cm-eIF4E*) and a single amino acid change at position 228 of the protein led to the resistance to MNSV. The recessive-resistance gene *zym* to ZYMV in watermelon was also shown to code for eIF4E (Ling et al., 2009). A single nucleotide polymorphism in that gene was associated to resistance (Ling et al., 2009), indicating that slight variations in eIF4E can confer resistance in cucurbits to viruses belonging to different families. More classical hypersensitive resistance has been observed in melon against PRSV; two alleles (Prv^1 and Prv^2) at the same locus have been described that can be distinguished by the reaction with different PRSV isolates (Pitrat & Lecoq, 1983).

Resistance may prevent short- or long-distance virus movement within the plant or impede virus multiplication. This type of resistance can be very efficient as resistance to ZYMV found in cucumber which is controlled by one recessive gene (*zym*; Provvidenti, 1987). Ullah and Grumet (2002) showed a localization of ZYMV in the veinal regions of the first leaves of resistant plants and suggested that resistance conferred by *zym* in cucumber is developmentally regulated and occurs at the level of systemic movement since no virus is detected in the apex of resistant plants. In other cases, such resistances are quantitative or partial and are often characterized as "intermediate" resistance. An interesting situation has been observed in melon, where resistance to WMV has been described in accessions PI 414723 with a monogenic-dominant (*Wmr*) genetic control (Gilbert, Kyle, Munger, & Gray, 1994). Resistance was correlated with reduced virus movement within leaves by reducing cell-to-cell movement, but did not affect vascular transport (Gray, Moyer, & Kennedy, 1988). More recently, Diaz-Pendon, Fernandez-Munoz, Gomez-Guillamon, and Moriones (2005) studied a new source of resistance to WMV in melon accession TGR-1551. The resistance resulted in a dramatic and significant reduction of virus titer and infected plants were asymptomatic or exhibited mild-disease symptoms. No effect was observed on virus accumulation in inoculated leaves, suggesting that the initial phases of infection were not affected. Resistance was determined to be monogenic recessive (*wmr*) but the relationship has not been established with *Wmr* (Diaz-Pendon et al., 2005). A resistance under a monogenic-dominant control (*Cys*) limiting CYSDV movement and multiplication has also been described in melon TGR-1551 (Marco, Aguilar, Abad, Gomez-Guillamon, & Aranda, 2003).

Finally, plants can be tolerant to viruses; in this case virus multiplication does occur normally but plants do not express symptoms and virus infection has no or only limited impact on yield. This is the case for ZYMV in zucchini squash. No resistance has been found in *C. pepo* or in *C. maxima* but accessions of several wild species or *C. moschata* (Nigerian local, Menina) are resistant to the virus (Paris, Cohen, Burger, & Yoseph, 1988; Provvidenti, 1986). One main dominant gene (*Zym-1*) and several complementary genes have been described in Menina and Nigerian local (Gilbert-Albertini, Lecoq, Pitrat, & Nicolet, 1993; Pachner, Paris, & Lelley, 2011; Paris et al., 1988). In commercial zucchini squash cultivars, *Zym-1* confers tolerance which allows good levels of fruit production even when ZYMV epidemics occur. Disadvantage of tolerant genotypes is that they constitute virus sources for susceptible genotypes.

4.3.2 Resistance durability, an important issue

Since breeding for resistance is a long and costly process, breeders are entitled to expect that resistant cultivars should provide a durable protection to the crop at least for the cultivar commercial life. However, resistances may be either strain specific (i.e., efficient only against some strains) or not stable (i.e., favor the emergence of new virus strains that overcome the resistance). Resistance durability will depend not only upon resistance specificity and stability but also upon the effect on virus fitness of the genetic change involved in the ability to overcome the resistance. Resistance specificity can be determined at the first steps of resistance characterization, when confronting resistant plants to a collection of isolates representative of the known virus variability. Stability of the resistance may be established after several years of cultivation of resistant varieties, if virulent variant emerges as a result of virus adaptation. Stability can sometimes be predicted from laboratory studies using different methods of inoculation on resistant genotypes and studying virus progenies (Gomez, Rodriguez-Hernandez, Moury, & Aranda, 2009; Lecoq et al., 2004).

An oligogenic-recessive resistance to CMV has been described in PI 161375 (*Songwhan charmi*) from Korea (Risser, Pitrat, & Rode, 1977), which also has the *Vat* gene (Pitrat & Lecoq, 1980). Leroux, Quiot, Lecoq, and Pitrat (1979) showed that 35% of 1124 CMV isolates collected among naturally infected vegetables and weeds in southeastern France were able to infect PI 161375 systemically; they were referred as CMV "Song strains". In order to evaluate the durability of this resistance when in association with *Vat* conferring a resistance to CMV transmission by *A. gossypii*, two types of observations were obtained (Lecoq et al., 2004). First, it was shown that the repeated cultivation of PI 161375 in the same plot did not significantly increase the occurrence of "Song strains" in weeds, which are natural reservoirs of CMV between successive melon crops. Second, the effectiveness of the CMV protection under natural epidemic conditions was similar in 1983 and 10 years later, in 1993 (Lecoq et al., 2004). Taken together, these observations indicated that there was no erosion with time of the protection conferred by this composite resistance to CMV and strongly suggested that although it is strain specific, it may be durable (Lecoq et al., 2004).

Durability of ZYMV resistance has been extensively studied in cucumber, squash, and melon with contrasting results. The *zym* resistance gene in cucumber Taichung Mou Gua (TMG) was efficient against several ZYMV reference strains. After inoculation, no systemic symptoms were observed except occasionally a mild-vein clearing limited to one or two leaves over the inoculated cotyledons. Virus was detected by DAS-ELISA only in these leaves with mild-vein clearing but not from asymptomatic leaves at the plant apex (Ullah & Grumet, 2002). Serial mechanical inoculations to TMG of extracts from leaves showing mild-vein clearing did not favor the emergence of virus variant overcoming the resistance and no ZYMV infection has been reported so far in fields of resistant cucumber cultivars. These observations strongly suggest that the resistance conferred by the *zym* gene in cucumber is stable and not specific (Lecoq, Desbiez, Wipf-Scheibel, Girard, & Pitrat, 2002).

ZYMV reference strains induce mild symptoms on *Zym-1*-tolerant zucchini squash commercial hybrids. Occasionally more severe mosaic patches are observed in restricted leaf areas. When these mosaic patches are used for sequential mechanical or aphid inoculation of tolerant hybrids, very severe symptoms may be observed, similar to those observed on susceptible controls (Desbiez, Gal-On, Girard, Wipf-Scheibel, & Lecoq, 2003). A single mutation at position 917 on P3 protein is responsible for this tolerance breakdown. However, this mutation resulted in a loss of virus fitness in susceptible cultivar and a rapid reversion toward the wild-type virus genotype. Alternating cultivation of tolerant and susceptible cultivars should therefore prevent the increase of isolates breaking down the tolerance and should contribute to tolerance durability (Desbiez et al., 2003).

When ZYMV reference strains were inoculated to melon PI 414723 possessing the *Zym*-resistance gene, two pathotypes were observed. Strains of pathotype 0 did not produce systemic symptoms, and the virus was not detected by DAS-ELISA in the plant apex. In contrast, strains of pathotype 1 induced systemic necrotic lesions leading sometimes to a generalized plant necrosis (Lecoq & Pitrat, 1984). It was easy by serial mechanical inoculation to PI 414723 of extracts from tissue with systemic necrotic lesions, to obtain variants that totally overcome the resistance. These variants grouped in pathotype 2 induced vein clearing and mosaic symptoms in PI 414723 very similar to those observed on susceptible plants (Lecoq & Pitrat, 1984). The same evolution was observed when instead of mechanical inoculation plants were inoculated by aphids. This demonstrated that the resistance conferred by *Zym* in melon is not stable since it may favor the rapid emergence of virus variants that break down the resistance. Therefore, the use of *Zym* has not been recommended in melon breeding programs.

Resistance breakdowns are not always associated to single or few mutations inducing changes in viral proteins. Two MNSV strains overcoming the *nsv* resistance gene in melon have been isolated in Spain, and in both cases the gain in virulence has been related to recombination that occurred in the 3′-noncoding region of the genome. Interestingly, one of these variants is the result of a recombination event between MNSV and CABYV, an unrelated cucurbit polerovirus (Diaz et al., 2004; Miras et al., 2014). However, both viruses frequently coinfect melons or cucumbers, what provide opportunities for recombination.

4.4. Transgenic resistance in cucurbits

The advantages of transgenic virus resistance compared to conventional virus resistance have been discussed by Gaba, Zelcer, and Gal-On (2004). Genetic engineering is a potentially easy way to achieve high levels of virus resistance not found in natural genetic resources. The feasibility of creating transgenic virus resistance in cucurbit cultivars depends (1) on the availability of sequence data for cloning a fragment of a viral gene, (2) on the ability to transform a given species or cultivar, and (3) on the political context of the country in which the transgenic crop has to be grown.

Sequence data of nearly all known cucurbit viruses are nowadays available. Transgenic virus resistance has been tested in laboratory conditions and proved to be efficient against major cucurbit viruses belonging to comoviruses, cucumoviruses, potyviruses, and tobamoviruses in the four major cucurbit crops: cucumber, melon, squash, and watermelon (Table 5.4). Besides classical single virus transgenic resistance, multiple transgenic virus resistances have been produced in a single operation, using several genes on a single construct (Tricoli et al., 2002). In this case, multiple transgenic virus resistance behaves as only a single gene in subsequent conventional crosses. Another interest of this approach is that the same single/multiple transgenic virus-resistance construct can be used to transform different species or cultivars (Gaba et al., 2004).

The practical limitations to the use of transgenic resistance are both technical and political. It depends on the efficiency of plant transformation systems and on the specificity of the transgenic virus resistance. Some species or cultivars (i.e., Charentais type melons) are particularly difficult to transform and some transgenic resistances were shown to be strain specific. There are also strong social limitations in some countries on the use of transgenic plants. This is particularly noticeable in Europe where parts of the farmer and consumer populations are strongly reluctant to the deployment of GMOs.

4.4.1 Diversity of transgenic resistances

The genes used for achieving the first transgenic virus resistance were mainly the viral coat proteins (CPs). CP-mediated resistance has been demonstrated in melon, cucumber, squash, and watermelon (Fuchs, McFerson, et al., 1997; Gonsalves & Slightom, 1993; Tricoli et al., 2002). Transgenic virus-resistance levels varied and conferred either complete resistance or only delayed symptom appearance. An important characteristic of this type of resistance is that increasing the level of CP expression (homozygote vs. heterozygote) enhances the virus-resistance response (Fuchs, McFerson, et al., 1997). Other viral sequences were used to obtain virus-resistant plants, including genes coding for movement proteins or polymerases.

Posttranscriptional gene silencing is another mechanism of pathogen-derived resistance that was found to be effective in melon carrying partial CP genes of ZYMV and PRSV (Wu et al., 2010). In this case, the level of resistance was inversely correlated to the level of expression of the transgene RNA, which contrasts with CP-mediated transgenic virus resistance.

Antisense-mediated resistance has been tested with ZYMV *CP* gene (Fang & Grumet, 1993) and a delay of several days in the appearance of viral symptoms in melon was observed in transgenic plants. More recently, a novel strategy has been developed based on the expression of a hairpin RNA transcript with virus genome fragments inserted in opposite orientation. This approach was used to obtain cucumber and melon resistant to ZYMV using the *HC-Pro* gene (Leibman et al., 2011). Interestingly, a cucumber line that accumulated high levels of small interfering-RNAs was resistant to WMV and PRSV, in addition to ZYMV. This opens the possibility of getting broad spectrum resistance to potyviruses (Leibman et al., 2011). Finally, ribozyme-mediated virus resistance was demonstrated in melon against ZYMV, WMV, and CMV (Huttner et al., 2001, Lenee, Perez, Gruber, Baudot, & Ollivo, 2001).

Several transgenic cucurbits obtained through these different strategies have been also tested in field conditions. The high levels of resistance observed in the laboratory were confirmed in the conditions of natural virus epidemics.

4.4.2 Commercial use of transgenic virus-resistant cultivars

Transgenic squash cultivars resistant to WMV and ZYMV were the first virus resistant transgenic plants deregulated in the United States in 1994 and cultivated for commercial production. They were followed by the release of commercial transgenic squash cultivars resistant to WMV, ZYMV, and CMV. Therefore, transgenic squashes constitute an interesting case study for analyzing their commercial impact, since they have been grown in the United States already for 20 years now.

The acreage of transgenic virus-resistant squash cultivated in the United States increases steadily, but still remains limited. It was close to 2000 ha in 2003 and reached 3200 ha in 2006 (Johnson, 2007) representing 13% of the total U.S. squash area. This relatively modest part of the market is probably due to various factors including (1) the availability of transgenic virus resistance in a few squash types only (crookneck, zucchini), (2) the added seed cost of transgenic cultivars in comparison with regular squash cultivars (approximately 380 U.S. $ per ha; Johnson, 2007), and (3) the availability of cheaper squash cultivars with good conventional intermediate resistance to CMV, WMV, and ZYMV.

The acreage of virus-resistant transgenic squash varies between States probably due to major differences in virus diseases pressure. In 2006, it reached 70% in New Jersey and 20% in Georgia and Florida—the two major squash producing states—but apparently none were cultivated in California (Johnson, 2007).

Interestingly, after 20 years of commercial use of transgenic virus-resistant squash in the United States, there is no report of breakage of the CP-mediated resistances to CMV, WMV, or ZYMV, which therefore appear to be durable.

4.4.3 Risk assessment studies

A major issue raised by the opponents to GMOs concerns their potential risks for the environment. Numerous field experiments conducted on transgenic virus-resistant cucurbits have been devoted to address these safety issues and, in this respect, cucurbits constitute a model for the evaluation of potential risks associated with virus-resistant transgenic plants (Fuchs & Gonsalves, 2007). Potential risks concern the interaction of transgenic plants with nontarget viruses (heteroencapsidation and recombination), with wild plants in the environment (gene flow) or with food safety issues (i.e., allergy, toxicity, etc.) (Fuchs & Gonsalves, 2007).

Heteroencapsidation should be considered only in the case of CP-mediated resistance. It is a phenotypic modification of viral particles that are encapsidated not only by their own CP, but also by the CP of another virus produced by the transgenic plant, as this may occur in mixed virus infections (Hammond, Lecoq, & Raccah, 1999). Heteroencapsidation was shown to occur very rarely in potyvirus-resistant transgenic squash and, since it concerns only a single generation (it is a phenotypic change that does not affect the viral genome), heteroencapsidation is considered as having a limited significance in terms of risks (Fuchs & Gonsalves, 2007). In addition, use of constructs with modified CP genes that would prevent heteroencapsidation is possible (Jacquet et al., 1998).

Recombination is a potentially more significant event since it could durably modify viral genomes. In the case of RNA viruses, recombination could occur between the transcript of a viral transgene and the genome of a challenge virus when replicating in a transgenic plant cell. However, so far no evidence of recombination has been observed in CMV, WMV, and ZYMV transgenic squash (Fuchs & Gonsalves, 2007), but probably more systematic surveys should be conducted to definitively exclude this possibility.

Gene flow from cultivated transgenic squash to a wild relative (*C. pepo* subsp. *ovifera* var. *texana*, a weed in Texas) has been extensively studied (Fuchs, Chirco, McFerson, & Gonsalves, 2004). The transfer of transgenes from transgenic squash to its wild relative was demonstrated and conferred virus resistance to the weed. The weeds carrying the transgenes were found to have a better fitness (seed production) under strong virus pressure than their nontransgenic counterpart. In contrast, under conditions of low virus pressure—what seems to be the rule in natural habitats of this weed—transgenic wild squashes were less fit than nontransgenic (Fuchs et al., 2004; Fuchs & Gonsalves, 2007).

Finally, numerous observations and sequence comparisons with known allergens suggest that CMV, WMV, or ZYMV CPs as expressed in transgenic plants does not pose a specific allergenic threat (Fuchs & Gonsalves, 2007). In addition, these CPs are abundantly found on the market in nontransgenic fresh cucurbits that are frequently virus infected, and no allergy or toxicity problem has ever been reported.

5. CONCLUDING REMARKS

Cucurbit viruses constitute an ever-changing pathosystem with new viruses or new virus strains emerging regularly in different cropping systems in various parts of the world (Lecoq & Desbiez, 2012; Navas-Castillo et al., 2011; Romay et al., 2014). The number of viruses infecting cucurbit crops has more than tripled in less than 35 years, with more than one new virus species reported each year. This evolution is very likely to continue and new severe viruses will be described in the coming years. Therefore, there is a need to maintain a careful surveillance of cucurbit crops in the main production areas and to detect early any major change in virus populations. This is of outmost importance to deploy control measures adapted to the actual prevailing virus species or virus strains in a specific region and to anticipate their spread to other areas. Surveys should be conducted regularly and diagnostic methods should be improved. Serological (ELISA) and molecular (PCR and RT-PCR) diagnostic tests have greatly improved, during last decades, our knowledge of the various viruses infecting major cucurbit crops worldwide (Lecoq, 2003). But for extensive surveys and to reduce their cost, there is the need for new simple and reliable mulitiplex techniques allowing detection of several viruses in a single operation in great numbers of samples (Boonham et al., 2014; Charlermroj et al., 2013; Lee et al., 2003; Wei, Pearson, Blohm, Nölte, & Armstrong, 2009).

Next generation sequencing (NGS) provides new opportunities to identify rapidly emerging viruses with a blind approach (Boonham et al., 2014). This will not only rapidly lead to decide whether a new encounter is a new strain of a known virus or a new virus species, but also provide important nucleotide sequence information. This will facilitate the development of specific RT-PCR primers and eventually, if the *CP* sequence is determined, the rapid production of specific antiserum through the heterologous expression of the CP in bacteria (Cotillon et al., 2005). However, the development of specific diagnostic tools should not exclude biological studies, which will enable to determine the actual host range of the new virus and it means of spread (transmission by seeds and vectors). However, the inclusion of a new virus to a specific genus or family may indeed suggest the nature of the vector. It should be emphasized that basic knowledge on virus biology is a prerequisite to the deployment of adapted control methods (Jones, 2004).

The diversity of control methods corresponds to the diversity of cucurbit viruses. Some methods will be virus specific or even strain specific such as cross-protection or breeding for resistance. In contrast, cultural practices intended to interfere with vector activities are more generic and will be equally effective to control viruses having the same vectors and sharing the same virus–vector interaction types (persistent, semipersistent, and nonpersistent).

None of the control method described in this chapter is sufficient by itself to control cucurbit viruses. To provide the optimal protection to the crop, they must be combined within an integrated management scheme (Jones, 2004). Numerous examples show the benefit of combining control methods. For instance, the combined use of insecticide and mulching resulted in suppression of *B. tabaci* populations and reduced the incidence of the persistently transmitted CuLCrV in zucchini crops (Nyoike & Liburd, 2010) and of the semipersistently transmitted SqVYV in watermelon (Kousik et al., 2012). Similarly, a combination of reflective mulch, mineral oil, and insecticide applications was very efficient in controlling the aphid-borne PRSV in zucchini (Pinese, Lisle, Ramsey, Halfpapp, & De Faveri, 1994). However, the use of mulches and row covers has a major drawback; this requires a lot of plastic material that farmers must dispose in an ecologically sound way after the crop cycle. Similarly, the extensive use of chemicals on crops (oil, insecticides) is negatively perceived by consumers (Perring et al., 1999). Therefore, there is a need to develop biodegradable mulches and new insecticides with limited if any impact on the environment.

Cultural and chemical control methods can also be combined with cross-protection and the use of resistant cultivars. Better protection was obtained in squash by combining ZYMV mild-strain protection and reflective mulches or floating row covers than by each of this method alone (Abou-Jawdah et al., 2000). Similarly, combining the use of insect-proof nets and CVYV or CYSDV partially resistant cucumber varieties provided an efficient virus control (Janssen et al., 2003) as did the combination of partial CMV resistance in melon and plastic mulches (Lecoq et al., 2004) or the use of tall border crops and ZYMV tolerance in squash (Coutts et al., 2011).

Combination of cultural practices and use of resistant cultivars not only may provide the best virus control but also can contribute to improve resistance durability. Indeed, Moury et al. (2011) described two contrasting scenarios for deploying resistance traits. A virtuous circle where farmers are encouraged to use resistant cultivars with cultural practices intended to interfere with vector activities. This reduces the virus "pressure" on resistant plants and decreases the probability of occurrence of virulent variants overcoming the resistance. A vicious circle would be if farmer, because of the good protection provided by the resistance, decide to stop using cultural control methods. This would thus increase the virus "pressure" on resistant plants which may facilitate the emergence of virulent isolates (Moury et al., 2011).

High levels of conventional resistance to major cucurbit viruses are still lacking (Table 5.4) and there is still a need to expand testing genetic resources available in public or private institutions. High throughput inoculation and evaluation methods have to be developed, particularly for the severe emerging viruses.

Cucurbits have been the first crops that demonstrated the potential of genetic engineering to obtain commercial resistant cultivars (Fuchs, Ferreira, et al., 1997). Transgenic squashes have also been a model for exhaustive risk assessment studies (Fuchs & Gonsalves, 2007). The use of different types of construct has been explored successfully. A novel strategy based on the expression of a hairpin RNA transcript with virus genome fragments inserted in opposite orientation is very promising since it may provide broad spectrum resistance (Leibman et al., 2011). Using shorter noncoding virus sequences will also contribute to limit the potential consequences of recombination events with nontarget viruses. A wider use of transgenic resistances still depends on legal issues for many countries. Hopefully, the development of virus-resistant cultivars using a minimum of foreign genetic information will contribute to facilitate their deregulation in countries still reluctant to their cultivation.

New control methods have to be explored. Some chemicals activate the defense response of plants, which is well known as induced systemic resistance or systemic acquired resistance, and induce an antiviral response. These chemicals do not act specifically against viruses, but they can also reduce losses caused by bacteria or fungi. Recently, application of acibenzolar-*S*-methyl resulted in decreased accumulation of CCYV in the treated plants and attenuated systemic symptoms in melon plants (Takeshita et al., 2013). In contrast, applications of the commercially available resistance-inducing treatment BioYield™ did not reduce WMV incidence in squash (Murphy, Eubanks, & Masiri, 2009). More experiments have to be conducted to check the potential against virus diseases of these resistance-inducing molecules.

A new generation of cross-protection might be at hands. Recently, Tamura et al. (2013) have reported that, using a latent virus in cucumber, *Apple latent spherical virus*, harboring part of ZYMV and/or CMV genome, they could obtain not only an effective protection against the target viruses but also a curative effect when inoculated after the target virus infection. If practically and legally applicable in cucurbit crops, this would be the first virus control method that could have a curative effect in cucurbit fields. In addition, it could be easily adapted to control any other cucurbit virus.

Finally, it is to the cucurbit farmer who is the end-user to decide which virus control method to apply to his crop. Control methods have to be cost effective, easy to apply with minimal extra investment and labor. The economic threshold will vary according to the income to be expected from the crop and to the fruit quality requirements by local consumers. Obviously, major differences are to be expected according to the country. Since we are lacking models predicting the intensity of virus epidemics in cucurbit crops, farmer's decision to use or not to use control methods will always be a kind of a bet. If the decision is to apply costly protective measures (mulches, row cover, cross-protection) and virus epidemics are early and severe, he will win the bet, but he loses if virus epidemics are late and mild.

REFERENCES

Abd-Rabou, S., & Simmons, A. M. (2012). Some cultural strategies to help manage *Bemisia tabaci* (*Hemiptera*: *Aleyrodidae*) and whitefly-transmitted viruses in vegetable crops. *African Entomology*, *20*, 371–379.

Abou-Jawdah, Y., Sobh, H., Fayad, A., Lecoq, H., Delecolle, B., & Trad-Ferre, J. (2000). Cucurbit yellow stunting disorder virus—A new threat to cucurbits in Lebanon. *Journal of Plant Pathology*, *82*, 55–60.

Aguilar, M. I., Guirado, M. L., Melero-Vara, J. M., & Gomez, J. (2010). Efficacy of composting infected plant residues in reducing the viability of *Pepper mild mottle virus*, *Melon necrotic spot virus* and its vector, the soil-borne fungus *Olpidium bornovanus*. *Crop Protection*, *29*, 342–348.

Antignus, Y. (2000). Manipulation of wavelength-dependent behaviour of insects: An IPM tool to impede insects and restrict epidemics of insect-borne viruses. *Virus Research*, *71*, 213–220.

Antignus, Y., Lapidot, M., Hadar, D., Messika, Y., & Cohen, S. (1998). Ultraviolet-absorbing screens serve as optical barriers to protect crops from virus and insect pests. *Journal of Economic Entomology*, *91*, 1401–1405.

Antignus, Y., Nestel, D., Cohen, S., & Lapidot, M. (2001). Ultraviolet-deficient greenhouse environment affects whitefly attraction and flight–Behavior. *Environmental Entomology*, *30*, 394–399.

Astier, S., Albouy, J., Maury, Y., Robaglia, C., & Lecoq, H. (2007). *Principles of plant virology*. Enfield: Science Publishers.

Atkins, S., Webster, C. G., Kousik, C. S., Webb, S. E., Roberts, P. D., Stansly, P. A., et al. (2011). Ecology and management of whitefly-transmitted viruses of vegetable crops in Florida. *Virus Research*, *159*, 110–114.

Blancard, D., Lecoq, H., & Pitrat, M. (1994). *A colour atlas of cucurbit diseases*. London: Manson Publishing.

Boonham, N., Kreuze, J., Winter, S., van der Vlugt, R., Bergervoet, J., Tomlinson, J., et al. (2014). Methods in virus diagnostic: From ELISA to next generation sequencing. *Virus Research*, *186*, 20–31.

Boubourakas, I. N., Hatziloukas, E., Antignus, Y., & Katis, N. I. (2004). Etiology of leaf chlorosis and deterioration of the fruit interior of watermelon plants. *Journal of Phytopathology*, *152*, 580–588.

Call, A. D., & Wehner, T. C. (2011). Gene list 2010 for cucumber. *The Cucurbit Genetics Cooperative Report*, *33–34*, 69–103.

Campbell, R. N., Wipf-Scheibel, C., & Lecoq, H. (1996). Vector-assisted seed transmission of *Melon necrotic spot virus* in melon. *Phytopathology*, *86*, 1294–1298.

Charlermroj, R., Himananto, O., Seepiban, C., Kumpoosiri, M., Warin, N., Oplatowska, M., et al. (2013). Multiplex detection of plant pathogens using a microsphere immunoassay technology. *PLoS One*, *8*(4), e62334. http://dx.doi.org/10.1371/journal.pone.0062344.

Choi, G.-S. (2001). Occurrence of two tobamovirus diseases in cucurbits and control measures in Korea. *Plant Pathology Journal*, *17*, 243–248.

Chyzik, R., Dobrinin, S., & Antignus, Y. (2003). Effect of a UV-deficient environment on the biology and the flight activity of *Myzus persicae* and its hymenopterous parasite *Aphidius matricariae*. *Phytoparasitica*, *31*, 467–477.

Cohen, R., Burger, Y., Herev, C., Koren, A., & Edelstein, M. (2007). Introducing grafted cucurbits to modern agriculture: The Israel experience. *Plant Disease*, *91*, 916–923.

Cotillon, A.-C., Desbiez, C., Bouyer, S., Wipf-Scheibel, C., Gros, C., Delecolle, B., et al. (2005). Production of a polyclonal antiserum against the coat protein of *Cucurbit yellow stunting disorder crinivirus* expressed in *Escherichia coli*. *Bulletin OEPP/EPPO*, *35*, 99–103.

Coudriet, D. L., Kishaba, A. N., & Bohn, G. W. (1981). Inheritance of resistance to *Muskmelon necrotic spot virus* in a melon aphid resistant breeding line of muskmelon. *Journal of the American Society for Horticultural Science*, *106*, 789–791.

Coutts, B. A., Kehoe, M. A., & Jones, R. A. C. (2011). Minimising losses caused by *Zucchini yellow mosaic virus* in vegetable cucurbit crops in tropical, sub-tropical and Mediterranean environments through cultural methods and host resistance. *Virus Research*, *159*, 141–160.

Coutts, B. A., Kehoe, M. A., & Jones, R. A. C. (2013). *Zucchini yellow mosaic virus*: Contact transmission, stability on surfaces, and inactivation with disinfectants. *Plant Disease*, *97*, 765–771.

Damicone, J. P., Edelson, J. V., Sherwood, J. L., Myers, L. D., & Motes, J. E. (2007). Effects of border crops and intercrops on control of cucurbit virus diseases. *Plant Disease*, *91*, 509–516.

Davis, A. R., Perkins-Veazie, P., Sakata, Y., Lopez-Galarza, S., Moroto, J. V., Lee, S.-G., et al. (2008). Cucurbit grafting. *Critical Reviews in Plant Sciences*, *27*, 50–74.

Desbiez, C., Gal-On, A., Girard, M., Wipf-Scheibel, C., & Lecoq, H. (2003). Increase in *Zucchini yellow mosaic virus* symptom severity in tolerant zucchini cultivars is related to a point mutation in P3 protein and is associated with a loss of relative fitness on susceptible plants. *Phytopathology*, *93*, 1478–1484.

Diaz, J. A., Nieto, C., Moriones, E., Truniger, V., & Aranda, M. A. (2004). Molecular characterization of a *Melon necrotic spot virus* strain that overcomes the resistance in melon and nonhost plants. *Molecular Plant–Microbe Interactions*, *17*, 668–675.

Diaz-Pendon, J. A., Fernandez-Munoz, R., Gomez-Guillamon, M. L., & Moriones, E. (2005). Inheritance of resistance to *Watermelon mosaic virus* in *Cucumis melo* that impairs virus accumulation, symptom expression, and aphid transmission. *Phytopathology*, *95*, 840–846.

Dogimont, C. (2011). Gene list 2011 for melon. *The Cucurbit Genetics Cooperative Report, 33–34*, 104–133.

Doolittle, S. P., & Walker, M. N. (1925). Further studies on the overwintering and dissemination of cucurbit mosaic. *Journal of Agricultural Research, 31*, 1–58.

Doring, T. F., & Chiitka, L. (2007). Visual ecology of aphids: A critical review on the role of colours in host finding. *Arthropod–Plant Interactions, 1*, 3–16.

Duffus, J. E. (1971). Role of weeds in the incidence of virus diseases. *Annual Review of Phytopathology, 9*, 319–340.

Fang, G. W., & Grumet, R. (1993). Genetic-engineering of potyvirus resistance using constructs derived from the *Zucchini yellow mosaic virus* coat protein gene. *Molecular Plant–Microbe Interactions, 6*, 358–367.

FAOSTAT. (2014). http://faostat.fao.org.

Fereres, A. (2000). Barrier crops as a cultural control measure of non-persistently transmitted aphid-borne viruses. *Virus Research, 71*, 221–231.

Fereres, A., & Moreno, A. (2011). Integrated control measures against viruses and their vectors. In C. Caranta, M. A. Aranda, M. Tepfer, & J. J. Lopez-Moya (Eds.), *Recent advances in plant virology* (pp. 237–261). Norfolk: Caister Academic Press.

Fuchs, M., Chirco, E. M., McFerson, J., & Gonsalves, D. (2004). Comparative fitness of a free-living squash species and free-living × virus-resistant transgenic squash hybrids. *Environmental Biosafety Research, 3*, 17–28.

Fuchs, M., Ferreira, S., & Gonsalves, D. (1997). Management of virus diseases by classical and engineered protection. *Molecular Plant Pathology*. Online, http://www.bspp.org.uk/mppol/, 1997/0116fuchs.

Fuchs, M., & Gonsalves, D. (2007). Safety of virus-resistant transgenic plants two decades after their introduction: Lessons from realistic field risk assessment studies. *Annual Review of Phytopathology, 45*, 173–202.

Fuchs, M., McFerson, J. R., Tricoli, D. M., McMaster, J. R., Deng, R. Z., Boeshore, M. L., et al. (1997). Cantaloupe line CZW-30 containing coat protein genes of *Cucumber mosaic virus, Zucchini yellow mosaic virus*, and *Watermelon mosaic virus-2* is resistant to these three viruses in the field. *Molecular Breeding, 3*, 279–290.

Gaba, V., Zelcer, A., & Gal-On, A. (2004). Cucurbit biotechnology—The importance of virus resistance. *In Vitro Cellular & Developmental Biology—Plant, 40*, 346–358.

Gal-On, A., & Raccah, B. (2000). A point mutation in the FRNK motif of the potyvirus helper component-protease gene alters symptom expression in cucurbits and elicits protection against the severe homologous virus. *Phytopathology, 90*, 467–473.

Gerling, D., Alomar, O., & Arno, J. (2001). Biological control of using predators and parasitoids. *Crop Protection, 20*, 779–799.

Gilbert, R. Z., Kyle, M. M., Munger, H. M., & Gray, S. M. (1994). Inheritance of resistance to watermelon mosaic virus in *Cucumis melo* L. *HortScience, 29*, 107–110.

Gilbert-Albertini, F., Lecoq, H., Pitrat, M., & Nicolet, J.-L. (1993). Resistance of *Cucurbita moschata* to *Watermelon mosaic virus* type 2 and its genetic relation to resistance to zucchini yellow mosaic virus. *Euphytica, 69*, 231–237.

Gomez, P., Rodriguez-Hernandez, A. M., Moury, B., & Aranda, M. A. (2009). Genetic resistance for sustainable control of plant virus diseases: Breeding, mechanisms and durability. *European Journal of Plant Pathology, 125*, 1–22.

Gonsalves, D., & Slightom, J. L. (1993). Coat protein-mediated protection—Analysis of transgenic plants for resistance in a variety of crops. *Seminars in Virology, 4*, 397–405.

Gray, S. M., Moyer, J. W., & Kennedy, G. G. (1988). Resistance in *Cucumis melo* to *Watermelon mosaic virus 2* correlated with reduced virus movement within leaves. *Phytopathology, 78*, 1043–1047.

Greer, L., & Dole, J. M. (2003). Aluminium foil, aluminium-painted plastic, and degradable mulches increase yields and decrease insect-vectored viral diseases of vegetables. *HortTechnology, 13*, 276–284.

Guner, N., & Wehner, T. C. (2003). Gene list for watermelon. *The Cucurbit Genetics Cooperative Report, 26*, 76–92.

Hammond, J., Lecoq, H., & Raccah, B. (1999). Epidemiological risks from mixed virus infections and transgenic plants expressing viral genes. *Advances in Virus Research, 54*, 189–314.

Herrera-Vasquez, J. A., Cordoba-Selles, M. C., Cebrian, M. C., Alfaro-Fernandez, A., & Jorda, C. (2009). Seed transmission of *Melon necrotic spot virus* and efficacy of seed-disinfection treatments. *Plant Pathology, 58*, 436–442.

Hilje, L., Costa, H. S., & Stansly, P. A. (2001). Cultural practices for managing *Bemisia tabaci* and associated viral diseases. *Crop Protection, 20*, 801–812.

Hooks, C. R. R., & Fereres, A. (2006). Protecting crops from non-persistently aphid-transmitted viruses: A review on the use of barrier plants as a management tool. *Virus Research, 120*, 1–16.

Huitron-Ramirez, M. V., Ricardez-Salinas, M., & Camacho-Ferre, F. (2009). Influence of grafted watermelon plant density on yield and quality in soil infested with melon necrotic spot virus. *HortScience, 44*, 1838–1841.

Huttner, E., Tucker, W., Vermeulen, A., Ignart, F., Sawyer, B., & Birch, R. (2001). Ribozyme genes protecting transgenic melon plants against potyviruses. *Current Issues in Molecular Biology, 3*, 27–34.

Jacquet, C., Delécolle, B., Raccah, B., Lecoq, H., Dunez, J., & Ravelonandro, M. (1998). Use of modified plum pox virus coat protein genes developed to limit heteroencapsidation-associated risks in transgenic plants. *Journal of General Virology, 79*, 1509–1517.

Janssen, D., Ruiz, L., Cano, M., Belmonte, A., Martin, G., Segundo, E., et al. (2003). Physical and genetic control of *Bemisia tabaci*-transmitted *Cucurbit yellow stunting disorder virus* and *Cucumber vein yellowing virus* in cucumber. *Bulletin OILB/SROP, 26*, 101–106.

Johnson, S. R. (2007). *Quantification of the impacts on US agriculture of biotechnology-derived crops planted in 2006*. Washington, DC, USA: National Center for Food and Agricultural Policy. www.ncfap.org.

Jones, R. A. C. (2004). Using epidemiological information to develop effective integrated virus disease management strategies. *Virus Research, 100*, 5–30.

Kosaka, Y., & Fukunishi, T. (1997). Multiple inoculation with three attenuated viruses for the control of cucumber virus disease. *Plant Disease, 81*, 733–738.

Kousik, C. S., Adkins, S., Turechek, W. W., Webster, C. G., Webb, S. E., Baker, C. A., et al. (2012). Progress and challenges in managing watermelon vine decline caused by whitefly-transmitted *Squash vein yellowing virus* (SqVYV). *Israel Journal of Plant Sciences, 60*, 435–445.

Kousik, C. S., Webster, C. G., Turechek, W. W., Stanley, P., & Roberts, P. D. (2010). Effect of reflective mulch and insecticidal treatments on development of watermelon vine decline caused by *Squash vein yellowing virus*. In J. A. Thies, S. Kousik, & A. Levi (Eds.), *Cucurbitaceae 2010 Proceedings* (pp. 237–239), Alexandria, VA: ASHS Press.

Kring, J. B. (1972). Flight behavior of aphids. *Annual Review of Entomology, 17*, 461–492.

Lecoq, H. (1992). Les virus des cultures de melon et de courgette de plein champ (II). *PHM—Revue Horticole, 324*, 15–25.

Lecoq, H. (1998). Control of plant virus diseases by cross protection. In A. Hadidi, R. K. Kheterpal, & H. Koganezawa (Eds.), *Plant virus disease control* (pp. 33–40). St. Paul, MN: APS Press.

Lecoq, H. (1999). Epidemiology of *Cucurbit aphid-borne yellows virus*. In H. G. Smith, & H. Barker (Eds.), *The Luteoviridae* (pp. 243–248). Wallingford, UK: CAB International.

Lecoq, H. (2003). Cucurbits. In G. Loebenstein, & G. Thottapilly (Eds.), *Virus and virus-like diseases of major crops in developing countries* (pp. 665–688). Dordrecht, The Netherlands: Kluwer Academic Publishers.

Lecoq, H., Cohen, S., Pitrat, M., & Labonne, G. (1979). Resistance to *Cucumber mosaic virus* transmission by aphids in *Cucumis melo*. *Phytopathology, 69,* 1223–1225.

Lecoq, H., & Desbiez, C. (2008). *Watermelon mosaic virus* and *Zucchini yellow mosaic virus*. In B. W. J. Mahy, & M. H. V. Van Regenmortel (Eds.), *Encyclopedia of Virology: Vol. 5.* (3rd ed., pp. 433–440). Oxford, UK: Elsevier.

Lecoq, H., & Desbiez, C. (2012). Viruses of cucurbit crops in the Mediterranean region: An ever-changing picture. *Advances in Virus Research, 84,* 67–126.

Lecoq, H., Desbiez, C., Wipf-Scheibel, C., & Girard, M. (2003). Potential involvement of melon fruit in the long distance dissemination of cucurbit potyviruses. *Plant Disease, 87,* 955–959.

Lecoq, H., Desbiez, C., Wipf-Scheibel, C., Girard, M., & Pitrat, M. (2002). Durability of *Zucchini yellow mosaic virus* resistances in cucurbits. In D. N. Maynard (Ed.), *Cucurbitaceae 2002* (pp. 294–300). Alexandria, VA: ASHS Press.

Lecoq, H., Labonne, G., & Pitrat, M. (1980). Specificity of resistance to virus transmission by aphids in *Cucumis melo*. *Annales de Phytopathologie, 12,* 139–144.

Lecoq, H., Lemaire, J.-M., & Wipf-Scheibel, C. (1991). Control of ZYMV in squash by cross protection. *Plant Disease, 75,* 208–211.

Lecoq, H., Moury, B., Desbiez, C., Palloix, A., & Pitrat, M. (2004). Durable virus resistance in plants through conventional approaches: A challenge. *Virus Research, 100,* 31–39.

Lecoq, H., Piquemal, J.-P., Michel, M.-J., & Blancard, D. (1988). Virus de la mosaïque de la courge: Une nouvelle menace pour les cultures de melon en France? *PHM—Revue Horticole, 289,* 25–30.

Lecoq, H., & Pitrat, M. (1984). Strains of *Zucchini yellow mosaic virus* in muskmelon (*Cucumis melo* L.). *Journal of Phytopathology, 111,* 165–173.

Lecoq, H., Wipf-Scheibel, C., Nozeran, K., Millot, P., & Desbiez, C. (2014). Comparative molecular epidemiology provides new insights in *Zucchini yellow mosaic* occurrence in France. *Virus Research, 186,* 135–143.

Lecoq, H., Wisler, G., & Pitrat, M. (1998). Cucurbit viruses: The classics and the emerging. In J. D. McCreight (Ed.), *Cucurbitaceae '98: Evaluation and enhancement of cucurbit germplasm* (pp. 126–142). Alexandria, VA: ASHS Press.

Lee, G. P., Min, B. E., Kim, C. S., Choi, S. H., Harn, C. H., Kim, S. U., et al. (2003). Plant virus cDNA chip hybridization for detection and differentiation of four cucurbit-infecting tobamoviruses. *Journal of Virological Methods, 110,* 19–24.

Lee, D. H., Nyrop, J. P., & Sanderson, J. P. (2009). Attraction of *Trialeurodes vaporariorum* and *Bemisia argentifolii* to eggplant, and its potential as a trap crop for whitefly management on greenhouse poinsettia. *Entomologia Experimentalis et Applicata, 133,* 105–116.

Legarrea, S., Betancourt, M., Plaza, M., Fraile, A., Garcia-Arenal, F., & Fereres, A. (2012). Dynamics of nonpersistent aphid-borne viruses in lettuce crops covered with UV-absorbing nets. *Virus Research, 165,* 1–8.

Legarrea, S., Weintraub, P. G., Plaza, M., Vinuela, E., & Fereres, A. (2012). Dispersal of aphids, whiteflies and their natural enemies under photoselective nets. *Biocontrol, 57,* 523–532.

Leibman, D., Wolf, D., Saharan, V., Zelcer, A., Arazi, T., Shiboleth, Y., et al. (2011). A high level of transgenic viral small RNA is associated with broad potyvirus resistance in cucurbits. *Molecular Plant–Microbe Interactions, 24,* 1220–1238.

Lenee, P., Perez, P., Gruber, V., Baudot, G., & Ollivo, C. (2001). Polyribozyme capable of conferring on plants resistance to *Cucumber mosaic virus* and resistant plants producing this polyribozyme. US Patent 6,265,634.

Leroux, J. P., Quiot, J.-B., Lecoq, H., & Pitrat, M. (1979). Mise en évidence et répartition dans le Sud-Est de la France d'un pathotype particulier du virus de la mosaïque du concombre. *Annales de Phytopathologie, 11,* 431–438.

Ling, K. S., Harris, K. R., Meyer, J. D. F., Levi, A., Guner, N., Wehner, T. C., et al. (2009). Non-synonymous single nucleotide polymorphisms in the watermelon eIF4E gene are closely related with resistance to Zucchini yellow mosaic virus. *Theoretical and Applied Genetics*, *120*, 191–200.

Loebenstein, G., & Raccah, B. (1980). Control of non-persistently aphid-borne viruses. *Phytoparasitica*, *8*, 221–235.

Lovisolo, O. (1980). Virus and viroid diseases of cucurbits. *Acta Horticulturae*, *88*, 33–82.

Marco, C. F., Aguilar, J. M., Abad, J., Gomez-Guillamon, M. L., & Aranda, M. A. (2003). Melon resistance to *Cucurbit yellow stunting disorder virus* is characterized by reduced virus accumulation. *Phytopathology*, *93*, 844–852.

Masmoudi, K., Duby, C., Suhas, M., Guo, J. Q., Guyot, L., Olivier, V., et al. (1994). Quality control of pea seed for pea seed borne virus. *Seed Science and Technology*, *11*, 491–503.

Medley, T. (1994). Availability of determination of nonregulated status for virus resistant squash. *Federal Register*, *59*, 64187–64188.

Miras, M., Sempere, R. N., Kraft, J. J., Miller, W. A., Aranda, M. A., & Truniger, V. (2014). Interfamilial recombination between viruses led to acquisition of a novel translation-enhancing RNA element that allows resistance breaking. *The New Phytologist*, *202*, 233–246.

Moury, B., Fereres, A., Garcia-Arenal, F., & Lecoq, H. (2011). Sustainable management of plant resistances to viruses. In C. Caranta, M. A. Aranda, M. Tepfer, & J. J. Lopez-Moya (Eds.), *Recent advances in plant virology* (pp. 219–236). Norfolk: Caister Academic Press.

Murphy, J. F., Eubanks, M. D., & Masiri, J. (2009). Reflective plastic mulch but not a resistance-inducing treatment reduced *Watermelon mosaic virus* incidence and yield losses in squash. *International Journal of Vegetable Science*, *15*, 3–12.

Navas-Castillo, J., Fiallo-Olive, E., & Sanchez-Campos, S. (2011). Emerging virus diseases transmitted by whiteflies. *Annual Review of Phytopathology*, *49*, 219–248.

Nieto, C., Morales, M., Orjeda, G., Clepet, C., Monfort, A., Sturbois, B., et al. (2006). An eIF4E allele confers resistance to an uncapped and non-polyadenylated RNA virus in melon. *Plant Journal*, *48*, 452–462.

Nischiguchi, M., & Kobayashi, K. (2011). Attenuated plant viruses: Preventing virus diseases and understanding the molecular mechanism. *Journal of General Plant Pathology*, *77*, 221–229.

Nolan, P. A., & Campbell, R. N. (1984). *Squash mosaic virus* detection in individual seeds and seed lots of cucurbits by enzyme-linked immunosorbent assay. *Plant Disease*, *68*, 971–975.

Nyoike, T. W., & Liburd, O. E. (2010). Effect of living (buckwheat) and UV reflective mulches with and without imidacloprid on whiteflies, aphids and marketable yields of zucchini squash. *International Journal of Pest Management*, *56*, 31–39.

Orozco-Santos, M., Perez-Zamora, O., & Lopez-Arriaga, O. (1995). Effect of transparent mulch on insect populations, virus diseases, soil temperature, and yield of cantaloup in a tropical region. *New Zealand Journal of Crop and Horticultural Science*, *23*, 199–204.

Pachner, M., Paris, H. S., & Lelley, T. (2011). Genes for resistance to zucchini yellow mosaic in tropical pumpkin. *Journal of Heredity*, *102*, 330–335.

Paris, H. S., & Brown, R. N. (2004). Gene list for *Cucurbita* species, 2004. *The Cucurbit Genetics Cooperative Report*, *27*, 77–97.

Paris, H. S., Cohen, S., Burger, Y., & Yoseph, R. (1988). Single gene resistance to zucchini yellow mosaic virus in *Cucurbita moschata*. *Euphytica*, *37*, 27–37.

Pauquet, J., Burget, E., Hagen, L., Chovelon, V., Le Menn, A., Valot, N., et al. (2004). Map-based cloning of the Vat gene from melon conferring resistance to both aphid colonization and aphid transmission of several viruses. In A. Lebeda, & H. S. Paris (Eds.), *Progress in cucurbit genetics and breeding research* (pp. 325–329). Olomouc: Palaki University in Olomouc.

Perring, T. M., Farrar, C. A., Mayberry, K., & Blua, M. J. (1992). Research reveals pattern of cucurbit virus spread. *California Agriculture*, *46*(2), 35–40.
Perring, T. M., Gruenhagen, N. M., & Farrar, C. A. (1999). Management of plant viral diseases through chemical control of insect vectors. *Annual Review of Entomology*, *44*, 455–481.
Pinese, B., Lisle, A. T., Ramsey, M. D., Halfpapp, K. H., & De Faveri, S. (1994). Control of aphid-borne papaya ringspot potyvirus in zucchini marrow (*Cucurbita pepo*) with reflective mulches and mineral oil-insecticide spray. *International Journal of Pest Management*, *40*, 81–87.
Pitrat, M., & Lecoq, H. (1980). Inheritance of resistance to cucumber mosaic virus transmission by *Aphis gossypii* in *Cucumis melo*. *Phytopathology*, *70*, 958–961.
Pitrat, M., & Lecoq, H. (1983). Two alleles for Watermelon Mosaic Virus 1 resistance in melon. *The Cucurbit Genetics Cooperative Report*, *6*, 52–53.
Powell, C. C., & Schlegel, D. E. (1970). Factors influencing seed transmission of squash mosaic virus in cantalupe. *Phytopathology*, *60*, 1466–1469.
Provvidenti, R. (1986). *Viral diseases of cucurbits and sources of resistance*. Taiwan: Food & Fertilizer Technology Center, Technical Bulletin, 16 pp.
Provvidenti, R. (1987). Inheritance of resistance to a strain of zucchini yellow mosaic virus in cucumber. *HortScience*, *22*, 102–103.
Provvidenti, R. (1996). In T. A. Zitter, D. L. Hopkins, & C. E. Thomas (Eds.), *Compendium of cucurbit diseases: Diseases caused by viruses* (pp. 37–45). St. Paul, MN: American Phytopathological Society.
Quiot, J. B., Labonne, G., & Marrou, J. (1982). Controlling seed and insect-borne viruses. In R. T. Plumb, & J. M. Thresh (Eds.), *Pathogens, vectors and plant diseases* (pp. 95–122). New York, NY: Academic Press.
Ratcliff, F. G., MacFarlane, S. A., & Baulcombe, D. C. (1999). Gene silencing without DNA: RNA-mediated cross-protection between viruses. *Plant Cell*, *11*, 1207–1215.
Raviv, M., & Antignus, Y. (2004). UV radiation effects on pathogens and insect pests of greenhouse-grown crops. *Photochemistry and Photobiology*, *79*, 219–226.
Risser, G., Pitrat, M., & Rode, J.-C. (1977). Etude de la résistance du melon (*Cucumis melo* L.) au virus de la mosaïque du concombre. *Annales d'Amélioration des Plantes*, *27*, 509–522.
Robinson, R. W., & Decker-Walters, D. S. (1997). *Cucurbits*. Wallingford, UK: CAB International.
Roditakis, E., Grispou, M., Morou, E., Kristoffersen, J. B., Roditakis, N., Nauen, R., et al. (2009). Current status of insecticide resistance in Q biotype *Bemisia tabaci* populations from Crete. *Pest Management Science*, *65*, 313–322.
Romay, G., Lecoq, H., & Desbiez, C. (2014). Cucurbit crops and their viral diseases in Latin America and the Caribbean islands: A review. *Journal of Plant Pathology*, *96*, 227–242.
Sakata, Y., Ohara, T., & Sugiyama, M. (2008). The history of melon and cucumber grafting in Japan. *Acta Horticulturae*, *767*, 217–228.
Sang-Min, K., Sang-Hyun, N., Jung-Myung, L., Kyu-Ock, Y., & Kook-Hyung, K. (2003). Destruction of cucumber green mottle mosaic virus by heat treatment and rapid detection of virus inactivation by RT-PCR. *Molecules and Cells*, *16*, 338–342.
Schuster, D. J., Stanley, P. A., Polston, J. E., Gilreath, P. R., & McAvoy, E. (2007). *Management of whiteflies, whitefly-vectored plant virus, and insecticide resistance for vegetable production in southern Florida*. IFAS Extension Publication ENY-735, http://edis.ifas.Ufl.edu/pdffiles/IN/IN695000.pdf.
Shelton, A. M., & Badenes-Perez, F. R. (2006). Concepts and application of trap cropping in pest management. *Annual Review of Entomology*, *51*, 285–308.
Simmons, A. M., Kousik, C. S., & Levi, A. (2010). Combining reflective mulch and host plant resistance for sweet potato whitefly (*Hemiptera*: *Aleyrodidae*) management in watermelon. *Crop Protection*, *29*, 898–902.

Stapleton, J. J., & Summers, C. G. (2002). Reflective mulches for management of aphids and aphid-borne virus diseases in late-season cantaloupe (*Cucumis melo* L. var. *cantalupensis*). *Crop Protection*, *21*, 891–898.

Sugiyama, M. (2013). The present status of breeding and germplasm collection for resistance to viral diseases of cucurbits in Japan. *Journal of the Japanese Society for Horticultural Science*, *82*, 193–202.

Summers, C. G., Mitchell, J. P., & Stapleton, J. J. (2004). Management of aphid-borne viruses and *Bemisia argentifolii* (*Homoptera*: *Aleyrodidae*) in zucchini squash by using UV reflective plastic and wheat straw mulches. *Environmental Entomology*, *33*, 1447–1457.

Takeshita, M., Okuda, M., Okuda, S., Hyodo, A., Hamano, K., Furuya, N., et al. (2013). Induction of antiviral responses by acibenzolar-S-methyl against *Cucurbit chlorotic yellows virus* in melon. *Phytopathology*, *103*, 960–965.

Tamura, A., Kato, T., Taki, A., Sone, M., Satoh, N., Yamagishi, N., et al. (2013). Preventive and curative effects of Apple latent spherical virus vectors harboring parts of the target virus genome against potyvirus and cucumovirus infection. *Virology*, *446*, 314–324.

Thresh, J. M. (1981). The role of weeds and wild plants in the epidemiology of plant virus diseases. In J. M. Thresh (Ed.), *Pests, pathogens and vegetation* (pp. 53–70). London: Pitman.

Tomlinson, J. A., & Thomas, B. J. (1986). Studies on melon necrotic spot virus disease of cucumber and on the control of the fungus vector (*Olpidium radicale*). *Annals of Applied Biology*, *108*, 71–80.

Tricoli, D. M., Carney, K. J., Russell, P. F., Quemada, H. D., McMaster, R. J., Reynolds, J. F., et al. (2002). Transgenic plants expressing DNA constructs containing a plurality of genes to impart virus resistance. US Patent 6,337,431.

Tricoli, D. M., Carney, K. J., Russell, P. F., Russell McMaster, J., Groff, D. W., Hadden, K. C., et al. (1995). Field evaluation of transgenic squash containing single or multiple virus coat protein gene for resistance to CMV, WMV2, and ZYMV. *Biotechnology*, *13*, 1458–1465.

Ucko, O., Cohen, S., & Ben-Joseph, R. (1998). Prevention of virus epidemics by a crop-free period in the Arava region of Israel. *Phytoparasitica*, *26*, 313–321.

Ullah, Z., & Grumet, R. (2002). Localization of *Zucchini yellow mosaic virus* to the veinal regions and role of viral coat protein in veinal chlorosis conditioned by the *zym* potyvirus resistance locus in cucumber. *Physiological and Molecular Plant Pathology*, *60*, 79–89.

Walters, S. A. (2003). Suppression of watermelon mosaic virus in summer squash with plastic mulches and rowcovers. *HortTechnology*, *13*, 352–357.

Webb, S. E., & Linda, S. E. (1993). Effect of oil and insecticide on epidemics of potyviruses in watermelon in Florida. *Plant Disease*, 77, 869–874.

Wei, T., Pearson, M. N., Blohm, D., Nölte, M., & Armstrong, K. (2009). Development of a short oligonucleotide microarray for the detection and identification of multiple potyviruses. *Journal of Virological Methods*, *162*, 109–118.

Wu, H. W., Yu, T. A., Raja, J. A. J., Christopher, S. J., Wang, S. L., & Yeh, S. D. (2010). Double-virus resistance of transgenic oriental melon conferred by untranslatable chimeric construct carrying partial coat protein genes of two viruses. *Plant Disease*, *94*, 1341–1347.

Yarden, G., Hemo, R., Livne, H., Maoz, F., Lev, E., Lecoq, H., et al. (2000). Cross-protection of Cucurbitaceae from zucchini yellow mosaic potyvirus. *Acta Horticulturae*, *510*, 349–356.

Zitter, T. A., & Simons, J. N. (1980). Management of viruses by alteration of vector efficiency and by cultural practices. *Annual Review of Phytopathology*, *18*, 289–310.

CHAPTER SIX

Virus Diseases of Peppers (*Capsicum* spp.) and Their Control

Lawrence Kenyon[1], Sanjeet Kumar, Wen-Shi Tsai, Jacqueline d'A. Hughes

AVRDC—The World Vegetable Center, Shanhua, Tainan, Taiwan, P.R. China

[1]Corresponding author: e-mail address: lawrence.kenyon@worldveg.org

Contents

Abstract

The number of virus species infecting pepper (*Capsicum* spp.) crops and their incidences has increased considerably over the past 30 years, particularly in tropical and subtropical pepper production systems. This is probably due to a combination of factors, including the expansion and intensification of pepper cultivation in these regions, the increased volume and speed of global trade of fresh produce (including peppers) carrying viruses and vectors to new locations, and perhaps climate change expanding the geographic range suitable for the viruses and vectors. With the increased incidences of diverse virus species comes increased incidences of coinfection with two or more virus species in the same plant. There is then greater chance of synergistic interactions between virus species, increasing symptom severity and weakening host resistance, as well as the opportunity for genetic recombination and component exchange and a possible increase in aggressiveness, virulence, and transmissibility. The main virus groups infecting peppers

Advances in Virus Research, Volume 90
ISSN 0065-3527
http://dx.doi.org/10.1016/B978-0-12-801246-8.00006-8

are transmitted by aphids, whiteflies, or thrips, and a feature of many populations of these vector groups is that they can develop resistance to some of the commonly used insecticides relatively quickly. This, coupled with the increasing concern over the impact of over- or misuse of insecticides on the environment, growers, and consumers, means that there should be less reliance on insecticides to control the vectors of viruses infecting pepper crops. To improve the durability of pepper crop protection measures, there should be a shift away from the broadscale use of insecticides and the use of single, major gene resistance to viruses. Instead, integrated and pragmatic virus control measures should be sought that combine (1) cultural practices that reduce sources of virus inoculum and decrease the rate of spread of viruliferous vectors into the pepper crop, (2) synthetic insecticides, which should be used judiciously and only when the plants are young and most susceptible to infection, (3) appropriate natural products and biocontrol agents to induce resistance in the plants, affect the behavior of the vector insects, or augment the local populations of parasites or predators of the virus vectors, and (4) polygenic resistances against viruses and vector insects with pyramided single-gene virus resistances to improve resistance durability.

1. INTRODUCTION

Peppers (*Capsicum* spp.) of both the pungent (chili or hot pepper) and non-pungent (sweet pepper) types are increasingly important spice and vegetable crops worldwide (Gniffke et al., 2013). The *Capsicum* genus is native to tropical Central and South America and comprises 27 species, though only five of these are domesticated (Ibiza, Blanca, Cañizares, & Nuez, 2012). Among these, *Capsicum annuum* with its diversity of market types of pungent and non-pungent fruits is the most widely commercially cultivated worldwide. *Capsicum frutescens* and *Capsicum chinense* are commercially cultivated at a very limited scale in different regions of Central America, West and Central Africa, and Oceania. The commercial cultivation of *Capsicum baccatum* and *Capsicum pubescens* is restricted to Central and South America and the Caribbean. *Capsicum* fruits are grown mainly for use as a fresh vegetable (hot and sweet peppers), for drying as spice (paprika, chili powder), for extraction of food colorings and flavorings, and extraction of other compounds for medicinal or industrial uses. The FAO production statistics (http://faostat.fao.org/) indicate that over the 30 years from 1982 to 2012, annual world production of fresh peppers increased by almost fourfold, from about 8.2 to about 31.2 million tons. Much of this increase was in East Asia, particularly China, where during the same period annual production increased by almost eightfold, from about 2.3 to about 16 million tons. However, these figures are only estimates and may not accurately reflect the increasing cultivation of peppers by smallholder growers in many tropical and subtropical countries for home consumption and/or sale at local markets.

Pests and diseases severely limit pepper cultivation throughout the world, and in some areas the measures used to control pests and diseases can account for a large proportion of the production costs. For example, some pepper growers in Java, Indonesia, reported spraying with pesticide to control diseases and pests about every 5 days, often with cocktails of different active ingredients (Mariyono & Bhattarai, 2009). Cultivated peppers are generally regarded as being relatively susceptible to infection by viruses. Green and Kim (1991) listed 35 different virus species able to infect peppers; 12 years later, the list had grown to 68 species (Pernezny, Roberts, Murphy, & Goldberg, 2003). In reviewing the most important viruses affecting pepper crops in the Mediterranean Basin region and the methods used to control them, Moury and Verdin (2012) suggested that only about 20 virus species belonging to 15 different taxonomic groups had been reported to cause damage to pepper crops. These authors also observed that there had been relatively few viral emergences in peppers in the Mediterranean Basin region in the previous 20 years compared to other crops such as tomato. Contrary to this, in other parts of the world the number of virus species recorded as infecting peppers has grown rapidly over this period. In this review, we take a broader look at the important viruses infecting peppers around the world and the methods used to control them, and we argue for the need to develop and adopt more integrated crop and pest management strategies that are more sustainable and safer for the environment, growers, and consumers.

2. THE MAIN VIRUSES INFECTING PEPPERS

2.1. Aphid-transmitted viruses

2.1.1 Potyviruses

The genus *Potyvirus* (Family *Potyviridae*) is one of the larger groups of plant viruses, and they cause diseases in a wide range of plant species. Potyviruses have a single-stranded, linear, positive-sense RNA genome of about 9.7 kb, and the particles are flexuous filaments 680–900 nm long. All are transmitted by one or more aphid species and can be transmitted mechanically and by grafting. Some potyviruses can be seed-borne, though this has not been conclusively proved from infected *Capsicum* species. At least 11 different species of potyvirus have been described causing disease in pepper crops in different parts of the world. Phylogenetic analysis of their genome sequences with other potyviruses separates the pepper-infecting potyviruses among three evolutionary clades (Moury & Verdin, 2012) (Table 6.1). The largest of these clades contains *Potato virus Y* (PVY; the type species of the genus), *Pepper mottle virus* (PepMoV), *Pepper severe mosaic virus* (PepSMV), *Pepper yellow mosaic virus* (PepYMV), *Peru tomato mosaic virus* (PTV), and the tentative species, Ecuadorian rocoto virus (EcRV). PVY has been reported from practically all geographic regions where peppers are grown, while all the other pepper-infecting species from the PVY clade, except PepMoV, are confined to South America, the center of origin of the Solanaceae. PepMoV has probably been recently introduced from South America into Japan (Ogawa et al., 2003), Korea (Han et al., 2006), and Taiwan (Cheng et al., 2011). PVY can cause plant stunting, systemic vein clearing, leaf mosaic or mottling, and dark green vein banding of pepper leaves. Fruits from infected pepper plants may have a mosaic pattern and be smaller and deformed (Pernezny et al., 2003). Depending upon the pepper cultivar, the strain of PVY, environmental conditions, and the time of infection, the plants may develop stem and apical bud necrosis, and necrotic spots may be found on fruits.

The second major clade of pepper-infecting potyviruses contains *Chilli veinal mottle virus* (ChiVMV), *Pepper veinal mottle virus* (PVMV), *Chilli ringspot virus* (ChiRSV), and *Wild tomato mosaic virus* (WTMV). ChiVMV is probably the most prevalent virus infecting peppers across South, East, and Southeast Asia (Tsai et al., 2008). Characteristic symptoms of ChiVMV infection are leaf mottle and dark green vein banding, and leaves may be small and distorted. If plants are infected when young, they may become stunted, with dark green streaks on their stems and branches. Most flowers will drop, resulting in considerable yield loss. Plants infected with PVMV may show chlorotic vein banding, mottling, mosaic, and leaf distortion. They may also be stunted with reduced and distorted fruit set. First identified from Ghana (Brunt & Kenten, 1971), PVMV was subsequently found to be widespread in countries of West and North Africa (Tsai et al., 2010). More recently, it has been detected in Taiwan, often in mixed infection with ChiVMV (Cheng et al., 2009).

Tobacco etch virus (TEV) is the only pepper-infecting virus in the third clade with *Potato virus A* (with which it has sometimes been confused) and *Tobacco vein mottling virus* (Moury & Verdin, 2012). TEV can cause mottle or severe mosaic and leaf distortion and stunting of pepper plants. Depending on the cultivar, isolate, and time of infection, TEV can cause flower drop, fruit mosaic and distortion, and root necrosis. Although

Table 6.1 Geographic distribution and evolutionary relationships between potyviruses infecting *Capsicum* species

Potyvirus name	Evolutionary clade[a]	Countries and/or regions	Key references
Ecuadorian rocoto virus[b] (EcRV)	1 (PVY)	Ecuador	Janzac, Fabre, Palloix, and Moury (2008)
Pepper mottle virus (PepMoV)	1 (PVY)	Mainly in the Americas and Caribbean, more recently in Japan, Korea, and Taiwan	Cheng et al. (2011), Han, Choi, Kim, Lee, and Kim (2006), Nelson and Wheeler (1978), Ogawa, Hagiwara, Iwai, Izumi, and Arai (2003), and Quiñones et al. (2011)
Pepper severe mosaic virus (PepSMV)	1 (PVY)	Argentina	Rabinowicz, Bravo-Almonacid, and Mentaberry (1993)
Pepper yellow mosaic virus (PepYMV)	1 (PVY)	Brazil	Inoue-Nagata et al. (2002)
Peru tomato virus (PTV)	1 (PVY)	Chile, Peru	Lucinda, da Rocha, Inoue-Nagata, and Nagata (2012), Melgarejo, Alminaite, Fribourg, Spetz, and Valkonen (2004), and Spetz et al. (2003)
Potato virus Y (PVY)	1 (PVY)	Worldwide where solanaceous crops are grown	Moury (2010)
Tobacco etch virus (TEV)	3 (TEV)	Americas, Caribbean, and Turkey	Buzkan et al. (2006), Nelson and Wheeler (1978), Spetz et al. (2003)
Chilli ringspot virus (ChiRSV)	2 (ChiVMV)	China, Vietnam	Gong et al. (2011), Ha, Revill, Harding, Vu, and Dale (2008), Wang et al. (2012)

Continued

Table 6.1 Geographic distribution and evolutionary relationships between potyviruses infecting *Capsicum* species—cont'd

Potyvirus name	Evolutionary clade	Countries and/or regions	Key references
Chilli veinal mottle virus (ChiVMV)	2 (ChiVMV)	South and Southeast Asia	Ha, Revill, et al. (2008), Nono-Womdim, Swai, Chadha, Gebre-Selassie, and Marchoux (2001), and Tsai et al. (2008)
Pepper veinal mottle virus (PVMV)	2 (ChiVMV)	North and West Africa, Taiwan	Brunt and Kenten (1971), Cheng, Wang, Chen, Chang, and Jan (2009), and Tsai, Abdourhamane, and Kenyon (2010)
Wild tomato mosaic virus (WTMV)	2 (ChiVMV)	Thailand, Vietnam	Ha, Revill, et al. (2008)

[a]Evolutionary clade as defined by Moury and Verdin (2012).
[b]Proposed name, not yet ratified by ICTV.

TEV has a relatively wide geographic distribution, the incidences of strains causing significant disease in pepper crops are rather localized, such as in Jamaica (McDonald, Halbert, Tolin, & Nault, 2003) or Turkey (Buzkan et al., 2006).

2.1.2 Cucumoviruses

Cucumber mosaic virus (CMV), the type member of the genus *Cucumovirus* in the family *Bromoviridae*, has a very wide host range and is one of the most prevalent viruses worldwide. CMV has a genome of three linear, positive-sense single-stranded RNAs, each separately encapsidated in 29-nm diameter icosahedral virions. Many different species of aphid have been described as CMV vectors, and transmission is in a nonpersistent manner. CMV probably affects pepper crops in all pepper-growing regions of the world. Based on serology and phylogenetic analysis of genome sequences, CMV isolates are classified into two subgroups (I and II). Subgroup I can be further divided into subgroups IA and IB. Peppers are more likely to be infected with subgroup I isolates since these are the more prevalent in warmer climates where peppers are grown (Moury & Verdin, 2012). The symptoms caused by CMV infection in pepper depend on cultivar, strain, time of infection, and growing conditions, and include mosaic and dull-colored leaves, shoe-stringing of leaves, and necrotic "oak-leaf" symptoms on older leaves. Infection can also prevent flower formation and set, as well as misshapen and patchily discolored fruit. Early infection with a severe strain can result in stem necrosis and plant death. Ali and Kobayashi (2010) demonstrated with seed grow-out tests that at least one strain of CMV could be seed-transmitted in pepper at a rate of 10–14%, though the importance of seed transmission in the epidemiology of CMV in pepper crops in the field remains unclear. CMV can occur in mixed infection with another virus, and there can be a synergistic interaction between the CMV and the other virus. For example, Murphy and Bowen (2006) demonstrated synergy in bell pepper plants coinfected with CMV and the potyvirus PepMoV, relative to each virus alone, for a decrease in stem height and above ground fresh weight. Similarly, Kim, Kim, Hong, Choi, and Ryu (2010) observed synergistic effects on symptom expression and reduced growth rate in plants infected with CMV prior to infection with PepMoV and/or PMMoV, and preliminary results at AVRDC—The World Vegetable Center indicate that coinfection with CMV reduces the effectiveness of resistance against ChiVMV.

Tomato aspermy virus (TAV), another member of the genus *Cucumovirus*, predominantly infects composites and solanaceous plants, including peppers, but is much less prevalent than CMV and only occasionally is a problem for pepper cultivation; Suzuki, Kuroda, Miura, and Murai (2003) in Japan considered it useful to screen pepper germplasm for resistance to TAV.

2.1.3 Poleroviruses

Poleroviruses (Family *Luteoviridae*) have nonenveloped icosahedral virions 26–30 nm in diameter, and a genome of positive-sense single-stranded RNA of about 6.2 kb. They can be transmitted by grafting and by certain aphids in a circulative, nonpropagative manner, but cannot be mechanically transmitted. Yonaha, Toyosato, Kawano, and Osaki (1995) named a virus that had been causing vein yellowing and leaf curling of bell pepper plants in Okinawa prefecture in Japan since the early 1980s *Pepper vein yellows virus* (PeVYV) and, by sequencing the entire genome, Murakami, Nakashima, Hinomoto, Kawano, and Toyosato (2011) showed that this was a distinct species of polerovirus with closest sequence similarity to *Tobacco vein distorting virus*. At about the same time, Dombrovsky, Glanz, Pearlsman, Lachman, and Antignus (2010) partially sequenced a polerovirus causing similar symptoms in pepper in Israel and suggested the name Pepper yellow leaf curl virus (PYLCV) for this virus. Dombrovsky et al. (2013) have recently shown that PeVYV and PYLCV are distinct, but related, poleroviruses and suggest that the speciation of these viruses occurred through putative recombination event(s) between poleroviruses coinfecting a common host(s). The use of better general (universal) polerovirus polymerase chain reaction (PCR) primers has resulted in PeVYV and PYLCV being detected in peppers from fields in Turkey and Tunisia (Buzkan, Arpaci, Simon, Fakhfakh, & Moury, 2012), India, Indonesia, Mali, Philippines, Thailand and Taiwan (Knierim, Tsai, & Kenyon, 2013), and Spain (Villanueva et al., 2013). PeVYV was also detected infecting a nightshade (*Solanum*) species in India (Knierim et al., 2013), but more information is required on the host range of this group of viruses to determine if weeds or other crop species could be overwintering reservoirs of infection for pepper. A virus infecting peppers in Australian glass houses in the late 1980s and named Capsicum yellows virus (Gunn & Pares, 1990), is most likely to have been a strain of *Potato leaf roll virus* (PLRV) (Knierim et al., 2013). Based on serological evidence, strains of *Beet western yellows virus* (BWYV) were associated with disease of solanaceous crops, including pepper, in the United States (Duffus, 1981), and partial genome sequences have recently suggested that BWYV is the predominant polerovirus infecting peppers in Turkey; in Tunisia, it is less important than PeVYV (Buzkan et al., 2012).

2.1.4 Other aphid-transmitted viruses of pepper

Alfalfa mosaic virus (AMV) (*Alfamovirus*, *Bromoviridae*) has a wide host range, is distributed worldwide, and can be transmitted mechanically, by grafting, and in the nonpersistent manner by several aphid species. It can also be transmitted through pepper seed. It causes a bright yellow or blotchy white mosaic on pepper leaves, and if plants are infected when young they may become stunted with misshapen and blotchy fruit. Infection of pepper crops is generally sporadic and of little importance. However, it can become prevalent in pepper and have local economic importance, for example, in Zambia (Ndunguru & Kapooria, 1999), Turkey (Buzkan et al., 2006), and Oklahoma (Abdalla & Ali, 2012).

Broad bean wilt virus 1 (BBWV1) and *Broad bean wilt virus 2* (BBWV2) are members of the genus *Fabavirus*, with genomes of two positive-sense single-stranded RNAs separately encapsidated in 25–30-nm diameter nonenveloped icosahedral particles. They have a very wide host range and are transmitted by aphids in a nonpersistent manner. BBWV1 was reported as the predominant species in Europe and North America causing mosaic and concentric rings on leaves and fruits of sweet or bell peppers (Rubio et al., 2002), while BBWV2 causing necrotic spots or streaks on leaves and stems with stunting and apical chlorosis in chili peppers was reported as occurring sporadically in East and Southeast Asia (Lee et al., 2000). However, Ryu et al. (2009) reported about 32% of paprika samples with symptoms of virus infection from Jeonnam Province in Korea as being infected with BBWV (species not specified), with about one-third of these being mixed infections with CMV and/or PepMoV. More recently, Svoboda and Leisova-Svobodova (2013) reported the relatively high incidence of both BBWV1 and BBWV2, either singly or as mixed infections in pepper crops in the Czech Republic.

2.2. Whitefly-transmitted viruses

2.2.1 Crinivirus

Tomato chlorosis virus (ToCV) is a typical *Crinivirus* (Family *Closteroviridae*) with a bipartite genome of positive-sense single-stranded RNA encapsidated in flexuous virions of 800–850 nm. It was first identified from tomato in Florida in 1989 (Wisler, Li, Liu, Lowry, & Duffus, 1998). It is transmitted in a semipersistent manner by three whitefly (Hemiptera: *Aleyrodidae*) species (*Bemisia tabaci*, *Trialeurodes vaporariorum*, and *Trialeurodes abutilonea*), and is emerging as a problem worldwide (Fortes, Moriones, & Navas-Castillo, 2012). In 1999, ToCV was identified as the cause of interveinal yellowing, mild upward leaf curling, and stunting of sweet pepper plants in greenhouses of Almería, Spain, that were heavily infested with *Bemisia tabaci* (Lozano, Moriones, & Navas-Castillo, 2004). ToCV has since been identified causing disease in pepper plants in Brazil (Barbosa, Teixeira, & Rezende, 2010), Costa Rica (Vargas et al., 2011), Tunisia, and Uruguay (sequences in GenBank, http://www.ncbi.nlm.nih.gov/ accessed in January 2014).

2.2.2 Begomoviruses

Members of the genus *Begomovirus* (Family *Geminiviridae*) infect dicotyledonous plants and are transmitted in a persistent manner by *B. tabaci* (Gennadius). They have twinned icosahedral (geminate) particles, and are either monopartite with a genome of a single circle of single-stranded DNA of about 2.7 kb, or bipartite with two circles of ssDNA both of about 2.7 kb. Begomoviruses are sometimes associated with satellite molecules of circular ssDNA; alphasatellites or/and betasatellites. The monopartite begomoviruses were originally considered to have originated in the Old World, while the bipartite begomoviruses were predominantly located in the New World. However, there is evidence that some bipartite begomoviruses were present in Asia (Vietnam) prior to continental separation (Ha, Coombs, et al., 2008), and that monopartite viruses are emerging in Latin America through convergent evolution of bipartite viruses (Melgarejo et al., 2013). The current International Committee on Taxonomy of Viruses (ICTV) criterion for distinguishing between begomovirus species is that there should be 89% or less nucleotide sequence identity in the DNA-A genome components (Brown et al., 2012). Strains of a species are defined by having 90–93% DNA-A nucleotide sequence identity (Fauquet et al., 2008).

The emergence of begomoviruses over the past 30 years as one of the most important groups of plant viruses affecting production of many different vegetable crops in tropical and subtropical countries has generally been associated with increases of populations of the vector whitefly. Pernezny et al. (2003) listed five begomoviruses as causing disease in peppers in the Americas, and only one, the chili leaf curl disease agent (listed as Tobacco leaf curl virus) from Asia. Since then, many more begomoviruses have been detected and their genome component(s) sequenced. Thus, a search of the GenBank database (http://www.ncbi.nlm.nih.gov/ accessed in January 2014) finds at least 37 ICTV ratified species and a further 6 ICTV candidate species of begomovirus detected from *Capsicum* species from different parts of the world (Table 6.2). However, not all of these cause serious diseases in pepper crops. More than half of these begomovirus species were detected from pepper plants growing in Asia, with serious disease of pepper crops due to begomoviruses being reported from Bangladesh, China, India, Indonesia, Thailand, and Vietnam. In Southeast and East Asia, the major contributory factors for the emergence and spread of pepper-infecting begomoviruses are probably expansion and intensification of production systems, leading to greater selection for more aggressive or crop-adapted begomovirus variants that have arisen through mutation, recombination, pseudorecombination, and acquisition of satellite DNA molecules. Intensification of cropping systems probably also resulted in increased populations of indigenous and introduced whitefly cryptic species, increasing the incidence and rate of spread of the viruses. There has been little introduction of begomovirus species into Southeast Asia from outside the region (except TYLCV mentioned below), and the recent spread of introduced and native species within the region has likely been through the movement of infected plant materials by man, and through spread in introduced whitefly biotypes that are more polyphagous and virus-vectoring competent (Kenyon, Tsai, Shih, & Lee, 2014). The number of begomoviruses reported to cause disease in pepper crops in the Americas also has increased over recent years, probably through similar causes as in Asia; a very recent example is the yet to be ratified *Pepper leafroll virus* in Peru (PepLRV) (Martínez-Ayala et al., 2014).

Table 6.2 Begomovirus species identified in GenBank[a] as naturally infecting pepper plants in different countries and regions

Begomovirus species[b]	Acronym	Country	Region
Okra yellow crinkle virus	OYCrV	Cameroon	Africa
Pepper yellow vein Mali virus	PepYVMLV	Burkina Faso, Mali	
Tomato yellow leaf curl Mali virus	TYLCMLV	Cameroon	
Euphorbia mosaic virus	EuMV	Mexico	America (North, Central, Caribbean)
*Merremia mosaic virus**	MerMV	Belize	
Pepper golden mosaic virus	PepGMV	Mexico, The United States	
Pepper huasteco yellow vein virus	PHYVV	Mexico	
Potato yellow mosaic virus	PYMV	Guadeloupe	
Squash yellow mild mottle virus	SYMMoV	Costa Rica	
*Tobacco yellow crinkle virus**	TbYCV	Cuba	
Tomato chino La Paz virus	ToChLPV	Mexico	
Tomato leaf curl Sinaloa virus	ToLCSiV	Nicaragua	
Tomato mottle virus	ToMoV	Belize	
Tomato severe leaf curl virus	ToSLCV	Mexico	
Pepper leafroll virus#	PepLRV	Peru	America (South)
*Tomato chlorotic leaf distortion virus**	TCLDV	Venezuela	
Tomato severe rugose virus	ToSRV	Brazil	
Chilli leaf curl India virus#	ChiLCINV	India	Asia
Chilli leaf curl Pakistan virus#	ChiLCPKV	China, Pakistan	

Continued

Table 6.2 Begomovirus species identified in GenBank as naturally infecting pepper plants in different countries and regions—cont'd

Begomovirus species	Acronym	Country	Region
Pepper leaf curl Varanasi virus#	PepLCVaV	India	
Tomato leaf curl Palampur virus#	ToLCPalV	Pakistan	
Tomato leaf curl Sulawesi virus#	ToLCSuV	Indonesia	
Chilli leaf curl virus	ChiLCV	India, Oman, Pakistan	
Indian cassava mosaic virus	ICMV	India	
*Jatropha mosaic India virus**	JMINV	India	
Malvastrum yellow vein Yunnan virus	MaYVYnV	China	
Pepper leaf curl Bangladesh virus	PepLCBV	Bangladesh, India, Pakistan	
Pepper leaf curl Lahore virus	PepLCLaV	Pakistan	
Pepper leaf curl virus	PepLCV	Malaysia, Thailand	
*Pepper leaf curl Yunnan virus**	PepLCYnV	China	
Pepper yellow leaf curl Indonesia virus	PepYLCIV	Indonesia	
Radish leaf curl virus	RaLCuV	India	
Sri Lankan cassava mosaic virus	SLCMV	India	
Tobacco curly shoot virus	TbCSV	China	
*Tomato leaf curl Cebu virus**	ToLCCeV	Philippines	
Tomato leaf curl Joydebpur virus	ToLCJoV	India	
Tomato leaf curl Karnataka virus	ToLCKV	India	

Table 6.2 Begomovirus species identified in GenBank as naturally infecting pepper plants in different countries and regions—cont'd

Begomovirus species	Acronym	Country	Region
Tomato leaf curl New Delhi virus	ToLCNDV	India, Pakistan	
Tomato leaf curl Philippines virus	ToLCPV	Philippines	
Tomato yellow leaf curl Kanchanaburi virus	TYLCKaV	Indonesia, Thailand	
Tomato yellow leaf curl Sardinia virus	TYLCSV	Jordan	
Tomato yellow leaf curl Thailand virus	TYLCTHV	Taiwan, Thailand	
Tomato yellow leaf curl virus	TYLCV	China, Mexico, Oman, Spain	Wide

[a]http://www.ncbi.nlm.nih.gov/, accessed in January 2014
[b]Species name followed by * were ratified as distinct species by the ICTV membership in early 2014. (http://talk.ictvonline.org/files/ictv_documents/m/msl/4825.aspx); species names followed by # are yet to be ratified; all other species are listed as ratified in Brown et al. (2012).

Tomato yellow leaf curl virus (TYLCV) causes devastating disease in tomato crops and its spread over the last 30 years from its center of origin in the Mediterranean Basin region to many tropical and subtropical countries has been documented extensively (Lefeuvre et al., 2010). Although strains of TYLCV also have been detected infecting peppers in some of these countries, they appear relatively unimportant in peppers. Morilla et al. (2005) showed that at least some *C. annuum* cultivars are dead-end hosts becoming infected with TYLCV, but from which whiteflies cannot acquire the virus. On the other hand, Polston, Cohen, Sherwood, Ben-Joseph, and Lapidot (2006) showed that other *Capsicum* cultivars were symptomless hosts and reservoirs of this virus.

2.3. Thrips-transmitted viruses

2.3.1 Tospoviruses

Tospoviruses (Family *Bunyaviridae*) are among the most damaging and economically important plant viruses, affecting a wide range of ornamental and food crops in many regions of the world. They have a genome consisting of one negative and two ambisense single-stranded RNAs that present a pseudocircular or panhandle conformation. The particles are 80–120 nm in diameter and pleomorphic. Multiple copies of the viral nucleocapsid (N) protein encapsidate each RNA to form ribonucleoprotein structures, which are enveloped in a host-derived membrane bilayer with the viral L protein and two viral glycoproteins forming projections from the envelope surface. Although tospoviruses can be transmitted mechanically in the lab with varying efficiency, in the field they are transmitted from plant to plant by specific thrips species (order Thysanoptera, Family *Thripidae*) in a persistent, propagative manner. Only larval thrips can acquire tospoviruses, but both larval and adult thrips can transmit them.

Tomato spotted wilt virus (TSWV), the type member of the Genus *Tospovirus*, was for a long time regarded as the only species in this genus, and it was not until the 1990s that other species of tospovirus started being identified. It was then realized that some of the earlier diagnoses of TSWV based on symptoms alone were probably other tospoviruses. Despite this, TSWV remains the most widespread of the tospoviruses. In peppers, TSWV causes chlorosis, often sudden yellowing and browning, and chlorotic rings on leaves. Necrotic streaks may appear on stems and extend to the terminal shoots. Fruit of infected plants may show chlorotic or necrotic spots and streaks with mosaic and concentric ring patterns.

Strains of *Impatiens necrotic spot virus* (INSV), the second of the tospoviruses to be identified and characterized (de Haan et al., 1992), are also able to cause a similar disease in peppers, but are less common than TSWV in this crop. INSV was initially confined to temperate regions of North America and Europe. However, the geographic range of this virus appears to have expanded in recent years. During 2009 and 2011, in the counties of Guanajuato and Querétaro, Central Mexico, pepper plants presenting symptoms of dark necrotic spots on some leaves and on the stems were diagnosed as being infected with a new strain of INSV (González-Pacheco & Silva-Rosales, 2013).

Groundnut ringspot virus (GRSV) was first recognized as a distinct species of tospovirus based on the serology (serogroup II) and phylogeny of nucleoprotein gene sequence of an isolate from groundnut in South Africa (de Ávila et al., 1993). Symptoms of GRSV in pepper closely resemble those induced by TSWV and include chlorotic and necrotic spots on leaves, inward rolling of leaves, and general plant stunting. Fruits of infected plants are generally discolored, often deformed and with chlorotic or necrotic ring patterns. *Tomato chlorotic spot virus* (TCSV, another serogroup II tospovirus) was identified as distinct from GRSV based on the nucleoprotein gene sequence difference of an isolate from tomato in Brazil (de Ávila et al.,

1993). Similar to infection with GRSV, pepper plants infected with TCSV exhibit necrotic spots or ringspots and severe leaf deformation; if infected at a young stage, they fail to grow and are severely stunted. The fruit from plants that are infected later on can be deformed, discolored, and show ringspot symptoms. For a long time, GRSV was thought to be confined to South Africa and South America, while TCSV was only reported from Brazil and Argentina. However, in 2009–2010, GRSV was detected for the first time in the United States, mostly at low incidence in tomato plantings across South Florida, and in 2011, it was confirmed infecting pepper, tomatillo, and eggplant at several locations in South Florida. Analysis of the genome composition of the GRSV isolates from Florida revealed that reassortment had occurred, with the large (L) and small (S) RNAs coming from GRSV and the medium (M) RNA coming from TCSV (Webster, Reitz, Perry, & Adkins, 2011); this was the first documented occurrence of RNA component reassortment between different tospovirus species. Since neither GRSV nor TCSV had been reported from the United States previously, it was suggested that this isolate had been introduced from an area where both viruses occurred (Webster et al., 2011). However, TCSV has since been identified from Florida (Londoño, Capobianco, Zhang, & Polston, 2012), the Dominican Republic (Batuman, Rojas, Almanzar, & Gilbertson, 2013), and Puerto Rico (Webster et al., 2013).

In Taiwan, *Watermelon silver mottle virus* (WSMoV) was identified as a distinct serogroup IV tospovirus primarily infecting cucurbits (Yeh & Chang, 1995) but was also observed to occasionally infect tomato and pepper, and germplasm of these species was screened for resistance to this virus (Green, Hwang, & Chou, 1996). WSMoV also was detected from pepper plants in the field in Thailand (Chiemsombat et al., 2008). In 1999, samples of pepper and tomato from the Bunderberg district of southern Queensland, Australia, presented symptoms suggesting tospovirus infection, and by serology and sequencing of the N gene, the causal agent was identified as a new serogroup IV tospovirus for which the name Capsicum chlorosis virus (CaCV) was proposed (McMichael, Persley, & Thomas, 2002). CaCV has since been detected with *Ceratothripoides claratris* as vector in Thailand (Premachandra, Borgemeister, Maiss, Knierim, & Poehling, 2005), China (Chen, Xu, Yan, & Wang, 2007), India (Kunkalikar et al., 2007), Taiwan (Zheng, Chen, Yang, Yeh, & Jan, 2008), and most recently in Hawaii (Melzer et al., 2014).

Satyanarayana et al. (1996) identified *Groundnut bud necrosis virus* (GBNV) (often also referred to as peanut bud necrosis virus) as a distinct tospovirus in

India. As the most widespread of the tospoviruses in India, GBNV has a relatively wide host range and has emerged as a serious constraint to tomato production. It also has been detected infecting peppers in several states of India. Although *Groundnut yellow spot virus* (GYSV; also known as Peanut yellow spot virus) was identified as a distinct tospovirus infecting groundnut in India (Satyanarayana et al., 1998), it has not been reported from peppers in India. However, two sequences of GYSV from pepper in China in 2007 and 2008 have been deposited in GenBank (http://www.ncbi.nlm.nih.gov/, accessed in January 2014). Similarly, Watermelon bud necrosis virus (WBNV), which has been known in watermelon in India since the early 1990s and was recently identified as a distinct species of tospovirus (Li et al., 2011), is sometimes detected infecting chili peppers in India (Mandal et al., 2012). Chiemsombat et al. (2010) proposed the name Tomato necrotic ringspot virus (TNRV) for a new tospovirus species causing symptoms of bud necrosis and necrotic ringspots on leaves, stem, and fruits of tomato and pepper in field production areas in Chiang Rai and Ratchaburi provinces of Thailand in 2007. Concurrently, Seepiban, Gajanandana, Attathom, and Attathom (2011) identified the same virus as the causal agent of necrotic spots, necrotic ringspots, and stem necrosis in tomato and pepper plants in Nakhon Pathom province, Thailand, and showed that it could be transmitted by both *Thrips palmi* and *C. claratris*. Pepper necrotic spot virus (PNSV) was then proposed as the name for a previously undescribed tospovirus species identified causing necrotic spots on leaves and stems of pepper and tomato in the La Joya valley, Arequipa, Peru, in 2010 (Torres, Larenas, Fribourg, & Romero, 2012). Similarly, pepper plants presenting symptoms of yellow mosaic, chlorosis, and necrosis in Tehran province, Iran, in 2010–2011 were identified as the first occurrence of Tomato yellow fruit ring virus (TYFRV) infecting pepper crops (Golnaraghi, Hamedi, Yazdani-Khameneh, & Khosroshahi, 2013). Most recently, Cheng et al. (2014) proposed the name Pepper chlorotic spot virus (PCSV) for a previously undescribed tospovirus species identified in Taiwan that causes mottling and deformation on leaves and fruits of sweet pepper.

2.3.2 Ilarviruses

Tobacco streak virus (TSV), the type species of the ilarviruses, is seed- and pollen-transmitted in certain host species, and there is often enhanced inoculation if both thrips and infected pollen are present on the inoculated plant (Sdoodee & Teakle, 1987). TSV was reported causing disease of peppers in

Argentina in 1974 (Gracia & Feldman, 1974) and in northern India in 2004 (Jain, Bag, & Awasthi, 2005). Also in 2004, the related Ilarvirus, *Parietaria mottle virus* (PMoV) was identified for the first time from samples of bell peppers being grown in south-east Spain (Janssen et al., 2005).

2.4. Viruses of pepper with other invertebrate vectors

Virus species of the genus *Tymovirus* (Family *Tymoviridae*) have isometric, nonenveloped particles of about 30 nm diameter and genomes of a single molecule of positive-sense single-stranded RNA of 6.0–7.5 kb in length. They are transmitted in a semipersistent manner by beetles of the families *Chrysomelidae* and *Curculionidae*; all are also mechanically transmissible, and a few are transmitted through seeds. The three species of Tymovirus cited as able to infect *Capsicum* species by Pernezny et al. (2003) are *Belladonna mottle virus*, *Eggplant mosaic virus*, and *Physalis mottle virus*. However, none of these currently appears important in pepper production, and only the more recently described (and as yet to be ratified by the ICTV Tymovirus species), *Chiltepín yellow mosaic virus* (ChiYMV) in Mexico shows the potential to become important in pepper production (Pagán et al., 2010). Similarly, although several nematode-transmitted viruses have been reported to infect peppers in the past, currently only the Nepovirus, *Tomato ringspot virus* in Iran, is reported as a potential concern (Sokhansanj, Rakhshandehroo, & Pourrahim, 2012).

2.5. Viruses not transmitted by invertebrate vectors

2.5.1 Tobamoviruses

Tobamoviruses are the most important viruses of pepper that are not transmitted by an invertebrate vector. They have a genome of a single linear positive-sense single-stranded RNA of about 6.3–66 kb encapsidated in ridged cylindrical particles about 18 nm in diameter and 300 nm long. The virus particles are produced to high titre in susceptible species of host and are very stable. Tobamoviruses are transmitted by contact and are often seed-transmitted to a significant degree. Seed from an infected plant can easily be externally contaminated with virus and, more rarely, can be infected internally (Genda, Sato, Nunomura, Hirabayashi, & Tsuda, 2011). Debris in the soil from a previous pepper crop or virus-carrying seed usually serve as the primary sources of infection, with subsequent spread within the crop being through contact between plants and mechanical transfer of virus on the hands and implements of those working with the crop. The international trade of pepper and tomato seed unknowingly contaminated with tobamoviruses has probably facilitated the spread of the viruses to different parts of the world.

The seven species of *Tobamovirus* known to be able to infect *Capsicum* species are *Tobacco mosaic virus* (TMV), *Tomato mosaic virus* (ToMV), *Tobacco mild green mosaic virus* (TMGMV), *Bell pepper mottle virus* (BPeMV), *Paprika mild mottle virus* (PaMMV), *Obuda pepper virus* (ObPV) and *Pepper mild mottle virus* (PMMoV). Symptom severity varies with virus strain and *Capsicum* genotype and usually includes leaf distortion with chlorotic mottle or mosaic, and small, misshapen, and discolored fruits. Often there are necrotic patches on fruits and leaves. These symptoms generally affect both quantity and quality of production. TMV, ToMV, TMGMV, and PMMoV are probably the predominant tobamoviruses affecting peppers in southern Europe (Moury & Verdin, 2012), while in most of Southeast Asia PMMoV and ToMV are the most common. The pepper-infecting tobamoviruses are classified into the different pathotypes, P_0, P_1, $P_{1.2}$, $P_{1.2.3}$, and $P_{1.2.3.4}$, based on the reaction of pepper cultivars carrying different *L*-gene alleles (Table 6.5; Mizumoto et al., 2012) to the pathotypes.

2.5.2 Tombusvirus

Moroccan pepper virus (MPV, *Tombusvirus*) was first described from pepper in Morocco and causes leaf deformation, leaf and flower abscission, stunting and lethal systemic necrosis. There is no known invertebrate vector and the virus appears to be spread through the soil. It is more prevalent in low-lying areas where water logging is common or where silt from rivers has been spread on farmland. With the recent reclassification of *Lettuce necrotic stunt virus* as MPV (Wintermantel & Hladky, 2013), it is apparent that the prevalence of this virus has been steadily increasing in central Asia and western North America. However, although this is a serious constraint to lettuce production in some areas, it remains to be determined how much of a threat MPV could be to pepper production.

3. MANAGEMENT OF VIRUSES INFECTING PEPPERS

3.1. Cultural and phytosanitary practices

There are three main means of managing virus diseases in pepper crops: (1) prevent or reduce the rate at which the virus reaches the crop, (2) prevent or reduce the rate at which the virus spreads once it reaches the crop, and (3) reduce the severity of the disease caused by the virus and thus reduce the economic impact on yield and quality of product. There are many cultural and phytosanitary measures that can be used to help address the first two routes, though the relative efficacy or importance of each measure will differ depending on the type of virus and the manner by which it is transmitted (Table 6.3). Most pepper production starts with sowing the seed in a nursery bed, pots, or plugs, and the young seedlings are subsequently transplanted to the net house/glass house/polytunnel (protected cultivation) or to the open field. Virus disease management has to start with seedling production and continue for most of the vegetative duration of the crop, with a slightly different set of practices for protected and open-field cultivation (Table 6.3).

Since the means for long-distance spread of most of the viruses without an invertebrate vector (and possibly some of the nonpersistent insect-transmitted viruses) is through contaminated seed, control of these viruses should start by only using seed from healthy pepper plants. If seed is suspected of being contaminated with PMMoV or any other *Tobamovirus*, then soaking the seed in 15% trisodium phosphate solution for 20 min, rinsing in clean tap water, and drying the seed will significantly reduce the level of contamination (Rast & Stijger, 1987). It is important not to recontaminate the seed by using previously used containers or handling the seed with contaminated hands or other implements. As the tobamoviruses can remain viable in plant debris in the soil for at least several months, it is important to sow seed in a well-rotted and steam-sterilized compost–soil mixture. This will also avoid problems with viruses that have vectors such as nematodes or thrips pupae that can be carried in the soil. Seedling production areas should be located away from field production areas and be kept free of weeds, volunteer pepper plants (e.g., Fig. 6.1A and B), and other plants that can serve as overwintering or alternative hosts to the viruses and/or vectors. Because the seedling stage is generally very susceptible to virus infection, in most locations it is advantageous to grow the seedlings in a fine-mesh net cage or net house that excludes flying or crawling virus vectors such as aphids, whiteflies, and thrips. Loebenstein, Alper, and Levy (1970) demonstrated that air-blast sprays with an emulsion of 1–2% mineral oil in water at intervals of 4–5 days in the seedling and early after transplant stages were effective at suppressing the spread of two aphid stylet-borne viruses, CMV and PVY, but were not economically beneficial when applied to later growth stages of the pepper crop. As some of the viruses, for example, the tobamoviruses, are readily mechanically transmitted on hands or other implements, seedlings should be handled as little as possible before transplanting, though any seedlings showing virus symptoms should be carefully removed from the seedbed or net cage with one or two adjacent seedlings all around to avoid infecting other seedlings.

Table 6.3 Relative importance of different cultural and phytosanitary measures for management of viruses with different modes of transmission

	Relative importance[a]		
Cultural or phytosanitary measure	**No invertebrate vector**	**Nonpersistent insect vector**	**Persistent or semipersistent insect vector**
Seedling production			
Use clean seed (treated and/or from healthy plants)	***	**	
Sow seed in well-rotted and steam-sterilized compost	***	**	*
Grow seedlings in insect-proof net cage		***	***
Avoid handling seedlings before transplanting to the field	***		
Remove infected seedlings and immediate neighbors from seedling bed as soon as seen	***	***	***
Root-drench seedlings with systemic insecticide before transplanting to net house or field		**	***
Protected production in net house/glass house/polytunnel			
Use fine-mesh netting and double-door airlocks to exclude vectors		***	***
Keep net house clean and tidy (free of other plants)	**	**	**
Use colored sticky traps to monitor vector insect numbers and/or for mass trapping of insects		**	***
Introduce biological control of vector insects using pathogens, parasites, and predators		***	***
Monitor plants closely and remove infected plants as soon as seen	***	***	***

Table 6.3 Relative importance of different cultural and phytosanitary measures for management of viruses with different modes of transmission—cont'd

	Relative importance		
Cultural or phytosanitary measure	**No invertebrate vector**	**Nonpersistent insect vector**	**Persistent or semipersistent insect vector**
Avoid touching/handling plants, and when plants have to be handled (e.g., for staking, pruning or harvesting), wash hands and clean implements frequently	***	**	
Open-field production			
Destroy plants and debris of previous crop immediately after final harvest	***	***	***
Transplant to a field some distance from other susceptible crops	*	**	**
Time transplanting to when vector populations are low and not migrating		***	***
Increase planting density	(–)[b]	**	**
Use reflective plastic mulch		**	**
Remove weeds (alternative hosts for virus and/or vectors) from in and around the crop	***	***	***
Monitor plants closely and rogue out and destroy infected plants as soon as possible	**	***	***
Use live mulches, border crops, or hedges around the crop that are nonhost for the viruses		***	*
Introduce/augment natural populations of pathogens, parasites, and predators of virus vectors		**	**
Institute "susceptible crop and weed-free period" (last resort)	**	***	***

[a]Relative importance where *, little importance; **, moderately important; and ***, very important (blank = no importance).
[b](–), may have effect to increase rate of spread of virus.

Figure 6.1 (A) Remnant plant from previous season left growing at the edge of the field because it was still producing a few fruits, but it was acting as a potent source of PepYLCIV for the new pepper crop in the adjacent field, Yogyakarta area, Indonesia. (B) Young volunteer pepper plant severely infected with PepYLCV and acting as a source of infection for the new field of peppers adjacent, Yogyakarta area, Indonesia. (See the color plate.)

In protected pepper production in net or glass houses, good crop hygiene, the proper use of appropriate screen netting and double-door air-locks, and the early roguing (removal) of infected plants are important in controlling virus diseases (Table 6.3). Colored sticky traps can be used within the protected environment for monitoring and/or mass-trapping virus vectors, while biocontrol agents against such vectors will be more effective in confined/protected environments than in the open field. Many whitefly predators, such as the predatory beetle, *Delphastus pusillus*, and the predatory mirid bug, *Macrolophus caliginosus*, pathogens such as the fungi *Verticillium lecanii* and *Paecilomyces fumosoroseus*, and parasitoids such as the parasitic wasps *Encarsia formosa* and *Eretmocerus eremicus*, have been identified and can be effective (though perhaps expensive) for controlling whitefly populations in glass or net houses. Similarly, predatory mites such as *Amblyseius cucumeris* or *Amblyseius swerskii*, predatory bugs such as *Orius* species (Bosco, Giacometto, & Tavella, 2008), or parasitic nematodes such as *Steinernema feltiae* may be effective in preventing buildup of vector thrips populations, and hence reduce the spread of thrips-vectored viruses. Pheromone lures can be used in conjunction with sticky traps in glass or net houses to manage Western flower thrips populations.

For crops that are transplanted to the open field, virus inoculum pressure should be minimized by destroying the remains of previous susceptible crops, and by good weed control in and near the crop, particularly solanaceous weeds and volunteer plants that may be alternative host to virus and vectors. If there is sufficient local information on the epidemiology and vector dynamics of the vectors, then it may be possible to time a crop such that it is transplanted to the field when vector numbers are low and vector populations are not migrating. Also, live mulches, border crops, or hedges around the fields that are more attractive to the vectors than the pepper crop, but not susceptible to the viruses, may function as depositories for the non-persistently transmitted viruses by viruliferous vectors entering the crop. Alternatively, if the hedge around the crop is dense enough it may act as a physical barrier, preventing or delaying the vectors reaching the crop. Using reflective plastic mulch may deter some vectors, such as whiteflies or aphids, from alighting on the pepper plants, feeding and transmitting virus.

Close and regular monitoring of the crop and removing diseased plants from the field as soon as virus symptoms are noticed will remove virus inoculum and reduce the spread within the field. Care should also be taken to avoid spreading viruses from plant to plant during the usual crop management operations in the field: work in areas where disease has been or is known to be present last, after working in unaffected areas; and regularly disinfect tools and equipments by soaking in bleach solutions or washing in strong detergent solutions. In the open field, inundative release of the biocontrol agents, as mentioned under protected cropping above, is not practical. Here, the more sustainable approach is to augment the natural populations of these agents and allow them to multiply to reach equilibrium with the target virus vectors by limiting the use of pesticides. Targeted application (e.g., seedling drenching) of pesticides with very narrow specificity avoids harming beneficial natural enemies.

3.2. Vector management with insecticides

The number of insecticide applications, including mixtures of different active ingredients, to pepper crops appears to have increased considerably over recent years in many countries. The reason often given for this is the attempt to control rapidly emerging leaf curl and yellowing diseases; for example, in Indonesia (Mariyono & Bhattarai, 2009). However, insecticide applications are generally ineffective for the control of the non-persistently vector-transmitted viruses, such as the potyviruses, AMV, and CMV, because transmission by the vectors occurs during the vector's probing of the plant surface prior to feeding and before insecticides can take effect. On the other hand, losses caused by some of these viruses may be reduced by an appropriately timed application of a mixture of mineral oil and a rapid-acting insecticide (e.g., a pyrethroid). For the more persistently transmitted viruses (e.g., poleroviruses, begomoviruses, and tospoviruses) insecticides may be more effective, since the vectors require rather longer acquisition and inoculation periods to transmit these viruses. Pandey, Mathur, and Srivastava (2010) showed that in India, managing whitefly populations was possible using either the biological pesticide, "neem seed kernel extract" or the synthetic insecticide Imidacloprid, and this reduced the incidence of leaf curl disease in the treated fields. "Spinosad," a novel mode-of-action insecticide derived from a family of natural products obtained by fermentation of *Saccharopolyspora spinosa*, had limited efficacy in controlling whiteflies and reducing leaf curl incidence in this study. However, whiteflies and many of the other virus vector species (e.g., thrips and aphids) multiply rapidly, have relatively broad host ranges, and can rapidly develop insecticide-resistant populations. For the thrips species, there is an additional limitation: their eggs are usually hidden within leaf folds or between floral parts, and the pupae are buried in the soil, so it is only the larval and adult forms that are exposed to contact insecticides. Because of these problems, the current thinking is that application of insecticides to pepper plants in the field to control virus vector populations is not a sustainable strategy. Effective insecticides are expensive, populations of beneficial insects (such as predatory beetles and parasitic wasps) may be adversely affected by the insecticides, and there are safety concerns for the environment, farmers, and pepper consumers. The use of insecticides in pepper production should probably be limited to a targeted seedling dip/root drench with a systemic compound, such as one of the neonicotinoids (assuming such is approved for use on peppers in that location) at the time of transplanting to the field.

3.3. Mild-strain cross-protection

The terms "mutual exclusion" and "cross-protection" are used for the antagonistic virus–virus interactions through which infection by a mild strain of virus inhibits infection or suppresses symptom development induced by a closely related severe strain of the virus in the same plant (Syller, 2012). In Japan in the early 1980s, attenuated isolates of PMMoV (at the time regarded as pepper strains of TMV) were successfully used to provide cross-protection against severe wild-type strain infection of pepper plants with no genes for resistance to PMMoV (Goto, Iizuka, & Komochi, 1984). However, with the widespread deployment of the L^3 resistance gene against tobamoviruses (see below) in pepper and the subsequent buildup of L^3 resistance-breaking strains of PMMoV, these original attenuated strains were no longer useful and new attenuated isolates of the L^3-breaking strains were developed and shown to be effective in controlling PMMoV in pepper cultivars carrying the L^3 resistance (Ogai, Kanda-Hojo, & Tsuda, 2013).

Some strains of CMV have been found in association with satellite RNAs (satRNAs). The satRNAs are linear molecules 332–405 nt long and dependent on their "helper" CMV for their replication, encapsidation, and dispersion; they are not necessary for the life cycle of the virus. Although the presence of some satRNA variants intensify the symptoms of CMV, many variants attenuate the symptoms, and some of these "mild-strains" of CMV have been found to cause cross-protection against severe strains. Recently, Nyana, Suastika, Temaja, and Suprapta (2012) screened 43 chili pepper samples showing very mild mosaic symptoms and identified two CMV isolates that had significant cross-protection action against severe strains of CMV and which may be useful as biological control agents for managing CMV in peppers in Indonesia. However, there is some concern that in some instances the use of mild strain cross-protection against one virus may have a synergistic effect on infection by another virus. For example, the recessive potyvirus resistance gene, *pvr3*, usually prevents the systemic movement of PepMoV up the stem of the plant to the young leaves, but this resistance can be reduced if CMV is present (Guerini & Murphy, 1999).

3.4. Host plant resistance against viruses

Over the years since the 1970s, many monogenic and polygenic virus resistance quantitative trait loci (QTL) have been detected, and many have been identified and mapped in *Capsicum* species (Table 6.4). Several of these different virus–capsicum pathosystems have been extensively reviewed elsewhere, so it is not our intention to make a very thorough review of them here but rather to highlight those that have been exploited commercially and to list some of the newer findings with potential for exploitation.

3.4.1 Resistance to CMV

Various sources of resistance to CMV in pepper have been identified, but most provide only partial resistance controlled quantitatively by multiple genes (Table 6.4). The action of these QTLs is either to restrict virus entry and uncoating (installation) in the host cells (Caranta, Palloix, Lefebvre, & Daubèze, 1997), restrict virus multiplication, or restrict the long-distance movement of the virus. Restricting the long-distance movement of the virus, although classed as partial resistance, was shown to be efficient in the field in Europe in the first resistant large-fruited pepper cultivars (Caranta et al., 2002). In Tunisia, surveys revealed that 90% of CMV isolates from pepper fields were reassortants between CMV subgroups IA and IB, and 55 of 57 of these isolates were able to overcome the polygenic resistance to CMV movement in the commercial cultivar "Milord" (Ben Tamarzizt et al., 2013). The only identified single dominant resistance gene against CMV is named *Cmr1* and was identified from the Korean commercial *C. annuum* cultivar "Bukang." *Cmr1* was mapped to the centromeric region of pepper chromosome P2, and its action is to inhibit the movement of CMV from the epidermal cell layer to the mesophyll cells (Kang et al., 2010). However, a new isolate of CMV designated CMV-P1 has emerged in Korea that can overcome this resistance. It is the helicase domain on RNA1 of CMV-P1, which is associated with the inhibition to virus movement in pepper plants carrying *Cmr1* (Kang, Seo, Chung, Kim, & Kang, 2012).

Table 6.4 Virus resistances identified in *Capsicum* spp.

Virus genus	*Gene (Q) = QTL*	Virus	Donor species	Donor cultivar or accession	Chromosomal location	Type of gene or action	References
Begomovirus	(*2 recessive*)	PepGMV, PHYVV	*C. chinense*	BG-3821	?	Virus replication and movement	García-Neria and Rivera-Bustamante (2011)
	(*1 recessive*)	PepLCVaV	*C. frutescens* × *C. chinense*	Bhut Jolokia	?		Rai et al. (2014)
Cucumovirus	*Cmr1*	CMV-FNY	*C. annuum*	Bukang	P2	Movement epidermal cell layer to mesophyll	Kang et al. (2010)
	cmv.hb-4.1 (Q)	CMV-HB	*C. annuum*	BJ0747	P5 (LG4)		Yao, Li, Wang, and Ye (2013)
	cmv.hb-8.2 (Q)				P11 (LG8)		
	cmv11.1 (Q)	CMV 1 Israel	*C. annuum*	Perennial	P11		Ben-Chaim, Grube, Lapidot, Jahn, and Paran (2001)
	cmv13.1 (Q)				?		
	cmv4.1 (Q)				P4		
	cmv6.1 (Q)				P6		

Continued

Table 6.4 Virus resistances identified in *Capsicum* spp.—cont'd

Virus genus	*Gene* (Q) = *QTL*	Virus	Donor species	Donor cultivar or accession	Chromosomal location	Type of gene or action	References
	cmv11.1 (Q)	CMV-N	*C. baccatum/annuum*	Vania (Pen 3-4 × Antibois)	P11	Long distance movement	Caranta et al. (2002)
	cmv11.2 (Q)	CMV-N			P11		
	cmv12.1 (Q)	CMV-MES/N			P12		
	cmv5.1 (Q)	CMV-MES	*C. annuum*	H3	P5		
	cmv2.1 (Q)	CMV	*C. annuum*		P2		Grube, Radwanski, and Jahn (2000)
	cmv3.1 (Q)				P3		
	cmv3.2 (Q)				P3		
Potyvirus	*pvr1* (=*pvr2*)	PepMoV, PVY, TEV	*C. chinense*, *C. annuum*	PI 159236, PI 152225, Dempsey	P4	elF4E–VPg, replication	Murphy, Blauth, Livingstone, Lackney, and Jahn (1998), Kang, Yeam, Frantz, Murphy, and Jahn (2005), and Lee, Jahn, and Yeam (2013)

	pvr3	PepMoV	*C. annuum*	Avelar	?	Movement	Murphy et al. (1998)
	Pvr4	PVY, PepMoV,	*C. annuum*	Criollo de Morelos 334	P10 (*Pvr7*, *Tsw*)	Extreme HR-like	Caranta, Thabuis, and Palloix (1999) and Grube, Blauth, Arnedo Andrés, Caranta, and Jahn (2000)
	pvr5	PepMoV	*C. annuum*	Criollo de Morelos 334	P4	Replication	Dogimont et al. (1996) and Parrella et al. (2002)
	pvr6		*C. annuum*	Perennial	P3	elFiso4E–VPg	Caranta et al. (1996)
	Pvr7	PepMoV	*C. chinense*	PI159236	P10 (*Pvr4*, *Tsw*)	Extreme HR-like	Grube, Blauth, et al. (2000)
	PVY P1.1 (Q)	PVY	*C. annuum*	Perennial	P1		Caranta, Lefebvre, and Palloix (1997) and Palloix, Ayme, and Moury (2009)
	PVY P1.2 (Q)				P1		
	PVY P6.1 (Q)				P6		

Continued

Table 6.4 Virus resistances identified in *Capsicum* spp.—cont'd

Virus genus	*Gene* (*Q*) = *QTL*	Virus	Donor species	Donor cultivar or accession	Chromosomal location	Type of gene or action	References
Tobamovirus	*Hk*	PaMMV(#)	*C. annuum*	Nanbu-Ohnaga	?	?	Sawada et al. (2005)
	$L^1 L^{1a} L^2 L^3 L^4$	Tobamoviruses	*C. annuum*		P11 (*cmv11.1*)	Localization and HR, CC-NBS-LRR	Sawada et al. (2005) and Yang et al. (2012)
Tospovirus	*Cac*	CaCV	*C. chinense*	PI290972	?		Persley, McGrath, Sharman, and Walker (2010)
	(*1 recessive*)	TNRV	*C. annuum*	PY-6423, PY-6424	?		Puangmalai, Potapohn, Akarapisarn, Cheewachaiwit, and Insuan (2013) and Puangmalai, Potapohn, Akarapisarn, and Pascha (2013)
	Tsw	TSWV	*C. chinense*	PI152225, PI159236	P10 (*Pvr4*, *Pvr7*)	HR-like localization, NBS-LRR?	Jahn et al. (2000) and Hoang et al. (2014)

3.4.2 Resistance to potyviruses

The *Capsicum*–potyvirus pathosystem has been well studied and used as a model reference for the study of similar pathosystems in other plant species. Because peppers can be infected by several different potyviruses, the initial naming of resistance genes was irregular and led to confusion, so Kyle and Palloix (1997) proposed the revision whereby all the potyvirus resistance genes are referred to as *pvr* genes. Two major gene *pvr* systems have been identified in pepper. Originally identified from different *Capsicum* species, the recessive genes *pvr1* and *pvr2* both map to the same position on chromosome P-4 and are now designated as different alleles at the *pvr1* locus, where a multiallelic series produces a complex of potyvirus resistance phenotypes (Lee et al., 2013). This was the first natural recessive plant virus resistance gene to be cloned (as *pvr2*), and was shown to encode a translation initiation factor (eIF4E; eukaryotic initiation factor 4E) that is essential to the translation of cellular mRNAs (Ruffel et al., 2002). As physical interaction between the potyvirus genome-linked protein (VPg) and the eIF4E is important for the multiplication of many of the potyviruses, a recessive resistance allele can be created if nucleotide substitutions in the *pvr1* gene lead to changes in the eIF4E that disrupt the interaction with the VPg. Charron et al. (2008) identified nine natural *pvr1* alleles that differed from each other and the wild-type $pvr1^{+}$ allele, and also showed that nonsynonymous substitutions in the central part of the VPg of PVY and TEV could restore the interaction between the mutated eIF4E and the virus VPg. By allele mining using high-resolution melting analysis, Jeong et al. (2012) identified, from a pool of 248 *Capsicum* genotypes, 13 other variants of this allele including one that conferred similar resistance to TEV-HAT as $pvr1^{2}$. Both $pvr1^{1}$ and $pvr1^{2}$ provide resistance to PVY, while only $pvr1^{2}$ is effective against TEV. Both alleles have been used extensively in commercial pepper varieties and $pvr1^{2}$ has remained durable for many years. Strains of PVY that break the $pvr1^{2}$ resistance are very rare in nature, probably because two consecutive nucleotide substitutions are required in the VPg of most PVY isolates; $pvr1^{2}$-breaking strains of TEV are more frequent (Moury & Verdin, 2012).

Caranta et al. (1996) identified the gene they named *pvr6* from the Indian *C. annuum* cultivar "Perennial" by showing that in combination with $pvr1^{2}$ it provided resistance to PVMV. Ruffel et al. (2006) showed that *pvr6* corresponds to the eIF(iso)4E locus and that since mutations in both *pvr1* and *pvr6* were required to provide resistance to PVMV, this virus can use both eIF4E and eIF(iso)4E isoforms. The same was shown for a Korean isolate of ChiVMV (Hwang et al., 2009). As with the *pvr1* locus, allele mining

(Rubio, Nicolai, Caranta, & Palloix, 2009) or ecoTILLING (Ibiza, Canizares, & Nuez, 2010) identified many natural variants of *pvr6* and may point to alternative sources of resistance to ChiVMV and PVMV. Shah et al. (2011) identified at least two pathotypes of ChiVMV causing differential responses on different local and exotic pepper genotypes in Pakistan, and preliminary findings at AVRDC—The World Vegetable Center indicate that there are several more different pathotypes of ChiVMV in different areas of South and Southeast Asia (Tsai et al., 2008). Thus, it may be difficult to find a combination of *pvr1* and *pvr6* alleles that will prove durable against ChiVMV across the whole region.

Another single recessive resistance, named *pvr3*, was identified from the *C. annuum* cultivar "Avelar" and conferred resistance to PepMoV but not to TEV (Murphy et al., 1998). This resistance inhibits the systemic movement of PepMoV into the phloem and up the stem to the younger leaves. However, as mentioned above, infection with CMV compensated PepMoV in its inability to enter the phloem and negated the *pvr3* resistance (Guerini & Murphy, 1999). The recessive *pvr5* was identified from double haploid lines of the Mexican *C. annuum* cultivar "Criollo de Morelos 334" (Dogimont et al., 1996). Although *pvr5* confers the same resistance phenotype against common strains of PVY as $pvr1^{1}$, and Parrella et al. (2002) mapped both *pvr2* (=*pvr1*) and *pvr5* to the same interval between molecular markers TG135 and CT31 on pepper chromosome P4, *pvr5* is apparently not the same as *pvr1* (Dogimont et al., 1996).

The second group of single-gene potyvirus resistance mechanisms comprises the two dominant genes *Pvr4* and *Pvr7*. *Pvr4* was identified from the Mexican *C. annuum* cultivar "Criollo de Morelos 334" and controls an extreme resistance to all PVY pathotypes, as well as to PepMoV, PepYMV, EcRV, PTV, and PepSMV, but not to TEV, PVMV, or ChiVMV (Janzac, Fabre, Palloix, & Moury, 2009). Because of this, *Pvr4* has been widely deployed in pepper cultivars for more than 20 years, during which time it has been highly durable with no natural *Pvr4* resistance-breaking strains of PVY emerging. Janzac, Montarry, Palloix, Navaud, and Moury (2010) used grafting to select for a *Pvr4*-breaking isolate of PVY and showed, although a single nucleotide substitution in the RNA-dependent RNA-polymerase of this isolate was responsible for breaking the resistance, that this mutation confers a high competitiveness cost in susceptible cultivars, thus explaining the durability of *Pvr4*. The dominant gene, *Pvr7*, from *C. chinense* "PI159236" confers resistance to PepMoV, is tightly linked to *Pvr4*, and maps to a cluster of dominant disease resistance genes (including

Tsw) at the telomere region of chromosome P10 (Grube, Blauth, et al., 2000). Kim et al. (2011) showed that a major QTL contributing to the control of trichome density on the pepper main stem also mapped to the R-gene cluster on P10, and that trichome density could be used as a morphological marker for *Pvr4* in pepper breeding.

As well as the single-gene resistance to potyviruses described above, a few polygenic QTL have been identified. The small-fruited Indian cultivar "Perennial," as well as being the source of QTL for resistance to CMV and the source of *pvr6*, was also the source for several QTL against PVY (Caranta, Lefebvre, et al., 1997). Although these QTL on their own provide only partial resistance to PVY (slowed symptom development and reduced symptom severity), they were found to improve the durability of $pvr1^3$ significantly (Palloix et al., 2009). The observation that the two-step adaptation of PVY first to $pvr1^3$ and then to the combination of $pvr1^3$ and the QTLs was still possible, led Palloix et al. (2009) to conclude that the components of a polygenic resistance are better used together rather than separately since "giving priority to major resistance factors may jeopardize the durability expected from polygenic resistance."

Allele-specific molecular markers at the *pvr1* locus (Rubio, Caranta, & Palloix, 2008; Yeam et al., 2005) and markers linked to the *Pvr4* gene (Arnedo-Andrés, Gil-Ortega, Luis-Arteaga, & Hormaza, 2002; Caranta et al., 1999) have been developed for use in marker-assisted breeding of peppers with potyvirus resistance.

3.4.3 Resistance to tospoviruses

Relatively few mechanisms of resistance to tospoviruses have been identified from *Capsicum* species (Table 6.4). Resistance to TSWV was found in several *C. chinense* accessions, including PI152225 and PI159236. The resistance is controlled by the dominant gene *Tsw* and is expressed as a hypersensitive-like localization of the virus (Boiteux & de Ávila, 1994). This gene has been fine mapped to a 259-kb region of the distal portion of chromosome P10 where 22 genes are predicted, among which eight show annotations of NBS-LRR resistance proteins (Hoang et al., 2014). This region is closely linked to, or may contain, the dominant potyvirus resistance genes *Pvr4* and *Pvr7* (see above). The resistance conferred by *Tsw* is broken at high temperatures; it depends on plant age, with young plants being more susceptible. *Tsw* only confers resistance to TSWV, not to any of the other pepper-infecting tospoviruses (Jahn et al., 2000). Unfortunately, within a few years of deployment in commercial pepper cultivars, strains of TSWV that could overcome the *Tsw* resistance emerged in Italy (Roggero, Masenga, & Tavella, 2002), Spain (Margaria, Ciuffo, & Turina, 2004), and Australia (Sharman & Persley, 2006).

As the tospovirus CaCV became important in pepper crops in Queensland, Australia, resistance to CaCV was identified and characterized in *C. chinense* PI 290972. This resistance named *Cac* is not effective against TSWV and segregates as a dominant gene, independently of the *Tsw* gene. CaCV-resistant bell capsicum breeding lines have been released, from which hybrid cultivars incorporating resistance to both CaCV and TSWV are being developed for commercial use (Persley et al., 2010). The only other tospovirus resistance identified in *Capsicum* spp. is the apparently single-gene recessive resistance against TNRV recently detected in a couple of *C. annuum* accessions in Thailand (Puangmalai, Potapohn, Akarapisarn, Cheewachaiwit, et al., 2013; Puangmalai, Potapohn, Akarapisarn, & Pascha, 2013).

3.4.4 Resistance against begomoviruses

Despite the increasingly devastating effect of whitefly-transmitted leaf curl viruses (begomoviruses) of pepper in much of Asia, Central America, and West Africa, there are as yet (as far as we are aware) no commercial pepper cultivars resistant to any of the viruses in this large group. In Mexico, combinations of field screening with natural infection and glasshouse screening with biolistic inoculation (Godínez-Hernández et al., 2001), and glasshouse screening with biolistic inoculation followed by graft inoculation (Anaya-López et al., 2003) with both of the bipartite begomoviruses *Pepper huasteco yellow vein virus* (PHYVV) and *Pepper golden mosaic virus* (PepGMV), identified populations of *C. chinense* and *C. annuum* with potentially useful levels of resistance. García-Neria and Rivera-Bustamante (2011) went on to show that the resistance to PepGMV in one of the *C. chinense* accessions (BG-3821) was probably controlled by two genes with either additive or duplicate recessive epistatic action. The resistance was associated with reduced virus replication and movement, and the induction of genes associated with systemic acquired resistance (SAR).

Meanwhile, in a stratified screening program (open field, viruliferous whiteflies in a glass house, grafting onto infected rootstocks) in India, from a starting collection of 307 *Capsicum* genotypes, eventually only three genotypes (GKC-29, BS-35, and EC-497636) were identified that were apparently highly resistant or immune to leaf curl (probably caused by the proposed monopartite species *Pepper leaf curl Varanasi virus*, PepLCVaV).

When grafted onto infected rootstocks, these genotypes did not develop symptoms of infection, and viral DNA could not be detected in them by PCR (Kumar, Kumar, Singh, Singh, & Rai, 2006). Unfortunately, inheritance studies with these genotypes could not be performed because of reproductive barriers. However, the same team in India identified a *C. frutescens* × *C. chinense* hybrid "Bhut Jolokia" that showed a resistant reaction on inoculation by viruliferous whiteflies, and inheritance studies indicated that this resistance was controlled by a single recessive gene (Rai et al., 2014).

3.4.5 Resistance to tobamoviruses

Holmes (1937; cited in Wang & Bosland, 2006) showed that resistance to TMV was controlled by a series of multiple alleles he designated as "*L*" since they were effective through the "localization of TMV." Subsequently, Boukema (1980) demonstrated that the resistance in 10 *C. chinense* accessions appeared to be allelic with Holmes's *L* genes. They thus designated the new allele L^3 and placed it with Holmes's redesignated alleles in the series $L^3 > L^2 > L^1 > L^+$, as defined by specificity toward the different species and isolates (pathotypes) of *Tobamovirus* (Table 6.5). The L^4 allele, identified from *Capsicum chalcoense* PI260429 providing resistance to all the *Tobamovirus* pathotypes known at the time, was then added to the series (Boukema, 1984). Tomita et al. (2008) mapped the L^3 of *C. chinense* to a region of chromosome P11 containing many repetitive sequences and R-gene homologs and some pseudogenes. Molecular markers linked to several of the different L-gene alleles were developed and proved useful in marker-assisted selection and breeding (Yang, Liu, Kang, Jahn, & Kang, 2009; Yang et al., 2012). Fine mapping and sequencing showed that the *L* gene alleles encode "coiled-coil, nucleotide-binding, leucine-rich repeat" (CC-NBS-LRR)-type resistance proteins, which through interaction with the viral coat protein (CP) (avirulence effector) can induce the resistance response (Tomita et al., 2011). The hierarchical specificity of the *L* genes to the virus pathotypes was determined by the interaction of multiple subregions of the L protein LRR domain with the different viral CPs, or protein complexes including them (Tomita et al., 2011). The resistance conferred by most of the *L* alleles is less at temperatures above 30 °C, though the L^{1a} in the homozygous or heterozygous states confers resistance to P_0 pathotypes at both 24 and 30 °C, and also to PaMMoV at 24 °C, but only in the homozygous state (Sawada et al., 2004). A single incompletely dominant gene designated *Hk*, which differs from the *L* gene was identified from the Japanese

Table 6.5 Reaction of different *Capsicum* genotypes carrying different resistance alleles to the different Tobamovirus species and pathotypes

		Virus species and pathotype				
		TMV, ToMV, TMGMV, BPeMV	**PaMMV, ObPV**	**PMMoV**	**PMMoV**	**PMMoV**
Resistance source	***Genotype***	P_0	P_1	$P_{1.2}$	$P_{1.2.3}$	$P_{1.2.3.4}$
C. annuum Early Cal Wonder	L^+/L^+	S	S	S	S	S
C. annuum Tisana	$L^1/-$	R	S	S	S	S
C. annuum Oonatsume	$L^{1a}/-$	R	R$^{\#}$	S	S	S
C. frutescens Tabasco	$L^2/-$	R	R	S	S	S
C. chinense PI159236	$L^3/-$	R	R	R	S	S
C. chacoense PI260429	$L^4/-$	R	R	R	R	S
C. annuum Nanbu-Ohnaga	*Hk*	S	R*	S	S	S

S, susceptible; R, resistant; R$^{\#}$, resistant at ≤24 °C; R*, resistant at ≥30 °C.
TMV, *Tobacco mosaic virus*; ToMV, *Tomato mosaic virus*; TMGMV, *Tobacco mild green mosaic virus*; BPeMV, *Bell pepper mottle virus*; PaMMV, *Paprika mild mottle virus*; ObPV, *Obuda pepper virus*; PMMoV, *Pepper mild mottle virus*.

C. annuum "Nanbu-Ohnaga," confers resistance to PaMMoV at 30 °C but not at 24 °C and is ineffective against any of the other pathotypes of *Tobamovirus* (Sawada et al., 2005).

Although no L^1- or L^2-breaking strains of pathotype P_0 (TMV, ToMV, TMGMV, and BPeMV), or L^2-breaking strains of pathotype P_1 (PaMMV or ObPV) have been detected, the useful life of L^1 and L^2 was cut short in many areas by the emergence of PMMoV strains, and PMMoV is now probably the most widespread *Tobamovirus* in pepper cultivation. Further to this, three distinct clades of PMMoV isolates exist based on phylogenetic analysis of their CP-coding sequences, and L^3-breaking strains (=$P_{1.2.3}$) have been detected from all three clades, while L^4-breaking strains (=$P_{1.2.3.4}$) have been detected from clades 2 and 3 (Moury & Verdin, 2012). All isolates in clade 3 carry the $M_{139}N$ substitution in their CP, making them $P_{1.2.3}$ pathotypes able to overcome L^3 (Berzal-Herranz et al., 1995; Velasco, Janssen, Ruiz-Garcia, Segundo, & Cuadrado, 2002). The few $P_{1.2.3}$ pathotypes in the other clades are due to different mutations in the CP, as were the L^4-breaking ($P_{1.2.3.4}$) isolates (Antignus, Lachman, Pearlsman, Maslenin, & Rosner, 2008; Genda et al., 2007).

3.5. Natural resistance to virus vectors

Since most of the important viruses of peppers are spread from plant to plant by an insect vector (primarily aphids, whiteflies, and thrips, which can be pests of pepper in their own right), a potentially useful approach to controlling the viruses is through the use of natural resistance against the vector species. However, apart perhaps for the aphids, until recently relatively little attention has been paid to such resistance in *Capsicum* genotypes.

When 50 *Capsicum* accessions and commercial cultivars were screened for resistance to the green peach aphid (*Myzus persicae*), only a few commercial cultivars and sources of virus resistance exhibited levels of tolerance to the aphid that could be a useful component in an integrated pest management program (Frantz, Gardner, Hoffmann, & Jahn, 2004). Working with the cotton aphid, *Aphis gossypii*, da Costa et al. (2011) identified pepper accessions that were less preferred by the aphids than were other accessions. The aphids may have been repelled by the volatile organic compounds (VOCs) produced by the nonpreferred cultivars and this trait could perhaps be used in breeding aphid-resistant cultivars. In relation to this, Dewhirst et al. (2012) observed that treating pepper plants with the naturally occurring plant defense activator, *cis*-jasmone, changed the VOCs produced by the plants such that alate *M. persicae* were no longer attracted to the plants, while the aphid parasitoid, *Aphidius ervi*, showed preference for *cis*-jasmone-treated plants by a significant increase in time foraging for aphids on them.

With the aim of identifying whitefly resistance that could be incorporated into IPM packages for managing virus diseases of pepper, Firdaus et al. (2011) screened 44 pepper accessions (from *C. annuum*, *C. frutescens*, *C. chinense*, and *C. baccatum*) for resistance to *B. tabaci* whiteflies. Based on adult survival and oviposition rates in no-choice tests, and whitefly, egg, and nymphal density in free-choice tests, several *C. annuum* genotypes were highly resistant to whiteflies. Whitefly density and oviposition rate generally correlated positively with trichome density and negatively correlated with leaf cuticle thickness; though there were genotypes that deviated from this, indicating that several different factors could be responsible for the resistance observed. Preliminary findings at AVRDC—The World Vegetable Center indicate that the slow rate of infection with TYLCTHV of certain hot pepper accessions may be because the viruliferous whiteflies are deterred from alighting and/or feeding on these accessions in a free-choice situation.

Fery and Schalk (1991) reported considerable variability within pepper germplasm for reaction to the thrips *Frankliniella occidentalis*, but the resistance they observed appeared to be due to tolerance mechanisms, not antixenosis (nonpreference) or antibiosis mechanisms. Maris, Joosten, Peters, and Goldbach (2003) found that a *Capsicum* line with resistance to *F. occidentalis*, defined by lack of reproduction, low preference, and minimal feeding damage, was inherently as susceptible to TSWV as a thrips-susceptible line, and under the controlled experimental conditions used, that the transmission of TSWV was little affected by the vector resistance. However, in larger scale glasshouse trials, probably because of impeded development of the thrips population, the introduction and spread of TSWV was restricted in the thrips-resistant line (Maris, Joosten, Goldbach, & Peters, 2003). More recently, Maharijaya et al. (2011) screened 32 pepper accessions (from *C. annuum*, *C. baccatum*, *C. chinense*, and *C. frutescens*) and identified several *C. annuum* accessions with reasonable levels of resistance against *F. occidentalis* and *Thrips parvispinus*. The resistance factors in these accessions were shown to have significant effects on oviposition rate, larval mortality, and life cycle period, indicating that the resistance is based on antibiosis.

3.6. Transgenic resistance

Compared to the other widely grown members of the Solanaceae (tomato, tobacco, eggplant, and potato), the cultivated species of *Capsicum* remain relatively difficult to transform, and transformation protocols tend to be cultivar-specific (Balázs et al., 2008). In Korea, Kim, Lee, Kim, and Paek (1997) transformed hot pepper with a construct containing a symptom-attenuating CMV satellite RNA and showed that the transformed plants developed only mild symptoms on infection with two different strains of CMV. Soon after, Shin, Han, Lee, and Peak (2002) produced transgenic pepper plants coexpressing the CP genes of CMV and ToMV and showed that viral propagation of both CMV-Kor and PMMoV was significantly retarded in several of the transformed lines. Subsequently, Lee et al. (2009) transformed pepper inbred lines with the CP gene of CMV-P0 strain using *Agrobacterium*-mediated transformation. The three transgenic lines selected were tolerant to infection by CMV-P0 and also to infection by CMV-P1, an emerging CMV pathotype that is able to overcome *Cmr1*, which is the widely deployed natural resistance to CMV-P0. As yet, the only *Capsicum* cultivar carrying transgenic resistance to a virus disease in commercial production is a line of sweet pepper in China carrying similar CP-mediated resistance to CMV (James, 2013).

Over recent years, pathogen-derived transgenic resistance strategies based on RNA interference (RNAi) have been shown to be very effective in various other crop–virus pathosystems (Duan, Wang, & Guo, 2012). Medina-Hernández, Rivera-Bustamante, Tenllado, and Holguín-Peña (2013) showed that this approach could work for a pepper-infecting begomovirus by agroinfiltrating tobacco with RNAi constructs designed to express double-stranded RNA (dsRNA) of the intergenic region (AC-IR-AVI) of *Pepper golden mosaic virus* (PepGMV) or *Tomato chino La Paz virus* (ToChLPV). The plants were challenged 4 days after agroinfiltration by biolistic inoculation with infectious clones of both the DNA-A and DNA-B of PepGMV. The plants expressing the homologous PepGMV dsRNA were highly resistant, whereas the plants expressing the heterologous ToChLPV dsRNA were still moderately resistant compared to the control plants. Lin, Ku, Tsai, Green, and Jan (2011) took the RNAi approach a step further by demonstrating that linking gene segments from two viruses with distinct genomic organization, one DNA (the begomovirus *Tomato leaf curl Taiwan virus* [ToLCTWV]) and the other RNA (the tospovirus TSWV), could confer multiple virus resistance in transgenic tomato or tobacco plants via gene silencing.

An alternative to the pathogen-derived resistance described above is to develop plant host-derived transgenic resistances. Shin, Park, An, and Paek (2002) demonstrated that overexpression of tobacco stress-induced gene 1 (*Tsi1*) in transgenic hot pepper plants induced constitutive expression of several pathogenesis-related genes in the absence of stress or pathogen treatment. The transgenic hot pepper plants expressing *Tsi1* exhibited resistance to PMMoV and CMV. Similarly, when PPI1 (pepper-PMMoV interaction1, a bZIP transcription factor gene) isolated from *C. chinense* was overexpressed in hot pepper under the control of the 35S promoter, some transgenic plants showed resistance to PMMoV (Lee et al., 2004). Later, with the cloning and sequencing of different alleles of the *pvr1* (eIF4E) locus from pepper (as mentioned above), it was possible to test constructs carrying these genes for ability to confer resistance to potyviruses in hosts less recalcitrant to transformation. Transgenic tomato progenies that overexpressed the *Capsicum pvr1[1]* allele showed dominant resistance to several TEV strains and other potyviruses, including PepMoV, similar to that observed in pepper homozygous for the *pvr1[1]* allele (Kang, Yeam, Li, Perez, & Jahn, 2007; Yeam, Cavatorta, Ripoll, Kang, & Jahn, 2007). However, in a similar approach, Cavatorta et al. (2011) observed that the *pvr1[1]* allele, when transgenically expressed in potato, permitted virus symptom development and protein accumulation, but overexpression of pepper allele *pvr1[2]* resulted in transgenic potato plants resistant to all three strains of PVY tested.

Another variation on the plant-derived transgenic resistance approach is to express or overexpress one of the resistance genes (*R* gene) that encodes a nucleotide-binding site-leucine-rich-repeat (NBS-LRR) protein. This approach, which has been shown to be effective with several NBS-LRR genes including from barley, rice, *S. bulbocastanum*, and soybean (Collinge, Jørgensen, Lund, & Lyngkjær, 2010), has not been tried with the *L* or *Tsw* genes from *Capsicum*. Thus, these newer transgenic approaches show great potential for producing pepper lines with resistance to multiple viruses, provided that the transformation system for peppers is stabilized, the prohibition on growing transgenic crops is lifted, and public opinion turns more in favor of such crops.

4. DISCUSSION AND CONCLUSIONS

The cultivated species of the genus *Capsicum* are relatively susceptible to infection by many different species of virus. The symptoms induced by virus infection in peppers ranges from very mild leaf chlorosis to severe leaf curl, plant stunting, necrosis, die-back and death, but it is seldom possible to make an accurate diagnosis by symptoms alone because two viral species can induce very similar symptoms (e.g., Fig. 6.2A and B). Also, the symptoms induced can resemble natural senescence or those caused by abiotic stresses such as nutrient deficiency or toxicity (e.g., Fig. 6.2C). Diagnostic antisera are commercially available for some of the more established pepper-infecting viruses, but the rapid emergence of new species and strains of these viruses means that other more generic tests such as those based on PCR with genus-specific universal primers are being more frequently employed for detection and identification of viruses.

In contrast to the situation in the Mediterranean Basin region where the diversity of viruses infecting peppers has remained relatively static over the last 20 years (Moury & Verdin, 2012), in the rest of the world, particularly

Figure 6.2 (See legend on next page.) *(Continued)*

the tropical and subtropical pepper-growing areas, there has been a rapid increase in the number of virus species identified infecting peppers in the field. Hanssen, Lapidot, and Thomma (2010) found only 49 virus species being able to infect pepper (in comparison to at least 136 virus species able to infect tomato), of which about 20 from 15 different taxonomic groups had been reported to cause damage to pepper crops. For this review, we observe that there have been many recent reports of known pepper-infecting viruses infecting peppers, and also of previously undescribed viruses being identified from peppers from diverse locations. Most of this increase is in virus species that are vectored by insects. For example, the number of begomoviruses (whitefly-transmitted) detected from pepper plants has increased from 6 in 2003 (Pernezny et al., 2003) to about 43 in January 2014 (Table 6.2). Similarly, whereas there were only three tospoviruses reported from peppers in 2003, now 13 different tospovirus species have been identified infecting the crop. The rapid rise in numbers of viruses identified as able to infect or emerging in new locations is not restricted to the peppers, but has also been observed for other crops such as tomato (Hanssen et al., 2010) and the cucurbits (Abrahamian & Abou-Jawdah, 2014; Juarez et al., 2013).

Most of these newly emerging viruses in pepper crops are transmitted by specific vector insects and their increased rate of emergence in different locations over the past 10–20 years is probably due to a combination of factors. Pepper production systems have become more intensive and extensive in many regions, including regions where peppers were not previously grown at any scale, and this has probably led to greater selection for more virulent or pepper-adapted variants of local, indigenous virus strains, or species. This intensification of pepper production may also have resulted in selection for, and increased populations of, indigenous and introduced virus vector species and variants adapted to feeding on pepper, thus increasing the incidence and rate of local spread of pepper-infecting viruses (Kenyon et al., 2014). The increased volume and speed of international trade of fresh produce and flowers through globalization has carried virus vector species and viruses to areas where they were not seen previously, and climate change may be shifting or broadening the range of locations favorable to some of

Figure 6.2—Cont'd (A) ChiVMV (Potyvirus) inducing relatively mild symptoms in pepper, Khon Kaen area, Thailand. (B) PVMV (Potyvirus) infecting pepper in Taiwan. (C) PeVYV (Polerovirus) infected plant with general chlorosis/yellowing that could easily be mistaken for nutrient deficiency or toxicity, Kamphaeng Saen area, Thailand. (See the color plate.)

the vectors and permitting the spread of some of the viruses to new locations (Bebber, Ramotowski, & Gurr, 2013). The recently reported increase in number of virus species infecting peppers is probably a reflection of the increased interest in this crop, and the development and use of advanced, more specific, and easier to use methods for detecting and identifying viruses (Kenyon et al., 2014).

Mixed infections of two or more different virus species in a pepper plant are relatively common in some areas, and as a greater diversity of viruses emerge and extend and merge their geographic ranges, such mixed infections will likely become more prevalent (e.g., Fig. 6.3). This could lead to more severe disease incidence since some virus species of the same genus can act synergistically when in mixed infection. Bag, Mitter, Eid, and Pappu (2012) showed that the tospovirus TSWV could facilitate the systemic movement of IYSV (another tospovirus) in *Datura stramonium*, an otherwise restrictive host of IYSV. The begomoviruses PHYVV and PepGMV can act synergistically in pepper (Renteria-Canett, Xoconostle-Cazares, Ruiz-Medrano, & Rivera-Bustamante, 2011). Sometimes, viruses from different genera also can act synergistically when coinfecting pepper, for example, CMV and PepMoV (Murphy et al., 1998; Murphy & Bowen, 2006), or CMV with PMMoV and/or PepMoV (Kim et al., 2010). From our own field observations, it seems likely that there is synergy in South Asia between a begomovirus and a tospovirus, and from experience with tomato, it is highly likely that there will be a synergistic interaction when the Crinivirus

Figure 6.3 Severe chlorosis and leaf curl caused by mixed infection of an unidentified begomovirus and ChiVMV, Khon Kaen area, Thailand. (See the color plate.)

ToCV and one of the tospoviruses (TSWV, PBNV, CaCV, or TNRV) coinfect a pepper plant (García-Cano, Resende, Fernández-Muñoz, & Moriones, 2006).

Traditionally, the deployment of natural virus resistance was regarded as the most easily adopted and sustainable means for controlling virus diseases in most crops. Resistance to some of the longer-established viruses of pepper have been identified, both from cultivated types and from close wild relatives (Table 6.4). The pepper breeding program at AVRDC—The World Vegetable Center has been using resistances from a wide selection of germplasm accessions to produce improved breeding lines, often with multiple virus resistance (Table 6.6). Some of the resistances identified elsewhere, such as some of the *pvr* alleles for recessive resistance against some of the potyviruses, the *L* alleles for resistance against some of the tobamoviruses, and the *Cmr1* for resistance to CMV, have been introgressed and proved useful in commercial pepper cultivars. However, resistance-breaking strains of the different viruses have evolved, or been selected for, from the virus population that can overcome these resistance mechanisms. With the increased chance of mixed infections, not only is there the possibility of synergistic interactions (as mentioned above), but there is also greater opportunity for virus genome reassortments, component capture, and recombination. The greater genetic diversity within the virus populations arising from greater mixing of isolates increases the likelihood that more virulent virus genotypes will be selected for that can overcome the deployed natural resistances. Ben Tamarzizt et al. (2013) identified 90% of the CMV isolates from Tunisian pepper crops as reassortants between subgroups IA and IB, with two predominant haplotypes, IB-IA-IA and IB-IA-IB (nomenclature according to the subgrouping of the three genomic RNAs). Moreover, most of these isolates were able to overcome the polygenic resistance to CMV movement in cultivar "Milord" (Caranta et al., 2002). Thus, the rate of discovery and use of natural resistance mechanisms has not kept pace with the rate of spread to, and emergence of, different pepper-infecting virus species and genera in different pepper-growing areas, including resistance-breaking strains. This also means that reliance on a single major gene resistance is no longer tenable, and it is becoming more important to deploy several different resistance genes in the same field, either by planting several different cultivars each carrying a different resistance, or by combining (pyramiding) several resistance genes into a single cultivar appropriate to the local situation (Djian-Caporalino et al., 2014). The more broad-based, multigenic or QTL, horizontal resistances against viruses are likely to be more durable,

Table 6.6 Virus-resistant pepper germplasm accessions and improved lines in use at AVRDC

Virus	Germplasm accessions[a]	Improved lines
Cucumovirus (CMV)	VI037522 (PBC495; Perennial HDV), VI037606 (PBC370), VI059330 (Bathaida Select), VI037649 (VC41a), VI041283 (PBC145; Tiwari)	HP: AVPP0105 SP: AVPP0602
Potyvirus (ChiVMV)	VI046904 (PBC602), VI037522 (PBC495; Perennial HDV), VI037634 (HDA 832), VI041283 (PBC145; Tiwari), VI038298 (PBC142; Pant C1), VI012911 (Pangalengan-2), VI012901 (Criollo de Morelos 334)	HP: AVPP0105 SP: AVPP0602
Potyvirus (PVY)	VI037518 (PBC66; MC-4), VI039374 (PBC385), VI037519 (PBC67; MC-5), VI037458 (PBC204), VI037548 (PBC384), VI037563 (PBC473), VI037629 (PBC743), VI037584 (PBC580), VI046904 (PBC602), VI062408 (PI 201234), VI062409 (PI 188478), VI012560 (Orias Kossarvu)	HP: AVPP9813, AVPP9905, AVPP0012, AVPP0105 SP: AVPP9807, AVPP0006, AVPP0204, AVPP0408, AVPP0502
Tobamovirus (ToMV, TMV, PMMoV)	VI012496 (Aji Blanco Cristal), VI014871 (Florida VR2)	HP: AVPP0606 SP: AVPP9906 (moderately resistant)
Begomovirus (TYLCThV)	VI041283 (PBC145; Tiwari) VI041282 (PBC144; Laichi), VI037522 (PBC495; Perennial HDV), VI012005 (Early Spring green), VI012907 (PI 159236), VI037599 (PBC143; Bangala green 1), VI037450 (PBC149; Lorai)	Reselections of symptom-free plants from crosses involving resistant/tolerant accessions (to left) in progress

[a]Vegetable Introduction Number (VI0xxxxx), followed by other accession number and/or cultivar n-ame; See AVRDC AVGRIS database at http://203.64.245.173/.
HP, hot pepper; SP, sweet pepper.

and if combined with major- (single-) gene resistance, are likely to increase the durability of the major gene resistance (Palloix et al., 2009), especially if also combined with similar multigenic resistances against the virus vector(s).

With the increasing diversity of pepper-infecting viruses and their vectors in many pepper-growing areas, it is now even more important to adopt more holistic and integrated approaches to controlling all the diseases and pests (including the viruses and their vectors) in this crop. These approaches will have to be tailored to each location and the spectrum of pathogens and pest present in each. The rapid emergence of resistance to many of the synthetic pesticides in many of the insect pests and vectors of viruses, as well as general concern about the effect of synthetic pesticides on the environment and the health of growers and consumers, means there should be less use of these compounds. However, for this to happen, growers need effective alternative control practices. Several natural products with insecticidal or antifeedant properties, such as neem extract, have relatively little activity against the beneficial biota, and so may prove to be useful components of an integrated control package, though they should probably be used prophylactically from early in the crop life to be effective, and this may render them too expensive for many small-scale farms. Similarly, very specific and effective predators, parasites, and pathogens have been identified that are already proving useful in the biological control of some of the insect pests and virus vectors in protected (glass house) production of peppers, but these are currently not very widely available, are expensive, and have not been demonstrated to be effective in the tropical or subtropical field situations where the majority of peppers are grown.

In conclusion, for the effective and sustainable control of virus diseases of pepper crops in tropical and subtropical field situations, we can no longer rely on the blanket use of insecticides to control vector insects or the deployment of single, major gene virus resistance. To control these diseases, we must formulate packages of integrated rational measures, based on understanding the ecology and epidemiology of the diseases, which are affordable and work with the local environment. The packages should include the use of natural resistance, but to increase their durability, major gene resistances should be pyramided and combined with polygenic, QTL resistances against viruses and virus vectors. Use of synthetic insecticides will not be abandoned completely but should be restricted to early growth stages when the crop is most vulnerable to infection—i.e., up to around the time of transplanting to the field—when the least amount of insecticide can provide the most beneficial effect with minimal effect on the environment and natural biological control agents or other beneficial organisms. Other components of the package could be the use of natural products that affect the virus vectors directly, or that interact to make the pepper plants less attractive as hosts for the vectors, or induce resistance in the pepper plants to the viruses and/or the vectors. Finally, all the components already described will be more durable if cultural practices concerning crop timing and location, field sanitation, and physical barriers are adopted that reduce the virus inoculum sources, limit the spread and increase in viruses and vectors to, and within, the pepper crop, and hence limit the extent of exposure of the crop to the viruses.

REFERENCES

Abdalla, O. A., & Ali, A. (2012). First report of *Alfalfa mosaic virus* associated with severe mosaic and mottling of pepper (*Capsicum annuum*) and white clover (*Trifolium repens*) in Oklahoma. *Plant Disease*, *96*, 1705–1706.

Abrahamian, P., & Abou-Jawdah, Y. (2014). Whitefly-transmitted criniviruses of cucurbits: Current status and future prospects. *Indian Journal of Virology*, *25*, 26–38.

Ali, A., & Kobayashi, M. (2010). Seed transmission of *Cucumber mosaic virus* in pepper. *Journal of Virological Methods*, *163*, 234–237.

Anaya-López, J. L., Torres-Pacheco, I., González-Chavira, M., Garzon-Tiznado, J. A., Pons-Hernandez, J. L., Guevara-González, R. G., et al. (2003). Resistance to geminivirus mixed infections in Mexican wild peppers. *HortScience*, *38*, 251–255.

Antignus, Y., Lachman, O., Pearlsman, M., Maslenin, L., & Rosner, A. (2008). A new pathotype of *Pepper mild mottle virus* (PMMoV) overcomes the L^4 resistance genotype of pepper cultivars. *Plant Disease*, *92*, 1033–1037.

Arnedo-Andrés, M. S., Gil-Ortega, R., Luis-Arteaga, M., & Hormaza, J. I. (2002). Development of RAPD and SCAR markers linked to the Pvr4 locus for resistance to PVY in pepper (*Capsicum annuum* L.). *TAG Theoretical and Applied Genetics*, *105*, 1067–1074.

Bag, S., Mitter, N., Eid, S., & Pappu, H. R. (2012). Complementation between two tospoviruses facilitates the systemic movement of a plant virus silencing suppressor in an otherwise restrictive host. *PLoS One*, 7, e44803.

Balázs, E., Bukovinszki, Á., Csányi, M., Csillèry, G., Divéki, Z., Nagy, I., et al. (2008). Evaluation of a wide range of pepper genotypes for regeneration and transformation with an *Agrobacterium tumefaciens* shooter strain. *South African Journal of Botany*, *74*, 720–725.

Barbosa, J. C., Teixeira, L. D. D., & Rezende, J. A. M. (2010). First report on the susceptibility of sweet pepper crops to *Tomato chlorosis virus* in Brazil. *Plant Disease*, *94*, 374.

Batuman, O., Rojas, M. R., Almanzar, A., & Gilbertson, R. L. (2013). First report of *tomato chlorotic spot virus* in processing tomatoes in the Dominican Republic. *Plant Disease*, *98*, 286.

Bebber, D. P., Ramotowski, M. A. T., & Gurr, S. J. (2013). Crop pests and pathogens move polewards in a warming world. *Nature Climate Change*, *3*, 985–988.

Ben Tamarzizt, H., Montarry, J., Girardot, G., Fakhfakh, H., Tepfer, M., & Jacquemond, M. (2013). *Cucumber mosaic virus* populations in Tunisian pepper crops are mainly composed of virus reassortants with resistance-breaking properties. *Plant Pathology*, *62*, 1415–1428.

Ben-Chaim, A., Grube, R. C., Lapidot, M., Jahn, M., & Paran, I. (2001). Identification of quantitative trait loci associated with resistance to cucumber mosaic virus in *Capsicum annuum*. *TAG Theoretical and Applied Genetics*, *102*, 1213–1220.

Berzal-Herranz, A., De La Cruz, A., Tenllado, F., Diaz-Ruiz, J. R., Löpez, L., Sanz, A. I., et al. (1995). The *Capsicum* L^3 gene-mediated resistance against the tobamoviruses is elicited by the coat protein. *Virology, 209*, 498–505.

Boiteux, L. S., & de Ávila, A. C. (1994). Inheritance of a resistance specific to tomato spotted wilt *tospovirus* in *Capsicum chinense* 'PI 159236'. *Euphytica, 75*, 139–142.

Bosco, L., Giacometto, E., & Tavella, L. (2008). Colonization and predation of thrips (Thysanoptera: Thripidae) by *Orius* spp. (Heteroptera: Anthocoridae) in sweet pepper greenhouses in Northwest Italy. *Biological Control, 44*, 331–340.

Boukema, I. W. (1980). Allelism of genes controlling resistance to TMV in *Capsicum* L. *Euphytica, 29*, 433–439.

Boukema, I. W. (1984). Resistance to TMV in *Capsicum chacoense* Hunz. is governed by an allele of the *L* locus. *Capsicum and Eggplant Newsletter, 3*, 47–48.

Brown, J. K., Fauquet, C. M., Briddon, R. W., Zerbini, M., Moriones, E., & Navas-Castillo, J. (2012). Family—*Geminiviridae*. In A. M. King, E. Lefkowitz, M. J. Adams, & E. B. Carstens (Eds.), *Virus taxonomy: Ninth report of the international committee on taxonomy of viruses* (pp. 351–373). San Diego, CA: Elsevier.

Brunt, A. A., & Kenten, R. H. (1971). Pepper veinal mottle virus—A new member of the potato virus Y group from peppers (*Capsicum annuum* L. and *C. frutescens* L.) in Ghana. *Annals of Applied Biology, 69*, 235–243.

Buzkan, N., Arpaci, B. B., Simon, V., Fakhfakh, H., & Moury, B. (2012). High prevalence of poleroviruses in field-grown pepper in Turkey and Tunisia. *Archives of Virology, 158*, 881–885.

Buzkan, N., Demir, M., Öztekin, V., Mart, C., Çağlar, B. K., & Yilmaz, M. A. (2006). Evaluation of the status of capsicum viruses in the main growing regions of Turkey. *EPPO Bulletin, 36*, 15–19.

Caranta, C., Lefebvre, V., & Palloix, A. (1997). Polygenic resistance of pepper to potyviruses consists of a combination of isolate-specific and broad-spectrum quantitative trait loci. *Molecular Plant-Microbe Interactions, 10*, 872–878.

Caranta, C., Palloix, A., Gebre-Selassie, K., Lefebvre, V., Moury, B., & Daubèze, A. M. (1996). A complementation of two genes originating from susceptible *Capsicum annuum* lines confers a new and complete resistance to Pepper veinal mottle virus. *Phytopathology, 86*, 739–743.

Caranta, C., Palloix, A., Lefebvre, V., & Daubèze, A. M. (1997). QTLs for a component of partial resistance to cucumber mosaic virus in pepper: Restriction of virus installation in host-cells. *TAG Theoretical and Applied Genetics, 94*, 431–438.

Caranta, C., Pflieger, S., Lefebvre, V., Daubèze, A. M., Thabuis, A., & Palloix, A. (2002). QTLs involved in the restriction of cucumber mosaic virus (CMV) long-distance movement in pepper. *TAG Theoretical and Applied Genetics, 104*, 586–591.

Caranta, C., Thabuis, A., & Palloix, A. (1999). Development of a CAPS marker for the Pvr4 locus: A tool for pyramiding potyvirus resistance genes in pepper. *Genome, 42*, 1111–1116.

Cavatorta, J., Perez, K. W., Gray, S. M., Van Eck, J., Yeam, I., & Jahn, M. (2011). Engineering virus resistance using a modified potato gene. *Plant Biotechnology Journal, 9*, 1014–1021.

Charron, C., Nicola, M., Gallois, J.-L., Robaglia, C., Moury, B., Palloix, A., et al. (2008). Natural variation and functional analyses provide evidence for co-evolution between plant eIF4E and potyviral VPg. *The Plant Journal, 54*, 56–68.

Chen, K., Xu, Z., Yan, L., & Wang, G. (2007). Characterization of a new strain of Capsicum chlorosis virus from peanut (*Arachis hypogaea* L.) in China. *Journal of Phytopathology, 155*, 178–181.

Cheng, Y. H., Deng, T. C., Chen, C. C., Liao, J. Y., Chang, C. A., & Chiang, C. H. (2011). First report of *Pepper mottle virus* in bell pepper in Taiwan. *Plant Disease, 95*, 617.

Cheng, Y. H., Wang, R. Y., Chen, C. C., Chang, C. A., & Jan, F. J. (2009). First report of *Pepper veinal mottle virus* in tomato and pepper in Taiwan. *Plant Disease*, *93*, 107.

Cheng, Y. H., Zheng, Y. X., Tai, C. H., Yen, J. H., Chen, Y. K., & Jan, F. J. (2014). Identification, characterisation and detection of a new tospovirus on sweet pepper. *Annals of Applied Biology*, *164*, 107–115.

Chiemsombat, P., Gajanandana, O., Warin, N., Hongprayoon, R., Bhunchoth, A., & Pongsapich, P. (2008). Biological and molecular characterization of tospoviruses in Thailand. *Archives of Virology*, *153*, 571–577.

Chiemsombat, P., Sharman, M., Srivilai, K., Campbell, P., Persley, D., & Attathom, S. (2010). A new tospovirus species infecting *Solanum esculentum* and *Capsicum annuum* in Thailand. *Australasian Plant Disease Notes*, *5*, 75–78.

Collinge, D. B., Jørgensen, H. J. L., Lund, O. S., & Lyngkjær, M. F. (2010). Engineering pathogen resistance in crop plants: Current trends and future prospects. *Annual Review of Phytopathology*, *48*, 269–291.

da Costa, J. G., Pires, E., Riffel, A., Birkett, M., Bleicher, E., & Sant'Ana, A. (2011). Differential preference of *Capsicum* spp. cultivars by *Aphis gossypii* is conferred by variation in volatile semiochemistry. *Euphytica*, *177*, 299–307.

de Ávila, A. C., de Haan, P., Kormelink, R., Resende, R. d. O., Goldbach, R. W., & Peters, D. (1993). Classification of tospoviruses based on phylogeny of nucleoprotein gene-sequences. *Journal of General Virology*, *74*, 153–159.

de Haan, P., de Ávila, A. C., Kormelink, R., Westerbroek, A., Gielen, J. J. L., Peters, D., et al. (1992). The nucleotide sequence of the S RNA of *Impatiens necrotic spot virus*, a novel tospovirus. *FEBS Letters*, *306*, 27–32.

Dewhirst, S. Y., Birkett, M. A., Loza-Reyes, E., Martin, J. L., Pye, B. J., Smart, L. E., et al. (2012). Activation of defence in sweet pepper, *Capsicum annum*, by cis-jasmone, and its impact on aphid and aphid parasitoid behaviour. *Pest Management Science*, *68*, 1419–1429.

Djian-Caporalino, C., Palloix, A., Fazari, A., Marteu, N., Barbary, A., Abad, P., et al. (2014). Pyramiding, alternating or mixing: Comparative performances of deployment strategies of nematode resistance genes to promote plant resistance efficiency and durability. *BMC Plant Biology*, *14*, 53.

Dogimont, C., Palloix, A., Daubze, A.-M., Marchoux, G., Selassie, K. G., & Pochard, E. (1996). Genetic analysis of broad spectrum resistance to potyviruses using doubled haploid lines of pepper (*Capsicum annuum* L.). *Euphytica*, *88*, 231–239.

Dombrovsky, A., Glanz, E., Lachman, O., Sela, N., Doron-Faigenboim, A., & Antignus, Y. (2013). The complete genomic sequence of *Pepper Yellow Leaf Curl Virus* (PYLCV) and its implications for our understanding of evolution dynamics in the genus *Polerovirus*. *PLoS One*, *8*, e70722.

Dombrovsky, A., Glanz, E., Pearlsman, M., Lachman, O., & Antignus, Y. (2010). Characterization of *Pepper yellow leaf curl virus*, a tentative new *Polerovirus* species causing a yellowing disease of pepper. *Phytoparasitica*, *38*, 477–486.

Duan, C.-G., Wang, C.-H., & Guo, H.-S. (2012). Application of RNA silencing to plant disease resistance. *Silence*. *3*. http://dx.doi.org/10.1186/1758-1907X-1183-1185.

Duffus, J. E. (1981). Beet western yellows virus—A major component of some potato leaf roll-affected plants. *Phytopathology*, *71*, 193–196.

Fauquet, C., Briddon, R., Brown, J., Moriones, E., Stanley, J., Zerbini, M., et al. (2008). Geminivirus strain demarcation and nomenclature. *Archives of Virology*, *153*, 783–821.

Fery, R. L., & Schalk, J. M. (1991). Resistance in pepper (*Capsicum annuum*-L) to Western flower thrips [*Frankliniella occidentalis* (Pergande)]. *HortScience*, *26*, 1073–1074.

Firdaus, S., Van Heusden, A., Harpenas, A., Supena, E. D. J., Visser, R. G. F., & Vosman, B. (2011). Identification of silverleaf whitefly resistance in pepper. *Plant Breeding*, *130*, 708–714.

Fortes, I. M., Moriones, E., & Navas-Castillo, J. (2012). *Tomato chlorosis virus* in pepper: Prevalence in commercial crops in southeastern Spain and symptomatology under experimental conditions. *Plant Pathology, 61*, 994–1001.

Frantz, J. D., Gardner, J., Hoffmann, M. P., & Jahn, M. M. (2004). Greenhouse screening of *Capsicum* accessions for resistance to green peach aphid (*Myzus persicae*). *HortScience, 39*, 1332–1335.

García-Cano, E., Resende, R. O., Fernández-Muñoz, R., & Moriones, E. (2006). Synergistic interaction between *Tomato chlorosis virus* and *Tomato spotted wilt virus* results in breakdown of resistance in tomato. *Phytopathology, 96*, 1263–1269.

García-Neria, M. A., & Rivera-Bustamante, R. F. (2011). Characterization of geminivirus resistance in an accession of *Capsicum chinense* Jacq. *Molecular Plant-Microbe Interactions, 24*, 172–182.

Genda, Y., Kanda, A., Hamada, H., Sato, K., Ohnishi, J., & Tsuda, S. (2007). Two amino acid substitutions in the coat protein of *Pepper mild mottle virus* are responsible for overcoming the L^4 gene-mediated resistance in *Capsicum* spp. *Phytopathology, 97*, 787–793.

Genda, Y., Sato, K., Nunomura, O., Hirabayashi, T., & Tsuda, S. (2011). Immunolocalization of *Pepper mild mottle virus* in developing seeds and seedlings of *Capsicum annuum*. *Journal of General Plant Pathology, 77*, 201–208.

Gniffke, P. A., Shieh, S. C., Lin, S. W., Sheu, Z. M., Chen, J. R., Ho, F. I., et al. (2013). Pepper research and breeding at AVRDC—The World Vegetable Center. In *XV EUCARPIA meeting on genetics and breeding of capsicum and eggplant, 2–4 September 2013, Turin, Italy* (pp. 305–311).

Godínez-Hernández, Y., Anaya-López, J. L., Díaz-Plaza, R., González-Chavira, M., Torres-Pacheco, I., Rivera-Bustamante, R. F., et al. (2001). Characterization of resistance to Pepper Huasteco Geminivirus in chili peppers from Yucatan, Mexico. *HortScience, 36*, 139–142.

Golnaraghi, A. R., Hamedi, A., Yazdani-Khameneh, S., & Khosroshahi, T. S. (2013). First report of a natural occurrence of tomato yellow fruit ring virus on pepper in Iran. *Plant Disease, 97*, 1259.

Gong, D., Wang, J.-H., Lin, Z.-S., Zhang, S.-Y., Zhang, Y.-L., Yu, N.-T., et al. (2011). Genomic sequencing and analysis of *Chilli ringspot virus*, a novel potyvirus. *Virus Genes, 43*, 439–444.

González-Pacheco, B. E., & Silva-Rosales, L. (2013). First report of *Impatiens necrotic spot virus* in Mexico in tomatillo and pepper plants. *Plant Disease, 97*, 1124.

Goto, T., Iizuka, N., & Komochi, S. (1984). Selection and utilization of an attenuated isolate of pepper strain of tobacco mosaic virus. *Annals of the Phytopathological Society of Japan, 50*, 221–228.

Gracia, O., & Feldman, J. M. (1974). Tobacco streak virus in pepper. *Journal of Phytopathology, 80*, 313–323.

Green, S. K., Hwang, J. T., & Chou, J. C. (1996). Evaluation of selected *Lycopersicon* and *Capsicum* germplasm for watermelon silver mottle tospovirus resistance. *Plant Disease, 80*, 824.

Green, S. K., & Kim, J. S. (1991). *Characteristics and control of viruses infecting peppers: A literature review*. Technical Bulletin No. 18 (p. 60). Shanhua, Taiwan: Asian Vegetable Research and Development Center.

Grube, R. C., Blauth, J. R., Arnedo Andrés, M. S., Caranta, C., & Jahn, M. K. (2000). Identification and comparative mapping of a dominant potyvirus resistance gene cluster in *Capsicum*. *TAG Theoretical and Applied Genetics, 101*, 852–859.

Grube, R. C., Radwanski, E. R., & Jahn, M. (2000). Comparative genetics of disease resistance within the Solanaceae. *Genetics, 155*, 873–887.

Guerini, M. N., & Murphy, J. F. (1999). Resistance of *Capsicum annuum* 'Avelar' to pepper mottle potyvirus and alleviation of this resistance by co-infection with cucumber mosaic cucumovirus are associated with virus movement. *Journal of General Virology, 80*, 2785–2792.

Gunn, L. V., & Pares, R. D. (1990). Capsicum yellows—A disease induced by a luteovirus in glasshouse peppers (*Capsicum annuum*) in Australia. *Journal of Phytopathology, 129*, 210–216.

Ha, C., Coombs, S., Revill, P., Harding, R., Vu, M., & Dale, J. (2008). Molecular characterization of begomoviruses and DNA satellites from Vietnam: Additional evidence that the New World geminiviruses were present in the Old World prior to continental separation. *Journal of General Virology, 89*, 312–326.

Ha, C., Revill, P., Harding, R. M., Vu, M., & Dale, J. L. (2008). Identification and sequence analysis of potyviruses infecting crops in Vietnam. *Archives of Virology, 153*, 45–60.

Han, J.-H., Choi, H.-S., Kim, D. H., Lee, H.-R., & Kim, B.-D. (2006). Biological, physical and cytological properties of *Pepper mottle virus*-SNU1 and its RT-PCR detection. *Plant Pathology Journal, 22*, 155–160.

Hanssen, I. M., Lapidot, M., & Thomma, B. P. H. J. (2010). Emerging viral diseases of tomato crops. *Molecular Plant-Microbe Interactions, 23*, 539–548.

Hoang, N. H., Kang, W.-H., Yang, H.-B., Kim, S.-b., Yeom, S.-I., Kwon, J.-K., et al. (2014). P405: Fine mapping of the *Tomato spotted wilt virus* resistance gene, *Tsw*, in *Capsicum*. In *International plant & animal genome XXII, San Diego, CA, USA/January 10–15*. https://pag.confex.com/pag/xxii/webprogram/Paper12010.htm.

Holmes, F. O. (1937). Inheritance of resistance to tobacco mosaic disease in the pepper. *Phytopathology, 27*, 637–642.

Hwang, J., Li, J., Liu, W.-Y., An, S.-J., Cho, H., Her, N., et al. (2009). Double mutations in eIF4E and eIFiso4E confer recessive resistance to *Chilli veinal mottle virus* in pepper. *Molecules and Cells, 27*, 329–336.

Ibiza, V., Blanca, J., Cañizares, J., & Nuez, F. (2012). Taxonomy and genetic diversity of domesticated *Capsicum* species in the Andean region. *Genetic Resources and Crop Evolution, 59*, 1077–1088.

Ibiza, V., Canizares, J., & Nuez, F. (2010). EcoTILLING in *Capsicum* species: Searching for new virus resistances. *BMC Genomics, 11*, 631.

Inoue-Nagata, A. K., Fonseca, M. E. N., Resende, R. O., Boiteux, L. S., Monte, D. C., Dusi, A. N., et al. (2002). *Pepper yellow mosaic virus*, a new potyvirus in sweet pepper, *Capsicum annuum*. *Archives of Virology, 147*, 849–855.

Jahn, M., Paran, I., Hoffmann, K., Radwanski, E. R., Livingstone, K. D., Grube, R. C., et al. (2000). Genetic mapping of the *Tsw* locus for resistance to the *Tospovirus Tomato spotted wilt virus* in *Capsicum* spp. and its relationship to the *Sw-5* gene for resistance to the same pathogen in tomato. *Molecular Plant-Microbe Interactions, 13*, 673–682.

Jain, R. K., Bag, S., & Awasthi, L. P. (2005). First report of natural infection of *Capsicum annuum* by *Tobacco streak virus* in India. *Plant Pathology, 54*, 257.

James, C. (2013). *Global status of commercialized biotech/GM crops: 2013*. ISAAA Brief No. 46, Ithaca, NY: ISAAA.

Janssen, D., Saez, E., Segundo, E., Martín, G., Gil, F., & Cuadrado, I. M. (2005). *Capsicum annuum*—a new host of *Parietaria mottle virus* in Spain. *Plant Pathology, 54*, 567.

Janzac, B., Fabre, M.-F., Palloix, A., & Moury, B. (2008). Characterization of a new potyvirus infecting pepper crops in Ecuador. *Archives of Virology, 153*, 1543–1548.

Janzac, B., Fabre, M. F., Palloix, A., & Moury, B. (2009). Phenotype and spectrum of action of the *Pvr4* resistance in pepper against potyviruses, and selection for virulent variants. *Plant Pathology, 58*, 443–449.

Janzac, B., Montarry, J., Palloix, A., Navaud, O., & Moury, B. (2010). A point mutation in the polymerase of *Potato virus Y* confers virulence toward the *Pvr4* resistance of pepper and a high competitiveness cost in susceptible cultivar. *Molecular Plant-Microbe Interactions, 23*, 823–830.

Jeong, H.-J., Kwon, J.-K., Pandeya, D., Hwang, J., Hoang, N., Bae, J.-H., et al. (2012). A survey of natural and ethyl methane sulfonate-induced variations of *eIF4E* using high-resolution melting analysis in *Capsicum*. *Molecular Breeding*, *29*, 349–360.

Juarez, M., Legua, P., Mengual, C. M., Kassem, M. A., Sempere, R. N., Gómez, P., et al. (2013). Relative incidence, spatial distribution and genetic diversity of cucurbit viruses in eastern Spain. *Annals of Applied Biology*, *162*, 362–370.

Kang, W.-H., Hoang, N., Yang, H.-B., Kwon, J.-K., Jo, S.-H., Seo, J.-K., et al. (2010). Molecular mapping and characterization of a single dominant gene controlling CMV resistance in peppers (*Capsicum annuum* L.). *TAG Theoretical and Applied Genetics*, *120*, 1587–1596.

Kang, W.-H., Seo, J.-K., Chung, B. N., Kim, K.-H., & Kang, B.-C. (2012). Helicase domain encoded by *Cucumber mosaic virus* RNA1 determines systemic infection of *Cmr1* in pepper. *PLoS One*, *7*, e43136.

Kang, B. C., Yeam, I., Frantz, J. D., Murphy, J. F., & Jahn, M. M. (2005). The *pvr1* locus in *Capsicum* encodes a translation initiation factor eIF4E that interacts with *Tobacco etch virus* VPg. *Plant Journal*, *42*, 392–405.

Kang, B.-C., Yeam, I., Li, H., Perez, K. W., & Jahn, M. M. (2007). Ectopic expression of a recessive resistance gene generates dominant potyvirus resistance in plants. *Plant Biotechnology Journal*, *5*, 526–536.

Kenyon, L., Tsai, W.-S., Shih, S.-L., & Lee, L.-M. (2014). Emergence and diversity of begomoviruses infecting solanaceous crops in East and Southeast Asia. *Virus Research*. http://dx.doi.org/10.1016/j.virusres.2013.12.026.

Kim, H., Han, J.-H., Kim, S., Lee, H., Shin, J.-S., Kim, J.-H., et al. (2011). Trichome density of main stem is tightly linked to PepMoV resistance in chili pepper (*Capsicum annuum* L.). *TAG Theoretical and Applied Genetics*, *122*, 1051–1058.

Kim, M., Kim, M., Hong, J., Choi, J., & Ryu, K. (2010). Patterns in disease progress and the influence of single and multiple viral infections on pepper (*Capsicum annuum* L.) growth. *European Journal of Plant Pathology*, *127*, 53–61.

Kim, S. J., Lee, S. J., Kim, B. D., & Paek, K. H. (1997). Satellite-RNA-mediated resistance to cucumber mosaic virus in transgenic plants of hot pepper (*Capsicum annuum* cv. Golden Tower). *Plant Cell Reports*, *16*, 825–830.

Knierim, D., Tsai, W.-S., & Kenyon, L. (2013). Analysis of sequences from field samples reveals the presence of the recently described pepper vein yellows virus (genus *Polerovirus*) in six additional countries. *Archives of Virology*, *158*, 1337–1341.

Kumar, S., Kumar, S., Singh, M., Singh, A. K., & Rai, M. (2006). Identification of host plant resistance to pepper leaf curl virus in chilli (*Capsicum* species). *Scientia Horticulturae*, *110*, 359–361.

Kunkalikar, S., Sudarsana, P., Rajagopalan, P., Zehr, U. B., Naidu, R. A., & Kankanallu, R. S. (2007). *First report of Capsicum chlorosis virus in tomato in India*. (online). http://dx.doi.org/10.1094/PHP-2007-1204-01-BR(online).

Kyle, M. M., & Palloix, A. (1997). Proposed revision of nomenclature for potyvirus resistance genes in *Capsicum*. *Euphytica*, *97*, 183–188.

Lee, U., Hong, J. S., Choi, J. K., Kim, K. C., Kim, Y. S., Curtis, I. S., et al. (2000). *Broad bean wilt virus* causes necrotic symptoms and generates defective RNAs in *Capsicum annuum*. *Phytopathology*, *90*, 1390–1395.

Lee, J. M., Jahn, M. M., & Yeam, I. (2013). Allelic relationships at the *pvr1* locus in *Capsicum annuum*. *Euphytica*, *194*, 417–424.

Lee, Y. H., Jung, M., Shin, S. H., Lee, J. H., Choi, S. H., Her, N. H., et al. (2009). Transgenic peppers that are highly tolerant to a new CMV pathotype. *Plant Cell Reports*, *28*, 223–232.

Lee, Y. H., Kim, H. S., Kim, J. Y., Jung, M., Park, Y. S., Lee, J. S., et al. (2004). A new selection method for pepper transformation: Callus-mediated shoot formation. *Plant Cell Reports*, *23*, 50–58.

Lefeuvre, P., Martin, D. P., Harkins, G., Lemey, P., Gray, A. J. A., Meredith, S., et al. (2010). The spread of tomato yellow leaf curl virus from the Middle East to the World. *PLoS Pathogens, 6*, e1001164.

Li, J.-T., Yeh, Y.-C., Yeh, S.-D., Raja, J. J., Rajagopalan, P., Liu, L.-Y., et al. (2011). Complete genomic sequence of watermelon bud necrosis virus. *Archives of Virology, 156*, 359–362.

Lin, C.-Y., Ku, H.-M., Tsai, W.-S., Green, S., & Jan, F.-J. (2011). Resistance to a DNA and a RNA virus in transgenic plants by using a single chimeric transgene construct. *Transgenic Research, 20*, 261–270.

Loebenstein, G., Alper, M., & Levy, S. (1970). Field tests with oil sprays for the prevention of aphid-spread viruses in peppers. *Phytopathology, 60*, 212–215.

Londoño, A., Capobianco, H., Zhang, S., & Polston, J. E. (2012). First record of *Tomato chlorotic spot virus* in the USA. *Tropical Plant Pathology, 37*, 333–338.

Lozano, G., Moriones, E., & Navas-Castillo, J. (2004). First report of sweet pepper (*Capsicum annuum*) as a natural host plant for *Tomato chlorosis virus*. *Plant Disease, 88*, 224.

Lucinda, N., da Rocha, W., Inoue-Nagata, A., & Nagata, T. (2012). Complete genome sequence of pepper yellow mosaic virus, a potyvirus, occurring in Brazil. *Archives of Virology, 157*, 1397–1401.

Maharijaya, A., Vosman, B., Steenhuis-Broers, G., Harpenas, A., Purwito, A., Visser, R., et al. (2011). Screening of pepper accessions for resistance against two thrips species (*Frankliniella occidentalis* and *Thrips parvispinus*). *Euphytica, 177*, 401–410.

Mandal, B., Jain, R. K., Krishnareddy, M., Krishna Kumar, N. K., Ravi, K. S., & Pappu, H. R. (2012). Emerging problems of tospoviruses (*Bunyaviridae*) and their management in the Indian subcontinent. *Plant Disease, 96*, 468–479.

Margaria, P., Ciuffo, M., & Turina, M. (2004). Resistance breaking strain of *Tomato spotted wilt virus* (*Tospovirus*; *Bunyaviridae*) on resistant pepper cultivars in Almeria, Spain. *Plant Pathology, 53*, 795.

Maris, P. C., Joosten, N. N., Goldbach, R. W., & Peters, D. (2003). Restricted spread of *Tomato spotted wilt virus* in thrips-resistant pepper. *Phytopathology, 93*, 1223–1227.

Maris, P. C., Joosten, N. N., Peters, D., & Goldbach, R. W. (2003). Thrips resistance in pepper and its consequences for the acquisition and inoculation of *Tomato spotted wilt virus* by the western flower thrips. *Phytopathology, 93*, 96–101.

Mariyono, J., & Bhattarai, M. (2009). *Chili production practices in Central Java, Indonesia: A baseline report.* Shanhua, Tainan, Taiwan: AVRDC—The World Vegetable Center. 75p, http://203.64.245.61/fulltext_pdf/EB/2001-2010/eb0118.pdf.

Martínez-Ayala, A., Sánchez-Campos, S., Cáceres, F., Aragón-Caballero, L., Navas-Castillo, J., & Moriones, E. (2014). Characterisation and genetic diversity of pepper leafroll virus, a new bipartite begomovirus infecting pepper, bean and tomato in Peru. *Annals of Applied Biology, 164*, 62–72.

McDonald, S. A., Halbert, S. E., Tolin, S. A., & Nault, B. A. (2003). Seasonal abundance and diversity of aphids (Homoptera: *Aphididae*) in a pepper production region in Jamaica. *Environmental Entomology, 32*, 499–509.

McMichael, L. A., Persley, D. M., & Thomas, J. E. (2002). A new tospovirus serogroup IV species infecting capsicum and tomato in Queensland, Australia. *Australasian Plant Pathology, 31*, 231–239.

Medina-Hernández, D., Rivera-Bustamante, R., Tenllado, F., & Holguín-Peña, R. (2013). Effects and effectiveness of two RNAi constructs for resistance to *Pepper golden mosaic virus* in *Nicotiana benthamiana* plants. *Viruses, 5*, 2931–2945.

Melgarejo, T. A., Alminaite, A., Fribourg, C., Spetz, C., & Valkonen, J. P. T. (2004). Strains of Peru tomato virus infecting cocona (*Solanum sessiliflorum*), tomato and pepper in Peru with reference to genome evolution in genus Potyvirus. *Archives of Virology, 149*, 2025–2034.

Melgarejo, T. A., Kon, T., Rojas, M. R., Paz-Carrasco, L., Zerbini, F. M., & Gilbertson, R. L. (2013). Characterization of a New World monopartite begomovirus causing leaf curl disease of tomato in Ecuador and Peru reveals a new direction in geminivirus evolution. *Journal of Virology*, *87*, 5397–5413. http://dx.doi.org/10.1128/JVI.00234-13.

Melzer, M. J., Shimabukuro, J., Long, M. H., Nelson, S. C., Alvarez, A. M., Borth, W. B., et al. (2014). First report of Capsicum chlorosis virus infecting waxflower (*Hoya calycina* Schlecter) in the United States. *Plant Disease*, *98*, 571.

Mizumoto, H., Nakamura, I., Shimomoto, Y., Sawada, H., Tomita, R., Sekine, K.-T., et al. (2012). Amino acids in *Tobamovirus* coat protein controlling pepper L^{1a} gene-mediated resistance. *Molecular Plant Pathology*, *13*, 915–922.

Morilla, G., Janssen, D., Garcia-Andres, S., Moriones, E., Cuadrado, I. M., & Bejarano, E. R. (2005). Pepper (*Capsicum annuum*) is a dead-end host for *Tomato yellow leaf curl virus*. *Phytopathology*, *95*, 1089–1097.

Moury, B. (2010). A new lineage sheds light on the evolutionary history of *Potato virus Y*. *Molecular Plant Pathology*, *11*, 161–168.

Moury, B., & Verdin, E. (2012). Viruses of pepper crops in the Mediterranean basin: A remarkable stasis. In G. Loebenstein, & H. Lecoq (Eds.), *Advances in virus research* (pp. 127–162). Burlington: Academic Press (Chapter 4).

Murakami, R., Nakashima, N., Hinomoto, N., Kawano, S., & Toyosato, T. (2011). The genome sequence of pepper vein yellows virus (family *Luteoviridae*, genus *Polerovirus*). *Archives of Virology*, *156*, 921–923.

Murphy, J. F., Blauth, J. R., Livingstone, K. D., Lackney, V. K., & Jahn, M. K. (1998). Genetic mapping of the *pvr1* Locus in *Capsicum* spp. and evidence that distinct potyvirus resistance loci control responses that differ at the whole plant and cellular levels. *Molecular Plant-Microbe Interactions*, *11*, 943–951.

Murphy, J. F., & Bowen, K. L. (2006). Synergistic disease in pepper caused by the mixed infection of *Cucumber mosaic virus* and *Pepper mottle virus*. *Phytopathology*, *96*, 240–247.

Ndunguru, J., & Kapooria, R. G. (1999). Identification and incidence of virus diseases of *Capsicum annuum* in the Lusaka Province of Zambia. *EPPO Bulletin*, *29*, 183–189.

Nelson, M. R., & Wheeler, R. E. (1978). Biological and serological characterization and separation of potyviruses that infect peppers. *Phytopathology*, *68*, 979–984.

Nono-Womdim, R., Swai, I. S., Chadha, M. L., Gebre-Selassie, K., & Marchoux, G. (2001). Occurrence of *Chilli veinal mottle virus* in *Solanum aethiopicum* in Tanzania. *Plant Disease*, *85*, 801.

Nyana, D. N., Suastika, G., Temaja, I. G. R. M., & Suprapta, D. N. (2012). Protective mild isolates of Cucumber mosaic virus obtained from chili pepper in Bali. *Agricultural Science Research Journal*, *2*, 280–284.

Ogai, R., Kanda-Hojo, A., & Tsuda, S. (2013). An attenuated isolate of *Pepper mild mottle virus* for cross protection of cultivated green pepper (*Capsicum annuum* L.) carrying the L^3 resistance gene. *Crop Protection*, *54*, 29–34.

Ogawa, Y., Hagiwara, K., Iwai, H., Izumi, S., & Arai, K. (2003). First report of *Pepper mottle virus* on *Capsicum annuum* in Japan. *Journal of General Plant Pathology*, *69*, 348–350.

Pagán, I., Betancourt, M., de Miguel, J., Piñero, D., Fraile, A., & García-Arenal, F. (2010). Genomic and biological characterization of chiltepín yellow mosaic virus, a new tymovirus infecting *Capsicum annuum* var. *aviculare* in Mexico. *Archives of Virology*, *155*, 675–684.

Palloix, A., Ayme, V., & Moury, B. (2009). Durability of plant major resistance genes to pathogens depends on the genetic background, experimental evidence and consequences for breeding strategies. *New Phytologist*, *183*, 190–199.

Pandey, S. K., Mathur, A. C., & Srivastava, M. (2010). Management of leaf curl disease of chilli (*Capsicum annuum* L.). *International Journal of Virology*, *6*, 246–250.

Parrella, G., Ruffel, S., Moretti, A., Morel, C., Palloix, A., & Caranta, C. (2002). Recessive resistance genes against potyviruses are localized in colinear genomic regions of the tomato (*Lycopersicon* spp.) and pepper (*Capsicum* spp.) genomes. *TAG Theoretical and Applied Genetics*, *105*, 855–861.

Pernezny, K. L., Roberts, P. D., Murphy, J. F., & Goldberg, N. P. (Eds.), (2003). *Compendium of pepper diseases*. St. Paul, MN: APS Press.

Persley, D. M., McGrath, D., Sharman, M., & Walker, I. O. (2010). Breeding for tospovirus resistance in a package of disease resistances for capsicum and tomato. (abstract). In *IXth international symposium on thysanoptera and tospoviruses, 31 August–4 September, 2009* (p. 37), Queensland, Australia: Sea World Resort. http://www.insectscience.org/10.166/abstract69.html.

Polston, J. E., Cohen, L., Sherwood, T. A., Ben-Joseph, R., & Lapidot, M. (2006). *Capsicum* species: Symptomless hosts and reservoirs of *Tomato yellow leaf curl virus*. *Phytopathology*, *96*, 447–452.

Premachandra, W. T. S. D., Borgemeister, C., Maiss, E., Knierim, D., & Poehling, H. M. (2005). *Ceratothripoides claratris*, a new vector of a Capsicum chlorosis virus isolate infecting tomato in Thailand. *Phytopathology*, *95*, 659–663.

Puangmalai, P., Potapohn, N., Akarapisarn, A., Cheewachaiwit, S., & Insuan, N. (2013). Screening *Capsicum* accessions for tomato necrotic ringspot virus resistance. *Chiang Mai University Journal of Natural Sciences*, *12*, 35–42.

Puangmalai, P., Potapohn, N., Akarapisarn, A., & Pascha, H. J. (2013). Inheritance of tomato necrotic ring virus resistance in *Capsicum annuum*. *Journal of Agricultural Science*, *5*, 129–133.

Quiñones, M., Arana, F., Alfenas-Zerbini, P., Soto, M., Ribeiro, D., Diaz, A., et al. (2011). First report of *Pepper mottle virus* in sweet pepper in Cuba. *New Disease Reports*, *24*, 16.

Rabinowicz, P. D., Bravo-Almonacid, F. F., & Mentaberry, A. N. (1993). cDNA Sequence of the pepper severe mosaic virus coat protein gene. *Plant Physiology*, *103*, 1023.

Rai, V. P., Kumar, R., Singh, S. P., Kumar, S., Kumar, S., Singh, M., et al. (2014). Monogenic recessive resistance to *Pepper leaf curl virus* in an interspecific cross of *Capsicum*. *Scientia Horticulturae*, *172*, 34–38.

Rast, A. T. B., & Stijger, C. C. M. M. (1987). Disinfection of pepper seed infected with different strains of capsicum mosaic virus by trisodium phosphate and dry heat treatment. *Plant Pathology*, *36*, 583–588.

Renteria-Canett, I., Xoconostle-Cazares, B., Ruiz-Medrano, R., & Rivera-Bustamante, R. (2011). Geminivirus mixed infection on pepper plants: Synergistic interaction between PHYVV and PepGMV. *Virology Journal*, *8*, 104.

Roggero, P., Masenga, V., & Tavella, L. (2002). Field isolates of *Tomato spotted wilt virus* overcoming resistance in pepper and their spread to other hosts in Italy. *Plant Disease*, *86*, 950–954.

Rubio, M., Caranta, C., & Palloix, A. (2008). Functional markers for selection of potyvirus resistance alleles at the *pvr2-eIF4E* locus in pepper using tetra-primer ARMS–PCR. *Genome*, *51*, 767–771.

Rubio, L., Luis-Arteaga, M., Cambra, M., Serra, J., Moreno, P., & Guerri, J. (2002). First report of *Broad bean wilt virus 1* in Spain. *Plant Disease*, *86*, 698.

Rubio, M., Nicolai, M., Caranta, C., & Palloix, A. (2009). Allele mining in the pepper gene pool provided new complementation effects between pvr2-eIF4E and pvr6-eIF(iso)4E alleles for resistance to pepper veinal mottle virus. *Journal of General Virology*, *90*, 2808–2814.

Ruffel, S., Dussault, M.-H., Palloix, A., Moury, B., Bendahmane, A., Robaglia, C., et al. (2002). A natural recessive resistance gene against potato virus Y in pepper corresponds to the eukaryotic initiation factor 4E (eIF4E). *The Plant Journal*, *32*, 1067–1075.

Ruffel, S., Gallois, J.-L., Moury, B., Robaglia, C., Palloix, A., & Caranta, C. (2006). Simultaneous mutations in translation initiation factors eIF4E and eIF(iso)4E are required to prevent pepper veinal mottle virus infection of pepper. *Journal of General Virology*, *87*, 2089–2098.

Ryu, J.-G., Ko, S.-J., Lee, Y.-H., Kim, M.-K., Kim, K.-H., Kim, H.-T., et al. (2009). Incidence and distribution of virus diseases on paprika (*Capsicum annuum* var. *grossum*) in jeonnam province of Korea. *The Plant Pathology Journal*, *25*, 95–98.

Satyanarayana, T., Gowda, S., Lakshminarayana Reddy, K., Mitchell, S. E., Dawson, W. O., & Reddy, D. V. R. (1998). Peanut yellow spot virus is a member of a new serogroup of Tospovirus genus based on small (S) RNA sequence and organization. *Archives of Virology*, *143*, 353–364.

Satyanarayana, T., Mitchell, S. E., Reddy, D. V. R., Kresovich, S., Jarret, R., Naidu, R. A., et al. (1996). The complete nucleotide sequence and genome organization of the M RNA segment of peanut bud necrosis tospovirus and comparison with other tospoviruses. *Journal of General Virology*, *77*, 2347–2352.

Sawada, H., Takeuchi, S., Hamada, H., Kiba, A., Matsumoto, M., & Hikichi, Y. (2004). A new tobamovirus-resistance gene, L^{1a}, of sweet pepper (*Capsicum annuum* L.). *Journal of the Japanese Society for Horticultural Science*, *73*, 552–557.

Sawada, H., Takeuchi, S., Matsumoto, K., Hamada, H., Kiba, A., Matsumoto, M., et al. (2005). A new *Tobamovirus*-resistance gene, *Hk*, in *Capsicum annuum*. *Journal of the Japanese Society for Horticultural Science*, *74*, 289–294.

Sdoodee, R., & Teakle, D. S. (1987). Transmission of tobacco streak virus by *Thrips tabach* a new method of plant virus transmission. *Plant Pathology*, *36*, 377–380.

Seepiban, C., Gajanandana, O., Attathom, T., & Attathom, S. (2011). Tomato necrotic ringspot virus, a new tospovirus isolated in Thailand. *Archives of Virology*, *156*, 263–274.

Shah, H., Yasmin, T., Fahim, M., Hameed, S., Haque, I.-U., Munir, M., et al. (2011). Reaction of exotic and indigenous *Capsicum* genotypes against Pakistani isolates of *Chili veinal mottle virus*. *Pakistan Journal of Botany*, *43*, 1707–1711.

Sharman, M., & Persley, D. M. (2006). Field isolates of *Tomato spotted wilt virus* overcoming resistance in capsicum in Australia. *Australasian Plant Pathology*, *35*, 123–128.

Shin, R., Han, J.-H., Lee, G.-J., & Peak, K.-H. (2002). The potential use of a viral coat protein gene as a transgene screening marker and multiple virus resistance of pepper plants coexpressing coat proteins of *Cucumber mosaic virus* and *Tomato mosaic virus*. *Transgenic Research*, *11*, 215–219.

Shin, R., Park, J. M., An, J.-M., & Paek, K.-H. (2002). Ectopic expression of Tsi1 in transgenic hot pepper plants enhances host resistance to viral, bacterial, and oomycete pathogens. *Molecular Plant-Microbe Interactions*, *15*, 983–989.

Sokhansanj, Y., Rakhshandehroo, F., & Pourrahim, R. (2012). First report of *Tomato ringspot virus* infecting pepper in Iran. *Plant Disease*, *96*, 1828.

Spetz, C., Taboada, A. M., Darwich, S., Ramsell, J., Salazar, L. F., & Valkonen, J. P. T. (2003). Molecular resolution of a complex of potyviruses infecting solanaceous crops at the centre of origin in Peru. *Journal of General Virology*, *84*, 2565–2578.

Suzuki, K., Kuroda, T., Miura, Y., & Murai, J. (2003). Screening and field trials of virus resistant sources in Capsicum spp. *Plant Disease*, *87*, 779–783.

Svoboda, J., & Leisova-Svobodova, L. (2013). First report of *Broad bean wilt virus-2* in pepper in the Czech Republic. *Plant Disease*, *97*, 1261.

Syller, J. (2012). Facilitative and antagonistic interactions between plant viruses in mixed infections. *Molecular Plant Pathology*, *13*, 204–216.

Tomita, R., Murai, J., Miura, Y., Ishihara, H., Liu, S., Kubotera, Y., et al. (2008). Fine mapping and DNA fiber FISH analysis locates the tobamovirus resistance gene L^3 of *Capsicum chinense* in a 400-kb region of R-like genes cluster embedded in highly repetitive sequences. *TAG Theoretical and Applied Genetics*, *117*, 1107–1118.

Tomita, R., Sekine, K.-T., Mizumoto, H., Sakamoto, M., Murai, J., Kiba, A., et al. (2011). Genetic basis for the hierarchical interaction between *Tobamovirus* spp. and *L* resistance gene alleles from different pepper species. *Molecular Plant-Microbe Interactions*, *24*, 108–117.

Torres, R., Larenas, J., Fribourg, C., & Romero, J. (2012). Pepper necrotic spot virus, a new tospovirus infecting solanaceous crops in Peru. *Archives of Virology*, *157*, 609–615.

Tsai, W. S., Abdourhamane, I. K., & Kenyon, L. (2010). First report of *Pepper veinal mottle virus* associated with mosaic and mottle diseases of tomato and pepper in Mali. *Plant Disease*, *94*, 378.

Tsai, W. S., Huang, Y. C., Zhang, D. Y., Reddy, K., Hidayat, S. H., Srithongchai, W., et al. (2008). Molecular characterization of the CP gene and 3′UTR of *Chilli veinal mottle virus* from South and Southeast Asia. *Plant Pathology*, *57*, 408–416.

Vargas, J. A., Hammond, R., Hernández, E., Barboza, N., Mora, F., & Ramírez, P. (2011). First report of *Tomato chlorosis virus* infecting sweet pepper in Costa Rica. *Plant Disease*, *95*, 1482.

Velasco, L., Janssen, D., Ruiz-Garcia, L., Segundo, E., & Cuadrado, I. M. (2002). The complete nucleotide sequence and development of a diferential detection assay for a pepper mild mottle virus (PMMoV) isolate that overcomes L3 resistance in pepper. *Journal of Virological Methods*, *106*, 135–140.

Villanueva, F., Castillo, P., Font, M. I., Alfaro-Fernández, A., Moriones, E., & Navas-Castillo, J. (2013). First report of *Pepper vein yellows virus* infecting sweet pepper in Spain. *Plant Disease*, *97*, 1261.

Wang, D., & Bosland, P. W. (2006). The genes of *Capsicum*. *HortScience*, *41*, 1169–1187.

Wang, J. H., Zhang, S. Y., Gong, D., Wu, Y. P., Zhang, Y. L., Yu, N. T., et al. (2012). First report of *Chilli ringspot virus* on chili pepper in China. *Plant Disease*, *96*, 462.

Webster, C. G., de Jensen, C. E., Rivera-Vargas, L. I., Rodrigues, J. C. V., Mercado, W., Frantz, G., et al. (2013). First report of Tomato chlorotic spot virus (TCSV) in tomato, pepper, and jimsonweed in Puerto Rico. *Plant Health Progress*. http://dx.doi.org/10.1094/PHP-2013-0812-01-BR.

Webster, C. G., Reitz, S. R., Perry, K. L., & Adkins, S. (2011). A natural M RNA reassortant arising from two species of plant- and insect-infecting bunyaviruses and comparison of its sequence and biological properties to parental species. *Virology*, *413*, 216–225.

Wintermantel, W. M., & Hladky, L. L. (2013). Complete genome sequence and biological characterization of *Moroccan pepper virus* (MPV) and reclassification of Lettuce necrotic stunt virus as MPV. *Phytopathology*, *103*, 501–508.

Wisler, G. C., Li, R. H., Liu, H. Y., Lowry, D. S., & Duffus, J. E. (1998). Tomato chlorosis virus: A new whitefly-transmitted, phloem-limited, bipartite closterovirus of tomato. *Phytopathology*, *88*, 402–409.

Yang, H.-B., Liu, W., Kang, W.-H., Jahn, M., & Kang, B.-C. (2009). Development of SNP markers linked to the *L* locus in *Capsicum* spp. by a comparative genetic analysis. *Molecular Breeding*, *24*, 433–446.

Yang, H.-B., Liu, W.-Y., Kang, W.-H., Kim, J.-H., Cho, H., Yoo, J.-H., et al. (2012). Development and validation of *L* allele-specific markers in *Capsicum*. *Molecular Breeding*, *30*, 819–829.

Yao, M., Li, N., Wang, F., & Ye, Z. (2013). Genetic analysis and identification of QTLs for resistance to *cucumber mosaic virus* in chili pepper (*Capsicum annuum* L.). *Euphytica*, *193*, 135–145.

Yeam, I., Cavatorta, J. R., Ripoll, D. R., Kang, B.-C., & Jahn, M. M. (2007). Functional dissection of naturally occurring amino acid substitutions in eIF4E that confers recessive Potyvirus resistance in plants. *Plant Cell*, *19*, 2913–2928.

Yeam, I., Kang, B.-C., Lindeman, W., Frantz, J., Faber, N., & Jahn, M. (2005). Allele-specific CAPS markers based on point mutations in resistance alleles at the *pvr1* locus encoding eIF4E in *Capsicum*. *TAG Theoretical and Applied Genetics*, *112*, 178–186.

Yeh, S.-D., & Chang, T.-F. (1995). Nucleotide sequence of the N gene of watermelon silver mottle virus, a proposed new member of the genus *Tospovirus*. *Phytopathology*, *85*, 58–64.
Yonaha, T., Toyosato, T., Kawano, S., & Osaki, T. (1995). Pepper vein yellows virus, a novel luteovirus from bell pepper plants in Japan. *Annals of the Phytopathological Society of Japan*, *61*, 178–184.
Zheng, Y.-X., Chen, C.-C., Yang, C.-J., Yeh, S.-D., & Jan, F.-J. (2008). Identification and characterization of a tospovirus causing chlorotic ringspots on *Phalaenopsis* orchids. *European Journal of Plant Pathology*, *120*, 199–209.

CHAPTER SEVEN

Control of Virus Diseases in Soybeans

John H. Hill[1], Steven A. Whitham
Department of Plant Pathology and Microbiology, Iowa State University, Ames, Iowa, USA
[1]Corresponding author: e-mail address: johnhill@iastate.edu

Contents

Advances in Virus Research, Volume 90
ISSN 0065-3527
http://dx.doi.org/10.1016/B978-0-12-801246-8.00007-X

Abstract

Soybean, one of the world's most important sources of animal feed and vegetable oil, can be infected by numerous viruses. However, only a small number of the viruses that can potentially infect soybean are considered as major economic problems to soybean production. Therefore, we consider management options available to control diseases caused by eight viruses that cause, or have the potential to cause, significant economic loss to producers. We summarize management tactics in use and suggest direction for the future. Clearly, the most important tactic is disease resistance. Several resistance genes are available for three of the eight viruses discussed. Other options include use of virus-free seed and avoidance of alternative virus hosts when planting. Attempts at arthropod vector control have generally not provided consistent disease management. In the future, disease management will be considerably enhanced by knowledge of the interaction between soybean and viral proteins. Identification of genes required for soybean defense may represent key regulatory hubs that will enhance or broaden the spectrum of basal resistance to viruses. It may be possible to create new recessive or dominant negative alleles of host proteins that do not support viral functions but perform normal cellular function. The future approach to virus control based on gene editing or exploiting allelic diversity points to necessary research into soybean–virus interactions. This will help to generate the knowledge needed for rational design of durable resistance that will maximize global production.

1. INTRODUCTION

Soybean, *Glycine max,* is processed to be the world's most important source of animal protein feed and vegetable oil. In the United States, soybeans are the most planted field crop after corn. The United States has the highest land acreage under soybean cultivation with 77.2 million acres (31.2 million hectares) devoted to production in 2012. Brazil is close behind with 68.2 million acres (27.6 million hectares) followed by Argentina at 47.9 million acres (19.4 million hectares). According to 2013 USDA data, the major global producers are the United States, Brazil, Argentina, China, India, Paraguay, Canada, Uruguay, and Bolivia (United States Department of Agriculture, Economics Research Service, 2013).

As one of the first bioengineered crops to achieve commercial success, soybeans have become very popular among producers because of reduced production costs. However, because of currently held restrictions against bioengineered soybeans in Europe and Japan, there are potential ramifications for producers for resource use, marketing, and international trade. Although initial focus has been on traits that reduce production costs (e.g., herbicide resistance), enhanced functionality characteristics such as pest resistance, and high oleic, high stearate, and increased omega-3 will likely emerge. Currently, 94% of the soybeans grown in the United States are bioengineered while 89% and 100% of the soybeans grown in Brazil and Argentina, respectively, are bioengineered (United States Department of Agriculture, Economics Research Service, 2013).

Historically, soybeans were likely first domesticated in the eastern half of north China around the eleventh century B.C. The five main plant foods of China included soybeans along with rice, wheat, barley, and millet. Production was localized in China until after the Chinese–Japanese war of 1894–1895 when soybean oil cake was imported by the Japanese for use as fertilizer. However, it is believed that some seed may have been sent from China as early as 1740 and planted in France. In 1908, shipments of soybean to Europe were made and the soybeans began to attract worldwide attention (Gibson & Benson, 2005).

Soybeans were first planted in the United States on a farm in Thunderbolt, Georgia by Henry Younge in 1765. During the 1880s, they were sometimes brought to the United States as inexpensive ballast that was later dumped to make room for cargo. Later, they were grown to make soy sauce or for use as "coffee berries" to brew coffee during the Civil War. The first reference that soybeans were tested by an agricultural school came from an 1879 report by the Rutgers Agricultural College in New Jersey. Early in the United States, soybeans were grown as a forage crop, but that changed in 1904 when George Washington Carver began studies of the soybean at the Tuskegee Institute in Alabama. Soybean acreage in the United States grew very slowly until World War II and soybean fields in China were devastated by an internal revolution in the 1940s. Henry Ford developed soybean-based plastic in the 1940s and even built, in 1941, a prototype vehicle titled the "Soybean Car" out of such plastics (Hill, 2003; Woodyard, 2010).

Disruption of trade routes during World War II forced the United States to look for alternatives to edible fats and oils. This resulted in rapid expansion of soybean production with the result that the United States is among the world's largest producer and exporter of soybeans along with Brazil and Argentina. Main importers include China, the European Union, Japan, Mexico, the Philippines, and several Latin American countries.

Soybeans constitute the leading oil seed crop constituting 57% of the world's production. Soybeans are grown primarily for oil (ca. 19%) and protein (ca. 40%) contained in soybean meal. The protein is considered as a complete protein, which means that it contains significant amounts of all essential amino acids necessary for humans. Processing of soybeans results in a wide variety of products including adhesives, asphalts, resins, cosmetics, inks, paints, plastics, polyesters, textile fibers, salad dressings, meat substitutes, beverage powders, nondairy creamer, infant formulas, breads, breakfast cereals, pastas, tofu, and pet foods.

Improved management practices, including improved seed varieties as well as fertilizer and pesticide applications, have resulted in improved varieties with disease resistance that have increased yields and encouraged expansion of soybean acreage in the United States. However, disease caused by virus infection continues to result in significant loss to producers. For example in 2010, loss caused by virus disease in the United States was calculated to approximate $35 million from data provided by the USDA Economic Research Service and disease loss estimates of Wrather and Koenning (United States Department of Agriculture, Economics Research Service, 2013; Wrather & Koenning, 2010). It is likely the estimated loss due to virus disease may be conservative because numerous viruses are latent while still causing loss. The loss estimate due to virus disease would likely be much higher without disease management practices when available. For virus diseases of soybean, these include management of potential for seed transmission, avoidance of planting adjacent to alternative virus host plans, deployment of resistance genes when available, and use of insecticides, which may also result in reduced populations of nontarget pests which happen to be virus vectors.

A large number of viruses have been identified that can infect soybean (Table 7.1). These are reported to occur either naturally or through laboratory inoculation of potential host plants. In this chapter, we focus on virus characteristics that are important for management of those viruses which are generally considered to be the most economically important or widespread.

2. SOYBEAN MOSAIC VIRUS

2.1. Biology

Soybean mosaic virus (SMV) is a member of the plant virus family *Potyvirideae* and is the cause of soybean mosaic disease, which occurs in all soybean production areas of the world. The virus was first reported from Connecticut in 1915 and described in about 1921(Gardner & Kendrick, 1921). Reported yield reductions range from 8% to 35%; however, losses as high as 94% have been recorded (Ross, 1977). Yield reductions vary with

Table 7.1 Virus diseases of soybean
Genus type

Family		
Genus	**Viruses in field**	**By Inoculation**
Single-stranded RNA, positive sense		
Secoviridae		
Comovirus	Bean pod mottle	Broad bean stain
	Broad bean true mosaic	Cowpea mottle
	Cowpea mosaic	Glycine mosaic
	Cowpea severe mosaic	Pea mild mosaic
	Quail pea mosaic	Red clover mottle
Fabavirus	Broad bean wilt	
Nepovirus	Soybean severe stunt	Arabis mosaic
	Tobacco ringspot	Cacao necrosis
	Tomato ringspot	Cherry leaf roll
		Grapevine fanleaf
		Mulberry ringspot
		Raspberry ringspot
		Tomato blackring
Potyviridae		
Potyvirus	Azuki bean mosaic	Beet mosaic
	Bean common mosaic	Cassia yellow spot
	Bean yellow mosaic	Chickpea distortion mosaic
	Blackeye cowpea mosaic	Clover yellow vein
	Passion fruit woodiness	Cowpea aphid-borne mosaic
	Peanut chlorotic ring mottle	Kennedya Y
	Peanut mottle	Peanut green mosaic
	Peanut stripe	Watermelon mosaic
	Soybean mosaic	Wisteria vein mosaic

Continued

Table 7.1 Virus diseases of soybean—cont'd
Genus type

Family		
Genus	**Viruses in field**	**By Inoculation**
Unnamed		
Sobemovirus	Southern bean mosaic	Subterranean clover mottle
	Soybean yellow common mosaic	
Luteoviridae		
Luteovirus	Bean leaf roll	Chickpea stunt
	Soybean dwarf (subterranean clover red leaf)	Pea leafroll
Polerovirus	Beat western yellows	
Enamovirus	Pea enation mosaic	
Unnamed		
Umbravirus		Groundnut rosette
Tombusviridae		
Carmovirus	Black gram mottle	Bean mild mosaic
	Soybean yellow mottle mosaic	Cowpea mottle
		Glycine mottle
		Pea stem necrosis
		Pelargonium line pattern
		Tephrosia symptomless
Tombusvirus		Cymbidium ringspot
Necrovirus		Tobacco necrosis
Dianthovirus		Red clover necrotic mottle
Virgaviridae		
Tobamovirus	Tobacco mosaic	
	Sunn-hemp mosaic	
Tobravirus	Tobacco rattle (soybean fleck)	Pea early browning
Pecluvirus		Peanut clump

Table 7.1 Virus diseases of soybean—cont'd

Genus type		
Family		
Genus	**Viruses in field**	**By Inoculation**
Bromoviridae		
Alfamovirus	Alfalfa mosaic	
Bromovirus	Cowpea chlorotic mottle (bean yellow stipple)	Broad bean mottle
		Cassia yellow blotch
		Spring beauty latent
Cucumovirus	Cucumber mosaic (soybean stunt)	Tomato aspermy
	Peanut stunt	
Ilarvirus	Tobacco streak	Asparagus virus 2
		Sunflower ringspot
Closterovirideae		
Closterovirus		Cassia severe mosaic
Tymoviridae		
Tymovirus		Clitoria yellow vein
		Ononis yellow mosaic
		Abelia latent
		Kennedya yellow mosaic
		Plantago mottle
Betaflexivirdae		
Carlavirus	Cowpea mild mottle	Cassia mild mottle
		Cole latent
		Pea streak
		Poplar mosaic
		Red clover vein mosaic

Continued

Table 7.1 Virus diseases of soybean—cont'd

Genus type		
Family		
Genus	**Viruses in field**	**By Inoculation**
Alphaflexiviridae		
Potexvirus		Asparagus 3 virus
		Clover yellow mosaic
		Commelina X
		Crotalaria yellow mosaic
		Foxtail mosaic
		Pea wilt
		Narcissus mosaic
		White clover mosaic
Unnamed		
	Soybean yellow vein	
Single-stranded RNA, ambisense		
Bunyaviridae		
Tospovirus	Tomato spotted wilt	Peanut yellow spot
	Soybean vein necrosis	
Single-stranded DNA		
Geminiviridae		
Begomovirus	Abutilon mosaic	Bean calico mosaic
	African soybean dwarf	Cotton leaf crumple
	Euphorbia mosaic	Tobacco leaf curl
	Horsegram yellow mosaic	
	Mungbean yellow mosaic-Sb	
	Rhynchosia mosaic	
	Soybean crinkle leaf	
Double-stranded DNA, reverse transcribing		
Caulimoviridae		
Caulimovirus	Soybean chlorotic mottle	Peanut chlorotic streak

Modified from Tolin (1999) and Hill (2003).

cultivar, virus strain, location, and time of inoculation. Early plant infection reduces pod set, increases seed coat mottling, and reduces seed size and weight, while late season infection has little effect on seed quality and yield. Additional effects of SMV include reduced seed size, oil content, and nodulation. SMV adversely affects nitrogen fixation and can reduce seed germination. Seed quality may be reduced as a result of seed coat mottling (Hill, Bailey, Benner, Tachibana, & Durand, 1987; Ross, 1968, 1969). Increased susceptibility to *Phomopsis* spp. resulting in poor seed quality and low vigor has been associated with SMV infection at or before soybean growth stage R2 (Hepperly, Bowers, Sinclair, & Goodman, 1979; Koning, TeKrony, Pfeiffer, & Ghabrial, 2001).

Symptoms vary with host genotype, virus strain, plant age at infection, and environment. Most infected cultivars are slightly stunted with fewer pods that are sometimes dwarfed and flattened, without hairs, and without seeds. Trifoliolate leaves have a mosaic of light and dark green areas that may later become raised or blistered, particularly along the main veins. Chlorosis may develop between the dark green areas. Leaf margins may be wavy or curled downward. Leaf veins do not grow together as is typical with herbicide injury. Primary leaves of some cultivars may show necrotic local lesions, which may merge into veinal necrosis followed by yellowing and leaf abscission. Some strains can induce severe stunting, systemic necrosis, leaf yellowing, petiole and stem necrosis, terminal necrosis, and defoliation usually leading to death. Different strains often produce different symptoms on the same cultivar. Rugosity is most severe in plants grown at about 18 °C. Symptoms are less severe at 24–25 °C and are largely masked above 30 °C. Temperature also influences the length of time between infection and symptom appearance, which ranges from 4 days at 29.5 to 14 days at 18.5 °C (Hill, 1999).

Depending on hilum color, seeds from infected plants may have a brown or black mottle that has been associated with suppression of posttranscriptional gene silencing of chalcone synthase by an SMV-encoded silencing suppressor protein. Seeds may be smaller and germination reduced as compared to seed from noninfected plants. Mottling does not indicate that the virus is present in seeds as not all mottled seeds contain virus and not all seeds from virus-infected plants are mottled. Further, plants not infected with SMV may also produce low levels of mottled seed. The correlation of percentage of infected seed with mottled seeds and severity of mottling varies from year-to-year, from cultivar-to-cultivar, and with virus strain (Hill et al., 1980) demonstrating the virus–soybean interaction, which occurs is governed by host genotype and virus strain (Domier et al., 2011).

Synergism has been reported to occur in plants infected with SMV and either BPMV or AMV. Reduction in yield of plants infected with both SMV and BPMV may be as high as 66–86% in susceptible cultivars; this compares with a reduction of 8–25% in cultivars inoculated with SMV. Disease symptoms may include severe dwarfing, foliage distortion, necrosis, and mottling of leaves in plants infected with both viruses. Doubly infected plants have fewer root nodules than singly infected plants (Calvert & Ghabrial, 1983; Malapi-Nelson, Wen, Ownley, & Hajimorad, 2009).

Numerous strains of the virus have been identified based upon reactions on a set of differential soybean cultivars. In the United States, strains are currently known as G1 through G7, G7a, and C14 (although the existence of G7a may be in doubt; Hajimorad and Hill, unpublished) (Buzzel & Tu, 1984; Cho & Goodman, 1979), and in Japan strains A–E (Takahashi, Tanaka, Wataru, & Tsuda, 1980). Subsequently, collection and analysis of SMV isolates from China resulted in the proposed establishment of 21 strains of SMV (Li, Yang, Yang, Zhi, & Gai, 2010) in addition to naturally occurring recombinant strains (Yang, Lin, Zheng, Zhang, & Zhi, 2014). Studies to investigate the correspondence between the Japanese and United States strains suggest that Japanese strain B is most similar to the United States strain G2 and that strain G3 may correspond to the Japanese strain A (Kanematsu, Eggenberger, & Hill, 1998). Other countries, including Korea and Brazil also report differentiation of SMV isolates using soybean cultivars (Almeida, 1981; Anjos, Lin, & Kitajima, 1985; Cho, Chung, & Lee, 1977; Lee, Kim, & Cho, 1992). The large number of reported strains reflects the variability of the virus, and most probably, the lack of understanding of virus effectors and resultant resistance mechanisms. It is highly probable that strains exist that have not been reported.

At least 32 aphid species, belonging to 15 different genera, transmit SMV in a nonpersistent manner. Some of the most important aphid species include *Acyrthosiphon pisum, Aphis fabae, A. glycines, Myzus persicae,* and *Rhopalosiphum maidis* (Edwardson & Christie, 1991). Virus isolates may show some vector specificity. The efficient transmission by the colonizing soybean aphid (*A. glycines*), recently introduced into North America, has been of concern. However, its addition to the array of migrating noncolonizing aphids that transmit SMV has not been documented to result in significantly increased SMV incidence (Pedersen, Grau, Cullen, Koval, & Hill, 2007).

Infected plants resulting from transmission through seed play an important role in SMV epidemiology. Such plants are primary inoculum sources for SMV. The spread within and among fields is mostly aggregated from a point source with secondary spread by aphids occurring at a moderately fast rate (Hill et al., 1980; Nutter, Schultz, & Hill, 1998).

Virus in seeds remains infective for a long period of time and viable virus can be recovered from seeds that no longer germinate. The level of seed transmission is dependent upon cultivar and plant stage at infection, with the incidence of seed transmission higher in plants infected before the onset of flowering. In most commercial cultivars presently grown, seed transmission is less than 5%, but no transmission occurs in some cultivars and some older cultivars can have levels as high as 75%.

SMV moves both up and down in plants and can be detected in all parts of a systemically infected plant, including the roots. Because of this, SMV has been used successfully as an expression vector (Wang, Eggenberger, Hill, & Bogdanove, 2006).

The most economically important host is soybean. Plant hosts infected by various virus isolates of SMV are limited to six plant families, which include members of the *Fabaceae*, *Amaranthaceae*, *Chenopodiaceae*, *Passifloraceae*, *Schrophulariaceae*, and *Solanaceae*. Numerous hosts differ in susceptibility based upon strain differences. Latent infection has been reported in several hosts.

2.2. Management

Management of disease caused by SMV should consider the use of SMV-free seed and timing of planting to avoid high vector populations when plants are young. Serological seed indexing techniques and/or grow-out tests can be used for virus detection in seed lots. Rouging, although generally impractical in the field, may not be very effective because of the tendency for symptoms in soybean to be masked above 30 °C. Efforts to achieve management of disease through control of aphid vectors have not been successful (Pedersen et al., 2007).

The most effective disease management should be based upon resistance At least three naturally occurring independent loci, designated *Rsv1, Rsv3,* and *Rsv4,* have been identified and mapped for resistance to SMV. At the *Rsv1* locus, nine variants, which confer differential reactions to strains G1 to G7, have been identified as *Rsv 1* from PI 96983, *Rsv1-y* from York, *Rsv1-m* from Marshall, *Rsv1-h* from Suweon 97, *Rsv1-t* from Ogden, *Rsv1-k* from Kwanggyo, *Rsv1-r* from Raiden, *Rsv1-s* from LR1, and *Rsv1-n* from PI 507389. Five alleles at the *Rsv3* locus have been identified with two reported for *Rsv4*. Because of the high variability of SMV, use of single resistance genes is potentially dangerous and pyramiding available sources of resistance is recommended to achieve disease control (Saghai Maroof et al., 2008; Shakiba, Chen, Shi, et al., 2012; Shi et al., 2009).

SMV holds a unique position among soybean viruses because of recent studies to understand the *R*-gene defense response. Most recent studies have focused on *Rsv1*-mediated resistance. The *Rsv1* resistance mechanism operates against all major SMV strains including the N isolate (SMV-N) of strain SMV-G2, but strain SMV-G7 overcomes *Rsv1* and induces a lethal systemic necrosis in PI 96983 (*Rsv1*). The mechanism of *Rsv1*-mediated resistance against SMV-N has been characterized as extreme resistance (ER) in which no virus can be detected in the inoculated leaf. However, the SMV-N–*Rsv1* interaction has the potential to produce a restricted hypersensitive response (HR) under certain conditions. The elicitors of *Rsv1*-mediated resistance have been mapped to the helper component proteinase (HC-Pro) and P3 cistrons of SMV. Interestingly, the *Rsv1* locus consists of a cluster of related resistance genes that appear to independently recognize HC-Pro and P3. This likely accounts for the fact that few naturally occurring strains that overcome *Rsv1*-mediated resistance have been reported and suggests the long-term durability of *Rsv1*-mediated resistance. Under certain conditions, however, SMV can mutate to evade *Rsv1*-mediated resistance. In addition, other studies using different isolates of the same strains report the cytoplasmic inclusion (CI) cistron may also be an elicitor of *Rsv1* in these isolates (Chowda-Reddy, Sun, Hill, Poysa, & Wang, 2011; Hajimorad & Hill, 2001; Hajimorad, Eggenberger, & Hill, 2005, 2006, 2008; Hajimorad, Wen, Eggenberger, Hill, & Saghai Maroof, 2011; Wen, Saghai Maroof, & Hajimorad, 2011; Wen, Khatabi, Ashfield, Saghai Maroof, & Hajimorad, 2013; Zhang, Grosic, Whitham, & Hill, 2012).

In contrast to *Rsv1*, *Rsv3* confers resistance to SMV strains G5, G6, and G7. The resistance conferred by *Rsv3* is phenotypically different from that conferred by *Rsv1* as it allows for a limited spread of avirulent SMV-G7 from the point of inoculation and does not induce a typical HR phenotype. The elicitors of *Rsv3*-mediated resistance have been mapped to the N- and C-termini of the CI protein (Chowda-Reddy, Sun, Hill, et al., 2011; Zhang et al., 2009).

The *Rsv4* resistance, effective against SMV strains G1–G7, is not associated with ER or HR but is characterized by delayed virus replication and movement or, otherwise, nonnecrotic late susceptibility (Gunduz, Buss, Chen, & Tolin, 2004). The elicitor for *Rsv4*-mediated resistance maps to

P3. Single nucleotide mutations in P3 compromise resistance and lead to the appearance of new resistance-breaking isolates. Therefore, the use of soybean cultivars that carry this single resistance gene is likely dangerous (Chowda-Reddy, Sun, Chen, et al., 2011)

The use of these genes in cultivars used for commercial production is unknown. However, Wang et al. (2005) report SMV resistance to two strains in 23 of 52 ancestral North American ancestral soybean lines. Of all public registration articles cited since 2002, only 15 reported resistance to SMV. While uncertain, the source of resistance is suggested to be one or more of the multiple alleles at the *Rsv1* locus. In a more extensive study using 5 SMV strains and 303 cultivars grown in the mid-south region of the United States, similar results revealed that the majority of lines tested do not carry resistance genes to SMV. However, 74 cultivars appeared to contain one of the *Rsv1* alleles, which was most likely believed to be *Rsv1-y*. Only a few cultivars appeared to carry *Rsv3* and *Rsv4* (Shakiba, Chen, Gergerich, et al., 2012a). Thus, at least in North America, SMV resistance may be more widespread than expected.

In the first field test of stable transgenic resistance against any pathogen of soybean, coat protein-mediated resistance was developed against SMV by using a non-aphid transmissible isolate to minimize the risk of spreading new pathogenic strains by potential of recombination. The resistance was effective against numerous SMV strains (Wang, Eggenberger, Nutter, & Hill, 2001). In a tactic intended to maximize the durability of resistance, the resistance could be successfully deployed in the field as rate-reducing resistance (Steinlage, Hill, & Nutter, 2002). Subsequently, posttranscriptional gene silencing was also reported from Japan to protect plants from virus infection but no field test was reported (Furutani et al., 2007). At the present time, transgenic resistance to SMV has not been incorporated into commercially sold varieties.

3. BEAN POD MOTTLE VIRUS

3.1. Biology

The first incidence of Bean pod mottle virus (BPMV) in the United States was reported on common bean (*Phaseolus vulgaris*) in South Carolina (Zaumeyer & Thomas, 1948) and the first reported incidence on soybean was from Arkansas (Walters, 1958). The incidence of BPMV is widespread in major soybean-growing regions in the south and southeastern United States and recently its incidence has significantly increased throughout the north central United States (Giesler, Ghabrial, Hunt, & Hill, 2002). The virus is a member of the genus *Comovirus* in the family *Secoviridae* (Sanfacon et al., 2009). Isolates from the United States have been classified into two subgroups (I and II) and one reassortant group (I/II), based on nucleic acid hybridization analysis. Isolates belonging to subgroup I induce moderate mottling symptoms while those from subgroup II induce mild symptoms. The reassortant (I/II) isolate produces severe symptoms (Gu et al., 2002). The most obvious symptoms occur in plants infected early in the season or during periods of rapid growth. Symptoms are more clearly shown in low-ambient temperatures (Gergerich, 1999). The virus host range is generally limited to the *Fabaceae* with soybean being the most economically important host. Soybean infected by BPMV produces 3–52% less yield depending on the soybean variety and the time of infection. Virus infection occurring in early vegetative stages of the soybean causes the highest yield reduction (Ziems et al., 2007). The seeds produced by BPMV-infected soybean may show mottled seed coats reducing the quality of the seed (Hill, Koval, Gaska, & Grau, 2007; Hobbs et al., 2003). Virus infection of soybean also increases the risk of Phomopsis seed infection (Abney & Ploper, 1994). Coinfection of BPMV and SMV leads to synergistic infection where SMV presence increases the titer of BPMV (Anjos, Jarlfors, & Ghabrial, 1992), which translates into dramatic reduction of yield (Ross, 1968) and root nodules (Tu, Ford, & Quinones, 1970) of co-infected plants.

The virus is transmitted by several leaf-feeding beetles in North America, including bean leaf beetle (*Cerotoma trifurcata*, Forster), grape colaspis (*Colaspis brunnea*, Fabricius), banded cucumber beetle (*Diabrotica undecimpunctata howardi*, Barber), striped blister beetle (*Epicauta vittata*, Fabricius), Japanese beetle (*Popillia japonica* Newman) and Mexican bean beetle (*Epilachna varivestis*, Mulsant) (Gergerich, 1999; Wickizer & Gergerich, 2007). The bean leaf beetle is an extremely efficient vector; up to 80% of beetles feeding on infected soybean were able to transmit the virus to healthy plants (Giesler et al., 2002). The bean leaf beetle also transmits Cowpea mosaic virus and Southern bean mosaic virus (Boethel, 2004). Of all the soybean viruses transmitted by bean leaf beetle, BPMV is considered to be the most prevalent and widespread in the United States soybean producing regions (Tolin & Lacy, 2004).

Theoretically, there are at least three sources of primary BPMV inoculum: infected soybean seeds, virus from previous year's epidemic retained in overwintering bean leaf beetles, and virus-infected leguminous weeds. Seed infection with BPMV occurs at a very low rate and seed coat mottling is not a good indicator of seed infection as mottling may occur without presence of the virus in the seed (Krell, Pedigo, Hill, & Rice, 2003). Overwintering bean leaf beetles have been shown to retain BPMV from the previous year albeit at a very low frequency (Krell et al., 2003). BPMV infects a range of mostly leguminous hosts including soybean and dry bean. Tick trefoil (*Desmodium* spp.), a perennial leguminous weed, was found to host BPMV in nature (Bradshaw, Rice, & Hill, 2007; Krell et al., 2003). Beetles emerging from overwintering habitats and feeding on infected perennial tick trefoil may acquire the virus. When soybean seedlings emerge, the beetles move to these plants transmitting the virus. The subsequent generation(s) of bean leaf beetles may continue to spread the virus within and between soybean fields during the growing season (Bradshaw, Zhang, Hill, & Rice, 2011).

3.2. Management

Management of the disease caused by BPMV should first focus on tactics that will reduce initial inoculum. These may include seed lot testing to exclude lots with BPMV-infected seed and production of certified soybean seed in areas predicted to have zero-to-low BPMV risk. The second consideration is that population density of the bean leaf beetle during the growing season is an important risk factor (Byamukama, Robertson, & Nutter, 2011). Therefore strategies directed at reducing bean leaf beetle populations have include delayed planting to reduce density of the bean leaf beetle, insecticide seed treatments by using neonicotinoid insecticides, and application of foliar pyrethroid insecticides between emergence and the first trifoliolate leaf followed by an additional application aimed at controlling the first generation of beetles (Bradshaw, Rice, & Hill, 2008; Krell, Pedigo, Hill, & Rice, 2004). The efficacy of all these strategies can be inconsistent. Ideally, resistance to insect feeding or disease resistance could provide optimal disease control. Unfortunately, the use of soybean varieties with antixenosis (i.e., physical resistance) against bean leaf beetles does not reduce spread of BPMV in the field (Redinbaugh et al., 2010). Screening of 52 North American ancestral soybean lines and examination of all public registration articles through 2002 resulted in no identifiable resistance against BPMV (Wang et al., 2005). Similar results were reported from screening of 303 soybean cultivars grown in the mid-southern region of the United States (Shakiba, Chen, Gergerich, et al., 2012b). Recently, however, some varieties have been reported to have tolerance (Haidi, Bradshaw, Rice, & Hill, 2012; Hill et al., 2007; Ziems et al., 2007).

4. SOYBEAN VEIN NECROSIS VIRUS

4.1. Biology

Soybean vein necrosis virus (SVNV) is an emerging virus that was first discovered in Tennessee and Arkansas in 2008. It is widespread in the soybean-growing areas of the United States and the fully sequenced virus has a characteristic *Tospovirus* genome organization (Zhou et al., 2011). Despite the widespread occurrence of the virus, strains do not appear prevalent and the population structure is homogeneous. Visual symptoms include early light green to yellow (chlorotic) patches near the main leaf veins. The chlorosis progresses to development of necrosis followed eventually by leaf death and desiccation on apparent susceptible varieties. Veins may appear clear, yellow, or dark brown. Lesions progress more slowly or are restricted on apparent resistant varieties. A large percentage of the plant canopy may be killed with widespread infections on highly susceptible varieties. Both nucleic acid and immunologically based detection methods have been developed (Khatabi et al., 2012; Zhou & Tzanetakis, 2013).

SVNV is efficiently transmitted by the soybean thrips [*Neohydatothrips variabilis* (Beach)]. Plants in several plant families including the *Asteraceae, Convulaceae, Cucurbitaceae, Fabaceae,* and *Solanaceae* have been shown to be infected, although symptoms may not always be evident. Although no resistant soybean varieties are known, the field variation in symptom severity is likely due to host genotype variation because of the apparent lack of virus strains.

4.2. Management

The efficient transmission by the soybean thrips is most likely principally responsible for apparent epidemics of the disease caused by the virus. However, presently no yield loss data has been reported so that the potential economic impact of a virus epidemic is unknown. It is probable that infected weed hosts provide a virus source for thrips transmission to soybean, and in the southern United States the natural distribution of the asymptomatic ivyleaf morning glory (*Ipomoea hederacea* (Jacq.)) is coincident with the incidence of the virus (Zhou & Tzanetakis, 2013). Although apparent resistance based upon field observation may exist, the existence of resistance for disease management is not yet documented. Thus, the best management of the disease, because of the lack of information concerning resistance in soybean varieties, is likely dependent upon attempting to manage the thrips vector using insecticides. Insecticide application should be considered only in fields with a known risk of yield loss; however, management of virus vectors has not generally been an effective tactic for control of disease caused by soybean-infecting viruses.

5. TOBACCO RINGSPOT VIRUS

5.1. Biology

Tobacco ringspot virus (TRSV), a member of the genus Nepovirus in the plant virus family *Secoviridae,* was first described in the United States in 1941 and has been reported in Canada, Cuba, Brazil, India, Australia, the former Soviet Union, and the Peoples Republic of China (Almeida, 1980; Demski, Kuhn, & Hartman, 1999; Fernandez-Suarez, 1984; Gupta, 1978). It causes a disease known as bud blight, which should not be confused with Brazilian bud blight, caused by the Ilarvirus tobacco streak virus. Plants infected in early growth stages are severely stunted and will likely remain green after a killing frost. The most characteristic symptom is the "shepherd's crook" caused by the curving of the terminal bud to form a crook. Later, other buds on the plant become brown, necrotic, and brittle. Adventitious leaf and floral buds may proliferate excessively. Petioles of the youngest trifoliolate leaves are often shortened, thickened, and may be curved to distort the shoot tips. Leaflets are dwarfed and may cup or roll. Leaf blades become more or less rugose and bronzed. Pods are generally underdeveloped and often aborted. Those that do set often develop dark blotches, generally do not produce viable seeds, and drop early (Demski et al., 1999).

Of the many diseases caused by TRSV, budblight of soybeans is the most severe and can cause the most significant yield loss. However, it is not generally widespread in soybeans. In general, most significant yield loss occurs when plants become infected before flowering. Yield of early inoculated soybeans has been reported to be reduced by as much as 79% (Demski, Harris, & Jellum, 1971) and yield loss as a result of natural infection has been reported as greater than 50% (Crittenden, Hastings, & Moore, 1966).

The virus, characterized by numerous strains, has been reported to infect a large number of plant hosts in at least 30 different plant families (Hill, 2003). Additional potential hosts are found in the compilation of DeZeeuw and Hooker (1965), which includes 54 plant families.

Global distribution of TRSV probably occurs through seed transmission. Although seed transmission can occasionally range as high as 100%, the potential of harvesting seed fields with such high infection rates in unlikely because infected plants produce little seed and remain green until killed by frost. This makes mechanical harvesting almost impossible (Athow & Bancroft, 1959; Athow & Laviolette, 1961). Because plants must become infected before bloom for seed transmission to be significant (Athow & Bancroft, 1959; Demski & Harris, 1974) and because plants infected before bloom produce few or no seeds, seed transmission is not believed to provide a major source of primary inoculum in the field.

The dagger nematode, *Xiphinema americanum,* is an efficient vector of TRSV to some host species such as *Cucumis sativus.* However, transmission efficiency to soybean is low (McGuire & Douthit, 1978). Observational reports, based upon rapid downwind directional spread, suggest the potential of efficient natural transmission by arthropod vectors (Hill, Epstein, McLaughlin, & Nyvall, 1973). Several arthropod vectors have been reported including aphids *Myzus persicae* and *Aphis gossypii;* the grasshoppers *Melanoplus differentialis, M. mexicanus,* and *M. femrrubrum;* the tobacco flea beetle, *Epitrix hirtipennis;* and the thrips, *Thrips tabaci* and probably *Frankliniella* sp. (summarized in Granillo & Smith, 1974). Conflicting reports suggest the potential for very low transmission efficiency by the newly introduced soybean aphid *Aphis glycines* (Clark & Perry, 2002; Wang, Kritzman, Hershman, & Ghabrial, 2006). In general, the reported arthropod vectors are not efficient vectors of TRSV. Additional research is required to discern how apparent virus can spread in soybeans more rapidly than expected through activity of inefficient arthropod or nematode vectors. Because seed transmission is low and the unequivocal demonstration of efficient transmission by an arthropod vector remains elusive, the epidemiology of the virus remains vague. Demonstration of a disease gradient on the downwind side of mixed legume (alfalfa and red clover) fields suggests spread more rapid than that associated with nematode transmission (Hill et al., 1973).

5.2. Management

Although not clear that it contributes in a major way to significant soybean disease problems, the use of virus-free soybean seed seems appropriate for disease management. Also, producers should avoid fields that are known to be infested with the dagger nematode (or treat them with appropriate nematacides) as well as avoid planting soybeans adjacent to mixed legume fields. Resistance to TRSV has been identified in *Glycine soja* but is not known in the cultivated soybean, *Glycine max* (Orellana, 1981). Recently, the screening of 303 soybean cultivars grown in the mid-south region of the United States showed no virus resistance (Shakiba, Chen, Gergerich, et al., 2012b). Transgenic resistance has been reported in tobacco utilizing the virus coat protein (Zadeh & Foster, 2004) and a TRSV satellite, which interferes with TRSV replication and can ameliorate symptoms (Gerlach, Llewellyn, & Haseloff, 1987). However, no transgenic resistance in soybean has been reported.

6. SOYBEAN DWARF VIRUS

6.1. Biology

Soybean dwarf virus (SbDV), a member of the plant virus family *Luteoviridae,* has been reported as causing significant yield losses in Japan that can range up to 80% but in the United States soybean-growing region it has been regarded as an emerging virus and at present has economic impact only in localized areas. The virus was reported for the first time from Hokkaido, Japan in 1969 (Tamada, Goto, Chiba, & Suwa, 1969) and has also been reported from Australia and New Zealand. It was first discovered in the United States in 2003 in Wisconsin and has since been reported from Iowa, Illinois, Kentucky, Virginia, and Mississippi. A synonym of the virus is subterranean clover red leaf virus. Symptoms on soybean include stunting, and leaf puckering, rugosity, and yellowing. Reduced seed set can occur if plants are infected at the seedling stage. Symptom severity is dependent upon the interaction between the soybean cultivar and virus strain. There is no report of seed transmission. The major economic host is soybean with most susceptible host species found in the *Fabaceae* and a few susceptible species found in the *Chenopodiaceae* and *Polemoiaceae* (Damsteegt, Hewings, & Sindermann, 1990). Diagnostic methods include RT-PCR and ELISA.

Two viral strains, known as the dwarfing strain (SbDV-D) and yellowing strain (SbDV-Y), were initially identified. More recently, four strains have been described in Japan, SbDV-DS (dwarfing-solani), SbDV-DP (dwarfing, pisum), SbDV-YS (yellowing-solani), and SbDV–YP (yellowing-pisum) (Terauchi et al., 2001). Natural occurrence of mixed infection of strains as well as recombinant SbDV has been described (Schneider et al., 2011). Strains YS and DS are transmitted by the foxglove aphid (*Aulacorthum solani* (Kaltenbach)) and strains YP and DP are transmitted by both the pea aphid

(*Acyrthosiphon pisum)* and the clover aphid (*Nearctaphis bakeri*) in a persistent manner. The virus is not mechanically transmissible.

The possibility that SbDV could become significant to soybean-growing regions of the United States illustrates how understanding the interaction between insect vector and virus strain is an important determinant for predicting disease possibility and, potentially, subsequent management. In the United States, populations of *A. solani* exist in many areas, but frequency of occurrence is limited in time and location and *A. solani* populations in the United States do not colonize soybeans as reported for Japanese biotypes. SbDV strains transmitted by *A. solani* are reported to be exotic in the United States and thus their frequency is likely to be limited. Natural infections of red clover (*Trifolium pratense* L.) as well as potentially white clover (*T. repens* L.), alsike clover (*T. hybridum* L.), subterranean clover (*Trifolium subterraneum* L.), and crimson clover (*T incarnatum* L.) are a major source of SbDV-like isolates. Reportedly, SbDV-Y strains infected white clover (*T. repens* L) but not red clover (*T. pratense* L.) while SbDV-D strains infect red clover but not white clover (Tamada, 1970). *N. bakeri*, which colonizes clover, transmits SbDV from clover to soybean at very low efficiencies. The colonizing soybean aphid (*Aphis glycines*), introduced into the United States in 2000, transmits SbDV very rarely (Damsteegt et al., 1990; Damsteegt, Stone, & Hewings, 1995; Harrison, Steinlage, Domier, & D'Arcy, 2005). Thus, factors affecting vector host preference, virus strain, and transmission efficiency and specificity combine to preclude SbDV becoming a major disease threat to the soybean-growing region of the United States. Previous detection of mixed infections and/or recombinants could result in changes in vector specificity. Accidental introduction of colonizing aphid biotypes that are efficient vectors could result in significant disease epidemics.

6.2. Management

Management options include not planting soybeans next to clover fields. Also, pesticide applications may be useful for vector control. Alternatively, the use of tolerant soybean varieties is important for disease control. A major QTL, *Rsdv1*, for resistance has been reported between SSR markers Sat_217 and Sat_211 on chromosome 5 (Yamashita, Takeuchi, Ohnishi, Sasaki, & Tazawa, 2013). Also, coat protein-mediated transgenic resistance has been reported (Tougou et al., 2007). No reports suggest these strategies have been adopted in commercial production.

7. PEANUT MOTTLE VIRUS

7.1. Biology

Peanut mottle virus (PMotV) is most common in the southern United States and has been identified in East Africa, northeast Australia, and possibly west Kenya, Uganda, Venezuela, Japan, Malaysia, and Bulgaria. It was first named in 1965 after isolation from peanuts in Georgia in the United States. Reduction in seed quality and mean yield reduction of 44% was reported for 21 susceptible cultivars in Virginia, where the virus is economically important because of the prevalence of peanut growing areas (Roane, Tolin, & Reid, 1978). The virus has characteristics typical of those belonging to the plant virus family *Potyviridae*. It has a rather limited host range infecting species belonging to the families *Amaranthaceae*, *Chenopodiaceae*, *Cucurbitaceae*, *Fabaceae*, *Pedaliaceae*, and *Solanaceae*. As expected, the virus is transmitted nonpersistently by a variety of aphids including *Aphis craccivora*, *A. gossypii*, *Hyperomyzus lactucae*, *Myzus persicae*, and *Rhopalosiphum padi* (Brunt et al., 1996 Onwards). Symptoms include mosaic, chlorotic patterns, concentric rings, vein chlorosis, and leaf deformity. Early infection may result in stunting (National Pest Surveillance, Monitoring Systems—SENASA, Argentina, n.d.). Virus detection can readily be accomplished through ELISA (Hobbs, Reddy, Rajewhwari, & Reddy, 1987; Sherwood, Sanborn, & Keyser, 1987).

The virus is seed transmitted in peanuts but not in soybeans. Global dissemination of the virus likely occurs through seed of peanut. Seed transmission in peanut likely provides the primary inoculum source for the virus in soybeans and consequently the virus is generally limited to peanut growing areas. Spread of the virus is more rapid in peanut than in soybeans (Adams & Kuhn, 1977; Demski, 1975).

7.2. Management

Management recommendations for the disease caused by the virus include not planting soybeans adjacent to peanut fields and use of resistant varieties. Two dominant resistance genes, *Rpv1* and *Rpv3* as well as the single recessive gene *rpv2* have been identified. These could permit development of soybeans with broader based resistance (Bagade & Buss, 1998; Gore et al., 2002; Shipe, Buss, & Tolin, 1979).

8. PEANUT STUNT VIRUS

8.1. Biology

Peanut stunt virus (PStuV), a member of the genus *Cucumovirus* in the family *Bromoviridae*, was first described in peanuts in North Carolina and Virginia in the United States in 1964. Subsequently, it has been reported in France, Japan, Korea, Morocco, Poland, the Sudan, Iran, China, and Spain (Baker, Blount, & Quesenberry, 1999; Blount et al., 2006). It is found infrequently in soybeans in the southern United States in soybeans where it is generally not considered to be a major problem. However, in Japan typical yield losses of 33% can occur due to reductions in seed number and size along with reductions in quality caused by seed coat mottling similar to that caused by SMV (Iizuka & Yunoki, 1974; Kosaka, 1997). The virus has a wide host range including while clover (*Trifolium repens* L.), which is believed to be the principal reservoir host of this virus (McLaughlin, Pederson, Evans, & Ivy, 1992). In addition to soybeans, clover, and peanuts the virus has been reported from tobacco, snapbeans, various curcubits, peppers, and tomatoes (Baker et al., 1999). The virus is nonpersistently transmitted by aphids including the cowpea aphid *(Aphis craccivora)*, spirea aphid (*A. spiraecola*), and green peach aphid (*Myzus persicae*) (Hebert, 1967). Strains PSV-K and PSV-T of the virus have been identified in Japan with PSV-K representing strains causing mild mosaic in soybeans after infection and PSV-T representing isolates causing systemic leaf curling and mosaic of the virus (Saruta et al., 2012). In the United States, strains have been differentiated as the Eastern strain found in the southeastern United States, and the Western strain found primarily in Washington (sometimes referred to as subgroups I and II, respectively). The Eastern strain produces necrotic primary lesions, systemic necrotic rings, and veinal necrosis in Tennessee Greenpod bean, and mosaic without necrosis in Perfected Wales pea. In contrast, the Western strain produces mosaic without necrosis in Tennessee Greenpod bean and mosaic with severe necrosis of stems and petioles in Perfected Wales pea (Mink, Hebert, & Silbernagel, 1967; Naidu, Hu, Pennington, & Ghabrial, 1995). Apparent naturally occurring reassortants between the strains can also be found (Hajimorad, Hu, & Ghabrial, 1999; Hu & Ghabrial, 1998).

8.2. Management

Management of the disease caused by PStuV is not generally practiced in soybeans. However, it is advisable to use cultural practices such as using good quality seed and avoiding planting of soybeans next to forages known to be infected with PStuV. Satellite RNAs, which can ameliorate symptoms, have been identified but have not been exploited for disease control in soybeans (Naidu et al., 1995). Disease resistance in soybeans has been reported to be conferred by the single dominant *Rpsv1* mapped to soybean chromosome 7 (Saruta et al., 2012.)

9. ALFALFA MOSAIC VIRUS

9.1. Biology

Alfalfa mosaic virus (AMV) is an apparent emerging problem in soybean in the United States. Although its incidence in soybean-growing regions in the United States is generally unknown, the virus was detected in 52% and 54% of soybean fields examined in Nebraska in 2001 and 2002, and incidence levels of 30% or more were sufficient to cause yield reductions of approximately 25% in Wisconsin in 2002 and 2003 (Mueller & Grau, 2007). Nevertheless, the distribution and therefore importance of the virus in soybean producing areas of the United States continues to be uncertain. The virus belongs to the plant virus family *Bromoviridae* and has a very broad host range consisting of 51 dicotyledonous families. The impact of the large number of potential weed hosts acting as an inoculum reservoir for inoculation of soybeans is not known. The virus is transmitted by at least 15 aphid species in a nonpersistent manner. Seed transmission has been shown up to a rate of 9% and is dependent upon host genotype and virus strain (He, Fajolu, Wen, & Hajimorad, 2010). However, the impact of seed transmission in commercial production is unknown. Symptoms in soybeans can be variable and depend upon soybean cultivar, environment, and virus strain. Some varieties of soybean are asymptomatic. In general, symptoms can include leaves that have mottled patterns of bright yellow and dark green tissues. Leaves that are newly emerged may be smaller than usual with bright yellow spots and brown discoloration. Symptoms may tend to decrease as the soybean plant matures. Plants may be stunted. Coinfection of soybean with AMV and SMV results in synergism with enhancement in the level of AMV and a reduction in the level of SMV leading to more pronounced stunting, chlorosis, mosaic, and leaf deformation (Malapi-Nelson et al., 2009). Plants infected by AMV do not produce seed with mottled seedcoats, unlike some other soybean viruses. However, seeds from the same cultivar infected with different AMV strains have been shown to exhibit size and color differences (He et al., 2010).

9.2. Management

In general, little effort has been directed towards management of this virus in soybean. Attempts to control the aphid vectors by insecticide application are considered to have little effect. A major resistance gene (*Rav1*) to AMV has been mapped as a quantitative trait locus in PI 153282 but the gene has not been incorporated into commercial varieties (Kopisch-Obuch et al., 2008). Nevertheless, based on degree of symptom expression it is likely that currently grown soybean varieties differ in tolerance to AMV. The use of symptoms for evaluation of AMV infection and potentially resistance screening can lead to potentially erroneous results, however. The symptom variation and ability to detect AMV is variable in soybean and is dependent upon virus strain, soybean genotype, and time of sample collection. For optimal virus detection of virus in infected plants by ELISA, leaves at the upper parts of the plants should be collected (He, Hill, & Hajimorad, 2011). Transgenic resistance to AMV has been developed in tobacco but has not been reported in soybeans (e.g., Loesch-Fries et al., 1987).

10. MANAGEMENT: PRESENT AND PROSPECTS

Numerous virus diseases of soybeans have been documented. However, difficulty in correct identification of the causal agent of virus-like diseases of soybeans is common because symptoms caused by widely different viruses are often not unique except for the recently emerged SVNV and, under some environmental conditions, TRSV. The large variety of soybean cultivars, and symptoms that vary with virus strain, cultivar, time of inoculation, environment, and latency makes this problem more vexing. For most viruses, symptoms are not diagnostic but require a laboratory test for unequivocal identification. This information is vital for disease management because knowledge of the pathogen biology and its vector(s) is essential for development of strategies intended to reduce the economic impact of the disease. This includes understanding the ecology of the virus together with the wide variety of insect vectors, vector relationships, and potential alternative hosts that exist among disease causing viruses. These issues obviously significantly impact development of disease management tactics that might be proposed for disease control.

For the soybean viruses discussed in this chapter, resistance and or tolerance is the best management tactic. Alternative options include vector management, various cultural practices, and planting clean seed. As discussed, vector management for disease control has been largely unsuccessful. Only in the case of BPMV, vectored by the bean leaf beetle, is vector control recommended for disease management. However, vector management, even though inconsistent, provides the best possible option for BPMV at the current time because resistance is not readily available

Certain cultural practices may also assist attempts to minimize disease. These include avoiding planting adjacent to known alternative hosts of the virus as exemplified for TRSV, SbDV, PMotV, and PStuV. For those viruses that are seed transmitted, and particularly for which seed may provide the primary inoculum source, attempts to minimize seed transmission are significant. This is clearly most important for SMV, although seed transmission may also play a role as a primary source for BPMV and AMV.

Genetic resistance, either natural or engineered, is an ideal way to protect crops from viruses (Maule, Caranta, & Boulton, 2007; Ritzenthaler, 2005) and will be the best approach for management of soybean viruses. Resistance genes have been identified for control of SMV (*Rsv1, Rsv3, Rsv4*), SbDV (*Rsdv1*), AMV(*Rav1*) and PMotV (*Rpv1, Rpv3, rpv2*). In the case of SMV, there is some evidence that they may have been incorporated into cultivars used for production.

For the future, we need to consider ways to augment genetic resistance through a variety of approaches that build upon knowledge of antiviral RNA silencing mechanisms, induced defense responses to pathogens, and detailed knowledge of virus–host interactions. These approaches are expected to enable rational design and selection of effective virus resistance for integration into elite soybean germplasm.

Virus resistance traits based on RNA silencing have already been used in other crop species, and their feasibility has been demonstrated in soybean. This form of pathogen-derived resistance has been induced by a variety of constructs against different viruses that are able to infect soybean. Transgenic virus resistance, not yet commercially adopted, was first shown to be effective for SMV, and would likely be effective for other soybean viruses as well. In the first example of stable, genetically engineered disease resistance in soybean, soybean plants was transformed with the coat protein (CP) gene and 3′-UTR under control of the CaMV 35S promoter from SMV. The T_3 generation yielded two homozygous transgenic lines that were highly resistant to the homologous strain of SMV, even at inoculum levels as high as 200 μg of virus per plant. This RNA-mediated resistance conferred resistance to four SMV strains tested and was highly effective in field tests (Steinlage et al., 2002; Wang et al., 2001). In a subsequent report, the CP of SMV was expressed under control of the constitutive CaMV 35S promoter and a soybean line was recovered that produced plants that could not accumulate SMV CP but did accumulate SMV CP short interfering RNAs (siRNAs) (Furutani et al., 2007). Resistance to SMV in this line was correlated with accumulation of siRNAs from the SMV CP coding sequence. This construct was not designed with the purpose of producing consistent and high levels of siRNAs, and therefore, levels of resistance varied among siblings and over the course of development.

Constructs expressing inverted repeat transcripts of target sequences are designed to produce constitutively high levels of siRNAs and more consistent RNA silencing. Tougou et al. (2006) created transgenic soybean expressing an inverted repeat of the SbDV CP under control of the CaMV 35S promoter. Production of SbDV CP siRNAs was correlated with resistance to SbDV infection for up to at least 2 months after inoculation using aphids.

An interesting application of antiviral RNA silencing is to use it to target multiple viruses in a single soybean line. Transgenic soybean lines were made that expressed a single construct to express short inverted repeats of 150 bp to target SMV, BPMV, and AMV (Zhang et al., 2011). These lines accumulated siRNAs corresponding to all three viruses and were resistant to simultaneous infection by them. This study shows that it is possible to stack resistance to multiple strains of the same virus as well as different viruses at a single locus. Advantages of RNA silencing include the ability to target any virus and durability of the resistance. The durability derives from that fact that a virus would have to mutate extensively to overcome it. A potential limitation to this approach is that it relies on sequence identity and for many viruses there are many strains that are related by various levels of sequence identity. Therefore, to use RNA silencing most effectively and enhance durability of virus resistant lines, it will be necessary to have a good understanding of viral sequence diversity.

Recent advances, such as the sequence of the Williams 82 reference genome (Schmutz et al., 2010) and application of virus-induced gene silencing as a relatively high throughput technique for testing gene function (Zhang, Bradshaw, Whitham, & Hill, 2010; Zhang & Ghabrial, 2006; Zhang et al., 2012), have led to the identification of many genes that play important roles in defense against soybean viruses and other pathogens (e.g., Pandey et al., 2011; Zhang et al., 2012). These genes function at various levels within the networks that mediate soybean immune responses.

The networks are rapidly and specifically activated in response to avirulent isolates of pathogens, and they may also be activated by conserved features of pathogens (pathogen-associated molecular patterns). The overall architecture of soybean defense gene networks is likely to be highly conserved with the model plant, *Arabidopsis thaliana*, for which there is a wealth of information. We have found that most genes involved in Arabidopsis defenses have one or more homologs in soybean. Despite this conservation, there are still many challenges to understand how the networks are regulated and how the genes within them interact with other plant and pathogen genes. In addition, there are likely to be genes and regulatory relationships that are soybean specific that remain largely undiscovered. Nevertheless, we are beginning to identify key genes that control the expression of many other genes and are required for soybean defense responses. These genes may represent key regulatory hubs that may be exploited in the future to enhance or broaden the spectrum of basal resistance to viruses and other pathogens in soybeans. Proteins encoded by these genes may function as positive or negative regulators, and so, it will be important to test effects of knocking down or increasing their expression. There may also be opportunities to exploit genetic variation in these genes to select alleles that confer improved resistance to viruses.

Plant viruses are obligate pathogens that are intimately associated with and completely reliant on factors, such as proteins, within their host cells (Whitham & Wang, 2004). Host factors, therefore, represent potential Achilles heels that can be exploited for virus control. A prime example of this is the recessive eIF4E/eIF(iso)4E-based resistance that occurs in many plants against potyviruses and other viruses (Maule et al., 2007; Truniger & Aranda, 2009; Wang & Krishnaswamy, 2012). In the case of potyviruses, the basic mechanism underlying this resistance is that interaction between the viral VPg protein and eIF4E/eIF(iso)4E is disrupted. Despite the prevalence of this type of resistance in monocot and dicot crop plants, there is to our knowledge no example of eIF4E/eIF(iso)4E-based resistance in soybean against the potyvirus SMV or any other virus. There is a wealth of information accumulating on the structure of eIF4E/eIF(iso)4E and positions of amino acid residues that influence interactions with different VPgs (Wang & Krishnaswamy, 2012). Recently, Kim et al. (2014) showed that it is possible to rationally design eIF(iso)4E variants that provide resistance against potyviruses. This may eventually be feasible for the SMV–soybean interaction. In addition to eIF4E/eIF(iso)4E, we expect that viruses interact with many other host proteins (e.g., Elena, Carreera, & Rodrigo, 2011; Elena & Rodrigo, 2012), and that potentially any of these may be candidates for designing plants with novel recessive resistance. To build a foundation for such approaches, we need to generate much more information on the soybean and viral proteins that interact. After generating such interaction maps, it will then be necessary to establish the importance of these interactions in the virus lifecycle and then to investigate the molecular and atomic details of the interactions. With such information, it may be possible to create new recessive or dominant negative alleles of host proteins that perform their normal cellular functions but do not support viral functions.

As discussed above, there may be many ways to generate novel virus resistance traits in the future. The possibilities presented represent biotechnology solutions to important disease problems, but in many cases, they need not rely upon creating transgenic plants. Recent advances using site-directed nucleases create opportunities to engineer novel resistance traits (Schornack, Moscou, Ward, & Horvath, 2013). This was demonstrated in principle for soybean through the use of zinc finger nucleases (ZFNs) to modify sequences of duplicated *DICER-like 4* genes (Curtin, Voytas, & Stupar, 2012; Curtin et al., 2011). The potential for using other site-directed nucleases, such as TAL effector nucleases (TALENs) and CRISPR/Cas9, are currently being explored. TALENs and CRISPRs are favored over the ZFNs, because their design is much more straightforward. In addition to the site-directed nucleases, soybean genomics researchers will be increasingly generating sequences of more soybean accessions and allied species (Chan, Qi, Li, Wong, & Lam, 2012; Kim et al., 2010; Lam et al., 2010; Varala, Swaminathan, Li, & Hudson, 2011). These sequences may be mined for alleles that might be useful in selecting new traits for virus resistance. These future approaches to virus control based on gene editing or exploiting allelic diversity point to a need for further research into soybean–virus interactions to generate the knowledge base needed for the rational design of resistance that will insure the stability and durability of quality soybean production.

ACKNOWLEDGMENTS

This work was supported, in part, by the Iowa Soybean Association, the United Soybean Board, the North Central Soybean Association, and Hatch Act and State of Iowa Funds. This is a journal paper of the Iowa Agriculture and Home Economics Experiment Station, Ames, IA, project number 3708.

REFERENCES

Abney, T. S., & Ploper, L. D. (1994). Bean pod mottle virus effects on soybean and seed maturation and seedborne *Phomopsis* spp. *Plant Disease, 78*, 33–37.

Adams, D. A., & Kuhn, C. W. (1977). Seed transmission of peanut mottle virus in peanuts. *Phytopathology, 67*, 1126–1129.

Almeida, A. M. R. (1980). Survey of soybean common mosaic and bud blight viruses in different regions of Parana State. *Fitopatologia Brasileira, 5*, 125–128.

Almeida, A. M. R. (1981). Identification of strains of soybean common mosaic in Prana state. *Fitopatologia Brasileira, 6*, 131–136.

Anjos, J. R., Jarlfors, U., & Ghabrial, S. A. (1992). Soybean mosaic potyvirus enhances the titer of two comoviruses in dually infected soybean plants. *Phytopathology, 82*, 1022–1027.

Anjos, J. R. M., Lin, M. T., & Kitajima, E. W. (1985). Characterization of an isolate of soybean mosaic virus. *Fitopatologia Brasileira, 10*, 143–157.

Athow, K. L., & Bancroft, J. B. (1959). Development and transmission of tobacco ringspot virus in soybean. *Phytopathology, 49*, 697–701.

Athow, K. L., & Laviolette, F. A. (1961). The relation of seed-transmitted tobacco ringspot virus to soybean yield. *Phytopathology, 51*, 341–342.

Bagade, P., Jr., & Buss, G. R. (1998). *Genetics of resistance to peanut mottle virus in soybean*. Available from, http://scholar.lib.vt.edu/theses/available/etd-32298-102824/, Accessed 12 March 2014.

Baker, C., Blount, A., & Quesenberry, K. (1999). *Peanut stunt virus infecting perennial peanuts in Florida and Georgia*. Florida Dept of Agriculture and Consumer Serv. Division of Plant Industry. Plant Pathology Circular No. 395. Available from, http://www.freshfromflorida.com/content/download/11401/144702/pp395.pdf/, Accessed 12 March 2014.

Blount, A. R., Sprenkel, R. K., Pittman, R. N., Smith, B. A., Morgan, R. N., Dankers, W., et al. (2006). *Peanut stunt virus reported on perennial peanut in north Florida and southern Georgia*. The Institute of Food and Agricultural Sciences, University of Florida Extension. SS-AGR-37. Available from, http://www.caes.uga.edu/commodities/fieldcrops/forages/events/SHC11/Peanut%20Stunt%20Virus%20Reported%20on%20Perennial%20Peanut.pdf/, Accessed 12 March 2014.

Boethel, D. J. (2004). Integrated pest management of soybean insects. In H. R. Boerma, & J. E. Specht (Eds.), *Soybeans: Improvement, production and uses*. (3rd ed., pp. 853–881). Madison, WI: ASA/CSSA/SSSA.

Bradshaw, J. D., Rice, M. E., & Hill, J. H. (2007). No-choice preference of *Cerotoma trifurcata* (Coleoptera: Chrysomelidae) to potential host plants of Bean pod mottle virus (Comoviridae) in soybean. *Journal of Economic Entomology, 100*, 808–814.

Bradshaw, J. D., Rice, M. E., & Hill, J. H. (2008). Evaluation of management strategies for bean leaf beetles (*Coleoptera: Chrysomelidae*) and bean pod mottle virus (Comoviridae) in soybean. *Journal of Economic Entomology, 101*, 1211–1227.

Bradshaw, J. D., Zhang, C., Hill, J. H., & Rice, M. E. (2011). Landscape epidemiology of bean pod mottle comovirus: Molecular evidence of heterogeneous sources. *Archives of Virology, 156*, 1615–1619.

Brunt, A. A., Crabtree, K., Dallwitz, M. J., Gibbs, A. J., Watson, L., & Zurcher, E. J. (Eds.), (1996 Onwards). *Plant viruses online: Descriptions and lists from the VIDE database*. Version: 20th August 1996. Available from, http://biology.anu.edu.au/Groups/MES/vide/, Accessed 12 March 2014.

Buzzel, R. I., & Tu, J. C. (1984). Inheritance of soybean resistance to soybean mosaic virus. *Journal of Heredity, 75*, 82.

Byamukama, E., Robertson, A. E., & Nutter, F. W., Jr. (2011). Quantifying the within-field temporal and spatial dynamics of bean pod mottle virus in soybean. *Plant Disease, 95*, 126–136.

Calvert, L. A., & Ghabrial, S. A. (1983). Enhancement by soybean mosaic virus of bean pod mottle virus titer in doubly infected soybean. *Phytopathology, 73*, 992–997.

Chan, C., Qi, X., Li, M.-H., Wong, F.-L., & Lam, H.-M. (2012). Recent developments of genomic research in soybean. *Journal of Genetics and Genomics, 39*, 317–324.

Cho, E.-K., Chung, B. J., & Lee, S. H. (1977). Studies on the identification and classification of soybean viruses in Korea. II. Etiology of a necrotic disease of *Glycine max*. *Plant Disease Reporter, 61*, 313–317.

Cho, E.-K., & Goodman, R. M. (1979). Strains of soybean mosaic virus: Classification based on virulence in resistant cultivars. *Phytopathology, 69*, 467–470.

Chowda-Reddy, R. V., Sun, H., Chen, H., Poysa, V., Ling, H., Gijzen, M., et al. (2011). Mutations in the P3 protein of soybean mosaic virus G2 isolates determine virulence on *Rsv4*-genotype soybean. *Molecular Plant-Microbe Interactions, 24*, 37–43.

Chowda-Reddy, R. V., Sun, H., Hill, J. H., Poysa, V., & Wang, A. (2011). Simultaneous mutations in multi-viral proteins are required for an avirulent Soybean mosaic virus isolate to gain virulence on Rsv1-, Rsv3-, and Rsv4-genotype soybeans. *PLoS One, 6*(11), e28342. http://dx.doi.org/10.1371/journal.pone.0028342.

Clark, A. J., & Perry, K. L. (2002). Transmissibility of field isolates of soybean viruses by *Aphis glycines*. *Plant Disease, 86*, 1219–1222.

Crittenden, H. W., Hastings, K. M., & Moore, D. M. (1966). Soybean losses caused by tobacco ringspot virus. *Plant Disease Reporter, 50*, 910–913.

Curtin, S. H., Voytas, D. R., & Stupar, R. M. (2012). Genome engineering of crops with designer nucleases. *The Plant Genome, 5*, 42–50.

Curtin, S. J., Zhang, F., Sander, J. D., Haun, W. J., Starker, C., Baltes, N. J., et al. (2011). Targeted mutagenesis of duplicated genes in soybean with zinc-finger nucleases. *Plant Physiology, 156*, 466–473.

Damsteegt, V. D., Hewings, A. D., & Sindermann, A. B. (1990). Soybean dwarf virus: Experimental host range, soybean germplasm reactions, and assessment of potential threat to U.S. soybean production. *Plant Disease, 74*, 992–995.

Damsteegt, V. D., Stone, A. L., & Hewings, A. D. (1995). Soybean dwarf, bean leaf roll, and beet western yellows luteoviruses in southeastern U.S. white clover. *Plant Disease, 79*, 48–50.

Demski, J. W. (1975). Source and spread of peanut mottle virus in soybean and peanut. *Phytopathology, 65*, 917–920.

Demski, J. W., & Harris, H. B. (1974). Seed transmission of viruses in soybean. *Crop Science, 14*, 888–890.

Demski, J. W., Harris, H. B., & Jellum, M. D. (1971). Effects of time of inoculation with tobacco ringspot virus on the chemical composition and agronomic characteristics of soybean. *Phytopathology, 61*, 308–311.

Demski, J. W., Kuhn, C. W., & Hartman, G. L. (1999). Tobacco ringspot. In G. L. Hartman, J. B. Sinclair, & J. C. Rupe (Eds.), *Compendium of soybean diseases*. (4th ed., pp. 66–68). St. Paul, MN: APS Press.

DeZeeuw, D. J., & Hooker, W. J. (1965). Additional hosts of the tobacco ringspot virus. *Quarterly Bulletin of the Michigan Agricultural Experiment Station, 48*, 76–80.

Domier, L. L., Hobbs, H. A., McCoppin, N. K., Bowen, C. R., Steinlage, T. A., Chang, S., et al. (2011). Multiple loci condition seed transmission of soybean mosaic virus (SMV) and SMV-induced seed coat mottling in soybean. *Phytopathology, 101*, 750–756.

Edwardson, J. R., & Christie, R. G. (1991). Soybean mosaic virus. The potyvirus group, vol. III. *University of Florida Monograph Series No. 16-111* (pp. 821–835).

Elena, S. F., Carreera, J., & Rodrigo, G. (2011). A systems biology approach to the evolution of plant-virus interactions. *Current Opinion in Plant Biology, 14*, 372–377.

Elena, S. F., & Rodrigo, G. (2012). Towards an integrated molecular model of plant-virus interactions. *Current Opinion in Virology, 2*, 719–724.

Fernandez-Suarez, R. (1984). Bibliographic review of some virus diseases of soya bean *(Glycine max):* Reporte de Investigacion del Instituto de Investigaciones Fundamentales. *Agricultural Tropical No. 16,* 31 pp.

Furutani, N., Yamagishi, N., Hidaka, S., Shizukawa, Y., Kanematsu, S., & Kosaka, Y. (2007). Soybean mosaic virus resistance in transgenic soybean caused by post-transcriptional gene silencing. *Breeding Science, 57,* 123–128.

Gardner, M., & Kendrick, J. B. (1921). Soybean mosaic. *Journal of Agricultural Research, 22,* 111–114.

Gergerich, R. C. (1999). Bean pod mottle. In G. L. Hartman, J. B. Sinclair, & J. C. Rupe (Eds.), *Compendium of soybean diseases.* (4th ed., pp. 61–62). St. Paul, MN: APS Press.

Gerlach, W. L., Llewellyn, D., & Haseloff, J. (1987). Construction of a plant disease resistance gene from the satellite RNA of tobacco ringsot virus. *Nature, 328,* 802–805.

Gibson, L., & Benson, G. (2005). *Origin, history and uses of soybean (Glycine max).* Available from, www.agron.iastate.edu/Courses/agron212/Readings/Soy_history.htm/, Accessed 12 March 2014.

Giesler, L. J., Ghabrial, S. A., Hunt, T. E., & Hill, J. H. (2002). Bean pod mottle virus: A threat to U.S. soybean production. *Plant Disease, 86,* 1280–1289.

Gore, M. A., Hayes, A. J., Jeong, S. C., Yue, Y. G., Buss, G. R., & Saghai Maroof, M. A. (2002). Mapping tightly linked genes controlling potyvirus infection at the *Rsv1* and *Rpv1* region in soybean. *Genome, 45,* 592–599.

Granillo, C. R., & Smith, S. H. (1974). Tobacco and tomato ringspot viruses and their relationships with *Tetranychus urticae. Phytopathology, 64,* 494–499.

Gu, H. C., Clark, A. J., de Sa, P. B., Pfeiffer, T. W., Tolin, S., & Ghabrial, S. A. (2002). Diversity among isolates of bean pod mottle virus. *Phytopathology, 92,* 446–452.

Gunduz, I., Buss, G. R., Chen, P., & Tolin, S. A. (2004). Genetic and phenotypic analysis of soybean mosaic virus resistance in PI 88788 soybean. *Phytopathology, 94,* 687–692.

Gupta, V. K. (1978). Further studies on bud blight disease of soybean. *Acta Botanica Indica, 6*(Suppl), 169–170.

Haidi, B., Bradshaw, J., Rice, M., & Hill, J. (2012). Bean leaf beetle (*Coleoptera: Chrysomelidae*) and bean pod mottle virus in soybean: Biology, ecology, and management. *Journal of Integrated Pest Management, 3*(1). http://dx.doi.org/10.1603/IPM11007, Accessed 12 March 2014.

Hajimorad, M. R., Eggenberger, A. L., & Hill, J. H. (2005). Loss and gain of elicitor function of soybean mosaic virus G7 provoking *Rsv1*-mediated lethal systemic hypersensitive response maps to P3. *Journal of Virology, 79,* 1215–1222.

Hajimorad, M. R., Eggenberger, A. L., & Hill, J. H. (2006). Strain-specific P3 of soybean mosaic virus elicits *Rsv1*-mediated extreme resistance, but absence of P3 elicitor function alone is insufficient for virulence on *Rsv1*-genotype soybean. *Virology, 345,* 156–166.

Hajimorad, M. R., Eggenberger, A. L., & Hill, J. H. (2008). Adaptation of soybean mosaic virus avirulent chimeras containing P3 sequences from virulent strains to *Rsv1*-genotype soybeans is mediated by mutations in HC-Pro. *Molecular Plant-Microbe Interactions, 21,* 937–946.

Hajimorad, M. R., & Hill, J. H. (2001). *Rsv1*-mediated resistance against soybean mosaic virus -N is hypersensitive response-independent at inoculation site, but has the potential to initiate a hypersensitive response-like mechanism. *Molecular Plant-Microbe Interactions, 14,* 587–598.

Hajimorad, M. R., Hu, C.-C., & Ghabrial, S. A. (1999). Molecular characterization of an atypical old world strain of peanut stunt virus. *Archives of Virology, 144,* 1587–1600.

Hajimorad, M. R., Wen, R.-H., Eggenberger, A. L., Hill, J. H., & Saghai Maroof, M. A. (2011). Experimental adaptation of an RNA virus mimics natural evolution. *Journal of Virology, 85,* 2557–2564.

Harrison, B., Steinlage, T. A., Domier, L. L., & D'Arcy, C. J. (2005). Incidence of soybean dwarf virus and identification of potential vectors in Illinois. *Plant Disease*, *89*, 28–32.

He, B., Fajolu, O. L., Wen, R.-H., & Hajimorad, M. R. (2010). Seed transmissibility of alfalfa mosaic virus in soybean. Online, *Plant Health Progress*. http://dx.doi.org/10.1094/PHP-2010-1227-01-BR.

He, B., Hill, J. H., & Hajimorad, M. R. (2011). Factors to improve detection of alfalfa mosaic virus in soybean. Online, *Plant Health Progress*. http://dx.doi.org/10.1094/PHP-2010-0926-02-RS.

Hebert, T. T. (1967). Epidemiology of the peanut stunt virus in North Carolina. *Phytopathology*, *57*, 461.

Hepperly, P. R., Bowers, G. R., Jr., Sinclair, J. B., & Goodman, R. M. (1979). Predisposition to seed infection by *Phomopsis sojae* in soybean plants infected by soybean mosaic virus. *Phytopathology*, *69*, 846–848.

Hill, J. H. (1999). Soybean mosaic. In G. L. Hartman, J. B. Sinclair, & J. C. Rupe (Eds.), *Compendium of soybean diseases*. (4th ed., pp. 70–71). St. Paul, MN: APS Press.

Hill, J. H. (2003). Soybean. In G. Loebenstein, & G. Thottappilly (Eds.), *Virus and virus-like diseases of major crops in developing countries* (pp. 386–389). Dordrecht, The Netherlands: Kluwer Academic Publishers.

Hill, J. H., Bailey, T. B., Benner, H. I., Tachibana, H., & Durand, D. P. (1987). Soybean mosaic virus: Effect of primary disease incidence on yield and seed quality. *Plant Disease*, *71*, 237–239.

Hill, J. H., Epstein, A. E., McLaughlin, M. R., & Nyvall, R. F. (1973). Aerial detection of tobacco ringspot virus-infected soybean plants. *Plant Disease Reporter*, *57*, 471–472.

Hill, J. H., Koval, N. C., Gaska, J. M., & Grau, C. R. (2007). Identification of field tolerance to bean pod mottle and soybean mosaic viruses in soybean. *Crop Science*, *47*, 212–218.

Hill, J. H., Lucas, B. S., Benner, H. I., Tachibana, H., Hammond, R. B., & Pedigo, L. P. (1980). Factors associated with the epidemiology of soybean mosaic virus in Iowa. *Phytopathology*, *70*, 536–540.

Hobbs, H. A., Hartman, G. L., Wang, Y., Hill, C. B., Bernard, R. L., Pedersen, W. L., et al. (2003). Occurrence of seed coat mottling in soybean plants inoculated with bean pod mottle virus and soybean mosaic virus. *Plant Disease*, *87*, 1333–1336.

Hobbs, H. A., Reddy, D. V. R., Rajewhwari, R., & Reddy, A. S. (1987). Use of direct antigen coating and protein A coating ELISA procedures for detection of three peanut viruses. *Plant Disease*, *71*, 747–749.

Hu, C.-C., & Ghabrial, S. A. (1998). Molecular evidence that strain BV-15 of peanut stunt cucumovirus is a reassortant between subgroup I and II strains. *Phytopathology*, *88*, 92–97.

Iizuka, M., & Yunoki, T. (1974). Peanut stunt virus isolated from soybeans, *Glycine max* Merr. *Bulletin of Tohoku National Agricultural Experimental Station*, *47*, 1–12.

Kanematsu, S., Eggenberger, A. L., & Hill, J. H. (1998). Comparison of soybean mosaic virus strains G2 and G3 with Japanese strains A and B. *Annals of the Phytopathological Society of Japan*, *64*, 607.

Khatabi, B., Wen, R. H., Hershman, D. E., Kennedy, B. S., Newman, M. A., & Hajimorad, M. R. (2012). Generation of polyclonal antibodies and serological analyses of nucleocapsid protein of soybean vein necrosis-associated virus; a distinct soybean infecting tospovirus serotype. *European Journal of Plant Pathology*, *133*, 783–790.

Kim, J., Kang, W.-H., Hwang, J., Yang, H.-B., Dosun, K., Oh, C.-S., et al. (2014). Transgenic Brassica rapa plants overexpressing eIF(iso)4E variants show broad-spectrum TuMV resistance. *Molecular Plant Pathology*, *15*, 615–626.

Kim, M. Y., Lee, S., Van, K., Kim, T. H., Jeong, S. C., Choi, I. Y., et al. (2010). Whole-genome sequencing and intensive analysis of the undomesticated soybean (*Glycine soja* Sieb. and Zucc.) genome. *Proceedings of the National Academy of Sciences of the United States of America*, *107*, 22032–22037.

Koning, G. R., TeKrony, D. M., Pfeiffer, T. W., & Ghabrial, S. A. (2001). Infection of soybean with soybean mosaic virus increases susceptibility to *Phomopsis* spp. seed infection. *Crop Science*, *41*, 1850–1856.

Kopisch-Obuch, F. J., Koval, N. C., Mueller, E. M., Paine, C., Grau, C. R., & Diers, B. W. (2008). Inheritance of resistance to alfalfa mosaic virus in soybean PI 153282. *Crop Science*, *48*, 933–940.

Kosaka, Y. (1997). Studies on causal viruses and control measures of soybean virus disease. *Bulletin of the Kyoto Prefecture Institute of Agriculture*, *20*, 1–100.

Krell, R. K., Pedigo, L. P., Hill, J. H., & Rice, M. E. (2003). Potential primary inoculum sources of bean pod mottle virus in Iowa. *Plant Disease*, *87*, 1416–1422.

Krell, R. K., Pedigo, L. P., Hill, J. H., & Rice, M. E. (2004). Bean leaf beetle (Coleoptera: Chrysomelidae) management for reduction of bean pod mottle virus. *Journal of Economic Entomology*, *97*, 192–202.

Lam, H.-M., Xu, X., Liu, X., Chen, W. B., Yang, G. H., Wong, F. L., et al. (2010). Resequencing of 31 wild and cultivated soybean genomes identifies patterns of genetic diversity and selection. *Nature Genetics*, *42*, 1053–1059.

Lee, Y. C., Kim, J. J., & Cho, E.-K. (1992). Classification of seed-borne SMV strains and resistance to SMV in leading soybean cultivars. *Korean Journal of Breeding*, *23*, 53–58.

Li, K., Yang, Q. H., Yang, H., Zhi, J., & Gai, J. Y. (2010). Identification and distribution of soybean mosaic virus strains in southern China. *Phytopathology*, *94*, 351–357.

Loesch-Fries, L. S., Merlo, D., Zinnen, T., Burhop, L., Hill, K., Krahn, K., et al. (1987). Expression of alfalfa mosaic virus RNA4 in transgenic plants confers virus resistance. *European Molecular Biology Organization Journal*, *6*, 1845–1851.

Malapi-Nelson, M., Wen, R.-H., Ownley, B. H., & Hajimorad, M. R. (2009). Co-infection of soybean with soybean mosaic virus and alfalfa mosaic virus results in disease synergism and alteration in accumulation level of both viruses. *Plant Disease*, *93*, 1259–1264.

Maule, A. J., Caranta, C., & Boulton, M. I. (2007). Sources of natural resistance to plant viruses: Status and prospects. *Molecular Plant Pathology*, *8*, 223–231.

McGuire, J. M., & Douthit, L. B. (1978). Host effect on acquisition and transmission of tobacco ringspot virus by *Xiphinema americanum*. *Phytopathology*, *68*, 457–459.

McLaughlin, M. R., Pederson, G. A., Evans, R. R., & Ivy, R. L. (1992). Virus diseases and stand decline in white clover pasture. *Plant Disease*, *76*, 158–162.

Mink, G. I., Hebert, T. T., & Silbernagel, M. J. (1967). A strain of peanut stunt virus isolated from beans in Washington. *Phytopathology*, *57*, 1400.

Mueller, E. E., & Grau, C. R. (2007). Seasonal progression, symptom development, and yield effects of alfalfa mosaic virus epidemics on soybeans in Wisconsin. *Plant Disease*, *91*, 266–272.

Naidu, R. A., Hu, C.-C., Pennington, R. E., & Ghabrial, S. A. (1995). Differentiation of eastern and western strains of peanut stunt cucumovirus based on satellite RNA support and nucleotide sequence homology. *Phytopathology*, *85*, 502–507.

National Pest Surveillance and Monitoring Systems—SENASA, Argentina. http://www.sinavimo.gov.ar/en/pest/peanut-mottle-virus-pmv/. Accessed 12 March 2004.

Nutter, F. W., Jr., Schultz, P. M., & Hill, J. H. (1998). Quantification of within-field spread of soybean mosaic virus in soybean using strain-specific monoclonal antibodies. *Phytopathology*, *88*, 895–901.

Orellana, R. G. (1981). Resistance to bud blight in introductions from the germplasm of wild soybean. *Plant Disease*, *65*, 594–595.

Pandey, A. K., Yang, C., Zhang, C., Graham, M. A., Horstman, H. D., Lee, Y., et al. (2011). Functional analysis of the Asian soybean rust resistance pathway mediated by *Rpp2*. *Molecular Plant-Microbe Interactions*, *24*, 194–206.

Pedersen, P., Grau, C., Cullen, E., Koval, N., & Hill, J. H. (2007). Potential for integrated management of soybean virus disease. *Plant Disease*, *91*, 1255–1259.

Redinbaugh, M. G., Molineros, J. E., Vacha, J., Berry, S. A., Hammond, R. B., Madden, L. V., et al. (2010). Bean pod mottle virus spread in insect-feeding-resistant soybean. *Plant Disease*, *94*, 265–270.

Ritzenthaler, C. (2005). Resistance to plant viruses: Old issue, new answers? *Current Opinion in Biotechnology*, *16*, 118–122.

Roane, C. W., Tolin, S. A., & Reid, P. H. (1978). Effects of peanut mottle virus (PMV) on twenty-five soybean cultivars. In *Potomac Division, American Phytopathological Society 35th Annual Meeting,* Abstract.

Ross, J. P. (1968). Effect of single and double infections of soybean mosaic and bean pod mottle viruses on soybean yield and seed characters. *Plant Disease Reporter*, *52*, 344–348.

Ross, J. P. (1969). Effect of time and sequence of inoculation of soybeans with soybean mosaic and bean pod mottle viruses on yields and seed characters. *Phytopathologische Zeitschrift*, *59*, 1404–1408.

Ross, J. P. (1977). Effect of aphid-transmitted soybean mosaic virus on yields of closely related and susceptible soybean lines. *Crop Science*, *17*, 869–872.

Saghai Maroof, M. A., Jeong, S. C., Gunduz, I., Tucker, D. M., Buss, G. R., & Tolin, S. A. (2008). Pyramiding of soybean mosaic virus resistance genes by marker-assisted selection. *Crop Science*, *48*, 517–526.

Sanfacon, H., Wellink, J., Le Gall, O., Karasev, A., van de Vlugt, R., & Wetzel, T. (2009). Secoviridae: A proposed family of plant viruses that combines the families Secquiviridae and Comoviridae, the unassigned genera Cheravirus and Sadwavirus, and the proposed genus Torradovirus. *Archives of Virology*, *154*, 899–907.

Saruta, M., Takada, Y., Kikucho, A., Yamada, T., Komatsu, K., Sayama, T., et al. (2012). Screening and genetic analysis of resistance to peanut stunt virus in soybean: Identification of the putative *Rpsv1* resistance gene. *Breeding Science*, *61*, 625–630.

Schmutz, J., Cannon, S. B., Schlueter, J., Ma, J., Mitros, T., Nelson, W., et al. (2010). Genome sequence of the palaeopolyploid soybean. *Nature*, *463*, 178–183.

Schneider, A. L., Damsteegt, V. D., Stone, A. L., Kuhlman, M., Bunyard, V. A., Sherman, D. J., et al. (2011). Molecular analysis of soybean dwarf virus isolates in the eastern United States confirms the presence of both D and Y strains and provides evidence of mixed infections and recombination. *Virology*, *412*, 46–54.

Schornack, S., Moscou, M. J., Ward, E. R., & Horvath, D. M. (2013). Engineering plant disease resistance based on TAL effectors. *Annual Review of Phytopathology*, *51*, 383–406.

Shakiba, E., Chen, P., Gergerich, R., Li, S., Dombek, D., & Shi, A. (2012a). Reactions of commercial soybean cultivars from the mid south to soybean mosaic virus. *Crop Science*, *52*, 1990–1997.

Shakiba, E., Chen, P., Gergerich, R., Li, S., Dombek, D., Shi, A., et al. (2012b). Reactions of mid-southern U.S. soybean cultivars to bean pod mottle virus and tobacco ringspot virus. *Crop Science*, *52*, 1980–1989.

Shakiba, E., Chen, P., Shi, A., Li, D., Dong, D., & Brye, K. (2012). Two novel alleles for resistance to soybean mosaic virus in PI 399091 and PI 61947. *Crop Science*, *52*, 2587–2594.

Sherwood, J. L., Sanborn, M. R., & Keyser, G. C. (1987). Production of monoclonal antibodies to peanut mottle virus and their use in enzyme-linked immunosorbent assay and dot-immunobinding assay. *Phytopathology*, 77, 1158–1161.

Shi, A., Chen, P., Li, D., Zheng, C., Zhang, B., & Hou, A. (2009). Pyramiding multiple genes for resistance to soybean mosaic virus in soybean using molecular markers. *Molecular Breeding*, *23*, 113–124.

Shipe, E. R., Buss, G. R., & Tolin, S. A. (1979). A second gene for resistance to peanut mottle virus in soybeans. *Crop Science*, *19*, 656–658.

Steinlage, T. A., Hill, J. H., & Nutter, F. W., Jr. (2002). Temporal and spatial spread of soybean mosaic virus (SMV) in soybeans transformed with the coat protein gene of SMV. *Phytopathology*, *92*, 478–486.

Takahashi, K., Tanaka, T., Wataru, I., & Tsuda, T. (1980). Studies on virus diseases and causal viruses of soybean in Japan. *Bulletin of Tohoku National Agricultural Experimental Station, 62*, 1–130.
Tamada, T. (1970). Aphid transmission and host rage of soybean dwarf virus. *Annals of the Phytopathological Society of Japan, 36*, 266–274.
Tamada, T., Goto, T., Chiba, I., & Suwa, T. (1969). Soybean dwarf, a new virus disease. *Annals of the Phytopathological Society of Japan, 35*, 282.
Terauchi, H., Kanematsu, S., Honda, K., Mikoshiba, Y., Ishiguro, K., & Hidaka, S. (2001). Comparison of complete nucleotide sequences of genomic RNA's of four soybean dwarf virus strains that differ in their vector specificity and symptom production. *Archives of Virology, 146*, 1885–1898.
Tolin, S. A. (1999). Diseases caused by viruses. In G. L. Hartman, J. B. Sinclair, & J. C. Rupe (Eds.), *Compendium of soybean diseases.* (4th ed., pp. 57–59). St. Paul, MN: APS Press.
Tolin, S. A., & Lacy, G. H. (2004). Viral, bacterial, and phytoplasmal diseases of soybean. In H. R. Boerma, & J. E. Specht (Eds.), *Soybeans; improvement, production and uses.* (3rd ed., pp. 765–819). Madison, WI: ASA/CSSA/SSSA.
Tougou, M., Furutani, N., Yamagishi, N., Shizukawa, Y., Takahata, Y., & Hidaka, S. (2006). Development of resistant transgenic soybeans with inverted repeat-coat protein genes of soybean dwarf virus. *Plant Cell Reports, 25*, 1213–1218.
Tougou, M., Yamagishi, N., Furutani, N., Shizukawa, Y., Takahata, Y., & Hidaka, S. (2007). Soybean dwarf virus-resistant transgenic soybeans with the sense coat protein gene. *Plant Cell Reports, 26*, 1967–1975.
Truniger, V., & Aranda, M. A. (2009). Recessive resistance to plant viruses. *Advances in Virus Research, 75*, 119–159.
Tu, J. C., Ford, R. E., & Quinones, S. S. (1970). Effects of soybean mosaic virus and/or bean pod mottle virus infection on soybean nodulation. *Phytopathology, 60*, 518–523.
United States Department of Agriculture, Economics Research Service. (2013). *Soybeans and Oil Crops.* Washington, DC: USDA-NASS. Available from, http://ers.usda.gov/topics/crops/soybeans-oil-crops.aspx#.UxTlryhy_Hg/, Accessed 12 March 2014.
Varala, K., Swaminathan, K., Li, Y., & Hudson, M. E. (2011). Rapid genotyping of soybean cultivars using high throughput sequencing. *PLoS One, 6*, e24811.
Walters, H. J. (1958). A virus disease complex in Arkansas. *Phytopathology, 48*, 346.
Wang, L., Eggenberger, A., Hill, J., & Bogdanove, A. J. (2006). *Pseudomonas syringae* effector *avrB* confers soybean cultivar-specific avirulence on soybean mosaic virus adapted for transgene expression but effector *avrPto* does not. *Molecular Plant-Microbe Interactions, 19*, 304–312.
Wang, X., Eggenberger, A. L., Nutter, F. W., Jr., & Hill, J. H. (2001). Pathogen-derived transgenic resistance to soybean mosaic virus in soybean. *Molecular Breeding, 8*, 119–127.
Wang, Y., Hobbs, H. A., Hill, C. B., Domier, L. L., Hartman, G. L., & Nelson, R. L. (2005). Evaluation of ancestral lines of U.S. soybean cultivars for resistance to four soybean viruses. *Crop Science, 45*, 639–644.
Wang, A., & Krishnaswamy, S. (2012). Eukaryotic translation initiation factor 4E-mediated recessive resistance to plant viruses and its utility in crop improvement. *Molecular Plant Pathology, 13*, 795–803.
Wang, R. Y., Kritzman, A., Hershman, D. E., & Ghabrial, S. A. (2006). *Aphis glycines* as a vector of persistently and nonpersistently transmitted viruses and potential risks for soybeans and other crops. *Plant Disease, 90*, 920–926.
Wen, R.-H., Khatabi, B., Ashfield, T., Saghai Maroof, M. A., & Hajimorad, M. R. (2013). The HC-Pro and P3 cistrons of an avirulent *Soybean mosaic virus* are recognized by different resistance genes at the complex *Rsv1* locus. *Virology, 26*, 203–215.

Wen, R. H., Saghai Maroof, M. A., & Hajimorad, M. R. (2011). Amino acid changes in P3, and not the overlapping pipo-encoded protein, determine virulence of soybean mosaic virus on functionally immune *Rsv1*-genotype soybean. *Molecular Plant Pathology, 12*, 799–807.
Whitham, S. A., & Wang, Y. (2004). Roles for host factors in plant viral pathogenicity. *Current Opinion in Plant Biology*, 7, 365–371.
Wickizer, S. L., & Gergerich, R. C. (2007). First report of Japanese beetle (*Popillia japonica*) as a vector of southern bean mosaic virus and bean pod mottle virus. *Plant Disease, 91*, 637.
Woodyard, C. (2010). *USA Today, June 29, 2010.* Available from, http://content.usatoday.com/communities/driveon/post/2010/06/mystery-car-40-henry-fords-soybean-car/1, Accessed 12 March 2014.
Wrather, J. A., & Koenning, S. (2010). *Soybean disease loss estimates for the United States, 1996–2010.* Available from, http://aes.missouri.edu/delta/research/soyloss.stm, Accessed 12 March 2014.
Yamashita, Y., Takeuchi, T., Ohnishi, S., Sasaki, J., & Tazawa, A. (2013). Fine mapping of the major soybean dwarf virus resistance gene *Rsdv1* of the soybean cultivar Wilis. *Breeding Science, 63*, 417–422.
Yang, Y., Lin, J., Zheng, G., Zhang, M., & Zhi, H. (2014). Recombinant soybean mosaic virus is prevalent in Chinese soybean fields. *Archives of Virology, 159*, 1793–1796.
Zadeh, A. H., & Foster, G. D. (2004). Transgenic resistance to tobacco ringspot virus. *Acta Virologica, 48*, 145–152.
Zaumeyer, W. J., & Thomas, H. R. (1948). Pod mottle, a virus disease of beans. *Journal of Agricultural Research*, 77, 81–96.
Zhang, C., Bradshaw, J. D., Whitham, S. A., & Hill, J. H. (2010). The development of an efficient multipurpose bean pod mottle virus viral vector set for foreign gene expression and RNA silencing. *Plant Physiology, 153*, 52–65.
Zhang, C., & Ghabrial, S. A. (2006). Development of bean pod mottle virus-based vectors for stable protein expression and sequence-specific virus-induced gene silencing in soybean. *Virology, 344*, 401–411.
Zhang, C., Grosic, S., Whitham, S. A., & Hill, J. H. (2012). The requirement of multiple defense genes in soybean *Rsv1*-mediated extreme resistance to soybean mosaic virus. *Molecular Plant-Microbe Interactions, 25*, 1307–1313.
Zhang, C., Hajimorad, M. R., Eggenberger, A. L., Tsang, S., Whitham, S. A., & Hill, J. H. (2009). Cytoplasmic inclusion cistron of soybean mosaic virus serves as a virulence determinant on Rsv3-genotpe soybean and a symptom determinant. *Virology, 391*, 240–248.
Zhang, Z., Sato, S., Ye, X., Dorrance, A. E., Morris, T. J., Clemente, T. E., et al. (2011). Robust RNAi-based resistance to mixed infection of three viruses in soybean plants expressing separate short hairpins from a single transgene. *Phytopathology, 101*, 1264–1269.
Zhou, J., Kantartzi, S. K., Wen, R. H., Newman, M., Hajimorad, M. R., Rupe, J. C., et al. (2011). Molecular characterization of a new tospovirus infecting soybean. *Virus Genes, 43*, 289–295.
Zhou, J., & Tzanetakis, I. E. (2013). Epidemiology of soybean vein necrosis-associated virus. *Phytopathology, 103*, 966–971.
Ziems, A. D., Giesler, L. J., Graef, G. L., Redinbaugh, M. G., Vacha, J. L., Berry, S. A., et al. (2007). Response of soybean cultivars to bean pod mottle virus infection. *Plant Disease, 91*, 719–726.

CHAPTER EIGHT

Control of Virus Diseases in Maize

Margaret G. Redinbaugh[*,1], **José L. Zambrano**[†]
[*]USDA, Agricultural Research Service, Corn, Soybean and Wheat Quality Research Unit and Department of Plant Pathology, Ohio State University-OARDC, Wooster, Ohio, USA
[†]Instituto Nacional Autónomo de Investigaciones Agropecuarias (INIAP), Programa Nacional del Maíz, Quito, Ecuador
[1]Corresponding author: e-mail address: redinbaugh.2@osu.edu

Contents

Abstract

Diseases caused by viruses are found throughout the maize-growing regions of the world and can cause significant losses for producers. In this review, virus diseases of maize and the pathogens that cause them are discussed. Factors leading to the spread

Advances in Virus Research, Volume 90
ISSN 0065-3527
http://dx.doi.org/10.1016/B978-0-12-801246-8.00008-1

of disease and measures for disease control are reviewed, as is our current knowledge of the genetics of virus resistance in this important crop.

1. INTRODUCTION

Maize is produced on nearly 177 million hectares in tropical, subtropical, and temperate regions. Worldwide production is about 875 million metric tons annually, more than wheat or rice production (Food and Agriculture Organizations, 2012). Most maize is used for animal feed, but in some parts of the world (e.g., sub-Saharan Africa) maize is a staple food crop. In the USA, almost 40% of the maize crop is used to produce biofuel. Maize is also a raw material used for manufacture of hundreds of other products and continued expansion and diversification of crop uses is expected. Because of its importance to agriculture and its critical role in food, feed, and fuel production, scientific resources for maize are well developed and include genome sequence, resources for high-density genotyping, and large and diverse germplasm collections, among many others (Schaeffer et al., 2011).

Plant viruses cause a significant proportion of crop diseases and economic losses around the world (Gomez, Rodriguez-Hernandez, Moury, & Aranda, 2009; Kang , Yeam, & Jahn, 2005), with losses in excess of $60 billion USD in crops annually (Hsu, 2002). In maize, Oerke and Dehne (2004) estimated annual losses at 3%, or almost $8 billion USD in 2012. Because the occurrence of virus diseases in the global maize crop is spotty, losses can be much higher locally or regionally. In 2011, smallholder farmers in severely affected areas of Kenya's South Rift Valley experienced more than 80% losses due to maize lethal necrosis (MLN), a disease caused by coinfection of plants with *Maize chlorotic mottle virus* (MCMV) and *Sugarcane mosaic virus* (SCMV) (see Table 8.1 for a summary of viruses and acronyms used; Wangai et al., 2012). Other viruses cause chronic losses in maize in the areas where they occur.

2. VIRUS DISEASES OF MAIZE

More than 50 viruses have been identified as infecting maize (Lapierre & Signoret, 2004). Typically, virus infection is first detected by the appearance of symptoms, including streaks, mosaics, and chlorosis. On older plants, leaves may become reddish or purple, and dwarfing or

Table 8.1 Virus diseases of maize discussed in this chapter

Disease	Pathogen	Abbrev.	Family	Dist.[a]	Primary vectors	Trans.[b]	Diag.[c]
Maize dwarf mosaic	*Maize dwarf mosaic virus*	MDMV	*Potyviridae*	Americas, Europe	Aphids	NP, M, S	CE, S, M
	Sugarcane mosaic virus	SCMV	*Potyviridae*	WW	Aphids		CE, S, M
Maize streak	*Maize streak virus*	MSV	*Geminiviridae*	Africa	*Cicadulina* spp.	P, V	CE, S, M
Maize chlorotic dwarf	*Maize chlorotic dwarf virus*	MCDV	*Secoviridae*	USA	*Graminella nigrifrons*	SP, V	CE, S, M
Maize mosaic	*Maize mosaic virus*	MMV	*Rhabdoviridae*	WW[d]	*Peregrinus maidis*	P, V	CE, S, M
Maize rough dwarf	*Maize rough dwarf virus*	MRDV	*Fijiviridae*	Europe	*Laodelphax striatellus*	P, V	CE, S, M
Mal de Rio Cuarto	*Rice black streaked dwarf virus*	RBSDV	*Fijiviridae*	Asia	*Laodelphax striatellus*		S, M
	Mal de Rio Cuarto virus	MRCV	*Fijiviridae*	S. Am.	*Delphacodes kuscheli*		
Maize stripe	*Maize stripe virus*	MSpV	*Tenuiviridae*	N. Africa	*Peregrinus maidis*	P, V	CE, S, M
Maize rayado fino	*Maize rayado fino virus*	MRFV	*Marafiviridae*	Americas	*Dalbulus maidis*	P, V	CE, S, M
Maize chlorotic mottle	*Maize chlorotic mottle virus*	MCMV	*Machlomovirus*	Americas, Hawaii, S.E. Asia, E. Africa	*Frankliniella williamsii* Beetles	NP, M, S	S, M
High Plains disease	*Wheat mosaic virus*	WMoV	*Emaravirus*	USA	Eriophyid mites	SP?	CE, S, M
Maize lethal necrosis	*Maize chlorotic mottle virus* Plus second virus[e]	MLN		Similar to MCMV		SP, M	na

[a]Distribution of the disease. WW, worldwide.
[b]Trans., modes of pathogen transmission. Vector transmission includes non-persistent (NP), semipersistent (SP), or persistent (P). "M" indicates that the virus is mechanically transmissible by leaf rub inoculation. "V" indicates obligate vector transmission. "S" indicates seed transmission.
[c]Available diagnostics. CE, commercial ELISA; S, antisera are available; M, virus genome sequence is available and/or PCR or RT-PCR-based assays have been developed.
[d]MMV is distributed throughout tropical and subtropical areas worldwide.
[e]Maize lethal necrosis occurs when maize infected by two viruses, generally MCMV plus a virus in the *Potyviridae*.

stunting is common in plants infected early in development. Symptoms on young plants are easier to distinguish than on plants after silking. Because maize-infecting viruses are transmitted primarily by insect vectors, diseased plants are usually either concentrated at the edges of fields or scattered throughout them. As in other crops, it is very difficult to diagnose virus diseases in maize based solely on symptoms, due to symptom variability based on plant genotype, time of infection, environmental conditions, and the potential for multiple infections. Therefore, visual diagnosis is verified using enzyme-linked immunosorbent assays (ELISA) or other serological tests, and molecular tests such as PCR or RT-PCR. These diagnostic tools are available for most maize-infecting viruses and are available for all of the diseases discussed in this chapter (Table 8.1).

At least a dozen viruses from eight families cause significant agronomic problems in maize worldwide (Louie, 1999; Redinbaugh & Pratt, 2008), and brief descriptions of the most widely prevalent maize virus diseases and the pathogens that cause them follow. Several other viruses are included because of their use as models in genetic studies: *Wheat streak mosaic virus* (WSMV), a tritimovirus in the *Potyviridae* that is transmitted by the wheat curl mite (*Aceria tosichella* Keifer); *Maize fine streak virus* (MFSV), a nucleorhabdovirus isolated from maize in southwest Georgia in the USA; and *Maize necrotic streak virus* (MNeSV), a virus in the *Tombusviridae* isolated from maize in Arizona (Louie, Redinbaugh, Gordon, Abt, & Anderson, 2000; McMullen, Jones, Simcox, & Louie, 1994; McMullen & Louie, 1991; Redinbaugh et al., 2002; Zambrano, Jones, Brenneret, et al., 2014). A number of other diseases that cause problems in limited geographic areas (e.g., *Maize white line*) and/or for which little information is available (e.g., Maize eye spot) are not included. Information on these diseases is described elsewhere (Lapierre & Signoret, 2004; Louie, 1999).

2.1. Maize dwarf mosaic

Maize dwarf mosaic burst into the news as an epidemic of maize in Ohio that destroyed the corn crop in 1963 and 1964 (Williams & Alexander, 1965). The disease is caused by viruses in the family *Potyviridae*, primarily MDMV and SCMV, and is arguably the most widespread disease of maize worldwide (Ali & Yan, 2012; Shulka, Wardand, & Brunt, 1994). MDMV is prevalent in North America and Europe and is also found in Central and South America (Lapierre & Signoret, 2004). SCMV (formerly known as MDMV-B) is found worldwide. Several other members of the family *Potyviridae* infect maize, including *Sorghum mosaic virus*, *Johnsongrass mosaic virus*, and WSMV, but these are of minor agronomic significance. All of these viruses incite mosaic symptoms and dwarfing in susceptible maize hybrids and cultivars and are naturally transmitted in a nonpersistent manner by about 25 aphid species. MDMV and SCMV also are transmitted through seed at a low rate (<0.5%). Like all potyviruses, MDMV and SCMV have single-stranded positive sense RNA genomes and flexuous rod-shaped virions of about 12×750 nm. The maize-infecting potyviruses are serologically distinct and have clearly distinct genome sequences (Adams, Antoniw, & Fauquet, 2005); they also can be distinguished based on their infection of oats (*Avena sativa* L.) and several sorghum (*Sorghum vulgare* L.) cultivars (Tosic, Ford, Shukla, & Jilka, 1990). Weeds and other poaceous hosts of MDMV and SCMV are important natural reservoirs of the viruses, with Johnsongrass (*Sorghum halepense* L.) being particularly important in MDMV epidemiology (Knoke, Louie, Madden, & Gordon, 1983). MDMV infects most sorghum cultivars, but SCMV infects just a few. SCMV, but not MDMV, infects sugarcane (*Saccharum* spp.).

All of the maize-infecting *Potyviridae* are easily, mechanically transmitted under greenhouse and field conditions (Louie, 1985; Louie, Knoke, & Reichard, 1983). The ease of transmission, coupled with the importance of the disease they cause, has led to extensive genetic studies of maize resistance to the potyviruses and their tritimovirus cousin, WSMV (Redinbaugh & Pratt, 2008).

2.2. Maize streak

Maize streak has been the most important and widespread virus disease of maize in sub-Saharan Africa and several Indian Ocean Islands for more than 100 years (Fuller, 1901; Lapierre & Signoret, 2004; Storey, 1925; Thottappilly, Bosque-Perez, & Rossel, 1993). The disease continues to cause significant losses and food insecurity across sub-Saharan Africa (Martin & Shepherd, 2009). The disease is caused primarily by *Maize streak virus*-strain A (MSV-A; family *Geminiviridae*; genus *Mastrevirus*), which is transmitted in a persistent manner by leafhoppers in the genus *Cicadulina*, especially *C. mbila* (Naudé). Typical symptoms include longitudinal chlorotic streaks along the leaf veins, and a reduction in plant growth and yield. Early planted maize crops serve as reservoirs of both virus and vectors in regions with staggered, overlapping growing seasons, and a number of wild grasses can also serve as virus reservoirs (Konate & Traore, 1992; Shepherd et al., 2010). Maize streak is managed using virus-resistant maize hybrids and cultivars, by removing virus reservoirs, and by reducing vector populations.

MSV can be transmitted using viruliferous vectors, and this approach is often used in genetic studies and in the development of resistant hybrids and cultivars. In the laboratory, MSV and virus clones can be transmitted using vascular puncture inoculation (VPI, described later) (Redinbaugh, 2003; Redinbaugh et al., 2001), and clones can be transmitted using agroinfiltration (Boulton, Pallaghy, Chatani, MacFarlane, & Davies, 1993).

2.3. Maize chlorotic dwarf

Among the virus diseases of maize that emerged rapidly and destructively in the 1960s in the USA was one caused by a "stunting agent" that was transmitted primarily by the black-faced leafhopper *Graminella nigrifrons* (Forbes) (Nault, Styer, Knoke, & Pitre, 1973; Rosencranz, 1969; Stewart, 2011). This disease is now called maize chlorotic dwarf caused by *Maize chlorotic dwarf virus* (MCDV; family *Secoviridae*, genus *Waikavirus*). By the 1970s, the disease had spread across the southeastern US west to Texas and north to Ohio. MCDV has a positive sense RNA genome and a ca. 30 nm in diameter icosahedral virion (Stewart, 2011). MCDV causes characteristic vein banding (a clearing of tertiary leaf veins) in leaves of infected plants. Leaf twist and tear, chlorosis, stunting, reddening or yellowing, and plant dwarfing are also common. The virus is transmitted by *G. nigrifrons* in a semipersistent manner (Nault et al., 1973), and there is no evidence for seed transmission. In addition to maize, the primary natural and reservoir hosts of MCDV are Johnsongrass, sorghum, *Setaria* spp., wheat, and crabgrass (*Digitaria saguinalis* L.). MCDV can be experimentally transmitted with vectors, but not by leaf rub inoculation.

Maize chlorotic dwarf is controlled through the use of resistant or tolerant hybrids, eradication of reservoir hosts, principally Johnsongrass, with selective sulfonylurea herbicides (Green & Ulrich, 1993; Kamau, 1990). MCDV is still present in Johnsongrass and can be detected in sweet corn in southern Ohio (Stewart et al., in press). Insecticide seed treatments are also useful for controlling the vector, and can provide some disease control (Kuhn, Jellum, & All, 1975).

2.4. Maize mosaic

Maize mosaic caused by *Maize mosaic virus* (MMV; family *Rhabdoviridae*; genus *Nucleorhabdovirus*) was first reported from Venezuela in 1960

(Herold, Bergold, & Weibel, 1960) and remains common in tropical and subtropical regions of Africa, the Americas, and Hawaii where its primary vector, the maize planthopper (*Peregrinus maidis* Ashmead), is present and sequential overlapping maize crops are grown (Brewbaker, 1979, 1981; Lapierre & Signoret, 2004). Typical maize mosaic symptoms include fine yellow stripes along leaf veins, but not leaf mosaic, despite the name. Stunting and yield loss also are common. MMV has a unipartite, single-stranded negative sense RNA genome and a bullet-shaped, membrane-bound virions (Jackson, Dietzgen, Goodin, Bragg, & Deng, 2005; Redinbaugh, Ammar, & Whitfield, 2012; Reed, Tsai, Willie, Redinbaugh, & Hogenhout, 2005). MMV is transmitted in a persistent propagative manner by *P. maidis* (Falk & Tsai, 1985), and the virus is not seed transmitted. Other natural hosts of MMV include teosinte (*Zea* spp.), itchgrass (*Rottboellia cochinchinensis* (Lour.) Clayton), and plains bristlegrass (*Setaria vulpiseta* (Lam.) Roemer & J.A. Schultes). Maize and teosinte are the major reservoir hosts. MMV can be transmitted using vectors, but not by leaf rub inoculation.

Maize mosaic is controlled with resistant or tolerant hybrids and cultivars. Controlling vector populations can also be helpful for disease control.

2.5. Maize stripe

Maize stripe was discovered in Florida in 1974 and is now known to occur in Africa, the Americas, Australia, and Asia, with its distribution overlapping that of maize mosaic and its vector *P. maidis* (Ammar, Gingery, & Madden, 1995; Falk & Tsai, 1998). The disease can cause large losses in Africa and is caused by *Maize stripe virus* (MSpV; genus *Tenuivirus*), a non-enveloped single-stranded RNA virus with five ambisense genome segments (Falk & Tsai, 1998). It is transmitted in a persistent propagative manner, and natural hosts include maize, sorghum, and itchgrass, similar to MMV. Maize stripe is easily confused with maize mosaic, because of similarities in symptoms, hosts, and vector species (Greber, 1981).

Maize stripe is managed through resistant hybrids and cultivars, and by controlling vector populations. A related virus, *Maize yellow stripe virus*, is transmitted by *C. chinai* Ghauri and found in Egypt (Ammar, Khlifa, Mahmoud, Abol-Ela, & Peterschmitt, 2007; Mahmoud et al., 2007).

2.6. Maize rayado fino (maize fine stripe)

Maize rayado fino was identified in the late 1960s in El Salvador and Costa Rica, and it continues to be one of the most important virus diseases of maize in parts of Mexico, Central, and South America (Gamez, 1969, 1976; Vasquez & Mora, 2007). Maize rayado fino is caused by *Maize rayado fino virus* (MRFV; family *Tymoviridae*; genus *Marafivirus*). The virus has a monopartite single-stranded RNA genome and a 30-nm icosahedral virion (Lapierre & Signoret, 2004). In nature, MRFV transmitted in a persistent propagative manner by the maize leafhopper, *Dalbulus maidis* (DeLong & Wolcott) (Rivera & Gamez, 1986), which is also an efficient vector for two other important pathogens, corn stunt spiroplasma (*Spiroplasma kunkelii*; CSS) and maize bushy stunt phytoplasma (MBSP; *Candidatus* Phytoplasma astri subgroup 16SrI-B). Together, these three pathogens form an aggressive disease complex known as "*achaparramiento*" or red stunt (Nault, 1990). Maize rayado fino symptoms include small chlorotic spots on young leaves that become elongated and more numerous along leaf veins as the plants age. Infected plants are stunted and grain yield is reduced (Bustamante, Hammond, & Ramirez, 1998; Toler, Skinner, Bockholt, & Harris, 1985; Vasquez & Mora, 2007). Maize rayado fino and achaparramiento are problems where overlapping maize crops provide a reservoir of pathogens and continuous food and breeding hosts for *D. maidis*.

The incidence of maize rayado fino can be reduced by decreasing vector populations with maize-free periods or insecticide treatments. MRFV-resistant lines and QTL for MRFV resistance have been identified (Bustamante et al., 1998; Vandeplas, 2003; Zambrano, Francis, & Redinbaugh, 2013; Zambrano, Jones, Francis et al., 2014), resistant hybrids and cultivars are not yet available.

2.7. Maize rough dwarf

Maize rough dwarf first appeared in Europe in the late 1940s, and it has been reported across southern Europe (Lapierre & Signoret, 2004). The disease emerged in Japan in the 1950s, and in China and Korea in the 1960s. Maize rough dwarf continues to be a major problem in China, where agronomic practices facilitate large populations of viruliferous vectors. The disease is caused by *Maize rough dwarf virus* (MRDV) and *Rice black streaked dwarf virus* (RBSDV), two closely related viruses (family *Reoviridae*; genus *Fijivirus*) (Lapierre & Signoret, 2004). The genomes of both viruses consist of 10 double-stranded RNA segments that are encased in complex, spherical ca. 70 nm diameter virions. Antisera for MRDV and RBSDV cross-react (Margaret Redinbaugh, unpublished results), and the capsid protein sequences are >85% identical (Fauquet, 2005). Both viruses are transmitted by the small brown planthopper (*Laodelphax striatellus* Fallen) in a persistent propagative manner, and both are transmitted transovirally to progeny from infected vectors. The diagnostic symptom of maize rough dwarf is small (<2 mm) irregular growths, or enations, on the underside of leaves that make the leaves rough to the touch. Infected plants also tend to be dark green in color, have shortened internodes leading to dwarfing, and poorly developed root systems. Yield losses can be significant (Li, Li, & Yuan, 1999). MRDV and RBSDV infect oats, wheat, crabgrass, bermudagrass (*Cyndon dactylon*), and cockspur (*Echinochloa crus-galli*).

Agronomic controls for the disease include reducing vector populations though insecticide applications, use of planting dates that avoid high vector populations, and planting tolerant lines (Li et al., 1999). Because MRDV and RBSDV are obligately transmitted by vectors, identification and development of highly resistant lines, hybrids, and cultivars have been slow. However, recent studies have identified strong quantitative trait loci (QTLs) for resistance in diverse germplasm (Luan, Wang, Li, Zhang, & Zhang, 2012; Shi et al., 2012; Tao, Liu, et al., 2013).

2.8. Mal de Río Cuarto

Mal de Río Cuarto first appeared in the Cordoba region of Argentina in the 1960s and has since been reported from Uruguay. Mal de Río Cuarto reduces yield, quality, and economic value of maize in parts of South America, mainly Argentina and Uruguay (Bonamico et al., 2010; Lapierre & Signoret, 2004), where it is a major concern for US companies that produce hybrid seed in Argentina. The disease is caused by *Mal de Río Cuarto virus* (MRCV; family *Reoviridae*; genus *Fijivirus*), which is closely related to MRDV and RBSDV. Antisera for any of the three viruses cross-react with all three viruses, and initially MRCV was proposed to be a strain of MRDV (Milne, Boccardo, Bo, & Nome, 1983). However, genome sequence analysis identifies MRCV as a separate virus species (Distefano et al., 2002). Disease symptoms are similar to those incited by MRDV and RBSDV and include leaf enations, dark green color, and shortened internodes. MRCV is transmitted in a persistent propagative manner by the planthoppers *Delphacodes kuscheli* Fennah and *D. haywardi* Muir in a persistent propagative manner (Arneodo, Guzman, Conci, Laguna, & Truol, 2002; Velazquez, Arneodo, Guzman, Conci, & Truol, 2003), and the virus is transovarially transmitted at a fairly high rate. Reservoir hosts of MRCV include wheat, oats, and several other grasses (Pardina, Gimenez, Laguna, Dagoberto, & Truol, 1998). Mal de Río Cuarto occurs in temperate areas where maize and winter wheat are both planted, with wheat serving as both a virus reservoir and a feeding and reproductive host of the vectors.

Similar to MRDV and RBSDV, MRCV can be managed by controlling vector populations with insecticides (March et al., 2002). Reducing weedy virus reservoirs can also help reduce disease. MRCV resistance has been fairly recently identified and mapped (Di Renzo et al., 2004; Martin et al., 2010), and it is not known whether resistant hybrids and cultivars are widely available.

2.9. Maize chlorotic mottle and corn (maize) lethal necrosis

Maize chlorotic mottle was initially described from Peru in 1971 (Castillo-Loayza, 1979). The disease is caused by MCMV (family *Tombusviridae*; genus *Machlomovirus*) (Lapierre & Signoret, 2004). MCMV has been identified in the central mainland USA, Hawaii, Mexico, Argentina, Thailand, and China (Lapierre & Signoret, 2004; Xie et al., 2011). Most recently, the virus has emerged in eastern sub-Saharan Africa (Wangai et al., 2012). Typical symptoms include mild chlorotic mottling and mosaic, although symptoms can be more severe depending on the maize genotype and plant age at the time of infection. MCMV has a single-stranded positive sense RNA genome that is encased in a 30-nm icosahedral virion. The virus is transmitted in a semipersistent manner by both Chrysomelid beetles and thrips (*Frankliniella williamsii* Hood and *F. occidentalis* Pergande) (Cabanas, Watanabe, Higashi, & Bressan, 2013; Nault et al., 1978; Zhao, Ho, Wu, He, & Li, 2014). Although there are a number of poaceous experimental hosts of MCMV, maize and sugarcane are the only known natural hosts (Lapierre & Signoret, 2004; Wu, Zhou, & Wu, 2014). Transmission of MCMV through seed and soil has been described (Jensen, Wysong, Ball, & Higley, 1991; Uyemoto, 1983).

A synergistic disease, MLN, occurs in maize simultaneously infected with MCMV and another virus, most frequently a potyvirus (i.e., MDMV or SCMV) (Niblett & Claflin, 1978). However, other viruses, including MRFV, MCDV, and MMV, can also cause synergistic reactions in coinfections with MCMV, and abiotic stresses can also exacerbate MCMV infections. With the exception of Hawaii, MLN is the primary disease problem wherever MCMV occurs, and MCMV and SCMV were present in disease epiphytotics that occurred in Kenya in 2011 and 2012 (Adams et al., 2013; Wangai et al., 2012). For MLN, field symptoms include a bright mosaic developing into tissue chlorosis and necrosis with plant stunting and death or a severe necrosis of terminal leaves (Niblett & Claflin, 1978; Uyemoto, Claflin, Wilson, & Raney, 1981a, 1981b).

In the US corn belt, outbreaks of corn lethal necrosis are fairly rare, with the last one occurring in 1988 (Jardine, 2014). The disease is managed by controlling vectors, crop rotation, and planting tolerant hybrids. In Hawaii, aggressive thrips management with insecticides, maintaining maize-free periods, crop rotation, and use of potyvirus-resistant and MCMV-tolerant germplasm are used for disease management (Nelson, Brewbaker, & Hu, 2011).

2.10. High Plains disease

High Plains disease was first reported in 1996 from the western USA (Jensen, Lane, & Seifers, 1996). Since then, the virus has spread as far east as Ohio and Florida (Stewart, Paul, et al., 2013). The disease is found on all types of maize, but is more prevalent on sweet corn and inbred lines used for hybrid development. In wheat, another economically important host, the disease frequently occurs with wheat streak mosaic. High Plains disease is caused by an ambisense-segmented RNA virus with high similarity to viruses in the family *Emaravirus* (Mielke-Ehret & Mühlbach, 2012). No formal species name has been assigned to this virus, and it is variously referred to as High Plains virus, wheat mosaic virus (WMoV), and maize red stripe virus (Skare et al., 2006). WMoV will be used to designate the virus in this chapter. Symptoms include small white spots along veins, mosaic, chlorosis and necrosis, reddening or red striping on leaves, and chlorosis at the leaf base. The virus is transmitted by the wheat curl mite (Seifers, Harvey, Martin, & Jensen, 1997), which also transmits WSMV. There is some evidence for a low level of seed transmission (Jensen et al., 1996). In addition to maize, WMoV infects wheat, barley, and several grasses.

High Plains disease occurs primarily in areas where winter wheat and maize are grown together. In spring, viruliferous vectors move from mature WMoV-infected wheat to maize seedlings to cause disease. Systemic insecticide treatments in wheat have limited effectiveness and are not generally economical. Planting maize 10–30 days after heading (spike emergence) in adjacent wheat plots reduced disease incidence (Fritts, Michels, & Rush, 1999). Resistance to WMoV is found in potyvirus-resistant maize germplasm, but it is inconsistently deployed in sweet corn and germplasm.

3. DISEASE EMERGENCE AND CONTROL

All of the economically important virus diseases in maize are transmitted by arthropod vectors, so three components, the virus, its vector, and a susceptible host, must come together in a suitable environment for disease to occur. Because of this complexity, the occurrence of virus diseases in the crop tends to be sporadic, with severe disease 1 year and little or moderate disease the next (Madden, Jeger, & Van Den Bosch, 2000).

From the descriptions of the virus diseases above, some factors arise as being frequently important in virus disease development in maize. The "sudden" appearance of emerging diseases can frequently be associated either with the emergence of a vector or with increased vector populations. For example, outbreaks of MCMV in Hawaii and Kenya occurred concurrently with outbreaks of maize thrips (Jiang, Meinke, Wright, Wilkinson, & Campbell, 1992; Jiang, Wilkinson, & Berry, 1990; Nyasani, Meyhöfer, Subramanian, & Poehling, 2012; Wangai et al., 2012). Similarly in Serbia, outbreaks of maize redness, caused by stolbur phytoplasma, occurred simultaneously with dramatically increased populations of a new stolbur vector (*Reptalus panzeri* Löw) that could transmit the pathogen to maize (Jovic et al., 2009). In other cases, chronic problems with endemic virus diseases can result from cropping practices that lead to emergence of high vector populations. For example, in temperate parts of China where annual dual cropping of winter wheat and maize leads to high populations of viruliferous *L. striatellus* populations, maize rough dwarf caused by RBSDV is a frequent disease problem (Li et al., 1999).

The most common approaches for management of virus diseases in maize include (1) breaking pathogen–vector and pathogen–maize interactions by removing virus reservoirs; (2) breaking vector–maize interactions by reducing vector numbers on susceptible maize or deploying insect resistant crops; and (3) breaking pathogen–maize interactions by deploying virus-resistant crops.

3.1. Removal of pathogen reservoirs

For maize virus diseases, vectors, weedy grasses, other poaceous crops, and maize itself can serve as reservoirs or sources of disease-causing viruses. In tropical regions such as Hawaii, Africa, and Central America, year-round maize production can result in sequential, overlapping crops with no maize-free periods. Here, the older maize crop can serve as a virus reservoir for a newly planted crop (Brewbaker, 1981; Hruska, Gladstone, & Obando, 1996; Shepherd et al., 2010). The mature maize crop also serves as a source of vectors for transmitting viruses. For seed-transmitted viruses such as the potyviruses, WMoV, and MCMV, maize can also serve as a pathogen source (Hohmann, Fuchs, Gruntzig, & Oertel, 1999; Jensen et al., 1991; Lebas, Ochoa-Corona, Elliott, Tang, & Alexander, 2005). It is important to note that the presence of a virus in maize seed does not always mean that the virus is transmitted (Hohmann et al., 1999), because of the possibility for phytosanitary conditions and restrictions on maize shipment that may be implemented in an effort to control diseases caused by seed-transmitted viruses.

In temperate regions, Johnsongrass is an important perennial reservoir of MDMV and MCDV (All, 1983). Weed eradication reduced the incidence of these diseases in the USA, but required very high levels of weed control. In maize–soybean rotations, control of Johnsongrass in the soybean crop has positive effects on virus disease control in the maize crop. Because MDMV and MCDV are acquired within minutes of feeding on virus-infected weeds, using post-emergent herbicides to control Johnsongrass after maize crops have emerged can make viruliferous vectors move from dying weeds to maize and increase disease incidence (Eberwine & Hagood, 1995; Eberwine, Hagood, & Tolin, 1998). Other poaceous crops and weeds can also serve as vector and virus reservoirs, with winter wheat serving as an important overwintering host for both the viruses and their vectors. Maize and wheat are frequently grown in close proximity in China and Argentina, leading to an increase in RBSDV and MRCV incidence (Grilli, 2008; Li et al., 1999). In these cases, delaying the maize planting date to avoid large populations of vectors can provide some disease control.

Vectors themselves can be reservoirs of propagatively transmitted pathogens. Recent outbreaks of maize rough dwarf in Europe were associated with high *L. striatellus* populations when the maize crop was at early developmental stages. The increased vector populations were thought to be the result, at least in part, of vector migration into the region (Achon, Subira, & Sin, 2013). In areas where maize is not grown in the winter, corn stunt-infected *D. maidis* overwinters on other crops such as alfalfa and returns to infect a new maize crop the next spring. In fact, corn stunt infection improved vector survival (Ebbert & Nault, 1994). Here, measures that reduce vector populations can be helpful.

3.2. Distrupting vector–maize interactions

Treatments that decrease vector populations reduce interactions between the vectors and maize and can be useful for controlling virus diseases. For persistently and semipersistently transmitted viruses, including MRDV, insecticide treatments directed at vector populations can reduce disease incidence (All, 1983; Antignus, Klein, & Ovadia, 2008; Perring, Gruenhagen, & Farrar, 1999). However, insecticide treatments are not generally effective for nonpersistently transmitted viruses that are transmitted within a few minutes of insect feeding (Jones, 2006). In other cases, adjusting planting dates and crop rotations have provided some disease control by limiting the access of viruliferous vectors to crops (Fritts et al., 1999; Gregory & Ayers, 1982; Hruska & Peralta, 1997). Barriers, such as tall plants, can also reduce vector access to maize and provide control for diseases caused by nonpersistently transmitted viruses (Jones, 2006). Another emerging area of research is to deploy vector-resistant crops, and maize inbred lines with resistance to *P. maidis*, *C. mbila*, and the aphid *Rhopalosiphum maidis* (Fitch) have been characterized (Dintinger, Boissot, Chiroleu, Hamon, & Reynaud, 2005; Dintinger, Verger, et al., 2005; Jones, 2006; Kairo, Kiduyu, Mutinda, & Empig, 1995; So, Ji, & Brewbaker, 2010). The rapidly developing genome information on maize virus vectors is expected to be key to identifying targets for development of vector-resistant maize (Cassone, Chen, Chiera, Stewart, & Redinbaugh, 2014; Chen, Cassone, Bai, Redinbaugh, & Michel, 2012; Whitfield, Rotenberg, Aritua, & Hogenhout, 2011).

3.3. Pathogen resistance in maize

Deploying virus-resistant crops is considered the most economically and environmentally sustainable approach to controlling disease. Maize is a highly diverse crop, and remarkable germplasm resources are available worldwide to maize breeders and geneticists (Romay et al., 2013). It is therefore not surprising that lines with moderate to strong resistance to the major maize-infecting viruses have been identified (Table 8.2). Genetic analyses of resistant lines indicate that virus resistance in maize is primarily monogenic or oligogenic and is either dominant or quantitative in character. Major and minor genes associated with virus resistance have been mapped using a number of different approaches by taking advantage of the array of genotyping tools that have been developed over the past 25 years and made available to the maize genetics community (Flint-Garcia, 2013; Schaeffer et al., 2011). Thus, the genetic materials and genotypic information needed

Table 8.2 Distribution of virus resistance genes and quantitative trait loci (QTL) in the maize genome

Chr[a]	Bin	Virus[b]	Locus[c]	Type	LOD	R^{2}[d]	Screen[e]	RS[f]	References
1	1.03	MRCV		QTL	3.8	30	NI	BLS14	Di Renzo et al. (2004)
	1.03	MMV		QTL	4.4	5	VI/GH	Oh1VI	Zambrano, Jones, Brenner, et al. (2014)
	1.05	MSV	*msv1*	QTL	36.8	59	VI/F	CML202	Welz, Schechert, Pernet, Pixley, and Geiger (1998)
								Tzi4	Kyetere et al. (1999)
								D211	Pernet, Hoisington, Dintinger, et al. (1999)
								CIRAD390	Pernet, Hoisington, Franco, et al. (1999)
2	2.01	MCDV		QTL	5.1	7	VI/GH	Oh1VI	Zambrano, Jones, Brenner, et al. (2014)
	2.02	MRCV		QTL	40.0	nr[g]	NI	Multiple	Martin et al. (2010)
	2.02	RBSDV	*qMRD2*	QTL	4.3	11	NI	90110	Luan et al. (2012)
	2.02	MSpV		QTL	nr	15	VI/F	Rev81	Dintinger, Boissot, et al. (2005)
	2.06	MSV		QTL	3.4	8	VI/F	CML202	Welz et al. (1998)
	2.08	MMV		QTL	22.9	32	VI/GH	Oh1VI	Zambrano, Jones, Brenner, et al. (2014)
	2.08	MFSV		QTL	40.5	60	VI/GH	Oh1VI	Zambrano, Jones, Brenner, et al. (2014)
	2.09	MSpV		QTL	nr	76	VI/GH	Rev81	Dintinger, Boissot, et al. (2005)
3	3.05	WSMV	*Wsm2*	Gene	nr	nr	MI/GH	Pa405	McMullen et al. (1994)
				Gene	7.0	10	MI/F	Oh1VI	Zambrano, Jones, Brenner, et al. (2014)
	3.05	SCMV	*Scm2*	QTL	nr	28	MI/F	D32	Xia, Melchinger, Kuntze, and Lübberstedt (1999)
					nr	15	MI/F	FAP1360A	Dussle, Melchinger, Kuntze, Stork, and Lubberstedt (2000)
					6.6	14	MI/F	Hongzao4	Zhang et al. (2003)
					10.4	13	MI/F	Oh1VI	Zambrano, Jones, Brenner, et al. (2014)

Continued

Table 8.2 Distribution of virus resistance genes and quantitative trait loci (QTL) in the maize genome—cont'd

Chr	Bin	Virus	Locus	Type	LOD	R^2	Screen	RS	References
	3.05	MMV	*mv1*	QTL	14.1	74	NI	Hi31	Ming, Brewbaker, Pratt, Musket, and McMullen (1997)
				QTL	5.3	6	VI/GH	Oh1VI	Zambrano, Jones, Brenner, et al. (2014)
	3.05	MCDV	*mcd1*	QTL	23.7	25	VI/GH	Oh1VI	Jones, Redinbaugh, Anderson, and Louie (2004)
					13.8	18	VI/GH	Oh1VI	Zambrano, Jones, Brenner, et al. (2014)
	3.05	MSpV		QTL	nr	63	VI/F	Rev81	Dintinger, Boissot, et al. (2005)
	3.05	WMoV		Gene	nr	nr	VI/GH	B73	Marçon et al. (1999)
	3.06	MSV		QTL	3.1	7	VI/F	CML202	Welz et al. (1998)
	3.09	MSV		QTL	3.9	8.3	VI/F	CIRAD390	Pernet, Hoisington, Franco, et al. (1999)
4	4.08	MSV		QTL	3.2	8	VI/F	CML202	Welz et al. (1998)
5	5.04	MSpV		QTL	nr	13	VI/F	Rev81	Dintinger, Boissot, et al. (2005)
	5.05	SCMV		QTL	3.6	6	MI/F	Hongzao4	Zhang et al. (2003)
6	6.01	WSMV	*Wsm1*	Gene	nr	nr	MI/GH	Pa405	McMullen et al. (1994)
					8.3	12	MI/GH	Oh1VI	Zambrano, Jones, Brenner, et al. (2014)
	6.01	MDMV	*Mdm1*	Gene	nr	nr	MI/GH	Pa405	McMullen and Louie (1989)
					93.5	79	MI/F	Oh1VI	Zambrano, Jones, Brenner, et al. (2014)
	6.01	SCMV	*Scm1*	QTL	nr	55	MI/F	D32	Xia et al. (1999)
					nr	64	MI/F	FAP1360A	Xu, Melchinger, Xia, and Lübberstedt (1999)
					4.0	8	MI/F	Hongzao4	Zhang et al. (2003)
					13.6	18	MI/F	Oh1VI	Zambrano, Jones, Brenner, et al. (2014)

	6.01	WMoV		Gene	nr	nr	VI/GH	B73	Marcon et al. (1999)
	6.01	MCDV		QLT	3.6	4	VI/GH	Oh1VI	Jones et al. (2004)
					9.5	12	VI/GH	Oh1VI	Zambrano, Jones, Brenner, et al. (2014)
	6.01	MFSV		QTL	4.4	5	VI/GH	Oh1IV	Zambrano, Jones, Brenner, et al. (2014)
	6.01	MMV		QTL	6.6	8	VI/GH	Oh1VI	Zambrano, Jones, Brenner, et al. (2014)
	6.01	MSV		QTL	3.9	29	VI/F	CIRAD390	Pernet, Hoisington, Franco, et al. (1999)
8	8.03	MRCV		QTL	3.8	13	NI	BLS14	Di Renzo et al. (2004)
	8.05	RBSDV	*qMrdd1*	QTL	nr	37	NI	CL1165	Tao, Jiang, et al. (2013)
	8.07		*qMRD8*		8.6	38	NI	X178	Shi et al. (2012)
					9.5	28	NI	90110	Luan et al. (2012)
9	9.02	MSV		QTL	nr	8	VI/F	Tzi4	Kyetere et al. (1999)
10	10.04	MRFV		QTL	12.2	23	VI/GH	Oh1VI	Zambrano, Jones, Francis, et al. (2014)
					10.9	23	VI/GH	Ki11	
	10.04	MNeSV		QTL	10.0	46	MI/GH	Oh1VI	José Zambrano et al. (unpublished results)
	10.05	MDMV		QTL	3.3	1	MI/F	Oh1VI	Zambrano, Jones, Brenner, et al. (2014)
	10.05	WSMV	*Wsm3*	Gene	nr	nr	MI/GH	Pa405	McMullen et al. (1994)
					5.4	7	MI/GH	Oh1VI	Zambrano, Jones, Brenner, et al. (2014)
	10.05	SCMV		QTL	6.0	15	MI/F	Hongzao4	Zhang et al. (2003)
	10.05	MCDV	*mcd2*	QTL	21.4	24	VI/GH	Oh1VI	Jones et al. (2004)

Continued

Table 8.2 Distribution of virus resistance genes and quantitative trait loci (QTL) in the maize genome—cont'd

Chr	Bin	Virus	Locus	Type	LOD	R^2	Screen	RS	References
	10.05	MSV		QTL	8.0	34	VI/F	D211	Pernet, Hoisington, Dintinger, et al. (1999)
					9.3	30	VI/F	CIRAD390	Pernet, Hoisington, Franco, et al. (1999)
	10.06	MSpV		QTL	Nr	18	VI/F	Rev81	Dintinger, Boissot, et al. (2005)

[a]Chr., chromosome.
[b]Virus acronyms are given in Table 8.1. MFSV, *Maize fine streak virus*; MNeSV, *Maize necrotic streak virus*.
[c]Locus names are provided where applicable.
[d]R^2, the proportion of the variance explained by the gene or QTL.
[e]The method used to screen for resistance. NI, natural infection under field conditions; VI, vector inoculation; MI, mechanical inoculation; GH, greenhouse conditions; F, field conditions.
[f]RP, resistance source; the inbred line in which the locus was identified.
[g]nr, not reported.

to develop resistant cultivars and hybrids using naturally occurring are virus resistance and are currently available for most diseases. Specific resistance virus genes will be discussed in more detail below.

Pathogen-derived resistance, achieved by expressing viral proteins or RNAs in transgenic plants, is another way to incorporate resistance into crop plants (Prins et al., 2008). For maize, plants transformed with MCDV coat protein(s), the MDMV P1, or the MSV replication associated protein genes were resistant to MCDV, MDMV, or MSV, respectively, under controlled conditions (Liu, Tan, Li, Zhang, & He, 2009; McMullen, Roth, & Townsend, 1996; Shepherd et al., 2007; Zhang, Fu, Gou, Wang, & Li, 2010). Interestingly, transgenic plants expressing SCMV (MDMV-B) were resistant to that virus, but also had pleiotropic resistance to MCMV (Murry et al., 1993). Further, transgenic maize expressing a bacterial dsRNA-specific nuclease (RNAse III) was resistant to RBSDV (Cao et al., 2013), but it is not yet known if this line is resistant to multiple viruses. The MSV-expressing transgenic maize are still being tested in confined field trials (Tomson & Shepherd, 2010), and lines expressing the MDMV P1 provided some virus resistance in the field (Zhang et al., 2013). Nonetheless, the agronomic utility of transgenic virus resistance in maize is not yet well understood, and commercial virus-resistant transgenic maize hybrids are not yet available.

4. DEVELOPMENT OF VIRUS-RESISTANT CROPS

Breeding maize for resistance to virus diseases initiated in 1965 in the USA when the first MDMV and MCDV outbreaks occurred (Findley, Louie, Knoke, & Dollinger, 1976). With little information on resistance genes or the pathogens themselves, maize breeders and cooperating pathologists and entomologists realized that the first step was to design efficient screening techniques to allow the identification of sources of resistance. While there have been many improvements in approach and technology over almost 50 years, development of resistant crops still requires identification and evaluation of resistant plants, and incorporation of favorable resistance traits into agronomically desirable genetic backgrounds. Because the identified virus resistance in maize is primarily mono- or oligogenic, most approaches for breeding virus resistance in maize include (i) screening germplasm collections to identify resistance sources; (ii) identification of markers linked to genes or QTLs conferring resistance; and (iii) the introgression of the resistance into elite breeding lines.

4.1. Screening methods

Selecting resistant plants is critical to identifying and characterizing virus resistance in maize, and a number of different approaches have been used to challenge plants with virus infection (Table 8.2). Natural infection, evaluated in disease nurseries or "hot spots" where the disease was naturally endemic or prevalent, was among the early approaches used (Findley et al., 1976). This method continues to be effective for identifying virus-resistant germplasm and characterizing resistance genes and QTLs in maize (Cao et al., 2013; Di Renzo et al., 2004, Ming et al., 1997; Tao, Liu, et al., 2013). However, this approach does not ensure gain under selection, because of unequal disease pressure from season to season and the high risk of escapes. In addition, the logistics of planting and evaluating trials in disease hot spots can make using these nurseries resource intensive. To circumvent these problems, different approaches for inoculating maize germplasm with defined virus cultures have been developed.

For viruses such as the potyviruses that can be transmitted mechanically, leaf rub inoculation, in which the virus is transmitted mechanically by hand rubbing virus-containing inoculum, usually with an abrasive agent (Louie, 1986), or air brush, backpack sprayer or solid stream protocols for mechanical transmission of viruses in the field (Louie et al., 1983; Meyer & Pataky, 2010) are effective. These protocols can be adjusted to provide very high (>90%) infection of susceptible plants under greenhouse and field conditions. VPI, in which the viral inoculum is introduced in germinating seeds with the aid of minute pins attached to an engraving tool (Louie, 1995), can be used to transmit all known maize-infecting viruses to maize. However, it is difficult to achieve high transmission efficiency with phloem limited viruses such as MCDV and MRDV (Louie & Abt, 2004).

Viruliferous insects can also be used for transmission of viruses, but this approach is resource and expertise intensive, requiring the ability to raise insects under controlled conditions. After feeding on virus-infected plants, virus-carrying vectors are then transmitted to test plants in the field or greenhouse for inoculation. By taking the virus' transmission mode into account, high efficiency inoculation conditions can be developed (Louie & Anderson, 1993; Zambrano et al., 2013). This approach has been effective for characterizing resistance QTLs under both greenhouse and field conditions (Table 8.2) (Jones et al., 2004; Kyetere et al., 1999; Welz et al., 1998; Zambrano, Jones, Brenner, et al., 2014).

Although high efficiency inoculation reduces the effects of disease "escapes" on resistance evaluation, plant responses to inoculation can vary among methods, environments, and experiments. Factors affecting inoculation efficiency, including plant development and genotype, can also influence plant response. For example, the inbred line Mo18W was resistant to MCDV in the field, but was susceptible in inoculations made under greenhouse conditions (Louie, Knoke, & Findley, 1990). Similarly, inbred lines from diverse breeding populations had low levels of MDMV infection when inoculated in the field, but were frequently rated as susceptible in the greenhouse (Scott & Louie, 1996). It has been suggested that greenhouse conditions provide high and evenly distributed inoculum pressure that is more favorable to symptom development and evaluation (Vandeplas, 2003).

4.2. Virus isolates and disease development

The number of genes involved in resistance and their mode of action has sometimes varied across germplasm and experiments (Bonamico et al., 2012; Louie, 1986; Pokorny & Porubova, 2006). Differential responses of resistant maize lines to different virus isolates or strains have only been documented in a few cases (Jones, Boyd, & Redinbaugh, 2011; Jones, Redinbaugh & Louie, 2007; Louie, 1986). In one case, a stronger resistance allele provided effective resistance to a more aggressive virus isolate (Jones et al., 2011).

Symptom development is most frequently used to evaluate maize germplasm for virus resistance. For most maize viruses, infected plants become symptomatic and serological assays do not detect the presence of virus in asymptomatic plants (Jones et al., 2011; Zambrano et al., 2013). Nonetheless, it is important to establish this empirically for each virus–host system, as virus tolerant (asymptomatic, virus-infected plants) inbred lines are known. For MCMV, lines with high levels of virus tolerance have been developed and are used widely in Hawaii, but the virus titers are similar in the tolerant and susceptible lines (J. Brewbaker, personal communication). For viruses controlled by dominant genes, disease incidence (the appearance of any symptom) alone is sufficient for mapping resistance genes (McMullen & Louie, 1989; Rufener, Balducci, Mowers et al., 1996). However, protocols that integrate disease incidence with a measure of symptom severity are frequently used (Brewbaker, 1981; Louie, 1986; Scott & Rosenkranz, 1981). Severity index ratings and area under the disease progress curve are also effective (Di Renzo et al., 2004; Madden, Hughes, & Van Den Bosch, 2007; Shi et al., 2012).

5. GENETICS OF RESISTANCE TO VIRUS DISEASES

An understanding of the genetic mechanisms involved in resistance to virus diseases is necessary to conduct efficient and effective breeding programs (Scott & Rosenkranz, 1981). Despite the difficulties in generating high disease pressure and understanding specific virus–vector–host relationships, the genetics of resistance to at least 11 viruses in 7 families has been studied (Table 8.2). The resources available for mapping and isolating genes in maize continue to evolve rapidly. Early studies relied on phenotypic markers, but the more recent studies outlined in Table 8.2 use various types of molecular markers. Emerging technologies for maize genotyping include genotyping by sequencing and high-density single nucleotide polymorphism platforms (Elshire et al., 2011; Ganal et al., 2011). Increasingly, genotyped association mapping populations, including the nested association mapping population, are available to researchers (Flint-Garcia et al., 2005; McMullen et al., 2009; Romay et al., 2013).

Both dominant genes, such as those for resistance to MDMV and WSMV (McMullen & Louie, 1989, 1991; McMullen et al., 1994), and QTLs, such as those for resistance to SCMV, MCDV, MRCV, or MMV (Jones et al., 2007), have been identified. Loci virus resistance in maize has been found on 9 of the 10 maize chromosomes (chr.) (Table 8.2). Resistance is both distributed and clustered in the maize genome, with clusters of resistance loci occurring on chr. 1, 2, 3, 6, 8, and 10.

Msv1, a gene that confers resistance to MSV, and QTLs that confer resistance to MRCV and MMV are found clustered on the short arm of chromosome 1 (Bonamico et al., 2012; Di Renzo et al., 2004; Kyetere et al., 1999; Welz et al., 1998; Zambrano, Jones, Brenner, et al., 2014). *Msv1* and the MRCV-resistance QTLs account for almost 60% and 30% of the phenotypic variation, respectively. Because these three QTLs were identified in different germplasm and have not yet been fine mapped, it is not yet possible to determine how closely linked the loci are.

Two clusters of virus resistance genes are present on chr. 2. At the distal end of the chromosome, QTLs for resistance to two maize-infecting fijiviruses and to the unrelated MCDV and MSpV were identified using different population structures and unrelated germplasm (Di Renzo et al., 2004; Luan et al., 2012; Martin et al., 2010; Zambrano, Jones, Brenner, et al., 2014). Overlapping QTLs for the nucleorhabdoviruses MMV and MFSV were identified in the inbred line Oh1VI in bin 2.08 (Zambrano,

Jones, Brenner, et al., 2014). A locus for resistance to MSpV, another membrane-bound virus, was identified nearby, but in different germplasm.

A major cluster of resistance loci (genes and QTLs) is located on chr. 3 near the centromere (Redinbaugh & Pratt, 2008; Wisser, Balint-Kurti, & Nelson, 2006). Resistance loci in this region include those for resistance to viruses in the *Potyviridae*, but also to viruses in other families, including those that use different genome, replication and movement strategies. This region contains *Wsm2* and *Scm2* (also known as *Scmv2* or *Rscmv2*), genes that confer resistance to WSMV and SCMV. The same region includes enhancers of resistance to MDMV and other potyviruses (De Souza et al., 2008; Ding et al., 2012; Dussle et al., 2000; Jones et al., 2011; McMullen et al., 1994; Stewart, Haque, Jones, & Redinbaugh, 2013; Wang, Zhang, Zhuang, Qin, & Zhang, 2007; Zambrano, Jones, Brenner, et al., 2014; Zhang et al., 2003). The region also contains QTLs that confer resistance to the waikavirus MCDV, the negative sense RNA rhabdovirus MMV, the ambisense RNA tenuivirus MSpV, the single-stranded DNA geminivirus MSV, and the ambisense RNA virus WoMV (Dintinger, Verger, et al., 2005; Jones et al., 2004; Marçon, Kaeppler, & Jensen, 1997; Marçon, Kaeppler, Jensen, Senior, & Stuber, 1999; Ming et al., 1997; Welz et al., 1998; Zambrano, Jones, Brenner, et al., 2014).

The second major cluster of resistance genes is located on the short arm of chromosome 6, where the dominant genes, *Mdm1* and *Wsm1*, colocalize with *Scm1* (also known as *Scmv1* and *Rscmv1*) (Jones et al., 2011, 2007; McMullen & Louie, 1989; Stewart, Haque, et al., 2013; Wang et al., 2007; Xia et al., 1999; Xu et al., 1999). As with the cluster on chr. 3, QTLs conferring resistance to viruses with different genome and replication strategies are found in this cluster, including MFSV, MCDV, WMoV, MMV, and MSV (Jones et al., 2004; Marçon et al., 1997, 1999; Pernet, Hoisington, Dintinger, et al., 1999; Zambrano, Jones, Brenner, et al., 2014).

The third major cluster of virus resistance genes and QTLs in the maize genome is located on chr. 10. In this region, *Wsm3*, which confers resistance to WSMV, colocalizes with QTLs that confer resistance to MDMV, MCDV, MRFV, MNeSV, MSV, and MSpV (Dintinger, Verger, et al., 2005; Jones et al., 2011, 2004; McMullen et al., 1994; Pernet, Hoisington, Dintinger, et al., 1999; Pernet, Hoisington, Franco, et al., 1999; Wang et al., 2007; Zambrano, Jones, Brenner, et al., 2014; Zambrano, Jones, Francis, Tomas, & Redinbaugh, 2014). This region of the maize genome also carries loci for bacterial and fungal resistance (Redinbaugh & Pratt, 2008), as well as putative R genes (Li et al., 2010; Xiao et al., 2007).

QTLs for resistance to the fijiviruses MRCV and RBSDV were identified in the same region of chr. 8 in South American and Chinese germplasm (Di Renzo et al., 2004; Tao, Jiang, et al., 2013).

For most studies defining loci for virus resistance in maize, sources of resistance were selected from locally adapted materials, and so are different between regions. The inbred line Oh1VI was developed from an open pollinated Virgin Island population as highly resistant to MCDV (Louie, Jones, Anderson, & Redinbaugh, 2002) and subsequently found to be highly resistant to MDMV, SCMV, and WSMV (Jones et al., 2007). Further study indicated that the line is highly resistant to MRFV, MMV, MFSV, and MNeSV (José Zambrano et al., unpublished results; Zambrano, Jones, Francis, et al., 2014). Although Oh1VI is susceptible to the fijiviruses and MSV (Margaret Redinbaugh et al., unpublished results), it would be of interest to characterize resistance to WMoV and MSpV in this line. Within Oh1VI, virus resistance was clearly clustered on chr. 2, 3, 6, and 10. The clustering of virus resistance genes within this line may provide an opportunity to examine the generally accepted "the birth-and-death" model proposed for resistance genes (Rosendahl & Taylor, 1997) for genes encoding virus resistance. In this model, new genes are created by gene duplication and genetic recombination, whereas others are deleted or become nonfunctional through deleterious mutations (Hulbert, Webb, Smith, & Sun, 2001; Nei & Rooney, 2005). Alternatively, a unique mechanism for resistance could be active in this line (Gomez et al., 2009).

5.1. Breeding methods and results

The development of resistant varieties requires thorough understanding of the crop host, the pathogen, and any vectors that may have coevolved and coexisted during domestication (Chahal & Gosal, 2002). *Mdm1* provides partial to complete resistance to MDMV and SCMV, depending on the source of resistance and virus isolate, and this gene has been used by breeders to develop lines and commercial hybrids with resistance to maize dwarf mosaic (Williams & Pataky, 2012). For MCDV, selection under natural disease pressure allowed some improvement of resistance (Eberhart, 1983). Later, development of MCDV-resistant inbred lines was achieved using marker-assisted selection (Rufener et al., 1996). The diseases caused by MDMV and MCDV are now largely controlled in the USA through the use of resistant hybrids and Johnsongrass removal (Redinbaugh & Pratt, 2008), and these viruses occur at very low rates in maize grown for grain Ohio even though they are present in the region (Stewart et al., in press).

Backcrosses have commonly been used to transfer virus resistance from resistant, but agronomically undesirable, lines to highly adapted elite lines. The recurrent parent is usually selected because of its good combining ability, high yield, and wide adaptability (Chahal & Gosal, 2002). This parent is crossed with a donor having virus resistance genes to produce F_1 and subsequent backcross progenies. For example, the *Mv1* gene conferring resistance to MMV was introduced into more than 100 inbreds and cultivars and 130 genetic stocks by at least five cycles of backcrossing and marker-assisted selection (Brewbaker, 1981; Brewbaker & Josue, 2007). Resistant inbred lines and cultivars were selected as MMV-resistant plants segregating in BC_5S_1 progenies in a disease nursery. Martin and coworkers (2010) have also used marker-based selection to successfully introgress the MRCV-resistance QTLs on chr. 2 into elite breeding lines.

For the past 30 years, maize breeding programs in sub-Saharan Africa have worked to incorporate *Msv1* into improved maize varieties and inbred lines (Shepherd et al., 2010). Originally, these programs combined backcrosses with disease screening that required transmitting MSV with viruliferous leafhoppers and selecting resistant lines. While this approach was effective, it is also required a lot of time and resources. Recent studies indicate the efficacy of combining backcrosses to elite lines with marker-assisted selection for development of resistant elite lines (Abalo, Tongoona, Derera, & Edema, 2009; Asea, Vivek, Lipps, & Pratt, 2012).

6. TOWARD UNDERSTANDING VIRUS RESISTANCE MECHANISMS IN MAIZE

None of the virus-resistance genes or loci mapped in genetic studies have been isolated to date, but recent reports describe fine mapping of genomic regions carrying the *Scmv1* and *Scm1/Scmv2/Rscmv2* genes that confer resistance to SCMV (Ding et al., 2012; Ingvardsen, Xing, Frei, & Lubberstedt, 2010; Tao, Jiang, et al., 2013). Tao and coworkers (2013) narrowed *Scmv1* to a region corresponding to a 59 kb of chr. 6 in the B73 genome and identified 10 candidate genes in this region. Of particular interest are genes previously associated with pathogen and abiotic stress responses such as putative thioredoxin h and cycloartenol synthase genes.Fine mapping of the *Scmv2* and *Rscmv2* genes identified a region of chr. 3 corresponding to 1.34 Mb (Ingvardsen et al., 2010) or 196 kb (Ding et al., 2012) of the B73 genome, respectively. Of the 20 ESTs homologous to portions of the *Scmv2* region, those similar to heat-shock protein

70, general vesicular transport factor p115, RhoGTPase-activating proteins and syntaxin proteins were considered candidate genes based on their involvement in plant responses to various stimuli (Ingvardsen et al., 2010). Ding and coworkers (2012) considered genes identified in the *Rscmv2* region of B73 that encode a RhoGTPase-activating protein and an auxin-binding protein-1 gene to be candidate genes. These two genes are also contained in the *Scmv2* region. Interestingly, no genes homologous to the NBS-leucine-rich repeat (LRR) containing class of R genes, which includes dominant genes conferring resistance to *Tobacco mosaic virus*, *Tomato mosaic virus*, and *Potato virus X* (Bendahmane, Kanyuka, & Baulcombe, 1999; Lanfermeijer et al., 2004; Whitham et al., 1994), were found in either the *Scmv1* or *Scmv2/Rscmv2* regions.

Within the larger genome region of chr. 3 in Oh1VI that carries QTLs for resistance to MCDV, MMV, and MRFV, the B73 genome encodes two genes for the translation factor eIF4e (Zambrano, Jones, Brenner, et al., 2014; Zambrano, Jones, Francis, et al., 2014). In several virus–host systems, recessive alleles of eIF4e confer resistance to several different viruses (Diaz-Pendon et al., 2004; Gomez et al., 2009). In these cases, the protein produced from the translation factor allele fails to interact with the virus and recruit the viral RNA to the cap-binding complex. eIF4e has also been shown to move from cell to cell, and some eIF4e alleles do not allow cell-to-cell movement of the potyvirus, *Pea seedborne mosaic virus* (Gao et al., 2004). However, these eIF4e genes are not within the *Scmv2/Rscmv2* regions identified in fine-mapping studies (Ding et al., 2012; Ingvardsen et al., 2010), so further research is needed to determine if the translation factors have any effect on virus resistance in maize.

6.1. Mechanisms of virus resistance in maize

MDMV levels in inoculated leaves of the resistant line Pa405 were similar to those in susceptible controls, and virus spread within inoculated leaves could be detected serologically (Law, Moyer, & Payne, 1989; Lei & Agrios, 1986). Similarly, when germinating kernels of Pa405 were inoculated with MDMV by VPI, bright mosaic symptoms developed and MDMV antigen could be detected in leaves that were present in the embryo at the time of inoculation, but no systemic infection occurred (Cassone et al., 2014; Louie & Abt, 2004). Together, these data indicate that MDMV can replicate and move from cell-to-cell in leaves of resistant plants, but that a barrier to systemic virus movement is present in resistant plants. Although candidate genes have been identified for SCMV resistance as discussed earlier, it remains unclear if a single pleiotropic gene or closely linked genes are responsible for resistance to multiple potyviruses in maize. In Arabidopsis, resistance to three different potyvirus species (*Tobacco etch virus*, *Lettuce mosaic virus*, and *Plum pox virus*) is conferred by three interacting dominant genes that restrict long-distance movement of the viruses: *RTM1*, which encodes a jacalin-like protein; *RTM2*, which encodes a small heat-shock protein; and *RTM3*, encodes a protein containing meprin and TRAF homology (MATH) domains (Chisholm, Mahajan, Whitham, Yamamoto, & Carrington, 2000; Cosson et al., 2010; Decroocq et al., 2006; Mahajan, Chisholm, Whitham, & Carrington, 1998; Revers et al., 2003; Whitham, Anderberg, Chisholm, & Carrington, 2000). Although the mechanism of resistance is not clear, *RTM1* and *RTM2* are expressed in phloem (Chisholm, Parra, Anderberg, & Carrington, 2001).

Microarray and proteome analysis have been used in a number of maize–virus systems to help gain information on pathways and genes important for virus resistance and susceptibility. Relatively few transcripts were upregulated in virus-resistant maize lines inoculated with potyviruses (Cassone et al., 2014; Shi, Thummler, Melchinger, Wenzel, & Lubberstedt, 2006; Uzarowska et al., 2009), but more were identified in the proteomic analyses (Wu, Han, et al., 2013; Wu, Wang, et al., 2013). The responses of susceptible maize lines to inoculation with RBSDV were also investigated (Jia et al., 2012). In the resistant inbred line Pa405 inoculated with MDMV, two transcripts annotated as potentially involved in virus resistance, pathogen response, or stress response were upregulated: an Hsp20/alpha crystalline-like gene and a cytochrome P450 (Cassone et al., 2014). Further, nine transcripts with putative involvement in signal transduction were identified as upregulated in MDMV-inoculated plants. Differences in maize genotypes, sampling times, the tissues, and virus isolates used, and experimental/bioinformatics platforms make it difficult to compare these experiments. However, a few transcripts were differentially regulated in different experiments. Increased accumulation of a transcript with similarity to both an LRR receptor protein kinase (brassinosteroid-insensitive 1 in rice) and an Hsp20/alpha crystalline-like gene was identified in MDMV-inoculated plants and was also upregulated RBSDV-infected maize (Jia et al., 2012). Different genes encoding homologous remorin proteins were upregulated in resistant and susceptible MDMV-inoculated plants (Cassone et al., 2014), and another remorin protein accumulated in SCMV-inoculated Mo17 (susceptible) plants, but not in the resistant line Siyi (Wu, Han, et al., 2013).

7. CONCLUSION

Thanks to advances in screening for resistance and the identification of molecular markers linked to resistance genes, virus-resistant hybrids are on the market. Resistance genes and QTLs for maize virus diseases have been successfully mapped and transferred from unadapted germplasm to elite lines by backcrossing and backcrossing with marker-assisted selection. Even though our knowledge about the genetic mechanisms for virus resistance in maize is currently limited and no cloned resistance genes have been reported, breeders have been able to effectively deploy resistance genes. Nonetheless, there is need for continuing efforts to deploy resistance to rapidly emerging virus diseases, such as MLN in sub-Saharan Africa, and to pyramid genes and QTLs for multiple virus diseases that threat maize production in some areas. Further characterization of multiply virus-resistant inbred lines and cloning of the maize genes conferring resistance to virus diseases will provide a better understanding of the biological and molecular processes that are important for the development of virus-resistant plants.

ACKNOWLEDGMENTS

The U.S. Department of Agriculture (USDA) prohibits discrimination in all its programs and activities on the basis of race, color, national origin, age, disability, and where applicable, sex, marital status, familial status, parental status, religion, sexual orientation, genetic information, political beliefs, reprisal, or because all or part of an individual's income is derived from any public assistance program. (Not all prohibited bases apply to all programs.) Persons with disabilities who require alternative means for communication of program information (Braille, large print, audiotape, etc.) should contact USDA's TARGET Center at (202) 720-2600 (voice and TDD). To file a complaint of discrimination, write to USDA, Director, Office of Civil Rights, 1400 Independence Avenue, S.W., Washington, DC 20250-9410, or call (800) 795-3272 (voice) or (202) 720-6382 (TDD). USDA is an equal opportunity provider and employer.

REFERENCES

Abalo, G., Tongoona, P., Derera, J., & Edema, R. (2009). A comparative analysis of conventional and marker-assisted selection methods in breeding *Maize streak virus* resistance in maize. *Crop Science*, *49*, 509–520.

Achon, M. A., Subira, J., & Sin, E. (2013). Seasonal occurrence of *Laodelphax striatellus* in Spain: Effect on the incidence of *Maize rough dwarf virus*. *Crop Protection*, *47*, 1–5.

Adams, M. J., Antoniw, J. F., & Fauquet, C. M. (2005). Molecular criteria for genus and species discrimination within the family *Potyviridae*. *Archives of Virology*, *150*, 459–479.

Adams, I. P., Miano, D. W., Kinyua, Z. M., Wangai, A., Kimani, E., Phiri, N., et al. (2013). Use of next-generation sequencing for the identification and characterization of *Maize chlorotic mottle virus* and *Sugarcane mosaic virus* causing maize lethal necrosis in Kenya. *Plant Pathology, 62*, 741–749.

Ali, F., & Yan, J. B. (2012). Disease resistance in maize and the role of molecular breeding in defending against global threat. *Journal of Integrative Plant Biology, 54*, 134–151.

All, J. (1983). Integrating techniques of vector and weed-host suppression into control programs for maize virus diseases. In D. T. Gordon, J. K. Knoke, L. R. Nault, & R. M. Ritter (Eds.), *International maize virus disease colloquium and workshop* (pp. 243–247). Wooster: Ohio State University, OARDC.

Ammar, E. D., Gingery, R. E., & Madden, L. V. (1995). Transmission efficiency of 3 isolates of Maize stripe tenuivirus in relation to virus titer in the planthopper vector. *Plant Pathology, 44*, 239–243.

Ammar, E. D., Khlifa, E. A., Mahmoud, A., Abol-Ela, S. E., & Peterschmitt, M. (2007). Evidence for multiplication of the leafhopper-borne maize yellow stripe virus in its vector using ELISA and dot-blot hybridization. *Archives of Virology, 152*, 489–494.

Antignus, Y., Klein, M., & Ovadia, S. (2008). The use of insecticides to control *Maize rough dwarf virus* (MRDV) in maize. *Annals of Applied Biology, 110*, 557–562.

Arneodo, J. D., Guzman, F. A., Conci, L. R., Laguna, I. G., & Truol, G. A. (2002). Transmission features of Mal de Rio Cuarto virus in wheat by its planthopper vector *Delphacodes kuscheli*. *Annals of Applied Biology, 141*, 195–200.

Asea, G., Vivek, B. S., Lipps, P. E., & Pratt, R. C. (2012). Genetic gain and cost efficiency of marker-assisted selection of maize for improved resistance to multiple foliar pathogens. *Molecular Breeding, 29*, 515–527.

Bendahmane, A., Kanyuka, K., & Baulcombe, D. C. (1999). The *Rx* gene from potato controls separate virus resistance and cell death responses. *Plant Cell, 11*, 781–791.

Bonamico, N. C., Balzarini, M. G., Arroyo, A. T., Ibañez, M. A., Díaz, D. G., Salerno, J. C., et al. (2010). Association between microsatellites and resistance to Mal de Rio Cuarto in maize by discriminant analysis. *Phyton-International Journal of Experimental Botany, 79*, 31–38.

Bonamico, N. C., Di Renzo, M. A., Ibañez, M. A., Borghi, M. L., Díaz, D. G., Salerno, J. C., et al. (2012). QTL analysis of resistance to Mal de Rio Cuarto disease in maize using recombinant inbred lines. *Journal of Agricultural Science, 150*, 619–629.

Boulton, M. I., Pallaghy, C. K., Chatani, M., MacFarlane, S., & Davies, J. W. (1993). Replication of Maize streak virus mutants in maize protoplasts—Evidence for a movement protein. *Virology, 192*, 85–93.

Brewbaker, J. L. (1979). Diseases of maize in the wet lowland tropics and the collapse of the classic Maya civilization. *Economic Botany, 33*, 101–118.

Brewbaker, J. L. (1981). Resistance to maize mosaic virus. In D. T. Gordon, J. K. Knoke, & G. E. Scott (Eds.), *Proceedings of the virus and viruslike diseases of maize in the United States, Wooster, OH* (pp. 145–151).

Brewbaker, J. L., & Josue, A. D. (2007). Registration of 27 maize parental inbred lines resistant to maize mosaic virus. *Crop Science, 47*, 459–461.

Bustamante, P. I., Hammond, R., & Ramirez, P. (1998). Evaluation of maize germplasm for resistance to *Maize rayado fino virus*. *Plant Disease, 82*, 50–56.

Cabanas, D., Watanabe, S., Higashi, C. H. V., & Bressan, A. (2013). Dissecting the mode of *Maize chlorotic mottle virus* transmission (*Tombusviridae*: *Machlomovirus*) by *Frankliniella williamsi* (Thysanoptera: Thripidae). *Journal of Economic Entomology, 106*, 16–24.

Cao, X. L., Lu, Y. G., Di, D. P., Zhang, Z., Liu, H., Zhang, A., et al. (2013). Enhanced virus resistance in transgenic maize expressing a dsRNA-specific endoribonuclease gene from *E. coli*. *PLoS One, 8*, e60829.

Cassone, B. C., Chen, Z., Chiera, J., Stewart, L. R., & Redinbaugh, M. G. (2014). Responses of highly resistant and susceptible maize to vascular puncture inoculation with *Maize dwarf mosaic virus*. *Physiological and Molecular Plant Pathology, 86*, 9–27.

Castillo-Loayza, J. (1979). Maize virus and virus-like diseases in Peru. In L. E. Williams, D. T. Gordon, & L. R. Nault (Eds.), *Proceedings of the international maize virus disease colloquium and workshop* (pp. 40–44). Wooster, OH: OARDC.

Chahal, G. S., & Gosal, S. S. (2002). *Principles and procedures of plant breeding*. United Kingdom: Alpha Science International Ltd.

Chen, Y. T., Cassone, B. J., Bai, X. D., Redinbaugh, M. G., & Michel, A. P. (2012). Transcriptome of the plant virus vector *Graminella nigrifrons*, and the molecular interactions of *Maize fine streak rhabdovirus* transmission. *PLoS One*, *7*, e40613.

Chisholm, S. T., Mahajan, S. K., Whitham, S. A., Yamamoto, M. L., & Carrington, J. C. (2000). Cloning of the Arabidopsis *RTM1* gene, which controls restriction of long-distance movement of *Tobacco etch virus*. *Proceedings of the National Academy of Sciences of the United States of America*, *97*, 489–494.

Chisholm, S. T., Parra, M. A., Anderberg, R. J., & Carrington, J. C. (2001). Arabidopsis *RTM1* and *RTM2* genes function in phloem to restrict long-distance movement of *Tobacco etch virus*. *Plant Physiology*, *127*, 1667–1675.

Cosson, P., Sofer, L., Le, Q. H., Léger, V., Schurdi-Levraud, V., Whitham, S., et al. (2010). *RTM3*, which controls long-distance movement of potyviruses, is a member of a new plant gene family encoding a meprin and TRAF homology domain-containing protein. *Plant Physiology*, *154*, 222–232.

De Souza, I. R. P., Schuelter, A. R., Guimaraes, C. T., Schuster, I., De Oliveira, E., & Redinbaugh, M. (2008). Mapping QTL contributing to SCMV resistance in tropical maize. *Hereditas*, *145*, 167–173.

Decroocq, V., Sicard, O., Alamillo, J. M., Lansac, M., Eyquard, J. P., Garcia, J. A., et al. (2006). Multiple resistance traits control Plum pox virus infection in *Arabidopsis thaliana*. *Molecular Plant-Microbe Interactions*, *19*, 541–549.

Di Renzo, M. A., Bonamico, N. C., Diaz, D. G., Ibañez, M. A., Faricelli, M. E., Balzarini, M. G., et al. (2004). Microsatellite markers linked to QTL for resistance to Mal de Rio Cuarto disease in *Zea mays* L. *Journal of Agricultural Science*, *142*, 289–295.

Diaz-Pendon, J. A., Truniger, V., Nieto, C., Garcia-Mas, J., Bendahmane, A., & Aranda, M. A. (2004). Advances in understanding recessive resistance to plant viruses. *Molecular Plant Pathology*, *5*, 223–233.

Ding, J. Q., Li, H. M., Wang, Y. X., Zhao, R. B., Zhang, X. C., Chen, J. F., et al. (2012). Fine mapping of *Rscmv2*, a major gene for resistance to *Sugarcane mosaic virus* in maize. *Molecular Breeding*, *30*, 1593–1600.

Dintinger, J., Boissot, N., Chiroleu, F., Hamon, P., & Reynaud, B. (2005). Evaluation of maize inbreds for *Maize stripe virus* and *Maize mosaic virus* resistance: Disease progress in relation to time and the cumulative number of planthoppers. *Phytopathology*, *95*, 600–607.

Dintinger, J., Verger, D., Caiveau, S., Risterucci, A. M., Gilles, J., Chiroleu, F., et al. (2005). Genetic mapping of maize stripe disease resistance from the Mascarene source. *Theoretical and Applied Genetics*, *111*, 347–359.

Distefano, A. J., Conci, L. R., Hidalgo, M. M., Guzman, F. A., Hopp, H. E., & Del Vas, M. (2002). Sequence analysis of genome segments S4 and S8 of *Mal de Rio Cuarto Virus* (MRCV): Evidence that the virus should be a separate fijivirus species. *Archives of Virology*, *147*, 1699–1709.

Dussle, C. M., Melchinger, A. E., Kuntze, L., Stork, A., & Lubberstedt, T. (2000). Molecular mapping and gene action of *Scm1* and *Scm2*, two major QTL contributing to SCMV resistance in maize. *Plant Breeding*, *119*, 299–303.

Ebbert, M. A., & Nault, L. R. (1994). Improved overwintering ability in *Dalbulus maidis* (Homoptera, Cicadellidae) vectors infected with *Spiroplasma kunkelii* (Mycoplasmatales, Spiroplasmataceae). *Environmental Entomology*, *23*, 634–644.

Eberhart, S. A. (1983). Developing virus resistant commercial maize hybrids. In D. T. Gordon, J. K. Knoke, L. R. Nault, & R. M. Ritter (Eds.), *International maize virus disease colloquium and workshop* (pp. 258–261). Wooster, OH: Ohio State University, OARDC.

Eberwine, J. W., & Hagood, E. S. (1995). Effect of johnsongrass (*Sorghum halepense*) control on the severity of virus diseases of corn (*Zea mays*). *Weed Technology, 9*, 73–79.

Eberwine, J. W., Hagood, E. S., & Tolin, S. A. (1998). Quantification of viral disease incidence in corn (*Zea mays*) as affected by johnsongrass (*Sorghum halepense*) control. *Weed Technology, 12*, 121–127.

Elshire, R. J., Glaubitz, J. C., Sun, Q., Poland, J. A., Kawamoto, K., Buckler, E. S., et al. (2011). A robust, simple genotyping-by-sequencing (GBS) approach for high diversity species. *PLoS One, 6*, e19379.

Falk, B. W., & Tsai, J. H. (1985). Serological detection and evidence for multiplication of *Maize mosaic virus* in the planthopper, *Peregrinus maidis*. *Phytopathology, 75*, 852–855.

Falk, B. W., & Tsai, J. H. (1998). Biology and molecular biology of viruses in the genus Tenuivirus. *Annual Review of Phytopathology, 36*, 139–163.

Fauquet, C. (2005). *Virus taxonomy: Classification and nomenclature of viruses, eighth report of the international committee on the taxonomy of viruses*. New York: Academic Press.

Findley, W. R., Louie, R., Knoke, J. K., & Dollinger, E. J. (1976). Breeding corn for resistance to virus in Ohio. In L. E. Williams, D. T. Gordon, & L. R. Nault (Eds.), *Proceedings of the international maize virus disease colloquium and workshop* (pp. 123–127), Wooster, OH: OARDC.

Flint-Garcia, S. A. (2013). Genetics and consequences of crop domestication. *Journal of Agricultural and Food Chemistry, 61*, 8267–8276.

Flint-Garcia, S. A., Thuillet, A. C., Yu, J. M., Pressoir, G., Romero, S. M., Mitchell, S. E., et al. (2005). Maize association population: A high-resolution platform for quantitative trait locus dissection. *Plant Journal, 44*, 1054–1064.

Food and Agriculture Organizations (2012). *FAO statistical yearbook. World food and agriculture*. Rome: Food and Agriculture Organization of the United Nations.

Fritts, D. A., Michels, G. J., & Rush, C. M. (1999). The effects of planting date and insecticide treatments on the incidence of high plains disease in corn. *Plant Disease, 83*, 1125–1128.

Fuller, C. (1901). *Mealie variegation. 1st Report of the Government Entomologist, Natal* 1899–1900.

Gamez, R. (1969). A new leafhopper-borne virus of corn in Central America. *Plant Disease Reporter, 53*, 929–932.

Gamez, R. (1976). Leafhopper-transmitted maize rayado fino virus in Central America. In L. E. Williams, D. T. Gordon, & L. R. Nault (Eds.), *Proceedings of the maize virus colloquium workshop* (pp. 15–19). Wooster, OH: OARDC.

Ganal, M. W., Durstewitz, G., Polley, A., Berard, A., Buckler, E. S., Charcosset, A., et al. (2011). A large maize (*Zea mays* L.) SNP genotyping array: Development and germplasm genotyping, and genetic mapping to compare with the B73 reference genome. *PLoS One, 6*, 28334.

Gao, Z. H., Johansen, E., Eyers, S., Thomas, C. L., Noel Ellis, T. H., & Maule, A. J. (2004). The potyvirus recessive resistance gene, *sbm1*, identifies a novel role for translation initiation factor eIF4E in cell-to-cell trafficking. *Plant Journal, 40*, 376–385.

Gomez, P., Rodriguez-Hernandez, A. M., Moury, B., & Aranda, M. A. (2009). Genetic resistance for the sustainable control of plant virus diseases: Breeding, mechanisms and durability. *European Journal of Plant Pathology, 125*, 1–22.

Greber, R. S. (1981). Maize stripe disease in Australia. *Australian Journal of Agricultural Research, 32*, 27–36.

Green, J. M., & Ulrich, J. F. (1993). Response of Corn (Zea mays L.) inbreds and hybrids to sulfonylurea herbicides. *Weed Science*, *41*, 508–516.

Gregory, L. V., & Ayers, J. E. (1982). Influence of planting date on yield reduction due to *Maize dwarf mosaic virus* in sweet corn. *Phytopathology*, *72*, 257.

Grilli, M. P. (2008). An area-wide model approach for the management of a disease vector planthopper in an extensive agricultural system. *Ecological Modelling*, *213*, 308–318.

Herold, F., Bergold, G. H., & Weibel, J. (1960). Isolation and electron microscopic demonstration of a virus infecting corn (*Zea mays* L.). *Virology*, *12*, 335–347.

Hohmann, F., Fuchs, E., Gruntzig, M., & Oertel, U. (1999). A contribution to the ecology of *Sugarcane mosaic potyvirus* (SCMV) and *Maize dwarf mosaic potyvirus* (MDMV) in Germany. *Zeitschrift fur Pflanzenkrankheiten und Pflanzenschutz—Journal of Plant Diseases and Protection*, *106*, 314–324.

Hruska, A. J., Gladstone, S. M., & Obando, R. (1996). Epidemic roller coaster: Maize stunt disease in Nicaragua. *American Entomologist*, *42*, 248–252, Winter.

Hruska, A. J., & Peralta, M. G. (1997). Maize response to corn leafhopper (Homoptera: Cicadellidae) infestation and Achaparramiento disease. *Journal of Economic Entomology*, *90*, 604–610.

Hsu, H. T. (2002). Biological control of plant viruses. In D. Pimentel (Ed.), *Encyclopedia of pest management* (pp. 68–70). Boca Raton, FL: CRC Press.

Hulbert, S. H., Webb, C. A., Smith, S. M., & Sun, Q. (2001). Resistance gene complexes: Evolution and utilization. *Annual Review of Phytopathology*, *39*, 285–312.

Ingvardsen, C. R., Xing, Y. Z., Frei, U. K., & Lubberstedt, T. (2010). Genetic and physical fine mapping of Scmv2, a potyvirus resistance gene in maize. *Theoretical and Applied Genetics*, *120*, 1621–1634.

Jackson, A. O., Dietzgen, R. G., Goodin, M. M., Bragg, J. N., & Deng, M. (2005). Biology of plant rhabdoviruses. *Annual Review of Phytopathology*, *43*, 623–660.

Jardine, D. (2014). Corn lethal necrosis. Department of Plant Pathology, Kansas State University, Fact Sheet, Manhattan, KS. http://www.plantpath.ksu.edu/doc695.ashx.

Jensen, S. G., Lane, L. C., & Seifers, D. L. (1996). A new disease of maize and wheat in the High Plains. *Plant Disease*, *80*, 1387–1390.

Jensen, S. G., Wysong, D. S., Ball, E. M., & Higley, P. M. (1991). Seed transmission of *Maize chlorotic mottle virus*. *Plant Disease*, *75*, 497–498.

Jia, M. A., Li, Y. Q., Lei, L., Di, D. P., Miao, H. Q., & Fan, Z. F. (2012). Alteration of gene expression profile in maize infected with a double-stranded RNA fijivirus associated with symptom development. *Molecular Plant Pathology*, *13*, 251–262.

Jiang, X. Q., Meinke, L. J., Wright, R. J., Wilkinson, D. R., & Campbell, J. E. (1992). *Maize chlorotic mottle virus* in Hawaiian-grown maize—Vector relations, host range and associated viruses. *Crop Protection*, *11*, 248–254.

Jiang, X. Q., Wilkinson, D. R., & Berry, J. A. (1990). An outbreak of *Maize chlorotic mottle virus* in Hawaii and possible association with thrips. *Phytopathology*, *80*, 1060.

Jones, R. A. C. (2006). Control of plant virus diseases. In J. M. Thresh (Ed.), *Vol. 67. Advances in virus research: Plant virus epidemiology* (pp. 206–238). San Diego: Academic Press.

Jones, M., Boyd, E., & Redinbaugh, M. (2011). Responses of maize (*Zea mays* L.) near isogenic lines carrying *Wsm1*, *Wsm2*, and *Wsm3* to three viruses in the *Potyviridae*. *Theoretical and Applied Genetics*, *123*, 729–740.

Jones, M. W., Redinbaugh, M. G., Anderson, R. J., & Louie, R. (2004). Identification of quantitative trait loci controlling resistance to *Maize chlorotic dwarf virus*. *Theoretical and Applied Genetics*, *110*, 48–57.

Jones, M. W., Redinbaugh, M. G., & Louie, R. (2007). The *Mdm1* locus and maize resistance to *Maize dwarf mosaic virus*. *Plant Disease*, *91*, 185–190.

Jovic, J., Cvrkovic, T., Mitrovic, M., Krnjajic, S., Petrovic, A., Redinbaugh, M. G., et al. (2009). Stolbur phytoplasma transmission to maize by *Reptalus panzeri* and the disease cycle of Maize redness in Serbia. *Phytopathology, 99*, 1053–1561.

Kairo, M. T. K., Kiduyu, P. K., Mutinda, C. J. M., & Empig, L. T. (1995). *Maize streak virus*—Evidence for resistance against *Cicadulina mbila* Naudé, the main vector species. *Euphytica, 84*, 109–114.

Kang, B. C., Yeam, I., & Jahn, M. M. (2005). Genetics of plant virus resistance. *Annual Review of Phytopathology, 43*, 581–621.

Kamau, G. M. (1990). *Use of sulfonylurea herbicides for the control of Johnsongrass (Sorghum halepense* L.) *in corn (Zea mays* L.). Texas: Texas A & M University.

Knoke, J. K., Louie, R., Madden, L. V., & Gordon, D. T. (1983). Spread of maize dwarf mosaic virus from Johnsongrass to corn. *Plant Disease, 67*, 367–370.

Konate, G., & Traore, O. (1992). Hosts of the *Maize streak virus* in the Sudan and the Sahel—Identification and spatiotemporal distribution. *Phytoprotection, 73*, 111–117.

Kuhn, C. W., Jellum, M. D., & All, J. N. (1975). Effect of carobfuran treatment on corn yield, *Maize chlorotic dwarf* and *Maize dwarf mosaic virus* diseases, and leafhopper populations. *Phytopathology, 65*, 1017–1020.

Kyetere, D. T., Ming, R., McMullen, M. D., Pratt, R. C., Brewbaker, J., & Musket, T. (1999). Genetic analysis of tolerance to *Maize streak virus* in maize. *Genome, 42*, 20–26.

Lanfermeijer, F. C., Jiang, G. Y., Ferwerda, M. A., Dijkhuis, J., de Haan, P., Yang, R. C., et al. (2004). The durable resistance gene *Tm-2(2)* from tomato confers resistance against ToMV in tobacco and preserves its viral specificity. *Plant Science, 167*, 687–692.

Lapierre, H., & Signoret, P. (2004). *Virus and virus diseases of Poaceae (Gramineae)*. Paris: Institut National de la Recherche Agronomique.

Law, M. D., Moyer, J. W., & Payne, G. A. (1989). Effect of host resistance on pathogenesis of maize dwarf mosaic virus. *Phytopathology, 79*, 757–761.

Lebas, B. S. M., Ochoa-Corona, F. M., Elliott, D. R., Tang, Z., & Alexander, B. J. R. (2005). Development of an RT-PCR for High Plains virus indexing scheme in New Zealand post-entry quarantine. *Plant Disease, 89*, 1103–1108.

Lei, J. D., & Agrios, G. N. (1986). Mechanisms of resistance in corn to *Maize dwarf mosaic* virus. *Phytopathology, 76*, 1034–1040.

Li, J., Ding, J., Zhang, W., Zhang, Y. L., Tang, P., Chen, J. Q., et al. (2010). Unique evolutionary pattern of numbers of gramineous NBS-LRR genes. *Molecular Genetics and Genomics, 283*, 427–438.

Li, J., Li, G., & Yuan, F. (1999). Control of the *Maize rough dwarf virus* disease. *Beijing Agricultural Science, 17*, 23–25.

Liu, X. H., Tan, Z. B., Li, W. C., Zhang, H. M., & He, D. W. (2009). Cloning and transformation of SCMV CP gene and regeneration of transgenic maize plants showing resistance to SCMV strain MDB. *African Journal of Biotechnology, 8*, 3747–3753.

Louie, R. (1985). Variations in maize inbred response to maize dwarf mosaic virus inoculation treatments. *Phytopathology, 75*, 1354.

Louie, R. (1986). Effects of genotype and inoculation protocols on resistance evaluation of maize to *Maize dwarf mosaic virus* strains. *Phytopathology, 76*, 769–773.

Louie, R. (1995). Vascular puncture of maize kernels for the mechanical transmission of *Maize white line mosaic virus* and other viruses of maize. *Phytopathology, 85*, 139–143.

Louie, R. (1999). Diseases caused by viruses. In D. G. White (Ed.), *Compendium of corn diseases* (pp. 49–55). St. Paul: APS Press.

Louie, R., & Abt, J. J. (2004). Mechanical transmission of maize rough dwarf virus. *Maydica, 49*, 231–240.

Louie, R., & Anderson, R. J. (1993). Evaluation of *Maize chlorotic dwarf virus* resistance in maize with multiple inoculations by *Graminella nigrifrons* (Homoptera, Cicadellidae). *Journal of Economic Entomology, 86*, 1579–1583.

Louie, R., Jones, M. W., Anderson, R. J., & Redinbaugh, M. G. (2002). Registration of maize germplasm Oh1VI. *Crop Science*, *42*, 991.

Louie, R., Knoke, J. K., & Findley, W. R. (1990). Elite maize germplasm—Reactions to *Maize dwarf mosaic* and *Maize chlorotic dwarf viruses*. *Crop Science*, *30*, 1210–1215.

Louie, R., Knoke, J. K., & Reichard, D. L. (1983). Transmission of *Maize dwarf mosaic virus* with solid-stream inoculation. *Plant Disease*, *67*, 1328–1331.

Louie, R., Redinbaugh, M. G., Gordon, D. T., Abt, J. J., & Anderson, R. J. (2000). *Maize necrotic streak virus*, a new maize virus with similarity to species of the family *Tombusviridae*. *Plant Disease*, *84*, 1133–1139.

Luan, J. W., Wang, F., Li, Y. J., Zhang, B., & Zhang, J. R. (2012). Mapping quantitative trait loci conferring resistance to rice black-streaked virus in maize (*Zea mays* L.). *Theoretical and Applied Genetics*, *125*, 781–791.

Madden, L. V., Hughes, G., & Van Den Bosch, F. (2007). *The study of plant disease epidemics*. St. Paul, MN: APS Press.

Madden, L. V., Jeger, M. J., & Van Den Bosch, F. (2000). A theoretical assessment of the effects of vector-virus transmission mechanism on plant virus disease epidemics. *Phytopathology*, *90*, 576–594.

Mahajan, S. K., Chisholm, S. T., Whitham, S. A., & Carrington, J. C. (1998). Identification and characterization of a locus (*RTM1*) that restricts long-distance movement of *Tobacco etch virus* in *Arabidopsis thaliana*. *The Plant Journal: For Cell and Molecular Biology*, *14*, 177–186.

Mahmoud, A., Royer, M., Granier, M., Ammar, E. D., Thouvenel, J. C., & Peterschmitt, M. (2007). Evidence for a segmented genome and partial nucleotide sequences of maize yellow stripe virus, a proposed new tenuivirus. *Archives of Virology*, *152*, 1757–1762.

March, G. J., Ornaghi, J. A., Beviacqua, J. E., Giuggia, J., Rago, A., & Lenardon, S. L. (2002). Systemic insecticides for control of *Delphacodes kuscheli* and the *Mal de Rio Cuarto virus* on maize. *International Journal of Pest Management*, *48*, 127–132.

Marçon, A., Kaeppler, S. M., & Jensen, S. G. (1997). Resistance to systemic spread of high plains virus and wheat streak mosaic virus cosegregates in two F2 maize populations inoculated with both pathogens. *Crop Science*, *37*, 1923–1927.

Marçon, A., Kaeppler, S. M., Jensen, S. G., Senior, L., & Stuber, C. (1999). Loci controlling resistance to *High Plains virus* and *Wheat streak mosaic virus* in a B73 X Mo17 population of maize. *Crop Science*, *39*, 1171–1177.

Martin, T., Franchino, J. A., Kreff, E. D., Procopiuk, A. M., Tomas, A., Luck, S., et al. (2010). Major QTLs conferring resistance of corn to fijivirus. US Patent 2010/0325750 Al.

Martin, D. P., & Shepherd, D. N. (2009). The epidemiology, economic impact and control of maize streak disease. *Food Security*, *1*, 305–315.

McMullen, M. D., Jones, M. W., Simcox, K. D., & Louie, R. (1994). 3 Genetic-loci control resistance to *Wheat streak mosaic virus* in the maize inbred Pa405. *Molecular Plant-Microbe Interactions*, *7*, 708–712.

McMullen, M. D., Kresovich, S., Villeda, H. S., Bradbury, P., Li, H. H., Sun, Q., et al. (2009). Genetic properties of the maize nested association mapping population. *Science*, *325*, 737–740.

McMullen, M. D., & Louie, R. (1989). The linkage of molecular markers to a gene controlling the symptom response in maize to *Maize dwarf mosaic virus*. *Molecular Plant-Microbe Interactions*, *2*, 309–314.

McMullen, M. D., & Louie, R. (1991). Identification of a gene for resistance to *Wheat streak mosaic virus* in maize. *Phytopathology*, *81*, 624–627.

McMullen, M. D., Roth, B. A., & Townsend, R. (1996). Maize chlorotic dwarf virus and resis-tance thereto. In Office PaT (Ed.) USA: Pioneer Hi-Bred International, Inc. and United States Department of Agriculture, 24. (5,569,828.).

Meyer, M. D., & Pataky, J. K. (2010). Increased severity of foliar diseases of sweet corn infected with *Maize dwarf mosaic* and *Sugarcane mosaic Viruses*. *Plant Disease*, *94*, 1093–1099.
Mielke-Ehret, N., & Mühlbach, H.-P. (2012). Emaravirus: A novel genus of multipartite, negative strand RNA plant viruses. *Viruses*, *4*, 1515–1536.
Milne, R. G., Boccardo, G., Bo, E. D., & Nome, F. (1983). Association of *Maize rough dwarf virus* with Mal de Rio Cuarto in Argentina. *Phytopathology*, *73*, 1290–1292.
Ming, R., Brewbaker, J. L., Pratt, R. C., Musket, T. A., & McMullen, M. D. (1997). Molecular mapping of a major gene conferring resistance to maize mosaic virus. *Theoretical and Applied Genetics*, *95*, 271–275.
Murry, L. E., Elliott, L. G., Capitant, S. A., West, J. A., Hanson, K. K., Scarafia, L., et al. (1993). Transgenic corn plants expressing MDMV strain-B coat protein are resistant to mixed infections of *Maize dwarf mosaic virus* and *Maize chlorotic mottle virus*. *Biotechnology*, *11*, 1559–1564.
Nault, L. R. (1990). Evolution of an insect pest—Maize and the corn leafhopper, a case-study. *Maydica*, *35*, 165–175.
Nault, L. R., Styer, W. E., Coffey, M. E., Gordon, D. T., Negi, L. S., & Niblett, C. L. (1978). Transmission of *Maize chlorotic mottle virus* by Chrysomelid beetles. *Phytopathology*, *68*, 1071–1074.
Nault, L. R., Styer, W. E., Knoke, J. K., & Pitre, H. N. (1973). Semipersistent transmission of leafhopper-borne *Maize chlorotic dwarf virus*. *Journal of Economic Entomology*, *66*, 1271–1273.
Nei, M., & Rooney, A. P. (2005). Concerted and birth-and-death evolution of multigene families. *Annual Review of Genetics*, *39*, 121–152.
Nelson, S., Brewbaker, J., & Hu, J. (2011). *Maize chlorotic mottle*. pp. 1–6. Honolulu, HI: College of Tropical Agriculture and Human Resources, University of Hawaii at Manoa.
Niblett, C. L., & Claflin, L. E. (1978). Corn lethal necrosis—New virus disease of corn in Kansas. *Plant Disease Reporter*, *62*, 15–19.
Nyasani, J. O., Meyhöfer, R., Subramanian, S., & Poehling, H.-M. (2012). Effect of intercrops on thrips species composition and population abundance on French beans in Kenya. *Entomologia Experimentalis et Applicata*, *142*, 236–246.
Oerke, E. C., & Dehne, H. W. (2004). Safeguarding production—Losses in major crops and the role of crop protection. *Crop Protection*, *23*, 275–285.
Pardina, P. E. R., Gimenez, M. P., Laguna, I. G., Dagoberto, E., & Truol, G. (1998). Wheat: A new natural host for the Mal de Rio Cuarto virus in the endemic disease area, Rio Cuarto, Cordoba province, Argentina. *Plant Disease*, *82*, 149–152.
Pernet, A., Hoisington, D., Dintinger, J., Jewell, D., Jiang, C., Khairallah, M., et al. (1999). Genetic mapping of maize streak virus resistance from the Mascarene source. II. Resistance in line CIRAD390 and stability across germplasm. *Theoretical and Applied Genetics*, *99*, 540–553.
Pernet, A., Hoisington, D., Franco, J., Isnard, M., Jewell, D., Jiang, C., et al. (1999). Genetic mapping of maize streak virus resistance from the Mascarene source. I. Resistance in line D211 and stability against different virus clones. *Theoretical and Applied Genetics*, *99*, 524–539.
Perring, T. M., Gruenhagen, N. M., & Farrar, C. A. (1999). Management of plant viral diseases through chemical control of insect vectors. *Annual Review of Entomology*, *44*, 457–481.
Pokorny, R., & Porubova, M. (2006). Heritability of resistance in maize to the Czech isolate of Sugarcane mosaic virus. *Cereal Research Communications*, *34*, 1081–1086.
Prins, M., Laimer, M., Noris, E., Schubert, J., Wassenegger, M., & Tepfer, M. (2008). Strategies for antiviral resistance in transgenic plants. *Molecular Plant Pathology*, *9*, 73–83.
Redinbaugh, M. G. (2003). Transmission of *Maize streak virus* by vascular puncture inoculation with unit-length genomic DNA. *Journal of Virological Methods*, *109*, 95–98.

Redinbaugh, M. G., Ammar, E. D., & Whitfield, A. E. (2012). Cereal rhabdovirus interactions with vectors. In R. G. Dietzgen, & I. Kuzmin (Eds.), *Rhabdoviruses: Molecular taxonomy, evolution, genomics, ecology, cytopathology and control* (pp. 147–163). London: Horizon Scientific Press.

Redinbaugh, M. G., Louie, R., Ngwira, P., Edema, R., Gordon, D. T., & Bisaro, D. M. (2001). Transmission of viral RNA and DNA to maize kernels by vascular puncture inoculation. *Journal of Virological Methods*, *98*, 135–143.

Redinbaugh, M. G., & Pratt, R. C. (2008). Virus resistance. In S. Hake, & J. L. Bennetzen (Eds.), *Handbook of maize: Its biology* (pp. 255–270). New York: Springer-Verlag.

Redinbaugh, M. G., Seifers, D. L., Meulia, T., Abt, J. J., Anderson, J. A., Styer, W. E., et al. (2002). *Maize fine streak virus*, a new leafhopper-transmitted rhabdovirus. *Phytopathology*, *92*, 1167–1174.

Reed, S. E., Tsai, C. W., Willie, K. J., Redinbaugh, M. G., & Hogenhout, S. A. (2005). Shotgun sequencing of the negative-sense RNA genome of the rhabdovirus *Maize mosaic virus*. *Journal of Virological Methods*, *129*, 91–96.

Revers, F., Guiraud, T., Houvenaghel, M. C., Mauduit, T., Le Gall, O., & Candresse, T. (2003). Multiple resistance phenotypes to Lettuce mosaic virus among Arabidopsis thaliana accessions. *Molecular Plant-Microbe Interactions*, *16*, 608–616.

Rivera, C., & Gamez, R. (1986). Multiplication of *Maize rayado fino virus* in the leafhopper vector *Dalbulus maidis*. *Intervirology*, *25*, 76–82.

Romay, M. C., Millard, M. J., Glaubitz, J. C., Peiffer, J. A., Swarts, K. L., Casstevens, T. M., et al. (2013). Comprehensive genotyping of the USA national maize inbred seed bank. *Genome Biology*, *14*, R55.

Rosencranz, E. (1969). A new leafhopper-transmissible corn stunt disease agent in Ohio. *Phytopathology*, *59*, 1293–1296.

Rosendahl, S., & Taylor, J. W. (1997). Development of multiple genetic markers for studies of genetic variation in arbuscular mycorrhizal fungi using AFLP. *Molecular Ecology*, *3*, 821–829.

Rufener, G. K., Balducchi, A. J., Mowers, R. P., Pratt, R. C., Louie, R., McMullen, M. D., et al. (1996). *Maize chlorotic dwarf virus* resistant maize and the production thereof. U.S.A. Patent No. 5,563,316.

Schaeffer, M. L., Harper, L. C., Gardiner, J. M., Andorf, C. M., Campbell, D. A., Cannon, E. K. S., et al. (2011). MaizeGDB: Curation and outreach go hand-in-hand. *Database—The Journal of Biological Databases and Curation*. 2011: bar022.

Scott, G. E., & Louie, R. (1996). Improved resistance to *Maize dwarf mosaic virus* by selection under greenhouse conditions. *Crop Science*, *36*, 1503–1506.

Scott, G. E., & Rosenkranz, E. (1981). Effectiveness of resistance to *Maize dwarf mosaic* and *Maize chlorotic dwarf viruses* in maize. *Phytopathology*, *71*, 937–941.

Seifers, D. L., Harvey, T. L., Martin, T. J., & Jensen, S. G. (1997). Identification of the wheat curl mite as the vector of the High Plains virus of corn and wheat. *Plant Disease*, *81*, 1161–1166.

Shepherd, D. N., Mangwende, T., Martin, D. P., Bezuidenhout, M., Thomson, J. A., & Rybicki, E. P. (2007). Inhibition of *Maize streak virus* (MSV) replication by transient and transgenic expression of MSV replication-associated protein mutants. *Journal of General Virology*, *88*, 325–336.

Shepherd, D. N., Martin, D. P., Van Der Walt, E., Dent, K., Varsani, A., & Rybicki, E. P. (2010). *Maize streak virus*: An old and complex 'emerging' pathogen. *Molecular Plant Pathology*, *11*, 1–12.

Shi, L. Y., Hao, Z. F., Weng, J. F., Xie, C. X., Liu, C. L., Zhang, D. G., et al. (2012). Identification of a major quantitative trait locus for resistance to *Maize rough dwarf virus* in a Chinese maize inbred line X178 using a linkage map based on 514 gene-derived single nucleotide polymorphisms. *Molecular Breeding*, *30*, 615–625.

Shi, C., Thummler, F., Melchinger, A. E., Wenzel, G., & Lubberstedt, T. (2006). Comparison of transcript profiles between near-isogenic maize lines in association with SCMV resistance based on unigene-microarrays. *Plant Science, 170*, 159–169.

Shulka, D. D., Wardand, C. W., & Brunt, A. A. (1994). *The Potyviridae.* Oxon, UK: CAB International.

Skare, J. M., Wijkamp, I., Denham, I., Rezende, J. A. M., Kitajima, E. W., Park, J. W., et al. (2006). A new eriophyid mite-borne membrane-enveloped virus-like complex isolated from plants. *Virology, 347*, 343–353.

So, Y. S., Ji, H. C., & Brewbaker, J. L. (2010). Resistance to corn leaf aphid (Rhopalosiphum maidis Fitch) in tropical corn (*Zea mays* L.). *Euphytica, 172*, 373–381.

Stewart, L. R. (2011). Waikaviruses: Studied but not understood. In *APSnet features.* APS Press, St. Paul, MN. https://www.apsnet.org/publications/apsnetfeatures/Pages/waikavirus.aspx.

Stewart, L. R., Haque, M. A., Jones, M. W., & Redinbaugh, M. G. (2013). Response of maize (*Zea mays* L.) lines carrying *Wsm1*, *Wsm2*, and *Wsm3* to the potyviruses *Johnsongrass mosaic virus* and *Sorghum mosaic virus*. *Molecular Breeding, 31*, 289–297.

Stewart, L. R., Paul, P. A., Qu, F., Redinbaugh, M. G., Miao, H., Todd, J., et al. (2013). *Wheat mosaic virus* (WMoV), the causal agent of High Plains Disease, is present in Ohio wheat fields. *Plant Disease, 97*, 1125.

Stewart, L. R., Teplier, R., Todd, J. C., Jones, M. W., Cassone, B. j., Wijeratne, S., Wijeratne, A. & Redinbaugh, M. G. Viruses in maize and Johnsongrass in southern Ohio. Phytopathology, in press.

Storey, H. H. (1925). The transmission of streak disease of maize by the leafhopper *Balclutha mbila* Naudé. *Annals of Applied Biology, 12*, 422–439.

Tao, Y. F., Jiang, L., Liu, Q. Q., Zhang, Y., Zhang, R., Ingvardsen, C. R., et al. (2013). Combined linkage and association mapping reveals candidates for *Scmv1*, a major locus involved in resistance to sugarcane mosaic virus (SCMV) in maize. *BMC Plant Biology, 13*, 162.

Tao, Y. F., Liu, Q. C., Wang, H. H., Zhang, Y. J., Huang, X. Y., Wang, B. B., et al. (2013). Identification and fine-mapping of a QTL, *qMrdd1*, that confers recessive resistance to maize rough dwarf disease. *BMC Plant Biology, 13*, 145.

Thottappilly, G., Bosque-Perez, N. A., & Rossel, H. W. (1993). Viruses and virus diseases of maize in tropical Africa. *Plant Pathology, 42*, 494–509.

Toler, R. W., Skinner, G., Bockholt, A. J., & Harris, K. F. (1985). Reactions of maize (*Zea mays*) accessions to *Maize rayado fino virus*. *Plant Disease, 69*, 56–57.

Tomson, J. A., & Shepherd, D. N. (2010). Developments in agricultural biotechnology in sub-Saharan Africa. *AgBioforum, 13*, 314–319.

Tosic, M., Ford, R. E., Shukla, D. D., & Jilka, J. (1990). Differentiation of sugarcane, maize dwarf, johnsongrass and sorghum mosaic viruses based on reactions of oat and some sorhum cultivars. *Plant Disease, 74*, 549–552.

Uyemoto, J. K. (1983). Biology and control of *Maize chlorotic mottle virus*. *Plant Disease, 67*, 7–10.

Uyemoto, J. K., Claflin, L. E., Wilson, D. L., & Raney, R. J. (1981a). Influence of *Maize chlorotic mottle* and *Maize dwarf mosaic viruses* inoculated at different plant growth stages on symptomatology and corn yields. *Phytopathology, 71*, 262–263.

Uyemoto, J. K., Claflin, L. E., Wilson, D. L., & Raney, R. J. (1981b). *Maize chlorotic mottle* and *Maize dwarf mosaic viruses*—Effect of single and double inoculations on symptomatology and yield. *Plant Disease, 65*, 39–41.

Uzarowska, A., Dionisio, G., Sarholz, B., Piepho, H. P., Xu, M. L., Ingvardsen, C. R., et al. (2009). Validation of candidate genes putatively associated with resistance to SCMV and MDMV in maize (*Zea mays* L.) by expression profiling. *BMC Plant Biology, 9*, 15.

Vandeplas, A. (2003). *Evaluation of sixty highland elite maize genotypes for resistance to maize rayado fino virus*. Leuven, Belgium: The Katholieke Universiteit Leuven, Ph.D.

Vasquez, J., & Mora, E. (2007). Incidence of and yield loss caused by maize rayado fino virus in maize cultivars in Ecuador. *Euphytica, 153*, 339–342.

Velazquez, P. D., Arneodo, J. D., Guzman, F. A., Conci, L. R., & Truol, G. A. (2003). *Delphacodes haywardi* Muir, a new natural vector of Mal de rio cuarto virus in Argentina. *Journal of Phytopathology, 151*, 669–672.

Wang, F., Zhang, Y. S., Zhuang, Y. L., Qin, G. Z., & Zhang, J. R. (2007). Molecular mapping of three loci conferring resistance to maize (*Zea mays* L.) rough dwarf disease. *Molecular Plant Breeding, 5*, 178–179.

Wangai, A. W., Redinbaugh, M. G., Kinyua, Z. M., Miano, D. W., Leley, P. K., Kasina, M., et al. (2012). First Report of *Maize chlorotic mottle virus* and Maize Lethal Necrosis in Kenya. *Plant Disease, 96*, 1582.

Welz, H. G., Schechert, A., Pernet, A., Pixley, K. V., & Geiger, H. H. (1998). A gene for resistance to the maize streak virus in the African CIMMYT maize inbred line CML202. *Molecular Breeding, 4*, 147–154.

Whitfield, A. E., Rotenberg, D., Aritua, V., & Hogenhout, S. A. (2011). Analysis of expressed sequence tags from Maize mosaic rhabdovirus-infected gut tissues of *Peregrinus maidis* reveals the presence of key components of insect innate immunity. *Insect Molecular Biology, 20*, 225–242.

Whitham, S. A., Anderberg, R. J., Chisholm, S. T., & Carrington, J. C. (2000). Arabidopsis RTM2 gene is necessary for specific restriction of *Tobacco etch virus* and encodes an unusual small heat shock-like protein. *Plant Cell, 12*, 569–582.

Whitham, S., Dineshkumar, S. P., Choi, D., Hehl, R., Corr, C., & Baker, B. (1994). The product of the *Tobacco mosaic virus* resistance gene N—Similarity to toll and the interleukin-1 receptor. *Cell, 78*, 1101–1115.

Williams, L. E., & Alexander, L. J. (1965). Maize dwarf mosaic, a new corn disease. *Phytopathology, 55*, 802–804.

Williams, M. M., & Pataky, J. K. (2012). Interactions between maize dwarf mosaic and weed interference on sweet corn. *Field Crops Research, 128*, 48–54.

Wisser, R. J., Balint-Kurti, P. J., & Nelson, R. J. (2006). The genetic architecture of disease resistance in maize: A synthesis of published studies. *Phytopathology, 96*, 120–129.

Wu, L. J., Han, Z. P., Wang, S. X., Wang, X. T., Sun, A. G., Zu, X. F., et al. (2013). Comparative proteomic analysis of the plant–virus interaction in resistant and susceptible ecotypes of maize infected with *Sugarcane mosaic virus*. *Journal of Proteomics, 89*, 124–140.

Wu, L. J., Wang, S. X., Chen, X., Wang, X. T., Wu, L. C., Zu, X. F., et al. (2013). Proteomic and phytohormone analysis of the response of maize (*Zea mays* L.) seedlings to *Sugarcane mosaic virus*. *PLoS One, 8*, e70295.

Wu, Q., Zhou, X., & Wu, J. (2014). First report of *Maize chlorotic mottle virus* infecting sugarcane (*Saccharum officinarum* Linn.). *Phytopathology, 98*, 572.

Xia, X. C., Melchinger, A. E., Kuntze, L., & Lübberstedt, T. (1999). Quantitative trait loci mapping of resistance to *Sugarcane mosaic virus* in maize. *Phytopathology, 89*, 660–667.

Xiao, W. K., Zhao, J., Fan, S. G., Li, L., Dai, J. R., & Xu, M. L. (2007). Mapping of genome-wide resistance gene analogs (RGAs) in maize (Zea mays L.). *Theoretical and Applied Genetics, 115*, 501–508.

Xie, L., Zhang, J. Z., Wang, Q. A., Meng, C. M., Hong, J. A., & Zhou, X. P. (2011). Characterization of Maize chlorotic mottle virus associated with maize lethal necrosis disease in China. *Journal of Phytopathology, 159*, 191–193.

Xu, M. L., Melchinger, A. E., Xia, X. C., & Lübberstedt, T. (1999). High-resolution mapping of loci conferring resistance to *Sugarcane mosaic virus* in maize using RFLP, SSR, and AFLP markers. *Molecular and General Genetics, 261*, 574–581.

Zambrano, J. L., Francis, D. M., & Redinbaugh, M. G. (2013). Identification of resistance to *Maize rayado fino virus* in maize inbred lines. *Plant Disease*, *97*, 1418–1423.
Zambrano, J. L., Jones, M. W., Brenner, E., Francis, D. M., Tomas, A., & Redinbaugh, M. G. (2014). Genetic analysis of resistance to six virus diseases in a multiple virus-resistant maize inbred line. *Theoretical and Applied Genetics*, *127*, 867–880.
Zambrano, J. L., Jones, M. W., Francis, D. M., Tomas, A., & Redinbaugh, M. G. (2014). Quantitative trait loci for resistance to *Maize rayado fino virus*. *Molecular Breeding*, *34*, 989–996.
Zhang, Z. Y., Fu, F. L., Gou, L., Wang, H. G., & Li, W. C. (2010). RNA interference-based transgenic maize resistant to *Maize dwarf mosaic virus*. *Journal of Plant Biology*, *53*, 297–305.
Zhang, S. H., Li, X. H., Wang, Z. H., George, M. L., Jeffers, D., Wang, F. G., et al. (2003). QTL mapping for resistance to SCMV in Chinese maize germplasm. *Maydica*, *48*, 307–312.
Zhang, Z. Y., Wang, Y. G., Shen, X. J., Li, L., Zhou, S. F., Li, W. C., et al. (2013). RNA interference-mediated resistance to maize dwarf mosaic virus. *Plant Cell, Tissue and Organ Culture*, *113*, 571–578.
Zhao, M., Ho, H., Wu, Y., He, Y., & Li, M. (2014). Western flower thrips (*Frankliniella occidentalis*) transmits *Maize chlorotic mottle virus*. *Journal of Phytopathology*, *162*, 532–536.

CHAPTER NINE

Tropical Food Legumes: Virus Diseases of Economic Importance and Their Control

Masarapu Hema*, Pothur Sreenivasulu†, Basavaprabhu L. Patil‡, P. Lava Kumar§, Dodla V.R. Reddy¶,1

*Department of Virology, Sri Venkateswara University, Tirupati, India
†Formerly Professor of Virology, Sri Venkateswara University, Tirupati, India
‡National Research Centre on Plant Biotechnology, IARI, Pusa Campus, New Delhi, India
§International Institute of Tropical Agriculture, Ibadan, Nigeria
¶Formerly Principal Virologist, ICRISAT, Patancheru, Hyderabad, India
[1]Corresponding author: e-mail address: dvrreddy7@yahoo.com

Contents

Abstract

Diverse array of food legume crops (*Fabaceae*: *Papilionoideae*) have been adopted worldwide for their protein-rich seed. Choice of legumes and their importance vary in different parts of the world. The economically important legumes are severely affected by a range of virus diseases causing significant economic losses due to

Advances in Virus Research, Volume 90
ISSN 0065-3527
http://dx.doi.org/10.1016/B978-0-12-801246-8.00009-3

reduction in grain production, poor quality seed, and costs incurred in phytosanitation and disease control. The majority of the viruses infecting legumes are vectored by insects, and several of them are also seed transmitted, thus assuming importance in the quarantine and in the epidemiology. This review is focused on the economically important viruses of soybean, groundnut, common bean, cowpea, pigeonpea, mungbean, urdbean, chickpea, pea, faba bean, and lentil and begomovirus diseases of three minor tropical food legumes (hyacinth bean, horse gram, and lima bean). Aspects included are geographic distribution, impact on crop growth and yields, virus characteristics, diagnosis of causal viruses, disease epidemiology, and options for control. Effectiveness of selection and planting with virus-free seed, phytosanitation, manipulation of crop cultural and agronomic practices, control of virus vectors and host plant resistance, and potential of transgenic resistance for legume virus disease control are discussed.

1. INTRODUCTION

Legumes belong to the family *Leguminasae* (*Fabaceae*), consisting of four subfamilies, the *Papilionoideae*, *Caesalpinoideae*, *Mimosoideae*, and *Swartzioideae* (Lewis, Schrire, MacKinder, & Lock, 2005). The *Papilionoideae* includes the major food legumes, soybean (*Glycine max*), groundnut (peanut or monkeynut, *Arachis hypogaea*), common bean (bean, French bean, or kidney bean, *Phaseolus vulgaris*), cowpea (southern pea, *Vigna unquiculata*), pigeonpea (red gram, arhar, *Cajanus cajan*), chickpea (Garbanzo or bengal gram, *Cicer arietinum*), pea (field pea, *Pisum sativum*), mungbean (green gram, *Vigna radiata*), urdbean (black gram, *Vigna mungo*), faba bean (*Vicia faba*), and lentil (*Lens culinaris*). They are usually cultivated in the tropical and subtropical areas of the world. Soybean, groundnut, common bean, cowpea, pigeonpea, mungbean, and urdbean are usually cultivated during the hot season, while chickpea, pea, faba bean, and lentil are cultivated during the cool season. Legume seeds (also called pulses or grain legumes) are second only to cereals as a source of human diet and animal feed. Nutritionally, legume seeds are two to three times richer in protein than cereal grains. Groundnut and soybean seeds are also rich in lipids. Diversity and importance of various food legumes vary in different parts of the world (Fig. 9.1). Groundnut is by far the most widely cultivated legume. Soybean is cultivated in a much larger area, dominating the legume production in both area and production (Fig. 9.1). It is extensively used in food industry and also as biofuel. Common beans and other legumes, viz., soybean, groundnut, cowpea, and chickpea, are the major source of food in Latin America, while lentil, pigeonpea, chickpea, mungbean, and urdbean are important in South Asia.

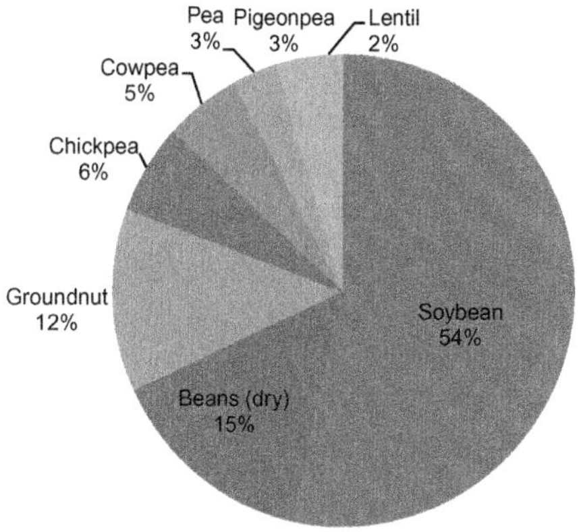

Crop	Area (ha)
soybean	106,625,241
beans (dry)	28,780,377
Groundnut	24,625,099
Chickpea	12,144,639
Cowpea	10,688,653
Pea	6,326,999
Pigeonpea	5,323,407
Lentil	4,249,725
Total	198,764,140

Figure 9.1 Percentage share of global production area of major food legumes. Total area = 198,764,140 ha. Source: *FAO Production Statistics of 2012.* (See the color plate.)

In the Middle East and North Africa, faba bean, lentil, and chickpea are particularly important. Groundnut, cowpea, and common bean are the most important food legumes in Africa. As per the 2012 production statistics (FAOSTAT, 2012), soybean, bean, groundnut, chickpea, pigeonpea, cowpea, lentil, and pea are cultivated in an area of 198.7 million hectares around the world, soybean dominating with 72% of production area.

Virus diseases are the major biotic constraints to legumes production, especially in the tropics and subtropics (Loebenstein & Thottappilly, 2003; Rao, Kumar, & Holguin-Peña, 2008; Sastry & Zitter, 2014). Cultivated food legumes are susceptible to natural infection by at least 150 viruses, belonging to different genera (ICTV, 2012).

The seed-transmitted viruses such as *Bean common mosaic virus* (BCMV), *Soybean mosaic virus* (SbMV), *Cucumber mosaic virus* (CMV), *Alfalfa mosaic virus* (AMV), *Peanut mottle virus* (PeMoV), *Peanut stripe virus* (PStV), *Pea seed-borne mosaic* virus (PSbMV), and *Bean yellow mosaic virus* (BYMV) are widely distributed and infect several legume crops. They have quarantine importance and also serve as primary source of inoculum in virus ecology and epidemiology (Albrechtsen, 2006; Sastry, 2013). Complex virus diseases like groundnut rosette are endemic only to Africa and chickpea stunt to Asia and Africa (Alegbejo & Abo, 2002; Kumar, Jones, & Waliyar, 2008). *Peanut bud necrosis virus* (PBNV), *Pigeonpea sterility mosaic virus* (PPSMV), *Mungbean yellow mosaic virus* (MYMV), and *Mungbean yellow mosaic India virus* (MYMIV) are confined to South East Asia (Kumar, Kumari, & Waliyar, 2008; Malathi & John, 2008; Mandal et al., 2012). Begomoviruses infecting common bean are prevalent in the Latin America (Navas-Castillo, Fiallo-Olive, & Sanchez-Campos, 2011). *Faba bean necrotic yellows virus*

(FBNYV) is predominant on cool season food legumes in West Asia and North Africa (Kumari & Makkouk, 2007; Makkouk & Kumari, 2009; Makkouk, Pappu, & Kumari, 2012). However, viruses such as BCMV, SbMV, PeMoV, AMV, and CMV are worldwide in distribution on legume crops (Loebenstein & Thottappilly, 2003). The virus that emerged as a great threat to groundnut during the last decade is *Tobacco streak virus* (TSV) (Kumar, Prasada Rao, et al., 2008). Many whitefly-transmitted legume begomoviruses have been characterized either as distinct species or as new pathotypes (Malathi & John, 2008; Qazi, Ilyas, Mansoor, & Briddon, 2007; Rey et al., 2012; Varma, Mandal, & Singh, 2011). The impact of virus diseases of food legumes on crop growth and yield is variable depending upon the crop cultivar, the virus strain or pathotype, and the time and duration of infection, season, location, and climate (Jones & Barbetti, 2012). The viruses infecting legumes have been reviewed (Bos, 2008; Hughes & Shoyinka, 2003; Malathi & John, 2008; Mishra, John, & Mishra, 2008).

In this chapter, the distribution, diversity of causal viruses, diagnosis, epidemiology, and control of virus diseases of economic significance of major annual food legumes are reviewed. In addition, begomoviruses infecting minor food legume crops are discussed.

2. VIRUS DISEASES OF MAJOR FOOD LEGUMES

The taxonomic position and modes of transmission of viruses causing or associated with diseases of major tropical food legumes are summarized in Table 9.1. For other biological and physicochemical characteristics of these viruses, refer to ICTVdB Management (2006) and ICTV (2012).

2.1. Soybean

Soybean is grown in tropical, subtropical, and temperate climates during warm, moist periods. The major soybean producing countries are the United States, Brazil, Argentina, China, India, Paraguay, Canada, Uruguay, and Ukraine (FAOSTAT, 2012). Nearly 70 viruses are known to naturally infect soybean worldwide (Hartman, Sinclair, & Rupe, 1999; Hill, 2003). Among them, diseases caused by SbMV, *Tobacco ring spot virus* (TRSV), PBNV, TSV, *Soybean dwarf virus* (SbDV), and begomoviruses are considered to be economically important (Wrather et al., 2010) (Table 9.1).

Table 9.1 Major virus diseases of tropical food legume crops and mode of transmission of their causal viruses

Crop	Disease	Causal virus (acronym)	Virus genus	Modes of transmission[a]
Soybean	Mosaic	*Soybean mosaic virus* (SbMV)	*Potyvirus*	Sap, seed, aphids (NP)
	Bud blight	*Tobacco ring spot virus* (TRSV)	*Nepovirus*	Sap, nematode, seed
		Cowpea severe mosaic virus (CPSMV)	*Comovirus*	Sap, beetles (SP)
		Peanut bud necrosis virus (PBNV)	*Tospovirus*	Sap, thrips (C)
	Brazilian bud blight	*Tobacco streak virus* (TSV)	*Ilarvirus*	Sap, seed, thrips-aided
	Dwarf	*Soybean dwarf virus* (SbDV)	*Luteovirus*	Aphids (C)
	Yellow mosaic	Several begomoviruses (e.g., *Mungbean yellow mosaic virus* (MYMV), *Bean golden mosaic virus* (BGMV))	*Begomovirus*	Whiteflies (C)
Groundnut	Rosette	*Groundnut rosette assistor virus* (GRAV)	*Luteovirus*	Aphids (C)
		Groundnut rosette virus (GRV)	*Umbravirus*	Sap, aphids(C), GRAV-dependent
		Satellite RNA		Aphids (C), GRAV-, and GRV-dependent
	Spotted wilt	*Tomato spotted wilt virus* (TSWV)	*Tospovirus*	Sap, thrips (CP)
	Bud necrosis	*Peanut bud necrosis virus* (PBNV)		
	Stem necrosis	TSV	*Ilarvirus*	Sap, seed, thrips aided
	Clump	*Peanut clump virus* (PCV)	*Pecluvirus*	Sap, seed, fungus
		Indian peanut clump virus (IPCV)		
	Mottle	*Peanut mottle virus* (PeMoV)	*Potyvirus*	Sap, seed, aphids (NP)
	Stripe	*Peanut stripe virus* (PStV)		
	Yellow mosaic	*Cucumber mosaic virus* (CMV)	*Cucumovirus*	Sap, seed, aphids (NP)

Continued

Table 9.1 Major virus diseases of tropical food legume crops and mode of transmission of their causal viruses—cont'd

Crop	Disease	Causal virus (acronym)	Virus genus	Modes of transmission
Common Bean	Common mosaic and black root	*Bean common mosaic virus* (BCMV) and *Bean common mosaic necrosis virus* (BCMNV)	*Potyvirus*	Sap, seed, aphids (NP)
	Golden mosaic	BGMV	*Begomovirus*	Whiteflies (C)
	Golden yellow mosaic	*Bean golden yellow mosaic virus* (BGYMV)		
	Dwarf mosaic	*Bean dwarf mosaic virus* (BDMV)		
	Mosaic due to CMV	CMV	*Cucumovirus*	Sap, seed, aphids (NP)
Cowpea	Mosaic due to potyviruses	*Cowpea aphid-borne mosaic virus* (CABMV) and *Bean common mosaic virus—blackeye cowpea mosaic strain* (BCMV-BICM)	*Potyvirus*	Sap, seed, aphids (NP)
	Mosaic due to comoviruses	*Cowpea mosaic virus* (CPMV) and CPSMV	*Comovirus*	Sap, seed, beetles (SP)
	Mosaic due to sobemovirus	*Southern bean mosaic virus* (SBMV)	*Sobemovirus*	Sap, seed, beetles (SP)
	Stunt	CMV	*Cucumovirus*	Sap, seed, aphids (NP)
		BCMV-BICM	*Potyvirus*	
	Golden mosaic and yellow mosiac	*Cowpea golden mosaic virus* (CGMV)	*Begomoviurus*	Whiteflies (C)
		Mungbean yellow mosaic India virus (MYMIV)		
		Dolichos yellow mosaic virus (DoYMV)		
	Mild mottle	*Cowpea mild mottle virus* (CMMV)	*Carlavirus*	Sap, seed, whiteflies (NP)
	Chlorotic mottle	*Cowpea chlorotic mottle virus* (CCMV)	*Bromovirus*	Sap, beetles (SP)
	Mottle	*Cowpea mottle virus* (CPMoV)	*Carmovirus*	Sap, seed, beetles (SP)

Pigeonpea	Sterility mosaic	Pigeonpea sterility mosaic virus (PPSMV)	*Emaravirus*	Sap, mites (SP)
	Yellow mosaic	MYMV, *Rhycosia mosaic virus* (RhMV) and *Tomato leaf curl New Delhi virus* (ToLCNDV)	*Begomovirus*	Whiteflies (C)
Mungbean and Urdbean	Yellow mosaic	MYMV, *Mungbean yellow mosaic India virus* (MYMIV) and *Horsegram yellow mosaic virus* (HgYMV)	*Begomovirus*	Whiteflies (C)
	Leaf curl	PBNV	*Tospovirus*	Sap, thrips (CP)
	Leaf crinkle	Urdbean leaf crinkle virus disease (ULCVD)	uncharacterized	Sap, seed, beetles (SP)
Chickpea	Stunt	*Bean leaf roll virus* (BLRV), SbDV, *Beet western yellows virus* (BWYV), *Legume yellows virus* (LYV) and Chickpea luteovirus (CpLV)	*Luteovirus*	Aphids (C)
	Chlorotic stunt	*Chickpea chlorotic stunt virus* (CpCSV)	*Polerovirus*	Aphids(C)
	Chlorotic dwarf	*Chickpea chlorotic dwarf virus* (CpCDV),	*Mastrevirus*	Leaf hoppers (C)
		Chickpea redleaf virus (CpRLV), *Chickpea yellows virus* (CpYV), *Chickpea chlorosis virus* (CpCV), *Chickpea chlorosis Australia virus* (CpAV)		
		Bean yellow dwarf virus (BeYDV)		

Continued

Table 9.1 Major virus diseases of tropical food legume crops and mode of transmission of their causal viruses—cont'd

Crop	Disease	Causal virus (acronym)	Virus genus	Modes of transmission
Pea	Mosaic due to potyviruses	*Pea seed-borne mosaic virus* (PSbMV)	*Potyvirus*	Sap, seed, aphids (NP)
		Bean yellow mosaic virus (BYMV)		
	Enation mosaic	*Pea enation mosaic virus*-1 (PEMV-1)	*Enamovirus*	Sap, aphids (C)
		Pea enation mosaic virus-2 (PEMV-2)	*Umbravirus*	
	Top yellows	BLRV	*Luteovirus*	Aphids (C)
Faba bean	Necrotic yellows	*Faba bean necrotic yellows virus* (FBNYV) and *Faba bean necrotic stunt virus* (FBNSV)	*Nanovirus*	Aphids (C)
	Leaf roll	BLRV	*Luteovirus*	
	Mosaic and necrosis	BYMV	*Potyvirus*	Sap, seed, aphids (NP)
	Mottle	*Broad bean mottle virus* (BBMV)	*Bromovirus*	Sap, seed, beetles (SP)
Lentil	Yellows and stunt	BLRV	*Luteovirus*	Aphids (C)
		FBNYV	*Nanovirus*	
	Mosaic and mottle	PSbMV	*Potyvirus*	Sap, seed, aphids (NP)
		BYMV		
		CMV	*Cucumovirus*	
		Broad bean stain virus (BBSV)	*Comovirus*	Sap, beetle (SP)

[a]NP, nonpersistent; SP, semipersistent; C, circulative; CP, circulative and propagative.

2.1.1 Mosaic

The yield losses due to mosaic disease caused by SbMV range from 8% to 50% under natural field conditions and can reach up to 100% in severe outbreaks (Arif, Stephen, & Hassan, 2002; Cui, Chen, & Wang, 2011; Hill, 2003) and when coinfected with *Bean pod mottle virus* (BPMV), *Cowpea mosaic virus* (CPMV), AMV, and TRSV (Cui et al., 2011; Hwang et al., 2011; Malapi-Nelson, Wen, Ownley, & Hajimorad, 2009). Symptoms induced by SbMV depend on the host genotype, the virus strain, the plant age, and the temperatures at which the plants are grown (Hill, 2003). Nine distinct strain groups, G1–G7, G7A, and C14 have been reported based on differential reactions on selected soybean cultivars (Cho & Goodman, 1979; Lim, 1985).

SbMV detection is based on ELISA, RT-PCR, real time-PCR, near infrared spectroscopy, and aquaphotomics (Cui et al., 2011; Jinendra et al., 2010). Plants grown from SbMV-infected seed often constitute the primary inoculum source with secondary spread, which occur rapidly, by aphids (Cui et al., 2011).

Control of SbMV is difficult because of its relatively broad host range, the number of aphid species that transmit the virus, and the frequency of seed transmission. Thus, utilization of virus-free seeds and control of aphid populations are effective management measures against SbMV. Rouging of early-infected plants in fields meant for seed production is recommended. Spraying with insecticides imidacloprid WP (wettable powder), benfuracarb EC (Commission Regulation), and acephate WP can reduce aphid population and SbMV incidence under field conditions (Kim, Roh, Kim, Im, & Hur, 2000). Soybean plants, cross-protected with an attenuated isolate of SbMV, showed negligible SbMV incidence (Kosaka & Fukunishi, 1994).

In order to achieve integrated control of SbMV in soybean, utilization of host–plant resistance was found to be the best option (Pedersen, Grau, Cullen, Koval, & Hill, 2007). A number of soybean accessions (germplasm) and cultivars carrying resistance to SbMV have been identified and used in the breeding programs (Cui et al., 2011). Since SbMV is genetically variable and continuously evolving via recombination and spontaneous mutations, strong directional breeding line selection can lead to the occurrence of resistance-breaking isolates (Gagarinova, Babu, Stromvik, & Wang, 2008).

Resistance against SbMV in soybean has been reported to be controlled by a single dominant gene (Wang, Gai, & Pu, 2003) or very closely linked genes against G1–G7 strain groups (Ma, Chen, Buss, & Tolin, 2004). Three independent dominant resistance loci *Rsv1*, *Rsv3*, and *Rsv4* conferring partial or complete genetic resistance to all SbMV strains have been identified.

Rsv2 was initially assigned to the resistance gene in the soybean cultivar OX670 and later dropped when it was known to actually possess two resistant genes *Rsv1* and *Rsv3* (Gunduz, Buss, Ma, Chen, & Tolin, 2001). *Rsv1* is a single locus, multiallelic gene (Zheng, Chen, & Gergerich, 2005). SbMV strain G7 overcomes the resistance conferred by *Rsv1* and results in systemic necrosis of virus-infected plants. The resistance conferring genes have been deployed in China, the United States, Canada, and other countries for developing SbMV-resistant soybean cultivars (Cui et al., 2011).

Incorporation of multiple resistance genes into soybean cultivars through gene pyramiding should become a high priority for soybean breeders to develop durable resistance to SbMV. Pyramiding viral resistance genes against SbMV is reported to be benefited by gene mapping and marker-assisted selection (MAS) with PCR-based markers for the *Rsv3* gene conferring resistance to three of the most virulent strains G5, G6, and G7 of SbMV (Jeong et al., 2002). Molecular markers of three resistance genes have been developed based on mapping with several molecular techniques such as restriction fragment length polymorphism, random amplified polymorphism DNA, amplified fragment length polymorphism (AFLP), simple sequence repeats (SSRs), and single nucleotide polymorphisms (SNPs) to assist in plant breeding programs (e.g., Hwang et al., 2006; Shi et al., 2009, 2011; Wang et al., 2011).

Arif and Hassan (2002) in Pakistan reported that PI 88788 germplasm line from China showed resistance to SbMV under field conditions, and the resistance in PI 88788 to SbMV-G1 was controlled by a single, partially dominant gene; however to SbMV-G7, the same gene was completely dominant (Gunduz, Buss, Chen, & Tolin, 2004). The inheritance and mapping of genes resistant to SbMV strain SC14 in soybean accessions in China have led to identification of markers for choosing resistance gene in soybean breeding programs and cloning of resistance genes (Li et al., 2006). *Rsv1*-mediated resistance to SbMV-G7 strain in soybean genotypes, Williams 82, PI 96983, and L78-379 was shown to be due to SbMV strain-specific protein P3 (Hajimorad, Eggenberger, & Hill, 2006). The resistance to SbMV-G7, governed by *Rsv3* gene, was attributed to cytoplasmic inclusion cistron of the virus modulating virulence and symptom expression (Zhang et al., 2009). Cylindrical inclusion protein of SbMV strains, G7H and G5H, was shown to be the pathogenic determinant in the two resistant cultivars L29 and Jinpumkong-2 in Korea (Seo et al., 2009). Genes governing resistance in 12 soybean genotypes in China to SbMV strains, SC4 and SC8, was found to be present at different loci (Wang et al., 2012).

Development of transgenic disease resistance to soybean, despite progress in other important crop plants, has advanced slowly. Soybean cv. 9341 transformed with coat protein (CP) gene showed resistance to SbMV strains G2, G6, and G7 (Steinlage, Hill, & Nutter, 2002; Wang, Eggenberger, Nutter, & Hill, 2001). Investigations on virus-induced gene silencing utilizing SbMV-G2 strain and soybean cultivars with *Rsv1*-resistant gene are likely to provide new insight into the soybean signaling network required for incorporating stable resistance (Zhang, Grosic, Whitham, & Hill, 2012). The complete sequence of soybean genome will facilitate the molecular cloning and characterization of three resistance (*R*) genes and elucidating their resistance signaling pathways that are likely to provide a better understanding of the co-evolutionary events of the *R* genes and SbMV genome (Schmutz et al., 2010). Information from such studies will help to develop novel strategies against SbMV and other genetically related viruses. Importance of RNA silencing as a tool to uncover gene function and engineer novel traits in soybean was reported (Kasai & Kanazawa, 2012). Robust RNAi-based resistance to mixed infection of SbMV, AMV, and BPMV in soybean plants was developed by expressing separate short hairpins from a single transgene (Zhang et al., 2011). This approach has a potential to develop multiple virus resistance in soybean and other legumes. Three microRNAs (miRNAs) known to regulate gene functions, involved in soybean's response to SbMV infection, were identified, and their interaction with genes conferring resistance is likely to elucidate mechanisms underlying pathogenesis by SbMV (Yin, Wang, Cheng, Wang, & Yu, 2013).

2.1.2 Dwarf

Soybean dwarf (SbD) disease, caused by SbDV, was reported from Japan, Indonesia, Africa, Australia, and New Zealand. SbDV host range includes pea, *Trifolium* species, bean, *Lupinus* speices, *Medicago* species, and *V. faba* (Hartman et al., 1999). Several distinct strains of SbDV were reported based on symptomatology in soybeans, aphid vector relationships, physicochemical properties, and molecular characteristics. The complete nucleotide sequences of the genomic RNAs of four SbDV strains were determined (Terauchi et al., 2001). Specific monoclonal antibodies (MAbs) can discriminate different strains of SbDV in ELISA (Mikoshiba, Honda, Kanematsu, & Fujisawa, 1994). Dot-blot hybridization test was developed for the detection and discrimination of S, P, Y, and D strains of SbDV (Yamagishi, Terauchi, Kanematsu, & Hidaka, 2006).

As primary and secondary spread of SbDV is carried by aphid vectors, organophosphorus insecticides have been used in Japan to minimize virus spread. Soybean cultivars "Adams" and "Yuuzuru" have shown tolerance to SbDV. The highly resistant cultivar "Tsurukogane" was released in Hokkaido, Japan, in 1994 (Hartman et al., 1999). The virus can be introduced successfully into soybeans without the aid of aphid vectors by transferring full-length cDNA clone with the aid of a gene gun, thus facilitating screening of genotypes under laboratory conditions (Yamagishi, Terauchi, Honda, Kanematsu, & Hidaka, 2006). In an attempt to produce SbDV-resistant transgenic soybean plants, a vector construct containing inverted repeat-SbDV-CP genes spaced by β-glucuronidase sequences was delivered into soybean somatic embryos via microprojectile bombardment. The T_2 plants 2 months after inoculation with SbDV by aphids showed negligible SbDV-specific RNA and remained symptomless. Additionally, they contained SbDV-CP-specific small interfering RNAs (siRNAs) suggesting that the T_2 plants acquired resistance to SbDV through RNA silencing-mediated process (Tougou et al., 2006). The T_2 progenies derived from soybeans transformed with positive-sense SbDV-CP gene remained symptomless after inoculation with SbDV through aphids and additionally showed little SbDV-specific RNA (Tougou et al., 2007). These results show good prospects for generating genetically engineered SbDV-resistant soybean cultivars.

2.1.3 Bud blight

In soybean, bud blight symptoms (curving of the terminal bud followed by necrosis) are induced during disease development by several taxonomically distinct viruses such as SbMV (Hartman et al., 1999), TRSV (Hartman et al., 1999), TSV (Arun Kumar, Lakshmi Narasu, Usha, & Ravi, 2008; Hartman et al., 1999), *Tomato spotted wilt virus* (TSWV) (Nischwitz, Mullis, Gitaitis, & Csinos, 2006), PBNV (Bhat, Jain, Varma, & Lal, 2002), *Cowpea severe mosaic virus* (CPSMV) (Anjos & Lin, 1984), and *Cowpea mild mottle virus* (CPMMV) (Almeida, Piuga, et al., 2005). Among these, bud blight caused by TRSV and PBNV are discussed next.

Bud blight caused by TRSV was reported from the United States, China, and India. Yield reduction was attributed to reduced pod set and seed formation (Hill, 2003). TRSV can be seed-transmitted to the extent of up to 100% (Frison, Bos, Hamilton, Mathur, & Taylor, 1990) and can remain viable in seeds for at least 5 years. Detection of TRSV in soybean seed can be done by duplex real time-PCR (Yi, Chen, & Yang, 2011). The virus is easily sap transmissible, and no efficient true arthropod vector of TRSV has been identified. However, it is reported to be transmitted nonspecifically by few insects and mites (*Aphis gossypii*, *Myzus persicae*, *Melanoplus differentialis*, *Epitrix hirtipennis*, *Thrips tabaci*, and *Tetranychus* species). The dagger nematode (*Xiphinema americanum*) is an inefficient vector, and the infection generally remains confined to roots (Hill, 2003).

In the absence of seed-borne inoculum, the disease first appears on the plants located at the periphery in soybean fields and advances inward as the season progresses. The disease spread depends on the availability of inoculum, from crops and weeds, adjacent to the field. More infection occurs in fields next to legume pastures and relatively less on those next to maize (*Zea mays*) fields. The hosts among legumes, which may act as reservoirs include *Crotalaria intermedia*, *Cyamopsis tetragonoloba*, *Lupinus* spp., *Melilotus* spp., *Phaseolus lunatus*, bean, pea, *Trifolium pretense*, and cowpea (Hill, 2003).

Virus-free soybean seed should be used in commercial fields. It may be desirable to avoid fields with dagger nematodes. Since the disease spread depends on the TRSV-susceptible crops and weeds next to soybean fields and to the presence of insect vector populations (Hill, 2003), location of soybean fields next to maize fields is recommended. One genotype (PI 407287) of wild soybean (*Glycine soja*) was shown to be resistant to the virus (Hartman et al., 1999).

PBNV causes chlorosis and necrosis of leaves, stems and buds, and stunting of soybean (Bhat et al., 2002). The disease was shown to cause severe yield losses to soybean crops in Maharashtra, India (Arun Kumar, Lakshmi Narasu, Usha, & Ravi, 2006). Unlike TRSV, PBNV is not transmitted through soybean seed.

2.1.4 Brazilian bud blight

Brazilian bud blight, caused by TSV, is currently known to occur in several countries that include Brazil, Argentina, the United States, and India (Arun Kumar et al., 2008; Rebedeaux, Gaska, Kurtzweil, & Grau, 2004). The characteristic symptoms are chlorosis and necrosis of leaves, stems and buds, and stunting. Bud blight symptoms caused by TSV are similar to those caused by TRSV and PBNV. A TSV isolate causing soybean bud blight disease in Brazil (TSV-BR) was reported to be a distinct strain which shared 81.3% and 80.7% nucleotide sequence homology with the CP gene of TSV-WC and TSV-MV (mungbean isolate from India), respectively (Almeida, Sakai, et al., 2005). The TSV isolate that caused necrosis in Maharashtra, India, was characterized by analyzing CP gene sequences and designated as TSV-SB (Arun Kumar et al., 2008). TSV was reported to be transmitted through soybean seed up to 90% (Frison et al., 1990). Thus, planting of virus-free seed is essential to minimize primary spread. Resistance of soybean to TSV has not been reported.

2.1.5 Yellow mosaic due to begomoviruses

Begomoviruses, viz., *Abutilon mosaic virus*, *Bean golden mosaic virus* (BGMV), *Euphorbia mosaic virus*, *Horsegram yellow mosaic virus* (HgYMV), MYMV, *Rhynchosia mosaic virus*, *Soybean crinkle leaf* virus, *Soybean golden mosaic virus*, *Tomato leaf curl Karnataka virus*, *Soybean mild mottle* virus (SbMMV), *Soybean chlorotic blotch virus* (SbCBV), *Soybean chlorotic spot virus* (SoCSV), and *Soybean yellow mosaic virus* have been reported to infect soybean under field conditions in different countries, and they are associated with yellow mosaic disease (YMD) (Alabi, Kumar, Mgbechi-Ezeri, & Naidu, 2010; Hartman et al., 1999; Malathi & John, 2008; Raj, Khan, Snehi, Srivastava, & Singh, 2006). Combined yield losses due to begomoviruses were estimated to exceed $300 million in black gram, mungbean, and soybean in India (Varma & Malathi, 2003). The identity of these begomoviruses is based on geminate virion morphology, whitefly transmission, and genome sequence analysis.

Sequence analysis of genomic components of the begomovirus isolates causing YMD in soybean from different locations of India revealed that they are the isolates of MYMV and MYMIV (Girish & Usha, 2005; Usharani, Surendranath, Haq, & Malathi, 2004, 2005). Recently, remote sensing technique was applied to determine the distribution of YMD in soybean (Gazala et al., 2013).

In Northwestern Argentina, sequence analysis of a begomovirus isolate from soybean indicated that it was closely related to *Sida mottle virus* (Rodriguez-Pardina, Zerbini, & Ducasse, 2006). In addition, BGMV, *Soybean blistering mosaic virus* (SbBMV), and *Tomato yellow spot virus* (ToYSV) were shown to infect soybean in Argentina (Alemandri et al., 2012). BGMV occurred at the highest incidence followed by SbBMV and ToYSV. Three distinct begomoviruses, i.e., BGMV, *Sida micrantha mosaic virus*, and *Okra mottle virus*, have been reported to naturally infect soybean sporadically in the Central Brazil based on phylogenetic analysis of DNA-A sequences (Fernandes, Cruz, Faria, Zerbini, & Aragao, 2009). Furthermore, the virus that induced chlorotic spots on soybean leaves in Brazil was identified as a novel begomovirus for which SoCSV name was proposed (Coco et al., 2013). In Nigeria, two distinct begomovirus isolates naturally infecting soybean, designated as SbMMV and SbCBV, were characterized (Alabi et al., 2010). The economic importance of these viruses is yet to be determined.

Since chemical control of whitefly vector that transmits YMD causal viruses is neither economical nor environment friendly, soybean germplasm was screened for YMD resistance in the Indian subcontinent under field conditions. Promising lines are yet to be utilized to develop YMD-resistant soybean cultivars (Akshay et al., 2013; Khan, Tyagi, & Dar, 2013; Kumar et al., 2014; Malek, Rahman, Raffi, & Salam, 2013). The cultivars Bossier and improved Pelican and the germplasm accessions TGm 119 and TGm 662 from the International Institute of Tropical Agriculture (IITA), Nigeria, were found to be resistant to *African soybean dwarf virus* (Hartman et al., 1999). Due to the wide genetic variation among the begomoviruses that infect soybeans, evaluation of resistance should be carried in multilocations to identify durable resistance. Screening of soybean germplasm against economically important begomoviruses by agroinoculation approach (as begomoviruses are not sap transmissible) is ideal to identify promising soybean cultivars under laboratory conditions.

2.2. Groundnut

The major groundnut-producing countries are China, India, Nigeria, the United States, Myanmar, Sudan, Argentina, Tanzania, and Indonesia (FAOSTAT, 2012). Thirty-two viruses have been reported to naturally infect this crop (Sreenivasulu, Subba Reddy, Ramesh, & Kumar, 2008). Of them, diseases caused by TSWV, PBNV, TSV, *Peanut clump virus* (PCV), *Indian peanut clump virus* (IPCV), PeMoV, PStV, CMV, and groundnut rosette disease (GRD) caused by a complex of three viral agents, *Groundnut rosette assistor virus* (GRAV), *Groundnut rosette virus* (GRV), and a satellite RNA, are considered to be economically important (Table 9.1).

2.2.1 Rosette

GRD is the most destructive disease in sub-Saharan Africa (SSA). It is known to occur in several African countries that include Malawi, Nigeria, Uganda, Senegal, Burkina Faso, Cotê d'Ivoire, South Africa, Niger, and Kenya (Wangai, Pappu, Pappu, Deom, & Naidu, 2001). Numerous epidemics of rosette have been reported in Africa resulting in substantial crop losses (Alegbejo & Abo, 2002; Naidu et al., 1999). Infection before flowering results in over 90% of crop loss. Based on symptoms, rosette is categorized into the chlorotic rosette, which is ubiquitous in SSA, while the green rosette occurs in West Africa, Uganda, and Angola; the third less frequently occurring type, the mosaic rosette, is recorded only in East and Central Africa. Among the three causal agents of GRD, GRAV is the helper virus involved in transmission of GRV and satellite RNA; GRV is the helper virus for satellite RNA replication, and disease symptoms are caused by satellite RNA. It also helps encapsidation of GRV RNA into GRAV particles through an unknown mechanism (Naidu et al., 1999). Thus, three components are intricately dependent on each other and play a crucial role in the biology and perpetuation of the disease (Alegbejo & Abo, 2002; Naidu et al., 1999; Taliansky, Robinson, & Murant, 2000). Sequence diversity of rosette disease causal agents from different geographic regions was studied (Deom, Naidu, Chiyembekeza, Ntare, & Subrahmanyam, 2000; Wangai et al., 2001). The causal viruses are readily transmitted by *Aphis craccivora* in a persistent manner.

The GRD can be tentatively diagnosed in the farmer's fields based on the characteristic symptoms. In the laboratory, GRD diagnosis is based on sap inoculation onto test plants such as *Chenopodium amaranticolor*, triple antibody sandwich-ELISA, dot-blot hybridization, gel electrophoresis for satellite RNA, and RT-PCR analysis (Blok et al., 1995; Breyel et al., 1988; Naidu, Robinson, & Kimmins, 1998; Rajeswari, Murant, & Massalski, 1987).

The GRD epidemiology is complex as it involves interactions between two distinct viruses and a satellite RNA, an aphid vector and the host plant in several areas of SSA. None of the three agents of rosette complex is seed-borne, and therefore, primary infection is introduced into the crop by viruliferous aphids likely to be derived from off season groundnut volunteers and self-sown plants. Secondary spread occurs from sources within the crop. The spread of GRD is complicated because a single aphid may not always transmit the three viral agents (Naidu, Miller, Mayo, Wesley, & Reddy, 2000). Plants that show symptoms but lack GRAV play no role in the spread of the disease because the CP of GRAV is required for encapsidation and transmission of GRV and satellite RNA. Therefore, the number of plants that possess all the three agents play a crucial role in the secondary spread of the disease in a given field, while the number of plants that show typical GRD symptoms influence yield.

Several approaches to manage GRD include application of pesticides to reduce vector populations, crop cultural practices to delay onset and spread of both the vector and the disease (reviewed by Alegbejo & Abo, 2002; Naidu et al., 1999; Ntare, Olorunju, & Waliyar, 2002; Olorunju & Ntare, 2003; Thresh, 2003). The control of aphid vectors by spraying the groundnut crop with insecticides could effectively control the disease (Davies, 1975). Many resource-poor farmers cannot afford pesticides and do not control GRD in the SSA. Early sowing may allow the crop to get established before aphid populations reach their peak and thus reduce the incidence of the disease. Dense plant stands discourage aphid infestation since aphids prefer light airy conditions. Nonetheless, only limited success has been achieved when these approaches were not combined.

Sources of resistance to GRD were first identified in groundnut landraces of late maturing Virginia type in West Africa. Resistance to this disease was also identified in the early maturing Spanish types. Resistant Virginia types were used in breeding program throughout SSA resulting in the development of several resistant cultivars (e.g., RMP 12, RMP 91, KH 241-D, and RG 1). Resistance among these cultivars was found to be effective against both chlorotic and green rosette and was governed by two independent recessive genes (Nigam & Bock, 1990). However, these cultivars being long duration were not widely grown. Early maturing chlorotic rosette-resistant Spanish types (90–110 days) suitable for diverse ecosystems of SSA were subsequently released (Bock, Murant, & Rajeswari, 1990; Naidu et al., 1999; Subrahmanyam, Hildebrand, Naidu, Reddy, & Singh, 1998). The majority of early maturing groundnut lines, evaluated in Nigeria in 2001 and 2002, showed resistance to GRD, early leaf spot, and late leaf spot (Iwo & Olorunju, 2009). Of the nine groundnut breeding lines possessing high yield and resistance to all the three diseases, ICGV-IS-96805 performed well at four locations and can be grown widely in SSA. These genotypes showed resistance to GRV but not to GRAV (Olorunju, Kuhn, Demski, Misari, & Ansa, 1991, 1992; Subrahmanyam et al., 1998). The resistance to GRV was shown to breakdown under high inoculum pressure and/or adverse environmental conditions. Most of the earlier studies on inheritance of disease resistance were based on visual symptoms and are applicable only to GRV and its satellite RNA, but not to GRAV. Immunity to all the three causative components of GRD was identified in wild *Arachis* species (Murant, Kumar, & Robinson, 1991; Subrahmanayam, Naidu, Reddy, Kumar, & Ferguson, 2001).

Dwivedi, Gurtu, Chandra, Upadhyaya, and Nigam (2003) determined AFLP diversity among selected rosette-resistant groundnut germplasm (ICGs 3436, 6323, 6466, 9558, 9723, 10347, 11044, 11968, and 12876) and one susceptible (ICG 7827) groundnut accession to identify DNA markers linked with resistance to GRD. Resistance to the aphid vector has been identified in groundnut genotype EC 36892 and in the breeding line ICG 12991 that was shown to be controlled by a single recessive gene. Herselman et al. (2004) first reported the identification of molecular markers closely linked to aphid resistance and GRD and constructed the first partial genetic map for cultivated groundnut using bulked segregant analysis and AFLP analysis.

Attempts have been made to exploit pathogen-derived resistance (PDR) (GRAV replicase and CP genes, and/or satellite RNA-derived sequences) to develop durable resistance to GRD (Taliansky, Ryabov, & Robinson, 1998). At present, it is not known that any transgenic groundnut lines, that possess GRD resistance, are in the pipeline for future deregulation (Reddy, Sudarshana, Fuchs, Rao, & Thottappilly, 2009).

2.2.2 Spotted wilt

Spotted wilt disease of groundnut caused by TSWV was reported from North and South Americas (Argentina, Brazil, and the United States), several African countries (South Africa, Nigeria, Kenya, Malawi, and Uganda), and Australia (Culbreath, Todd, & Brown, 2003). Losses up to 100% have been reported due to this virus (Culbreath & Srinivasan, 2011). In Asia, a similar disease was shown to be caused by a distinct tospovirus, later named as PBNV (Reddy, 1998). In addition to groundnut, TSWV was reported to naturally infect other legumes, viz., soybean, pea, *Tephrosia purpurea*, urdbean, mungbean, cowpea, *Crotalaria juncea*, *Canavalia gladiata*, faba bean, chickpea, and lentil (EPPO, 1997). TSWV produces a variety of symptoms, viz., concentric ring spots on leaflets, terminal bud necrosis, and severe plant stunting and mottled seed.

ELISA and Western blot analysis were used to differentiate various strains of tospoviruses into serogroups (Adam, Yeh, Reddy, & Green, 1993; Sreenivasulu, Demski, Reddy, Naidu, & Ratna, 1991). TSWV and PBNV are distinct virus species and TSWV belongs to the serogroup I and PBNV to the serogroup IV in the genus *Tospovirus* (Satyanarayana et al., 1998). The relationships between the isolates of different serogroups were studied by molecular hybridization and nucleotide sequence comparison (Tsompana & Moyer, 2008). Field diagnosis of TSWV infections based on symptoms may mislead as the virus induces a variety of symptoms, often influenced by environment. ELISA, dot-blot hybridization, RT-PCR, immunocapture-RT-PCR (IC-RT-PCR), and real-time fluorescent RT-PCR were applied for the detection of TSWV and other tospoviruses (Bandla et al., 1994; Boonham et al., 2002; Huguenot et al., 1990; Jain, Pappu, Pappu, Culbreath, & Todd, 1998; Resende, de Avila, Goldbach, & Peters, 1991; Rice, German, Mau, & Fujimoto, 1990; Ronco et al., 1989; Weekes, Mumford, Barker, & Wood, 1996).

Global status of tospovirus epidemics in diverse cropping systems and control measures has been reviewed by Pappu, Jones, and Jain (2009). Transmission by thrips appears to be the only means of virus spread (Riley, Joseph, Srinivasan, & Diffie, 2011). The principal vectors of TSWV, *Frankliniella fusca*, and *F. accidentalis* occur on groundnut throughout the Southeastern United States. Since *F. fusca* is the predominant species that reproduces on groundnut, it is considered to be the most important vector.

Adjustment to planting dates and planting at high density are valuable practices to reduce TSWV incidence (Culbreath, Branch, Holbrook, & Tilman, 2009, Culbreath, Tillman, Gorbet, Holbrook, & Nischwitz, 2008). The use of conservation tillage (no-tillage, minimum tillage, or strip tillage) in groundnut results in a lower incidence of spotted wilt and reduced severity of foliar fungal diseases compared to conventional tillage (Cantonwine et al., 2006; Monfort, Culbreath, Stevenson, Brenneman, & Perry, 2007). In general, chemical control of thrips does not significantly reduce the incidence of spotted wilt of groundnut. In some cases, the application of insecticides increased the incidence. Seed treatment or in-furrow application of the neonicotinoid insecticide imidacloprid increased incidence of the disease compared to the nontreated control susceptible cultivars (Todd & Culbreath, 1995).

Field resistance to TSWV was observed in the cultivars Southern Runner, Georgia Browne, Georgia Green, UF MDR 98, Tamrun 96, C-99R, and ViruGard (Culbreath et al., 2003). Intensive screening of breeding lines in multiple breeding programs has resulted in the identification of several sources with moderate to high levels of field resistance to TSWV than that in Georgia Green. They include breeding lines F NC 94002 and F NC 94022, which are used in generating TSWV-resistant groundnut cultivars (Culbreath & Srinivasan, 2011). Newly identified field-resistant groundnut genotypes have recently been evaluated under laboratory conditions against TSWV or thrips to understand the mechanism of resistance (Sundaraj, Srinivasan, Culbreath, Riley, & Pappu, 2014). Thrips feeding and survival were suppressed on some resistant genotypes compared with susceptible genotypes. In Brazil, three peanut breeding lines (IC-1, IC-34, and ICGV 86388) showed resistance to TSWV under glass house and field tests (Nascimento et al., 2006).

An integrated genetic linkage map of cultivated groundnut, constructed from the populations of two recombinant inbred lines (RILs), was employed to map TSWV resistance trait (Qin et al., 2012). Two major quantitative trait loci (QTLs) for TSWV resistance were identified for each RIL.

Wang et al. (2013) studied the genetic mapping and QTL analysis for TSWV and leaf spot disease resistance using F_2 and F_5 generations based on genetic maps derived from Tifrunner × GT-C20 cross.

Genetic engineering methods have been attempted to incorporate resistance to TSWV in groundnut (Li, Jarret, & Demski, 1997). They used nucleocapsid gene of TSWV Hawaiian L isolate to transform the groundnut cultivar New MEXICO Valencia A. The engineered plants showed resistance to TSWV. Groundnut cv. MARC I transformed with CP gene of TSWV showed lower spotted wilt incidence than the field-resistant cv. Georgia Green. Cultivar AT 120 transformed with the antisense nucleocapsid gene of TSWV also showed lower incidence of spotted wilt than comparable controls (Culbreath et al., 2003).

Integration of multiple disease suppressive factors is necessary for controlling TSWV in groundnut. The adaption of genetic, chemical, and cultural practices for disease management was enhanced greatly by the development and use of spotted wilt risk index. Virus index has become an important tool by which growers can assess the relative risk of spotted wilt incidence in a particular field and for the identification of suitable disease suppressive factors that best apply to their situation. Application of the various options such as planting at high density with field-resistant cultivars such as Georgia Green and C-11-2-39, application of phorate and conserved tillage had contributed to substantial reduction of TSWV incidence in Georgia (Culbreath & Srinivasan, 2011).

2.2.3 Bud necrosis

Bud necrosis disease (BND) of groundnut was first reported from India. It is known to occur in Nepal, Sri Lanka, Myanmar, Thailand, and parts of China (Reddy, 1998). BND was shown to be caused by a distinct tospovirus, PBNV (synonym *Groundnut bud necrosis virus*, GBNV). The virus also infects mungbean, urdbean, cowpea, pea, soybean, and lablab bean under field conditions (Mandal et al., 2012). The symptoms of BND caused by PBNV and spotted wilt caused by TSWV are similar on groundnut.

PBNV can be identified by characteristic chlorotic and/or necrotic ring spot symptoms on cowpea, cv. C-152. Various formats of ELISA, RT-PCR, and IC-RT-PCR have been used for detection of PBNV in plant samples (Hobbs, Reddy, Rajeshwari, & Reddy, 1987; Thein, Bhat, & Jain, 2003). Disease diagnosis becomes more difficult when there is coinfection of PBNV and TSV because both the viruses produce terminal bud necrosis. The incidence and progress of BND is dependent on several environmental factors and cropping practices which influence multiplication and spread of the vector, *Thrips palmi*. Primary as well as the secondary spread occurs through inoculum derived from alternate hosts that include mungbean, pepper (*Capsicum annuum*), potato (*Solanum tuberosum*), and the weed *Ageratum conyzoides* (Reddy, Amin, McDonald, & Ghanekar, 1983). Management of BND depends upon the control of *T. palmi*. Even though several weeds have been identified as sources of virus and vector thrips (Reddy et al., 1983), their eradication in the tropics is not practical. Rouging of early-infected plants in the fields can create gaps, which can lead to increased incidence. Insecticidal control of thrips was not effective in reducing virus incidence (Wightman & Amin, 1988). Botanical pesticides, neem leaf and seed extracts, castor cake, and monochrotophos, were found to lower BND incidence (Gopal, Muniayappa, & Jagadeeshwar, 2011).

Depending on the arrival of *T. palmi*, the sowing dates need to be adjusted to avoid them. A good crop canopy results in lower disease incidence (Reddy, 1998). For example, in Southern India, groundnut crops sown early with the onset of rains (mid to late June) escaped PBNV infection as thrips vector infestation usually occurred in July and August (Reddy, Buiel, et al., 1995). In contrast, Thira, Cheema, and Kang (2004) observed maximum PBNV infection in groundnut crops sown during May in Northern India. Further, maintenance of optimum plant density and intercropping with cereal crops such as pearl millet (*Pennisetum glaucum*), sorghum (*Sorghum bicolor*), and maize have contributed to lower percentage of PBNV incidence (Reddy, Buiel, et al., 1995a).

Tolerance or resistance to PBNV and/or thrips vector has been identified in germplasm and breeding lines. Robut 33-1, a cultivar commonly grown by marginal farmers in Asia and Africa, showed field resistance (Reddy, 1998). Several wild *Arachis* germplasm lines showed resistance to PBNV (Reddy et al., 2000). Genotypes ICGV 86388, IC 34, and IC 10 were found to be resistant to PBNV in Thailand (Pensuk, Daengpluang, Wongkaew, Jogloy, & Patanothai, 2002). It was subsequently shown that multiple genes governed resistance to PBNV (Pensuk, Jogloy, Wongkaew, & Patanothai, 2004). Groundnut breeding lines ICGV 90009, ICGV 86999, ICGV 86329, ICGV 91177, ICGV 91234, ICGV 94252, and TG 26 were found to be resistant to PBNV in India (Gopal et al., 2010).

The popular Spanish groundnut cultivar, JL 24, was engineered with *N* gene and the plants from T_1 and T_2 generations showed partial resistance to PBNV (Chander Rao, Bhatnagar-Mathur, Kumar, Reddy, & Sharma, 2013). The promising transgenic lines are yet to be deregulated.

2.2.4 Stem necrosis

Stem necrosis disease (SND) caused by TSV in India was first reported by Reddy et al. (2001). The disease may have been existing for a long time and attributed to PBNV because of striking similarity in symptoms that include terminal bud necrosis (Prasada Rao et al., 2003). The disease occurred on nearly 225,000 ha in Andhra Pradesh, India, in the year 2000 causing crop losses that exceeded $65 million. TSV is known to occur on groundnut in India, Pakistan, South Africa, and Brazil. The virus has a wide natural host range and infects many crops that include sunflower (*Helianthus annuus*), soybean, cowpea, mungbean, sunnhemp (*Crotalaria juncea*), green gram, and black gram (Jain, Vemana, & Sudeep, 2008; Kumar, Prasada Rao, et al., 2008), and it is widely expanding its host range. Characteristic symptoms are necrosis on stems and petioles and necrotic spots on pod shells. Genomes of various isolates of TSV have been partially sequenced (Jain et al., 2008). PBNV and TSV can be distinguished by assays on selected hosts that include cowpea, and *P. vulgaris* cv. Topcrop. TSV can be detected by ELISA, nucleic acid hybridization, and RT-PCR (Prasada Rao et al., 2003).

Three thrips species *Megalurothrips usitatus*, *F. schultzei*, and *Scritothrips dorsalis* assist the virus to transmit through pollen. Sunflower and marigold (*Tagetes patula*) could act as sources of inoculum. Primary source of inoculum is likely to be provided by numerous weeds. The disease is often found near fields surrounded by the parthenium (*Parthenium hysterophorus*) weed, suggesting its role in providing the primary source of inoculum.

Measures suggested for SND control include the destruction of virus sources, installation of barrier crops, and maintenance of optimum plant population and controlling of thrips through seed treatment. Removal of parthenium from the vicinity of the groundnut fields is expected to reduce the disease incidence. Border and intercropping with maize, pearl millet, or sorghum around the groundnut fields may decrease the disease incidence by obstructing thrips movement. Rouging of early-infected plants may not limit or restrict further spread of the disease. Cultivation of groundnut near sunflower and marigold should be discouraged because they act as a source of virus inoculum and/or thrips. Maintenance of sufficient plant density is important to discourage landing of thrips. Seed treatment with systemic insecticides (imidacloprid) may prevent vector infestation at early stages of crop growth. Limited germplasm screening revealed low disease incidence in ICGV 92267, 99029, 01276, ICG 94379, Kadiri 7, and Kadiri 9 groundnut genotypes (Kumar, Prasada Rao, et al., 2008; Vemana & Jain, 2013).

Efforts to produce engineered groundnut plants with viral genes yielded encouraging results. Groundnut cultivars Kadiri 6 and Kadiri 134 were genetically engineered with TSV-CP gene, and integration of the gene was confirmed in T_1, T_2, and T_3 generations. Engineered plants did not produce symptoms when sap inoculated with TSV (Mehta et al., 2013), but they should be evaluated under field conditions.

2.2.5 Clump

Peanut clump disease (PCD) is known to occur in India, Pakistan, and West Africa (Bragard, Doucet, Dieryck, & Delfosse, 2008; Reddy, Bragard, Sreenivasulu, & Delfosee, 2008). The causal virus of PCD that occurs in the Indian subcontinent is referred to as IPCV, whereas the virus that occurs in Africa is referred to as PCV. IPCV and PCV are causing indistinguishable symptoms. Both PCV and IPCV are shown to infect other economically important poaceous crops and pigeonpea. The annual loss due to PCD on global scale has been estimated to exceed US $38 million (Reddy & Thirumala-Devi, 2003). Furthermore, the disease also has quarantine importance, because the casual viruses are seed transmissible in groundnut, cereals, and millets (Delfosse et al., 1999; Reddy et al., 2008). Several strains of IPCV and PCV have been identified based on serological and genomic diversity. IPCV isolates have been grouped into three distinct serotypes, viz., IPCV-H (Hyderabad), IPCV-D (Durgapura), and IPCV-L (Ludhiana) (Nolt, Rajeshwari, Reddy, Barathan, & Manohar, 1988), whereas the PCV isolates are placed into five distinct groups by using MAbs (Huguenot, Givord, Sommermeyer, & Van Regenmortel, 1989; Manohar, Dollet, Dubern, & Gargani, 1995).

PCD can be readily identified in the farmers' fields by the characteristic symptoms and patchy distribution of infected plants. *C. amaranticolor* is a good diagnostic host for PCV and IPCV. ELISA and nucleic acid hybridization-based tests as well as RT-PCR have been used for the detection of these viruses (Dieryck, Delfosse, Reddy, & Bragard, 2010; Huguenot et al., 1989; Manohar et al., 1995; Reddy et al., 2008). Both PCV and IPCV are soil-borne and transmitted by the plasmodiophorid obligate biotrophic parasite, *Polymyxa graminis* (Reddy et al., 2008).

For the control of PCD, the following cultural practices are suggested: (a) early sowing of groundnut crop before the onset of monsoon rains; (b) use of pearl millet as a bait crop to reduce the inoculum load in the soil. To achieve this, bait crop should be planted soon after the onset of monsoon preferably under irrigation and uprooted in 3 weeks after germination;

(c) avoid rotation with highly susceptible cereal crops such as maize and wheat (*Triticum aestivum*); (d) sowing groundnut crops during postrainy season; (e) rotation with dicot hosts to reduce the inoculum in the soil; and (f) soil solarization during hot summer months. This is achieved by covering well-irrigated soils with a transparent polythene sheet (preferably biodegradable) for at least 3 weeks (Reddy et al., 2008).

No resistance to IPCV was found in nearly 9000 *Arachis* germplasm lines. Resistance was identified in wild *Arachis* species, and it is yet to be incorporated into cultivated groundnut (Reddy & Thirumala-Devi, 2003). Four genetically engineered groundnut lines with IPCV-H-CP gene and two with IPCV-H Rep gene have been developed (Sharma et al., 2006). These events were reported to show resistance to PCD based on contained field trials conducted between 2002 and 2005 on the experimental farm of International Crops Research Institute for Semi-arid Tropics (ICRISAT), Patancheru, India. However, these genotypes have never been evaluated for biosafety, and information on durability of resistance is not available.

2.2.6 Mottle and stripe

PeMoV and PStV are among the several potyviruses reported to infect groundnut naturally. They cause mottle and stripe symptoms, and it is difficult to make a distinction between the infections of these two viruses under field conditions. Symptoms induced by necrotic strains of these two viruses mimic the infections of TSWV and PBNV in groundnut (Sreenivasulu et al., 1988). They are seed-transmitted and relatively widely distributed. These two viruses are also reported to infect other legumes, viz., common bean, soybean, cowpea, peas, and white lupin. In Georgia, losses due to mottle were estimated up to 70%, and in India, it may reach 40% in susceptible groundnut cultivars (Reddy & Thirumala-Devi, 2003). Yield reductions by PStV in groundnut in Georgia were about 7% in experimental plots (Lynch, Demski, Branch, Holbrook, & Morgan, 1988) but can reach up to 70% in early-infected plants. In Northern China, annual yield losses due to PStV are estimated at over 200,000 tons of pods (McDonald, Reddy, Sharma, Mehan, & Subrahmanayam, 1998).

ELISA, dot-blot hybridization, RT-PCR, and IC-RT-PCR-based tests were employed for detection of both PeMoV and PStV in leaf and seed samples (Bijaisoradat & Kuhn, 1988; Dietzgen et al., 2001; Gillaspie, Pittman, Pinnow, & Cassidy, 2000; Gillaspie, Wang, Pinnow, & Pittman, 2007; Hobbs et al., 1987).

The transmission of PeMoV through the seed in groundnut (0–8.5%) and other grain legumes (cowpea, mungbean, and common bean) is contributing to the primary spread of PeMoV. In nature, PeMoV and PStV are transmitted by aphids in a nonpersistent manner. Alternate crops (cowpea, soybean, clover, pea, navy bean, French bean, and white lupin) and weeds (*Centrosena pubescence*, *Catasetum macrocarpum*, *Calopogonium caeruleum*, *Crotalaria straita*, *Desmodium siliquiosum*, *Pueraria phaseoloides*, and beggarweed, *Desmodium* spp.) are facilitating the survival and spread of PeMoV by aphids (Demski & Reddy, 1997).

Since the primary source of PeMoV or PStV inoculum is groundnut seed, planting should be done with seed lots obtained from disease-free areas. Genotypes in which PeMoV is not transmitted through the seed, such as ICG 2716 (EC 76446-292), ICG 7013 (NCAC 17133), and ICG 1697 (NCAC 17090), are useful in containing the spread of PeMoV. These lines were used in conventional breeding program to transmit the nonseed transmissible trait to high yielding groundnut cultivars. The seed of advanced breeding lines from these crosses has been tested for frequency of the virus transmission. Two nonseed transmitting high yielding groundnut genotypes (ICGS 65 and ICGS 76) were identified (Reddy & Thirumala-Devi, 2003). High yielding groundnut genotype ICG 89336 was found to be tolerant to PeMoV. *Arachis chacoense* and *Atriplex pusilla* as well as wild *Arachis* species have been reported to be resistant to PeMoV (Demski & Sowell, 1981). However, these resistant sources are yet to be utilized in breeding programs.

Enforcing of strict quarantine regulations in countries where PStV is known to be restricted to certain locations is important to avoid introduction of the virus into the virus-free locations. Only certified groundnut seed is to be moved between the locations and to countries with no record of PStV occurrence. Attempts made to control PStV by using 10% milk suspension, metasystox or milk alternated with metasystox or pyrimidine carbamate (systemic aphicide) were unsuccessful. Application of plastic film mulch in groundnut field in China was found to reduce PStV incidence (Demski et al., 1993). Attempts to identify genotypes which do not transmit PStV through seed were unsuccessful. Resistance to PStV could not be found in cultivated groundnut but was identified in some wild *Arachis* species (Culver, Sherwood, & Melouk, 1987). However, this resistance is yet to be transferred to cultivated high yielding groundnut cultivars.

Groundnut plants transformed with PStV-CP gene exhibited high levels of resistance to the virus (Higgins, Hall, Mitter, Cruickshank, & Dietzgen, 2004). Groundnut plants transformed with full-length untranslatable form of

CP gene and 3′ untranslated region (UTR) of an Indonesian blotch strain of PStV showed resistance to PStV infection, and the resistance was stably inherited over at least five generations (Dietzgen et al., 2004). However, these genetically engineered plants are yet to be commercialized.

2.2.7 *Yellow mosaic*

Yellow mosaic on groundnut, caused by CMV, was first reported from Northern parts of China (Xu & Barnett, 1984) and later from Argentina (De Breuil, Giolitti, & Lenardon, 2005). Crop losses up to 40% were reported from China, and the symptoms caused by CMV are described by Xu and Barnett (1984). Of the two strains of CMV reported to naturally infect groundnut in China, CMV-CA was the predominant strain and CMV-CS was of minor importance. Bioassays on cowpea, cucurbits, tobacco, *Datura stramonium*, and *Chenopodium* species were useful for the initial diagnosis of CMV isolates (Palukaitis & Garcı´a-Arenal, 2003). ELISA, dot-blot hybridization, and RT-PCR were optimized for routine detection of the virus (Dietzgen et al., 2001; Palukaitis & Garcı´a-Arenal, 2003). The CMV-CA isolate was seed transmissible (2–4%) in groundnut. Therefore, the primary spread is presumably initiated through the seed-borne inoculum, whereas the aphids may contribute to its secondary spread. Planting with CMV-free groundnut seed reduced disease incidence. Cultural practices like mulching with transparent plastic sheets and rouging of diseased seedlings at early stages of crop growth reduced disease incidence in China (Reddy & Thirumala-Devi, 2003). No resistance to CMV in the cultivated groundnut was reported.

2.3. Common bean

Common bean is an important grain legume crop cultivated in Myanmar, India, Brazil, China, the United States, Mexico, Tanzania, Kenya, and Rwanda (FAOSTAT, 2012). Among the large number of viruses infecting common bean, diseases caused by certain species of *Potyvirus*, *Begomovirus*, and *Cucumovirus* genera were considered to be economically important (Table 9.1). The symptoms caused by these viruses on common bean were described by Morales (2003) and Schwartz, Steadman, Hall, and Foster (2005).

2.3.1 *Common mosaic and black root*

BCMV and *Bean common mosaic necrosis virus* (BCMNV) cause common mosaic and black root diseases, respectively. BCMV predominates in the Western hemisphere, whereas BCMNV occurs in Eastern Africa, Dominican

Republic, and Haiti. They cause significant yield losses that can reach up to 80% (Morales, 2003). Substantial yield losses were reported in mixed infections with BCMV and *Bean rugose mosaic virus* (BRMV) (Castillo-Urquiza, Maia, Carvalho, Pinto, & Zerbini, 2006).

Several BCMV isolates have been distinguished on the basis of symptoms produced on *P. vulgaris* cultivars. BCMV and BCMNV strains cause two main types of symptoms on beans: "common mosaic" characterized by green vein banding and leaf malformation and "black root" characterized by systemic necrosis and plant death (Morales, 2003; Schwartz et al., 2005). BCMV and BCMNV strains, occurring on lima beans from Peru, were characterized using differential bean varieties and RT-PCR (Melgarejo, Lehtonen, Fribourg, Rannali, & Valkonen, 2007).

BCMV is seed-transmitted up to 83% in *P. vulgaris* and 7–22% in Tapari bean plants. The necrosis-inducing strains of BCMV and BCMNV are seed and pollen transmitted in common bean genotypes that lack the dominant *I* gene. Seed-borne inoculum contributes to primary spread, and secondary spread occurs through aphid vectors which transmit both viruses in a non-persistent manner. The planting of dominant *I* gene and recessive I^+ common bean cultivars side by side had resulted in major epidemics of black root because of the availability of seed-transmitted inoculum and the existence of several aphid species that can transmit these viruses (Morales, 2003).

Losses due to BCMV and BCMNV may be in principle curtailed by planting with certified seeds, control of aphid vectors by oil application, timely sowing of crops, use of optimum plant densities, and intercropping with maize. However, growing resistant cultivars was considered to be the best option for reducing crop losses by these viruses (Morales, 2003). Symptom severity could be reduced in French bean through seed treatment with Sanosil (a commercial formulation containing hydrogen peroxide and silver which can mask the expression of BCMV symptoms) and *Pseudomonos fluorescens* (plant growth-promoting bacterium) (Bhuvanendra Kumar, Uday Shankar, Prakash, & Shekar Shetty, 2005). Treatment with 15 mA electricity for 15 min resulted in substantial reduction of transmission of BCMV through bean seed (Hormozi-Nejad, Mozafari, & Rakhshandehroo, 2010).

The occurrence of numerous BCMV strains has important implications for the development of resistant cultivars. In the New World and Europe, where mosaic-inducing strains of BCMV are mostly prevalent, cultivars with the "*I*" gene provided effective protection against the virus for more than 50 years (Morales, 2003). Since this gene also prevents seed transmission, it has provided valuable means of eliminating quarantine risk. Gene combinations of *bc-u* (strain-nonspecific epistatic gene) plus any of the *bc-1*, *bc-1²*, *bc-2*, or *bc-2²* genes confer recessive resistance. The combination of genes *bc-u*, *bc-2²*, *bc-3*, and "*I*" gives durable resistance to all the known strains of BCMV. Nonetheless, it has been shown to be a considerable breeding task to incorporate these genes into beans (Morales, 2003; Schwartz et al., 2005). A number of BCMV/BCMNV resistance genes have been tagged including the dominant "*I*" gene and the recessive *bc-3*, *bc-2*, and *bc-1²* genes. These genes can be distinguished by inoculation with different virus isolates and by a range of molecular marker tags that are available for each gene (reviewed by Blair, Beaver, Nin, Prophete, & Singh, 2006; Kelly, Gepts, Miklas, & Coyne, 2003; Miklas, Kelly, Beebe, & Blair, 2006). The genetic and molecular characterization of the "*I*" locus of the *P. vulgaris* was studied (Eduardo Vallejos et al., 2006). In India, the presence of "*I*" gene was confirmed in some of the bean accessions of diverse origin when they were evaluated for resistance against BCMV strains and for inheritance patterns in selected cultivars against the strain NL-1 (Sharma et al., 2008). In Spain, the introgression and pyramiding of genes conferring genetic resistance to BCMV and anthracnose local races into breeding lines A25 and A3308 was reported (Ferreira, Campa, Perez-Vega, Rodriguez-Suarez, & Giraldez, 2012). The prospects of MAS for common bean diseases including BCMV and BCMNV were reviewed (Tryphone et al., 2013). Resistance to fungal angular leaf spot and BCMNV diseases was incorporated into adapted common bean genotype in Tanzania using molecular markers (Chilangane et al., 2013).

2.3.2 Golden mosaic, golden yellow mosaic, and dwarf mosaic

Several begomoviruses, viz., BGMV, *Bean golden yellow mosaic virus* (BGYMV), *Bean dwarf mosaic virus* (BDMV), *Bean summer death virus*, *Bean yellow dwarf virus*, and *Bean calico mosaic virus* (BCaMV) have been reported to naturally infect common bean in the tropical countries (Morales, 2006; Navas-Castillo et al., 2011; Schwartz et al., 2005). Later, natural occurrence of several other begomoviruses on common bean has been reported, and their economic significance is yet to be determined (e.g., Fiallo-Olive et al., 2013; Jyothi, Nagaraju, Padmaja, & Rangaswamy, 2013; Kamaal, Akram, Pratap, & Yadav, 2013; Shahid, Ikegami, & Natsuaki, 2012; Venkataravanappa et al., 2012). These examples indicate the emergence of new begomoviruses capable of infecting common bean.

Symptoms of golden yellow mosaic resemble BGMV infection and hence were considered to be caused by the same virus. However, advances in molecular techniques have revealed that the golden mosaic and golden YMDs are caused by related but distinct begomovirus species, BGMV and BGYMV, respectively (Morales, 2003).

Studies in Brazil on the begomoviruses infecting legume crops and weeds revealed that the recombination of viral genomes in weeds such as *Macroptilium lathyroides* (Rodriguez-Pardina et al., 2006; Silva et al., 2011) has contributed to their rapid evolution. DNA hybridization and PCR-based methods were applied for specific detection of BGMV, BGYMV, BCaMV, and BDMV in field-collected bean samples (Karkashian, Ramos-Reynoso, Maxwell, & Ramirez, 2011; Potter, Nakhla, Mejia, & Maxwell, 2003). Field spread of these viruses is by *Bemisia tabaci* in a persistent manner. Female whiteflies were found to be more efficient vectors of BGMV than males. The epidemics of these viruses depended on the presence of suitable whitefly reproductive hosts such as tomato (*Solanum lycopersicum*), eggplant (*Solanum melongina*), soybean, and tobacco (*Nicotiana tabacum*). Additionally, the development of pesticide-resistant whitefly populations reported in different countries has contributed to the spread of the viruses they transmit (Morales, 2003, 2006).

The incidence of BGMV, BGYMV, and BDMV decreases with increasing distance from preferred hosts of the vector. There are also opportunities for reducing severity of diseases by manipulating sowing time (during the rainy season) so as to escape peak infestation by *B. tabaci*. In the Dominican Republic, planting in November significantly reduced the disease. Also, seed treatment and spraying with systemic insecticides like carbofuran or aldicarb in combination with mineral oil were found to reduce begomovirus incidence (Morales, 2003).

Resistance to BGMV was reported in black-seeded common bean cultivars in South America. *P. vulgaris* accessions possessing partial resistance or tolerance to BGMV include Porrillo Sintetico, Porrillo 70, Turrialba 1, and ICA-Pijao. They have been used successfully in breeding black-seeded cultivars such as ICTA Quetzal and Negro Huastoco in Guatemala and Mexico, respectively. Additional sources of resistance were identified in 188 accessions selected from a set of 1660 accessions of bean germplasm. Sources of resistance to BGYMV have been identified in black-seeded Mesoamerican genotypes, which are at best tolerant and have the ability to escape infection under field conditions. Other BGYMV resistance genes discovered are *bgm* or *bgm-1* in bean genotypes of Mexican "Durango" race or the *bgm-2* of Andean origin (Morales, 2003). These genes were successfully used in the development of BGYMV-resistant cultivars. However, the presence of these genes resulted in deformed pods that could be avoided through incorporation of *Bgp-1* gene that contributed to normal pod development (Roman, Castaneda, Sanchez, Munoz, & Beaver, 2004). Blair, Rodriguez, Pedraza, Morales, and Beebe (2007) mapped *bgm-1* conferring resistance to BGYMV and linkage with *bc-1* conferring strain-specific resistance to BCMV. Resistant cultivars to BDMV infection include Porillo Sintetico, DOR 41, Red Mexican UI 35, and pinto UI 114. Furthermore, the sources of resistance to BGMV were also effective against BDMV. A single dominant gene, *Bdm*, conferred BDMV resistance in crosses between Othello and Topcrop bean cultivars (Seo, Gepts, & Gilbertson, 2004).

Scientists at the Brazilian Agricultural Research organization (EMBRAPA) produced genetically engineered bean cv. Olathe with a hairpin (hp) construct containing the Rep gene from BGMV. Transformed plants showed resistance to BGMV even under mixed infection with BCMNV and BRMV (Bonfim, Faria, Nogueira, Mendes, & Aragao, 2007). Two lines that showed high degree of resistance to BGMV in field trials under high vector pressure were identified. Further evaluations under field conditions for agronomic traits have been carried out in three regions of Brazil. The results confirmed superior performance of engineered plants in multilocation tests (Aragao & Faria, 2009). This engineered bean has received approval from Brazilian National Technical Commission on Biosafety, thus became the first deregulated native crop cultivar in Latin America (Tollefson, 2011). It is expected to receive wide acceptance from farmers in Brazil. Specific PCR-based test was developed for the detection of engineered BGMV-resistant common bean Embrapa 5.1 in Brazil (Dinon et al., 2012). Such tests could be useful to assess the performance of this crop.

2.3.3 Mosaic due to CMV

Several strains of CMV occurring worldwide can induce different symptoms in common bean ranging from mild mosaic to severe plant malformation and yield losses varying from 5% to 75% depending on the cultivar, age of infection, virus strain, and environmental conditions (Morales, 2003; Schwartz et al., 2005). ELISAs, dot-immunobinding assays (DIBAs), rapid immunofilter assay, and IC-RT-PCR were used for distinguishing CMV isolates (Palukaitis & Garcı´a-Arenal, 2003; Zein, Nakazawa, Ueda, & Miyatake, 2007). At least six bean infecting strains of CMV are seed borne (up to 10%) in beans and thus can be disseminated to long distances in seed shipments. Both seed and aphid transmission of CMV may be erratic and influenced by several factors (Schwartz et al., 2005).

In Chile, Turkey, and Iran, legume crop losses are caused by CMV warrant application of specific management measures. Use of virus-free bean seeds is probably the least costly management measure. It may also be useful to destroy reservoir hosts or to isolate the crop from such hosts. Management of aphid vector populations in crops and inoculum reservoir hosts may reduce the natural spread of CMV. This can be achieved by planting barrier crops that are immune to CMV infection, applying sticky traps, and mulching with aluminum foil. Developing new bean crop varieties resistant to CMV either by conventional breeding methods or by genetic engineering is gaining momentum (Makkouk et al., 2012). Several species of the genus *Phaseolus* such as *P. acutifolius*, *P. adenanthus*, *P. leptostachyus*, *P. palyanthus*, *P. trilobus*, and some accessions of *P. coccineus* are resistant to CMV. However, the resistance genes are yet to be transferred to *P. vulgaris* (Schwartz et al., 2005). The different strategies of PDR in developing transgenic plants resistant to CMV infection are reviewed by Morroni, Thompson, and Tepfer (2008) and need to be utilized to develop CMV-resistant common bean. Genes involved in resistance response in common bean cv. Othello were identified by inoculating the geminivirus (BDMV) reporter in transgenic *Nicotiana benthamiana*. The identified *RT4-4* gene did not confer resistance to the reported geminivirus, but it activated a resistance related response (systemic necrosis) to seven strains of CMV from tomato and pepper but not to a strain from common bean (Seo et al., 2006).

2.4. Cowpea

Cowpea is the most widely cultivated indigenous legume in SSA, especially by smallholder farmers, because of its tolerance to drought and ability to thrive in poor soils. Over 80% of cowpea production is confined to West Africa with Niger, Nigeria and Burkina Faso alone contributing 83% of the production area and 77% of the total production (FAOSTAT, 2012). It is the second most important food legume after groundnut (Boukar, Bhattacharjee, Fatokun, Kumar, & Gueye, 2013). Cowpea is susceptible to over 140 viruses and about 20 of these viruses are known to have widespread distribution (reviewed in Hampton & Thottappilly, 2003). They include BCMV-blackeye cowpea mosaic strain (BCMV-BlCM), *Cowpea aphid-borne mosaic virus* (CABMV), *Cowpea chlorotic mottle virus* (CCMV), CMV, *Cowpea golden mosaic virus* (CGMV), *Cowpea mottle virus* (CPMoV), CPMV, CPSMV, and *Southern bean mosaic virus* (SBMV; Table 9.1). BCMV-BlCM, CABMV, and CMV were detected in the majority of cowpea producing countries.

The viruses infecting cowpea are vectored by aphids (BCMV-BlCM, CABMV, and CMV), by beetles (CCMV, CPMV, CPSMV, CPMoV, and SBMV), or by whiteflies (CPMMV and CGMV). With the exception of CGMV and CCMV, all the other viruses are known to be seed-transmitted at a variable rate between none to 55% depending on virus strain, cowpea genotype, and time of infection (Bashir & Hampton, 1996; Salem, Ehlers, Roberts, & Ng, 2010).

Establishment of cowpea infection is dependent on the volunteer plants including cultivated and noncultivated species, insect vector density, and cultivar susceptibility. Therefore, measures targeting vector control, cultural management to eliminate volunteer sources including use of virus-free seed stocks (Biemond et al., 2013; Sharma & Varma, 1984) and resistant cultivars (reviewed in Hampton & Thottappilly, 2003) can significantly reduce the incidence, spread, and damage to cowpea crops. However, insect control measures by using pesticides and cultural measures other than growing virus-free seed is seldom feasible for smallholder farmers in SSA.

Protocols for screening germplasm for resistance by mechanical inoculation and/or natural infection facilitated by vectors under field conditions have been reported (Gillaspie, 2007; Goenaga, Quiles, & Gillaspie, 2008). Additionally, serological and nucleic acid-based diagnostic tools (Amayo et al., 2012; El-Kewey, Sidaros, Abdelkader, Emeran, & EL-Sharkawy, 2007; Naimuddin, 2010; Ojuederie, Odu, & Ilori, 2009; Roy & Malathi, 2004; Salem et al., 2010) have contributed to the sensitive detection and rapid evaluation of cowpea germplasm, leading to identification of acceptable levels of resistance by a number of research groups around the world (reviewed in Hampton & Thottappilly, 2003). However, these intensive efforts are limited to a few viruses, viz., BCMV-BlCM, CABMV, CMV, and CGMV. In the majority of cases, resistance identified is not immunity but tolerance. Cowpea genetics and list of resistant genes utilized for developing cultivars with resistance to one or more cowpea viruses are reviewed (Hampton & Thottappilly, 2003). At IITA, research work is being continued to identify durable resistant cowpea varieties and also to determine the genetic determinants of virus resistance in cowpea germplasm. Resistance to two potyviruses was found in germplasm accessions, TVu401, TVu1453, and TVu1948, and in breeding lines, IT82D-885, IT28D-889, and IT82E-60 (Gumedzoe, Rossel, Thottappilly, Asselin, & Huguenot, 1998). However, resistance to multiple virus infection is scarce in cowpea, and recent studies are putting greater emphasis on multiple virus resistance. These efforts have already resulted in identification of multiple virus resistance to three virus species

(BCMV-BlCM, CMV, and SBMV) in breeding lines, IT98K-1092-1 and IT97K-1042-3 (Ogunsola, Fatokun, Boukar, Ilori, & Kumar, 2010). RILs are being established for mapping studies to identify DNA markers linked to multiple virus resistance. For instance, resistance to CGMV in crosses between IT97K-499-25 × Canapu T16 was shown to be controlled by a single dominant gene (Rodrigues, Santos, & Santana, 2012). CABMV resistance was conditioned by a single recessive gene or more than one recessive gene (Orawu, Melis, Laing, & Derera, 2013). Resistance in cowpea to BCMV-BlCM was attributed to a single recessive gene pair (Arshad, Bashir, Sharif, & Malik, 1998). Very limited efforts have focused on breeding for vector resistance. Recently at IITA, 92 wild cowpea accessions were evaluated in a greenhouse for resistance to cowpea aphid, *A. craccivora*, which is a prominent pest and a vector for at least 11 viruses in SSA (Souleymane, Aken'Ova, Fatokun, & Alabi, 2013). This resulted in the identification of only TVNu1158 as the best aphid-resistant accession for incorporation into cultivated cowpea, while other accessions were found to be susceptible to aphid infestation and severe damage (Souleymane et al., 2013).

Progress in the development of varieties with superior performance depends on the availability of germplasm with desired traits. IITA conserves over 14,000 accessions of cultivated cowpea in its gene bank, and the USDA has duplicates of most of the IITA cowpea lines for safe keeping (Boukar et al., 2013; Dumet et al., 2012). Besides cultivated cowpea lines, some accessions of wild cowpea relatives are also conserved in the gene bank at IITA. However, there are not many reports in literature on the use of wild cowpea relatives for the genetic improvement of cultivated varieties for pest and disease resistance (reviewed in Boukar et al., 2013).

Application of DNA marker technologies for cowpea improvement has been very slow when compared to many other crops. However, recent advances in molecular biological techniques, genomics, and bioinformatics have been opening new vistas for molecular breeding in cowpea (Boukar et al., 2013; Diouf, 2011; Ehlers et al., 2012). In addition, SNP genotyping platforms, high-density consensus genetic map with more than 1000 markers, and QTL(s) linked to important biotic and abiotic resistance traits including resistance to foliar thrips, *Fusarium* wilt (FW), root-knot nematode, bacterial blight, ashy stem blight, and *Striga* have been established (Ehlers et al., 2012). Similar genomic resources are required to fast track development of virus-resistant cowpea.

High levels of resistance to several cowpea viruses, especially multiple virus infections are limited in cowpea germplasm. Efficient and stable transformation of cowpea and stable transmission of transgenes to progeny have been a major bottleneck for the development of cowpea transgenics. Good protocols for *Agrobacterium*-mediated transformation (Chaudhury et al., 2007; Popelka, Gollasch, Moore, Molvig, & Higgins, 2006) and also biolistic transformation (Ivo, Nascimento, Vieira, Campos, & Aragao, 2008) of cowpea have been developed and showed stable inheritance of the transgenes. Cowpea transformed with *Bt* gene (Popelka et al., 2006) is under field tests in Nigeria, Burkina, and Ghana. Similar transformation techniques combined with transgenically induced PTGS (posttranscriptional gene silencing) or RNAi can be exploited for the control of both DNA and RNA viruses infecting cowpea. The concept of using RNAi construct to silence the CPSMV proteinase cofactor gene and CABMV-CP gene is explored in transgenic cowpea. In the symptomless resistant lines, the resistance was homozygosis dependent. Only homozygous plants remained uninfected while heterozygous plants presented relatively mild symptoms (Cruz & Aragao, 2014). In Arlingtom line, cowpea extreme resistance to CPMV is controlled by a dominant locus designated *Cpa*. Using the transgenic approach and mutational analysis of the 24 k Pro gene, Fan, Niroula, Fildstein, and Bruening (2011) demonstrated the participation of protease in eliciting extreme resistance. These recent developments in cowpea transformation form a basis for strong cowpea genetic improvement program to enhance cowpea productivity.

2.5. Pigeonpea

Pigeonpea is an important grain legume crop predominately grown in the Indian subcontinent, covering an area of ~3.53 million hectares in India alone. It is also grown in southern and eastern Africa, the Caribbean and China. Fifteen viruses are known to naturally infect pigeonpea (Kumar, Kumari, et al., 2008). Of these, diseases caused by the PPSMV and very recently whitefly-transmitted bipartitate begomoviruses have been shown to be economically important (Table 9.1) (Jones et al., 2004; Kumar, Jones, & Reddy, 2003; Reddy, Raju, & Lenne, 1998).

2.5.1 Sterility mosaic

Sterility mosaic disease (SMD) is a serious constraint for pigeonpea cultivation in India, Bangladesh, Nepal, Thailand, Myanmar, and Sri Lanka with an estimated annual loss of over US$ 300 million in India alone (Kumar et al., 2007). SMD is characterized by partial or complete cessation of flower production (sterility), excessive vegetative growth, stunting, chlorotic ring spots or mosaics on the leaves, and reduction in the size of leaves. The yield losses due to SMD depend on at what stage the crop is infected; early infection (<45 days) can lead to a yield loss from 95% to 100%, whereas late infections can lead to 26–97% yield losses. In addition, infection from SMD predisposes the plants to subsequent infection by fungal diseases and spider mites attack (Kumar et al., 2003).

SMD was shown to be caused by PPSMV, with a negative-strand segmented RNA genome, classified under the newly created genus, *Emaravirus* (Kumar et al., 2003; Mielke-Ehret & Mühlbach, 2012). PPSMV is transmitted by the eriophyid mite *Aceria cajani* in a semipersistent manner (Kulkarni et al., 2003). Recently, an isolate of PPSMV from ICRISAT-Patancheru (India) was fully sequenced and was shown to contain five segments of RNA (Elbeaino, Digiaro, Uppala, & Sudini, 2014). The PPSMV isolates collected from SMD-affected plants in Coimbatore (C), Bangalore (B), Dharwad (D), Gulbarga (G), Varanasi (V) from India, and Nepalgunj (N) from Nepal have shown differences in virulence and biochemical properties (Kumar, Jones, & Waliyar, 2004, Kumar, Kumari, et al., 2008; Kumar, Latha, Kulkarni, Raghavendra, & Saxena, 2005). DAS-ELISA, DIBA, and PCR have been developed to detect PPSMV (Kumar et al., 2003, 2007). Very recently, degenerate PCR primers for amplification of partial RdRp sequences of emaraviruses have been shown to detect PPSMV isolates (Elbeaino, Whitefield, Sharma, & Digiaro, 2013).

Natural occurrence of PPSMV and its vector has been recorded only on cultivated and wild pigeonpea. PPSMV is not known to be seed-transmitted. Seed treatment or soil and foliar application of a number of organophosphorous-based insecticides or acaricides (carbofuran and aldicarb for soil application, and oxythioquinox, kelthane, dinocap, monocrotophos, tedion, and metasystox as foliar sprays) recommended for the management of the vector mites is seldom practiced by farmers due to high cost and environmental issues (Ghanekar, Sheila, Beniwal, Reddy, & Nene, 1992; Rathi, 1997). Destruction or rouging of SMD-infected stubbles, ratoon, weeds, or wild species of pigeonpea that support PPSMV and *A. cajani* multiplication are effective in the eradication of SMD inoculum, but rarely practiced under small holder subsistence farming systems (Reddy, Sharma, & Nene, 1990). Thus, major emphasis has been placed on the development of SMD-resistant cultivars and hybrids as the most effective and realistic approach to reduce the crop losses.

Sources of SMD resistance have been identified in the pigeonpea germplasm collection at ICRISAT (Ghanekar et al., 1992; Nene, 1995).

However, the occurrence of distinct strains/isolates of PPSMV in different locations makes it difficult to incorporate broad-spectrum resistance. Resistance to diverse isolates of PPSMV has been reported in very few cultivars such as ICP7035, and it has been released for cultivation (Rangaswamy et al., 2005). Wild *Cajanus* species were shown to have resistance to multiple isolates of PPSMV.

Screening for resistance to three PPSMV isolates from South India was conducted for 115 wild *Cajanus* accessions belonging to six species, *C. albicans*, *C. platycarpus*, *C. cajanifolius*, *C. lineatus*, *C. scarabaeoides*, and *C. sericeus*. Accessions, ICP 15614, 15615, 15626, 15684, 15688, 15700, 15701, 15725, 15734, 15736, 15737, 15740, 15924, 15925, and 15926 showed resistance to all the three isolates (Waliyar, 2005). These accessions are cross-compatible with cultivated pigeonpea. As a result, it should be possible to incorporate this resistance through conventional breeding. Most of the wild species did not support multiplication of mites, and the majority of the accessions inoculated with viruliferous mites were resistant to PPSMV, but were susceptible by graft inoculation, suggesting that the observed resistance was vector mediated. Some of the wild species of pigeonpea resistant to infestation by mites have a thicker leaf cuticle and epidermal cell wall, which hinders mite's stylet penetration into the leaf epidermal cells (Reddy, Sheila, Murthy, & Padma, 1995). The wild species, *C. scarabaeoides* (ICPW 94), which is resistant to all the isolates of PPSMV, was used in the crossing program, and the progeny was tested for SMD resistance, resulting in both resistant and moderately resistant plants. Hybrid lines derived from interspecific crosses involving *C. acutifolius* and *C. platycarpus* have shown resistance to Patancheru isolate of PPSMV under field conditions (Mallikarjuna et al., 2011). Recently, new sources of resistance to FW and SMD were identified in a minicore collection of pigeonpea germplasm at ICRISAT (Sharma, Rameshwar, & Pande, 2013; Sharma et al., 2012). High level of resistance to SMD was found in 24 accessions, which originated from India, Italy, Kenya, Nepal, Nigeria, the Philippines, and the United Kingdom. Combined resistance to FW and SMD was found in five accessions of pigeonpea (ICPs 6739, 8860, 11015, 13304, and 14819), and these diverse accessions should be utilized in the breeding programs.

Studies on the inheritance of SMD resistance trait in various cultivars of pigeonpea against several isolates of PPSMV have led to different interpretations on the genetics of inheritance of SMD. Studies by Srinivas, Reddy, Jain, and Reddy (1997) showed that the resistance to SMD was dominant in the two crosses (ICP 7035 × ICP 8863 and ICP 7349 × ICP 8863), and recessive in another cross (ICP 8850 × ICP 8863). The disease reaction for the Patancheru isolate of PPSMV appeared to be governed by a single gene with three alleles, with one resistance allele exhibiting dominance, and the other being recessive over the allele for susceptibility. However, monogenic inheritance of SMD resistance was noticed in the cross ICP 8850 × ICP 8863. The nature of inheritance of SMD resistance in two crosses involving resistant and susceptible pigeonpea cultivars revealed that the resistance is controlled by recessive gene and appeared to be monogenetic in one cross (TTB 7 × BRG 3) and governed by two independent nonallelic genes exhibiting complimentary epistasis in another cross (ICP 8863 × ICP 7035) (Ganapathy et al., 2012).

Ganapathy et al. (2009) generated two AFLP primer pairs, comprising four markers, polymorphic for SMD-resistant and -susceptible bulks of the F_2 population derived from TTB7 (susceptible) and BRG3 (resistant) parents. After screening over 3000 SSR markers on parental genotypes of each mapping population, intraspecific genetic maps comprising of 11 linkage groups and 120 and 78 SSR loci were developed for ICP 8863 × ICPL 20097 and TTB 7 × ICP 7035 populations, respectively (Gnanesh, Bohra, et al., 2011). The genotypes with high polymorphism as revealed by SSR markers were identified, and they were recommended for developing mapping populations (Naik et al., 2012). Composite interval mapping-based QTL analysis by using genetic mapping and phenotyping data provided four QTLs for Patancheru PPSMV isolate and two QTLs for Bangalore PPSMV isolate (Gnanesh, Ganapathy, Ajay, & Byre, 2011). Identification of different QTLs for resistance to Patancheru and Bangalore SMD isolates is an indication of involvement of different genes conferring the resistance to these two isolates (Gnanesh, Ganapathy, et al., 2011).

Complete sequence of pigeonpea genotype ICPL 87119 genome, an inbred line tolerant to SMD, popularly known as Asha was published (Varshney et al., 2012). Hence, it should be possible in the near future to identify the *R* genes, NBS-LRR, which contains a nucleotide-binding site and a leucine-rich repeat, involved in conferring resistance to PPSMV, using modern genomics and bioinformatics-based approaches such as next-generation sequencing and microarrays.

2.5.2 Yellow mosaic

YMD of pigeonpea, caused by whitefly-transmitted begomoviruses, occurs in Sri Lanka, India, Jamaica, Nepal, and Puerto Rico (Reddy et al., 2012). Although the incidence of YMD in pigeonpea is low, the late sown pigeonpea can show higher incidence resulting in a yield loss up to 40% (Beniwal, Deena, & Nene, 1983). Various begomovirus species that include MYMV, *Rhyncosia mosaic virus*, and *Tomato leaf curl New Delhi virus* have shown to be associated with YMD (Biswas & Varma, 2000).

2.6. Mungbean and urdbean

Mungbean is a nutritious grain legume crop containing 23.6% of easily digestible protein and 51% carbohydrates. It is cultivated in China, Thailand, the Philippines, Vietnam, Indonesia, Myanmar, Bangladesh, India, and in the hot and dry regions of Southern Europe and Southern United States (Nair et al., 2013). Urdbean is widely cultivated in India, Myanmar, Thailand, the Philippines, and Pakistan. Among several viruses reported to naturally infect these two pulse crops (Biswas, Tarafdar, & Biswas, 2012; Makkouk, Kumari, Huges, Muniyappa, & Kulkarni, 2003), yellow mosaic caused by begomoviruses, leaf curl caused by PBNV, and leaf crinkle caused by Urdbean leaf crinkle virus (ULCV) are considered to be economically important (Table 9.1) (Biswas et al., 2012; Malathi & John, 2008; Mandal et al., 2012; Qazi et al., 2007). Mixed infections by PBNV, MYMIV, and ULCV are common in both the crops in India and are often synergistic resulting in crop losses exceeding 90% (Biswas et al., 2012).

2.6.1 Yellow mosaic

Three begomoviruses, MYMV, MYMIV, and HgYMV, were shown to cause mungbean yellow mosaic disease (MYMD) symptoms (Islam, Sony, & Borna, 2012; Malathi & John, 2008). MYMIV occurs in Northern India, Pakistan, Nepal, Bangladesh, and Indonesia, while MYMV is confined to Thailand, Vietnam, and Peninsular region of India (Ilyas, Qazi, Mansoor, & Briddon, 2010; Islam et al., 2012; Shahid et al., 2012; Tsai et al., 2013). HgYMV occurs only in South India (Borah & Dasgupta, 2012; Varma & Malathi, 2003). Enzyme immunoassays, ISEM, nucleic acid hybridization, and PCR-based tests were employed for the detection of begomoviruses associated with YMD of mungbean and urdbean (Malathi & John, 2008). Hyperspectral remote sensing of yellow mosaic severity and associated chlorophyll losses in MYMD affected urdbean was analyzed and developed logistic regression models with spectral ratios for disease assessment (Prabhakar et al., 2013).

Since MYMV is not seed-transmitted, primary source of inoculum is contributed by numerous alternate hosts of the virus and the whitefly vector. Management approaches for MYMD have been reviewed (Malathi & John,

2008). Usually farmers do not use pesticides to control the vector. However, seed treatment and spaying with imidacloprid contributed to relatively low disease incidence (Malathi & John, 2008; Sethuraman, Manivannan, & Natarajan, 2001). Marigold as a trap crop along with yellow sticky traps (8/ha) recorded reduced disease incidence (Salam, Patil, & Byadgi, 2009). Integrated management of major mungbean diseases (wet root rot, leaf spots, and yellow mosaic due to MYMV) by using different combinations of an insecticide, fungicide, and bioformulation as seed treatment, with or without foliar sprays was recently reported (Dubey & Singh, 2013). Such studies should always include their cost-effectiveness.

Of over 10,000 mungbean germplasm lines screened for resistance to MYMV from 1977 to 2003, in Punjab, India, 31 germplasm lines were found to be resistant (Singh, Sharma, Shanmugasundaram, Shih, & Green, 2003). Subsequently, several reports have been published on the identification of germplasm lines/cultivars exhibiting varied levels of resistance to local isolates of causal viruses (e.g., Ahmad et al., 2013; Akthar et al., 2011; Habib, Shad, Javaid, & Iqbal, 2007; Paul, Biswas, Mandal, & Pal, 2013; Shad, Mughal, Farooq, & Bashir, 2006). Several mungbean and urdbean MYMV-resistant germplasm lines were identified in multilocation tests (Biswas et al., 2012; Malathi & John, 2008; Mondol, Rahman, Rashid, Hossain, & Islam, 2013). The inheritance of MYMV resistance in mungbean has been reported to be conferred by a single recessive gene (Reddy, 2009), a dominant gene (Sandhu, Brar, Sandhu, & Verma, 1985), two recessive genes (Ammavasai, Phogat, & Solanki, 2004), and complimentary recessive genes (Shukla & Pandya, 1985). Use of different source of MYMV resistance for genetic studies and infection by different strains of the same virus or distinct begomovirus species may have led to these conflicting results. A report on the genetic analysis of resistance to MYMV suggested that the resistance is governed by two recessive genes (Dhole & Reddy, 2012). Due to the divergence in begomoviruses, these studies need to be supported with precise virus identification.

Genetic markers that will aid in selecting MYMD-resistant breeding lines have been identified. A SCAR marker linked to MYMV resistance gene was identified with a distance of 6.8 cm (Souframanien & Gopalakrishna, 2006). Two marker loci, YR4 and CYR1 (CYR1 was completely linked with MYMIV resistance), were employed in multiplex PCR reaction to screen quickly and reliably urdbean germplasm and breeding lines for resistance to MYMV (Maiti, Basak, Kundagrami, Kundu, & Pal, 2011). A single dominant gene was shown to govern resistance in a cross that involved mungbean genotypes DPU 88-31 (resistant) × AKU 9904 (susceptible). Resistance genes were mapped using SSR markers. Out of 361 markers, 31 were found to be polymorphic between the parents (Gupta, Gupta, Anjum, Pratap, & Kumar, 2013). Markers CEDG 180, mapped at a distance of 12.9 cm, were found to be linked with resistance gene in analyses on bulked segregants. Mapping of QTL for MYMIV resistance in mungbean using an F_8 RIL generated in Thailand from a cross between NM10-12-1 (MYMIV resistance) and KPS2 (MYMIV susceptible) was done and field evaluated in India and Pakistan (Kitsanachandee et al., 2013).

Agroinoculation, using dimeric constructs of DNA-A and DNA-B, of sprouted seeds and seedling was shown to be a useful tool for screening crop germplasm for virus resistance. This technique was applied for screening of mungbean progenies (F_2) derived from a cross between Vamban (Gg) 2 (susceptible) × KMG 189 (MYMV resistant) (Karthikeyan et al., 2011). Later, a similar approach was used for screening of 78 mungbean germplasm lines for resistance to MYMV (Sudha et al., 2013).

In order to develop genetically engineered MYMV-resistant mungbean genotypes, MYMV-Vig CP, replication-associated protein (Rep-sense, Rep-antisense, truncated Rep, nuclear shuttle protein, and movement protein) genes were agroinoculated with partial dimers of MYMV-Vig and analyzed for viral DNA accumulation. Both mungbean and tobacco model systems have shown that engineered plants containing replicase gene and *AC4* hp RNA gene showed resistance to MYMV (Haq, Ali, & Malathi, 2010; Shivaprasad, Thillaichidambaram, Balaji, & Veluthambi, 2006; Sunitha, Shanmugapriya, Balamani, & Veluthambi, 2013). None of the laboratory-generated resistant plants has been tested under contained field trials, and hence, the deregulation of genetically engineered MYMV-resistant mungbean and urdbean is less likely to happen in the near future.

2.6.2 Leaf curl

Leaf curl caused by PBNV is widespread on mungbean and urdbean in the Indian subcontinent. Symptoms of the disease include necrosis of terminal bud, leaves, petioles, stems, and pods. Early infection results in crop losses of up to 90% (Biswas et al., 2012). Based on *N* gene sequence analysis, PBNV strains were classified into eight different evolutionary clusters irrespective of their geographical origin or host (Mandal et al., 2012). Sowing of mungbean during second half of May to the first half of June in summer, late sowing in spring and intercropping with pearl millet (at 2:1 ratio) resulted in relatively low disease incidence. Among the various insecticides (imidacloprid, thiamethoxam, acetamiprid fipronil, dimethoate, fenvalerate, and azadirachtin), imidachloprid gave the most satisfactory control of *T. palmi* and low PBNV incidence in mungbean (Sreekanth, Sriramulu, Rao, Babu, & Babu, 2004a, 2004b). Of 39 mungbean genotypes screened for resistance to PBNV under field conditions, accessions LGG 460, 480, 491, and 582 consistently showed lower disease incidence than the susceptible genotypes (Sreekanth, Sriramulu, Rao, Babu, & Babu, 2002).

2.6.3 Leaf crinkle

Leaf crinkle disease on urdbean is widely distributed in India and Pakistan and reported to cause crop losses up to 100% (Biswas et al., 2012; Reddy, Tonapi, Navi, & Jayaram, 2005). Based on the symptoms and transmission, the causal agent of this disease is considered to be a virus; however, there are no reports of its isolation or characterization.

2.7. Chickpea

Chickpea is the third most important pulse crop. It is extensively grown in India, Australia, Turkey, Myanmar, Ethiopia, Iran, Canada, the United States, Pakistan, and Tanzania (FAOSTAT, 2012). There are two distinct types of cultivated chickpea: *desi* and *kabuli*. Several viruses have been reported to naturally infect chickpea in different parts of the world (Abraham, Menzel, Lesemann, Varrelmann, & Vetten, 2006, Abraham, Menzel, Varrelmann, & Vetten, 2009; Kanakala, Sakhare, Verma, & Malathi, 2012, Kanakala, Verma, Vijay, Saxena, & Malathi, 2013; Kumar, Jones, et al., 2008; Mumtaz, Kumari, Mansoor, Martin, & Briddon, 2011; Nahid et al., 2008; Schwinghamer, Knights, Breeder, & Moore, 2009). Among them, viruses causing or associated with stunt and chlorotic dwarf diseases are considered to be economically important (Table 9.1).

2.7.1 Stunt

Chickpea stunt disease was first reported from Iran, and it occurs in North Africa, the Middle East, South Africa, Australia, Indian subcontinent, Spain, Turkey, and the United States (Kumar, Jones, et al., 2008). The causal agent of stunt was originally attributed to *Pea leaf roll virus* (syn. *Bean leaf roll virus*, BLRV) (Horn & Reddy, 1996; Nene & Reddy, 1987; Reddy & Kumar, 2004). Subsequently, BLRV and *Chickpea stunt luteovirus* (CpLV) were regarded as causal agents. They indicated that CpLV was probably a strain of BLRV but no comparative studies were reported. Later, studies showed that a leafhopper-transmitted geminivirus, *Chickpea chlorotic dwarf virus*

(CpCDV), was also capable of producing symptoms similar to those referred to as chickpea stunt in India and Pakistan (Horn, Reddy, Roberts, & Reddy, 1993). Surveys of chickpea with stunt symptoms in both India and Pakistan and follow-up serological and electron microscopic studies showed that the etiology of stunt disease was more complex than that was previously thought. The relative prevalence of the luteoviruses appeared to vary in the different chickpea growing areas of the Indian subcontinent. CpCDV- and CpLV-like isolates were widely distributed in India and Pakistan, whereas BLRV-like and *Beet western yellows virus*-like isolates were of minor importance (Horn, Reddy, van den Heuvel, & Reddy, 1996). The stunt disease is more common in the *kabuli* genotypes in Pakistan than in the *desi* types. The yield loss was nearly total if infection occurred in the early stage of growth; if infection occurred at the flowering stage, the yield losses could be as high as 75% (Horn, Reddy, & Reddy, 1995).

A new member of the genus *Polerovirus* named *Chickpea chlorotic stunt virus* has been reported to naturally infect chickpea (Abraham et al., 2006). It is persistently transmitted by *A. craccivora* and *Acyrthosiphon pisum*.

2.7.2 Chlorotic dwarf

CpCDV produces symptoms very similar to those of stunt that include leaf rolling, yellowing, necrosis, and stunting and was shown to be caused by a mastrevirus. It was recorded in India (Horn et al., 1993), Pakistan (Horn et al., 1995; Nahid et al., 2008), Iran and Sudan (Makkouk et al., 2003), Egypt, Iraq, Syria, and Yemen (Kumari, Makkouk, & Attar, 2006, Kumari et al., 2008). When the infection occurs before flowering, the yield loss was reported to be 100% in chickpea (Horn et al., 1993). Molecular characterization of the mastrevirus isolates associated with stunt disease in several countries was reported (Kanakala et al., 2012; Mumtaz et al., 2011; Nahid et al., 2008). The mastrevirus associated with severe stunting, reduction in leaf size, drying and eventual death of chickpea cultivars around Delhi, India, was characterized, and the differences between mastreviruses originating from Africa, the Middle East, Asia, and Australia were compared (Kanakala et al., 2012). Recently, five other mastrevirus species, *Chickpea red leaf virus*, *Chickpea yellows virus*, *Chickpea chlorosis virus*, *Chickpea chlorosis Australia virus*, and *Tobacco yellow dwarf virus* were found in Australia (Hadfield et al., 2012; Thomas, Parry, Schwinghamer, & Dann, 2010). All the mastrevirus isolates infecting chickpea in Africa, Australia, and Asia were reclassified on the basis of 78% nucleotide identity in the genomic DNA and were grouped into one species, CpCDV (Muhire et al., 2013). CpCDV was also found to infect faba bean, lentil, French bean, pigeonpea, and lablab bean (Makkouk et al., 2003). Tissue-blot immunoassay (TBIA) was applied for the detection of CpCDV in chickpea and faba bean (Kumari, Najar, Attar, Loh, & Vetten, 2010).

The management of stunt disease relies upon identification of resistant sources and introgression of resistance genes into desired chickpea genotypes. Over 10,000 germplasm lines have been screened for resistance to stunt at Hisar, India, which is a hot spot for CpCDV, and the lines GG 669 and ICCC 10 were found to be field-resistant. Resistance was expressed as slower symptom development, compared to the susceptible line WR 315. Chickpea lines identified as resistant at Hisar showed 40–70% infection when screened at Junagadh, India (Horn et al., 1996). Nine entries were found to be resistant, and 33 entries were moderately resistant in field screening tests conducted at Junagadh during the years 2008 and 2009 (Chickpea research highlights, IIPR, Kanpur, India, 2009). Various strategies to control virus infections on chickpea were evaluated in Australia (Schwinghamer et al., 2009). Effective field screening for resistance to chickpea stunt viruses should include serological assaying of both susceptible and resistant genotypes and evaluation under greenhouse conditions against virus types and strains.

At present, evaluation of CpCDV resistance is conducted on the basis of natural infection in the field, which is bound to be erroneous due to vagaries in vector population and similar symptoms by unrelated viruses. Kanakala et al. (2013) devised an agroinoculation technique that involves the delivery of CpCDV genomic DNA through *Agrobacterium tumefaciens*, thus facilitating precise screening for CpCDV resistance. They screened 70 chickpea genotypes both under field conditions and by agroinoculation. The genotype SCGP-WR-29 showed resistance under field conditions but exhibited 80% incidence under agroinoculation, indicating that the resistance was presumably due to nonpreference by the vector. The true virus resistance was identified in the genotypes L-550, GNG-1499 (Gauri), and IPC 09-07, which did not express any symptoms and did not show the presence of virus DNA in PCR tests. This type of resistance needs to be exploited for generating CpCDV-resistant chickpea cultivars.

The experiments conducted by International Center for Agricultural Research in the Dry Areas (ICARDA, Syria) in collaboration with Agriculture Research organization in Sudan showed that delayed planting of chickpea cultivars Shendi and ICCV-2 (late October, early or late November) and irrigation at short intervals resulted in reduced CpCDV incidence in Northern Sudan.

2.8. Pea

Pea is widely grown in the temperate regions. Its cultivation in tropics is restricted to cool season. Diseases caused by PSbMV, BYMV, *Pea enation mosaic virus* (PEMV)-1, PEMV-2, and BLRV are economically important (Table 9.1) (Kraft & Pfleger, 2001; Makkouk et al., 2003, 2012).

2.8.1 Mosaic caused by *PSbMV* and *BYMV*

PSbMV is widely distributed largely due to its high levels of seed transmission, thus facilitating entry through international exchange of germplasm. PSbMV is transmitted in a nonpersistent manner by the aphid species *A. pisum*, *M. persicae*, and *A. gossypii* (Kraft & Pfleger, 2001). ELISA, dot-immunobinding assay, and PCR-based tests were reported for virus detection and characterization (Makkouk et al., 2012). Managing the aphid vectors could offer some protection by reducing the aphid-mediated secondary spread from the initial disease foci that resulted from planting of PSbMV infected seed. Genetic resistance for seed transmission of PSbMV provides the best option for reducing the virus spread and its subsequent establishment especially in new areas where the virus was not previously reported (Kraft & Pfleger, 2001). Resistance, characterized as immunity, is controlled by a single recessive gene *sbm*. The *sbm-1* gene confers resistance to P-1 and P-2 isolates, and *sbm-2* and *sbm-3* to P-2 isolates, while only *sbm-4* to P-4 isolates of PSbMV (Johansen, Keller, Dougherty, & Hampton, 1996; Provvidenti & Alconero, 1988). The *sbm-1* gene is linked to *wlo* on chromosome 6 of the *Pisum* genome while *sbm-2* is linked to *mo* on chromosome 2. Recessive genes *sbm-3* and *sbm-4* are also found on chromosome 6, but linkages have not been established. Two homologous genes, translation initiation factor *eIF4E* and *eIF(iso)4E*, governed resistance to PSbMV and BYMV, respectively, at the *sbm-1* and *sbm-2* locus (Bruun-Rasmussen et al., 2007; Gao et al., 2004). In Czech Republic, *eIF4E*-specific molecular markers for PSbMV resistance were developed (Smýkal, Safarova, Navratil, & Dostalova, 2010). The genome of pea was fully sequenced thus permitting marker-assisted breeding (Smýkal et al., 2012).

BYMV was reported on pea in Syria, Egypt, Italy, and Libya and is seed-transmitted in peas, faba beans, lentils, lupins, and a number of forage legumes. ELISA and RT-PCR were used for the detection of BYMV (Makkouk et al., 2012). Spatial isolation of pea fields from virus reservoir hosts limited the spread of the virus by aphids. Several BYMV-resistant pea cultivars and breeding lines were developed using a single recessive gene, *mo* (Kraft & Pfleger, 2001).

2.8.2 Enation mosaic

Pea enation mosaic disease is caused by PEMV-1 and PEMV-2 (Makkouk et al., 2012). PEMV-1 is transmitted efficiently in a persistent manner by at least eight aphid species. *A. pisum* is considered to be the most efficient vector. ELISA- and RT-PCR-based tests for detection of PEMV were reported (Chomic et al., 2010; D'Arcy, Torrance, & Martin, 1989). Resistance to PEMV infection in peas and lentil was reported (Aydin, Muehlbauer, & Kaiser, 1987). PDR- and RNAi-mediated strategies have potential for introducing resistance to PEMV.

2.8.3 Top yellows

Top yellows disease is caused by BLRV. Yield is affected due to misshapen and/or poorly filled pods (Kraft & Pfleger, 2001; Makkouk et al., 2003, 2012). *Medicago*, *Trifolium*, and *Vicia* species act as sources of virus inoculum. Extensive secondary spread was reported if infestation by aphids was not controlled. Therefore, judicious application of insecticides coinciding with aphid monitoring can reduce incidence and spread of BLRV. Spatial isolation of pea fields from virus sources may not help as the virus is persistently transmitted by aphids. Rouging of initially infected pea plants reduced the secondary spread. Resistance to BLRV is inherited as a single recessive gene, designated *lr*. Another recessive gene, *lrv* confers tolerance to BLRV in pea. Several resistant/tolerant pea cultivars available from seed companies are listed (Kraft & Pfleger, 2001).

2.9. Faba bean

Faba bean is extensively cultivated in West Asia and North African (WANA) countries. Among several viruses reported to naturally infect this crop, FBNYV, *Faba bean necrotic stunt virus* (FBNSV), BLRV, BYMV, and *Broad bean mottle virus* (BBMV) are economically important (Table 9.1) (Jellis, Bond, & Boulton, 1998; Makkouk et al., 2012). Faba bean breeding for resistance to different types of diseases including viruses was reviewed (Sillero et al., 2010).

2.9.1 Necrotic yellows and necrotic stunt

Necrotic yellows and necrotic stunt diseases were reported to be caused by two distinct species of the genus *Nanovirus*, FBNYV and FBNSV, respectively. Of these two, FBNYV was widely distributed (Algeria, Egypt, Lebanon, Libya, Spain, Tunisia, Ethiopia, Jordan, Morocco, Syria, and Turkey). In addition to faba bean, it naturally infects other food legume crops pea, chickpea, cowpea, common bean, and lentil. ELISA, TBIA, dot-blot hybridization, and PCR-based tests were employed to differentiate FBNYV from other viruses infecting faba bean (Makkouk & Kumari, 2009). FBNYV is transmitted efficiently by aphids *A. pisum*, *A. craccivora*, and relatively less efficiently by *Aphis fabae*, in a circulative persistent manner. FBNYV is not known to be transmitted by seeds or mechanically (Makkouk & Kumari, 2009). The recommended integrated virus management practices consisted of (a) seed treatment with imidacloprid before planting, (b) judicious application of aphicides, (c) planting at an appropriate time to avoid peak number of viruliferous aphids, (d) planting to provide high-density crop stand, and (e) planting with resistant genotypes (Makkouk & Kumari, 2009).

2.9.2 Leaf roll

Leaf roll disease in faba bean is caused by BLRV. The occurrence of this disease has been reported in the Mediterranean countries (Egypt, Morocco, Tunisia, Lebanon, Syria, and Spain) and Australia. BLRV can be diagnosed by ELISA, TBIA, and RT-PCR (El-Beshehy & Farag, 2013; Freeman et al., 2013; Makkouk et al., 2012; Ortiz, Castro, & Romero, 2005). BLRV is transmitted by aphids (*A. pisum*, *A. craccivora*, *A. fabae*, and *M. persicae*) in a circulative, nonpropagative manner. Seed treatment with imidacloprid before planting and adjustment to sowing dates can reduce the disease incidence. Application of aphicides (organophosphorus, carbamate, and pyrethroid) was shown to reduce disease incidence (Jellis et al., 1998). Genotypes resistant to BLRV have been reported in faba bean (Makkouk, Kumari, & van Leur, 2002) and lentil (Makkouk, Kumari, Sarker, & Erskine, 2001). At ICARDA, faba bean accessions BPL 756, BPL 757, BPL 758, BPL 769, BPL 5278, and BPL 5279 were found to be resistant to BLRV (Kumari & Makkouk, 2003; Makkouk et al., 2002).

2.9.3 Mosaic and necrosis

Mosaic and necrosis in faba bean are caused by BYMV. It was reported from Israel, Italy, Lebanon, Libya, Morocco, Syria, Tunisia, Greece, and Turkey. A high incidence, up to 100%, has been reported in some regions of Egypt, Iraq, and Sudan with relatively warm winters (Makkouk et al., 2012). BYMV is seed-transmitted in faba bean up to 2.4% (Kaiser, 1973). BYMV was detected in several commercial faba bean seed samples up to 9.2% (Sayasa, Iwasaki, & Yamamoto, 1993). Primary source of inoculum was shown to come from seed (Jellis et al., 1998). Therefore, use of virus-free seed was recommended for planting to minimize disease incidence.

Adjustment to sowing dates, spraying with mineral oils, soil mulching with reflective polythene sheets, and ensuring that faba bean crops were not grown in the vicinity of known over wintering virus sources were recommended for reducing BYMV incidence (Mahdy, Fawzy, Hafez, Mohamed, & Shahwan, 2007). Faba bean accessions, BPL 1351, BPL 1363, BPL 1366, and BPL 1371, were found to be resistant to BYMV (Kumari & Makkouk, 2003; Makkouk & Kumari, 2009; Makkouk et al., 2012). In Egypt, resistance in faba bean to BYMV infection was analyzed through diallel mating scheme including reciprocals of six faba bean genotypes with varied resistance and susceptibility to BYMV infection. Resistance was inherited polygenically (El-Bramawy & El-Beshehy, 2012).

2.9.4 Mottle

Mottle disease in faba bean is caused by BBMV. Depending on time of infection, grain yield losses of up to 55% have been reported (Makkouk, Bos, Azzam, Kumari, & Rizkallah, 1988, Makkouk, Bos, Rizkallah, Azzam, & Katul, 1988). High incidence of BBMV was recorded in Morocco, Sudan, Tunisia, Syria, Egypt and Algeria (Makkouk et al., 2012). BBMV was detected in seed and plant tissues by employing ELISA or TBIA (Makkouk, Bos, Azzam, et al., 1988). BBMV was transmitted by mechanical inoculation and by beetle vectors (*Acalymma trivittata, Diabrotica undecimpunctata*, and *Spodoptera exigua*). Seed transmission up to 1.4% was reported, especially when BBMV occurred in mixed infection with BYMV. Planting with virus-free seed was recommended (Makkouk, Bos, Rizkallah, et al., 1988). Faba bean genotypes resistant to BBMV are currently not available.

2.10. Lentil

Lentil, one of the world's oldest cultivated plants, originated in the Middle East and spread through Western Asia to the Indian subcontinent (Erskine, Muehlbauer, Sarker, & Sharma, 2009). Of the several viruses reported to naturally infect lentil (Makkouk et al., 2003), diseases caused by BLRV, FBNYV, PSbMV, CMV, BBSV, and BYMV are of economic significance (Table 9.1).

2.10.1 Yellows and stunt

Yellowing and stunting diseases in lentil, caused by BLRV and FBNYV, were already discussed under faba bean. BLRV is an important virus reported on lentil from Bangladesh, Ethiopia, Iran, Iraq, Syria, Tunisia, and the United States. When plants were infected at the preflowering stage, yield reductions up to 91% were reported. FBNYV was reported on lentil from Ethiopia, Iran, Iraq, Pakistan, Syria, and Turkey (Kumari, Attar, Mustafayev, & Akparov, 2009). No sources of resistance have been recorded for these two viruses.

2.10.2 Mosaic and mottle

Mosaic and mottling diseases in lentil are caused by PSbMV, BYMV, BBSV, and CMV. The natural occurrence of PSbMV was reported from Algeria, Egypt, Ethiopia, Iran, Iraq, Jordan, Morocco, New Zealand, Pakistan, Syria, Tunisia, and Turkey (Kumari et al., 2009). Crop losses of up to 61% were reported from Pakistan (Kumari & Makkouk, 1995; Kumari, Makkouk, & Ismail, 1996). Seed transmission rates of PSbMV in lentil varied widely (0–44%) depending on the cultivar and virus isolate (Kumari et al., 2009). Under field conditions, the virus can over-winter in hairy vetch (*Vicia villosa*) and volunteer peas. From these sources, the virus is transmitted by aphids to nearby lentil crops.

Even though CMV occurs worldwide, its natural occurrence on lentil was reported only from Australia, Ethiopia, India, Iran, Nepal, New Zealand, Pakistan, and Syria. It is easily sap transmissible and nonpersistently vectored by more than 60 different aphid species. Its transmission through lentil seed varied from 0.05% to 37% (Kumari et al., 2009).

BBSV was reported on lentils from Ethiopia, Iran, Jordan, Syria, and Turkey. Grain yield losses varied from 14% to 61%, and the seed transmission rates were found to range from 0.2% to 32.4% when 19 lentil genotypes were inoculated at flowering stage. Infection of lentil plants at preflowering, flowering, and pod stages resulted in a seed-transmission rates of 20.6%, 19.1%, and 1.5%, respectively. BBSV is sap transmissible and by beetles *Apion aestivum*, *Apion arrogans*, *Sitona crinite*, *Sitona limosa*, and *Sitona lineatus* (Kumari et al., 2009). ELISA, TBIA, and PCR were employed for the detection of lentil viruses (Kumari et al., 2009; Kumari & Makkouk, 2007).

Five viruses affecting lentil were seed-borne. For such viruses, planting with virus-free seed was recommended, especially when the virus was also vectored by insects. Seed transmission of BBSV was reduced to zero when seeds were exposed to 70 °C for 28 days; but this treatment caused an unacceptable reduction (57%) in seed germination (Kumari & Makkouk, 1996). Thermotherapy may be useful to eliminate seed-borne viruses from germplasm accessions, meant for conservation. The use of lentil cultivars resistant to the virus or to seed transmission is an effective control option (Kumari et al., 2009).

Over-wintering or over-summering crops, which could act as sources of infection for such nonpersistent viruses as BYMV, PSbMV, and CMV, should be avoided through spatial isolation that will adequately reduce virus spread. In contrast, persistently transmitted viruses such as BLRV and FBNYV can be carried from lucerne fields over long distances making it more difficult to avoid virus spread from these sources (Kumari et al., 2009).

Field experiments at ICARDA showed that seed treatment with the systemic insecticide imidacloprid significantly improved yields of moderately resistant and susceptible lentil genotypes, but had no effect on the yield of resistant genotypes. Seed treatment was also effective in increasing yields from BLRV and FBNYV-inoculated plots, but had no effect in SbDV-inoculated plots (Makkouk & Kumari, 2001).

Four lentil accessions (PI 212610, PI 251786, PI 297745, and PI 368648) were found to be resistant to PSbMV (Makkouk & Kumari, 2009). The genotype ILL 7163 was shown to be highly resistant to BYMV. Additionally, ILL 75 showed resistance to BLRV, FBNYV, and SbDV and ILL 74, ILL 85, ILL 213, ILL 214, and ILL 6816 genotypes were resistant to FBNYV and BLRV. Two cultivars from the United States, "Redchief" and "Palouse," were tolerant to PSbMV infection, expressed as very low grain yield loss in addition to low seed-transmission rate (Kumari & Makkouk, 1995). "Redchief" was also reported to be tolerant to BBSV with low seed-transmission rates (Makkouk & Kumari, 1990).

3. VIRUS DISEASES OF MINOR FOOD LEGUMES

Minor food legume crops that have regional or local importance are hyacinth bean (*Lablab purpureus*, Syn. field bean, lablab bean, pole bean, dolichos bean, or Indian bean), horse gram (*Macrotyloma uniflorum*) and lima bean (*P. lunatus*). Viruses infecting minor food legume crops have been summarized (Makkouk et al., 2003; Odedara, Hughes, Odebode, & Odu, 2008). Only the diseases caused by begomoviruses on these crops have been shown to have the potential to cause significant crop losses in tropical environments and are briefly discussed.

3.1. Hyacinth bean

Hyacinth bean is thought to have originated in India, and spread to South and Eastern Asia, Africa, and the Americas (Murphy & Colucci, 1999; Shivashankar, Kulkarni, Shashidhar, & Mahishi, 1993). For over six decades, dolichos yellow mosaic disease attributed to *Dolichos yellow mosaic virus*

(DoYMV), was considered to be the major constraint of hyacinth bean production (Maruthi et al., 2006). DoYMV was identified as a geminivirus based on electron microscopy (Raj, Aslam, Srivastava, & Singh, 1988), serological and nucleic acid hybridization tests (Swanson, Varma, Muniyappa, & Harrison, 1992). DoYMV was poorly transmitted (to a maximum of 18.3%) by *B. tabaci* and had a narrow host range restricted to *L. purpureus* and *L. purpureus* var. *typicum*. MYMIV was also reported to be the causal agent of YMD of *L. purpureus* based on genome sequence analysis (Singh, Chakroborthy, Singh, & Pandey, 2006). YMD of pole bean in South India could be effectively managed by adopting integrated management practices that include border cropping with African tall maize, seed treatment with imidacloprid 70% WSW@ 5.0 kg, use of reflective mulches, spraying with triazophos 30 days after sowing (DAS) and with thiomethoxam 45 DAS. These measures contributed to yield of 32.2 tons/ha with a cost:benefit ratio of 1:3.17 (Jyothi et al., 2013).

Of the 300 *L. purpureus* genotypes screened under laboratory and field conditions, genotypes VRSEM 894, VRSEM 860, and VRSEM 887 showed no overt symptoms and did not show the virus in PCR tests (Singh, Kumar, Rai, & Singh, 2012). These genotypes have the potential for generating DoYMV-resistant *L. purpureus* cultivars.

3.2. Horse gram

Horse gram is mainly cultivated in the Indian subcontinent and Africa (Jayan & Maya, 2001). YMD was characterized by yellow mosaic on leaves, coupled with reduction of leaf size and plant height in severely infected plants. This virus was initially identified as HgYMV (Muniyappa, Rajeshwari, Bharathan, Reddy, & Nolt, 1987). In addition to horse gram, HgYMV infects French bean, groundnut, lima bean, mungbean, pigeonpea, soybean and bambara groundnut in India. HgYMV was identified as a distinct species of Old World bipartite begomoviruses (Barnabas, Radhakrishnan, & Ramakrishnan, 2010). *Indigofera hirsuta*, a legume weed, was shown to serve as a natural reservoir of HgYMV. Horse gram genotypes tolerant/resistant to HgYMV were identified (Muniyappa, Reddy, & Mustak Ali, 1978), and they are yet to be exploited in breeding programs. A wild relative of horse gram, *Macrotyloma axillare*, was found to be immune to virus infection.

3.3. Lima bean

Lima bean is native to Central America. Among several viruses reported to infect this crop, begomoviruses, viz., BCaMV (North America and

Mexico), BGMV (Latin America and the Caribbean) and *Lima bean golden mosaic virus* (Nigeria) are considered to be important (Makkouk et al., 2003). Lima bean crop is widely grown in Nepal. The frequently observed yellow mosaic symptoms are shown to be caused by MYMIV based on genome sequence analysis (Shahid et al., 2012). In SSA, lima bean is naturally infected by Lima bean golden mosaic begomovirus (Hughes, Naidu, & Shoyinka, 2001). The symptoms are golden mosaic and yellowing. The virus isolates associated with this disease in Nigeria were identified as strains of SbCBV based on sequence analysis of DNA-A component (Alabi et al., 2010).

4. CONCLUSIONS AND FUTURE PROSPECTS

Virus diseases have been shown to have significant impact on the production of major tropical and subtropical food legume crops soybean, groundnut, common bean, cowpea, pigeonpea, mungbean, urdbean, chickpea, pea, faba bean, and lentil (Table 9.1). Among the causal viruses, SbMV, TSWV, PBNV, PCV, IPCV, GRV, CMV, PSbMV, BCMV, BCMNV, BLRV, BGMV, BGYMV, BYMV, PeMoV, PStV, FBNYV, CpCDV, TSV, MYMV, and MYMIV are economically important. They infect more than one legume crops. Viruses transmitted through the seed of food legumes have quarantine importance and also serve as primary virus source in virus ecology and disease epidemiology (Sastry, 2013). Intensive cropping and changes to cropping systems as a result of increased access to irrigation facilities and abuse of pesticides are some of the factors aiding vector multiplication and spread. Occurrence of tospoviruses and TSV on numerous hosts, other than their natural hosts, is of major concern (e.g., Jones, 2009; Rojas & Gilbertson, 2008; Varma et al., 2011).

For the management of legume viruses, several control options available are selection and planting of virus-free seed, adjustment of crop cultural (agronomic) practices, chemical, physical, and biological control of virus vectors, and planting of virus-resistant crops developed through conventional and/or nonconventional breeding methods. These measures were practiced well to minimize the occurrence of important legume virus diseases (e.g., Hema, Gogoi, Dasgupta, & Sreenivasulu, 2014; Hooks & Fereres, 2006; Kumar, Jones, et al., 2008, Kumar, Kumari, et al., 2008; Malathi & John, 2008; Reddy et al., 2009; Sreenivasulu et al., 2008). With the exception of seed treatment to minimize spread of seed-transmitted viruses, pesticide use should be avoided as far as possible for controlling vectors of especially nonpersistently and semipersistently transmitted viruses. Large-scale screening of germplasm should be ideally done under field conditions, preferably at hotspot locations. Laboratory screening for nonsap transmissible viruses, such as some of the begomoviruses, agroinoculation methods are available (Kanakala et al., 2013). Bacterial artificial chromosome (BAC) libraries of pulse crops have the potential to accelerate gene discovery and enhance molecular breeding in these crops (Yu, 2012). Advances in the development of transgenic pulse crops have been reviewed (Eapen, 2008). Groundnut (e.g., Chander Rao et al., 2013), soybean (e.g., Grossi-de-Sa, Pelegrini, & Fragoso, 2011), common bean (Tollefson, 2011), and mungbean (Haq et al., 2010; Sunitha et al., 2013; Yadav, Shukla, & Chattopadhyay, 2009) have been genetically engineered with virus genome-derived genes/sequences and resistance against targeted viruses has been evaluated. Of these, only genetically engineered common bean resistant to BGMV has been commercialized in Brazil (Tollefson, 2011). RNAi-based approaches are being exploited to develop virus resistance in chickpeas (Nahid, Amin, Briddon, & Mansoor, 2011). Despite the progress in genomics of legumes (Sharma, Upadhyaya, Varshney, & Gowda, 2013; Varshney, Mohan et al., 2013; Varshney, Song, et al., 2013) and availability of tools for transforming the plants, incorporation of resistance to economically important legume viruses by genetic engineering is yet to be accomplished on a commercial scale.

Sensible integration of the various options available for control remains the best choice for virus disease control. It should, however, be emphasized that selecting the best measures for each virus–crop combination and production system requires knowledge of the epidemiology of the causal virus in a given agroecosystem and the mode of action and effectiveness of each individual control measure (Jones, 2009; Jones & Barbetti, 2012). Each strategy must be affordable to the farmer and fulfill the requirements of being environmentally friendly and socially acceptable.

ACKNOWLEDGMENTS

M. Hema acknowledges Council of Scientific and Industrial Research (CSIR), New Delhi, and Department of Biotechnology (DBT), New Delhi, for providing financial assistance.

REFERENCES

Abraham, A. D., Menzel, W., Lesemann, D. E., Varrelmann, M., & Vetten, H. J. (2006). Chickpea chlorotic stunt virus: A new polerovirus infecting cool-season food legumes in Ethiopia. *Phytopathology*, *96*, 437–446.

Abraham, A. D., Menzel, W., Varrelmann, M., & Vetten, H. J. (2009). Molecular, serological and biological variation among chickpea chlorotic stunt virus isolates from five countries of North Africa and West Asia. *Archives of Virology*, *154*, 791–799.

Adam, G., Yeh, S. D., Reddy, D. V. R., & Green, S. K. (1993). Serological comparison of tospovirus isolates from Taiwan and India with *Impatiens necrotic spot virus* and different *Tomato spotted wilt virus* isolates. *Archives of Virology, 130*, 237–250.
Ahmad, M., ud-Din, S. B., Muhammad, R. W., Liaqat, S., Qayyum, A., Hamza, A., et al. (2013). Numerical evaluation of different mungbean (*Vigna radiata* L., Wilezek) genotypes against yellow mosaic virus (YMV) under the ecological northern irrigated plain. *Journal of Food, Agriculture and Environment, 11*, 594–597.
Akshay, T., Harish, G. D., Shivakumar, M., Kumar, B., Verma, K., Lal, S. K., et al. (2013). Genetics of Yellow mosaic virus (YMV) resistance in cultivars soybean [*Glycine max* (L.) Merr.]. *Legume Research, 36*, 263–267.
Akthar, K. P., Sarwar, G., Abbas, G., Asghar, M. J., Sarwar, N., & Shah, T. M. (2011). Screening of mungbean germplasm against *Mungbean yellow mosaic India virus* and its vector *Bemisia tabaci*. *Crop Protection, 30*, 1202–1209.
Alabi, O. J., Kumar, P. L., Mgbechi-Ezeri, J. U., & Naidu, R. A. (2010). Two new 'legumoviruses' (genus *Begomovirus*) naturally infecting soybean in Nigeria. *Archives of Virology, 155*, 643–656.
Albrechtsen, S. E. (2006). *Testing methods for seed transmitted viruses: Principles and protocols* (pp. 1–268). Wallingford, CT: CABI Publishing.
Alegbejo, M. D., & Abo, M. E. (2002). Etiology, ecology, epidemiology and control of groundnut rosette disease in Africa. *Journal of Sustainable Agriculture, 20*, 17–29.
Alemandri, V., Rodriguez Pardina, P., Izaurralde, J., Medina, S. G., Caro, E. A., Mattio, M. F., et al. (2012). Incidence of begomoviruses and climatic characterization of *Bemisia tabaci*-geminivirus complex in soybean and bean in Argentina. *AgriScientia, 26*, 31–39.
Almeida, A. M. R., Piuga, F. F., Marin, S. R. R., Kitajima, E. W., Gaspar, J. O., de Oliveira, T. G., et al. (2005). Detection and partial characterization of a carlavirus causing stem necrosis of soybean in Brazil. *Fitopatologia Brasileira, 30*, 191–194.
Almeida, A. M. R., Sakai, J., Hanada, K., de Oliveira, T. G., Belintani, P., Kitajima, E. W., et al. (2005). Biological and molecular characterization of an isolate of *Tobacco streak virus* obtained from soybean in Brazil. *Fitopatologia Brasileira, 30*, 366–373.
Amayo, R., Arinaitwe, A. B., Mukasa, S. B., Tusiime, G., Kyamanywa, S., Rubaihayo, P. R., et al. (2012). Prevalence of viruses infecting cowpea in Uganda and their molecular detection. *African Journal of Biotechnology, 11*, 14132–14139.
Ammavasai, S., Phogat, D. S., & Solanki, I. S. (2004). Inheritance of resistance to *Mungbean yellow mosaic virus* (MYMV) in green gram (*Vigna radiata* L. Wilczek). *Indian Journal of Genetics, 64*, 146.
Anjos, J. R. N., & Lin, M. T. (1984). Budblight of soybeans caused by cowpea severe mosaic virus in central Brazil. *Plant Disease, 68*, 405–407.
Aragao, F. J. L., & Faria, J. C. (2009). First transgenic geminivirus-resistant plant in the field. *Nature Biotechnology, 27*, 1086–1088.
Arif, M., & Hassan, S. (2002). Evaluation of resistance in soybean germplasm to soybean mosaic potyvirus under field conditions. *Journal of Biological Sciences, 2*, 601–604.
Arif, M., Stephen, M., & Hassan, S. (2002). Effect of soybean mosaic potyvirus on growth and yield components of commercial soybean varieties. *Pakistan Journal of Plant Pathology, 1*, 54–57.
Arshad, M., Bashir, M., Sharif, A., & Malik, B. A. (1998). Inheritance of resistance in cowpea (*Vigna unguiculata* L. Walp.) to blackeye cowpea mosaic potyvirus. *Pakistan Journal of Botany, 30*, 263–270.
Arun Kumar, N., Lakshmi Narasu, M., Usha, B. Z., & Ravi, K. S. (2006). Natural occurrence and distribution of *Tobacco streak virus* in south India. *Indian Journal of Plant Protection, 34*, 54–58.

Arun Kumar, N., Lakshmi Narasu, M., Usha, B. Z., & Ravi, K. S. (2008). Molecular characterization of *Tobacco streak virus* causing soybean necrosis in India. *Indian Journal of Biotechnology*, 7, 214–217.

Aydin, H., Muehlbauer, F. J., & Kaiser, W. J. (1987). *Pea enation mosaic virus* resistance in lentil (*Lens culinaris*). *Plant Disease*, *71*, 635–638.

Bandla, M. D., Westcot, D. M., Chenault, K. D., Ullman, D. E., German, T. L., & Sherwood, J. L. (1994). Use of monoclonal antibody to the nonstructural protein encoded by the small RNA of tomato spotted wilt tospovirus to identity viruliferous thrips. *Phytopathology*, *84*, 1427–1431.

Barnabas, A. D., Radhakrishnan, G. K., & Ramakrishnan, U. (2010). Characterization of a begomovirus causing horsegram yellow mosaic disease in India. *European Journal of Plant Pathology*, *127*, 41–51.

Bashir, M., & Hampton, R. O. (1996). Detection and identification of seed-borne viruses from cowpea (*Vigna unguiculata* (L.) Walp.) germplasm. *Plant Pathology*, *45*, 54–58.

Beniwal, S. P. S., Deena, E., & Nene, Y. L. (1983). Effect of yellow mosaic on yield and its components in post-rainy season pigeonpea. *International Pigeonpea Newsletter*, *2*, 48.

Bhat, A. I., Jain, R. K., Varma, A., & Lal, S. K. (2002). Nucleocapsid protein gene sequence studies suggest that soybean bud blight is caused by a strain of *Groundnut bud necrosis virus*. *Current Science*, *82*, 1389–1392.

Bhuvanendra Kumar, H., Uday Shankar, A. C., Prakash, H. S., & Shekar Shetty, H. (2005). Effect of Sanosil and *Pseudomonas fluorescens* on bean common mosaic potyvirus incidence in French bean. *International Journal of Botany*, *1*, 163–167.

Biemond, P. C., Oguntade, O., Kumar, P. L., Stomph, T. J., Termorshuizen, A. J., & Struik, P. C. (2013). Does the informal seed system threaten cowpea seed health? *Crop Protection*, *43*, 166–174.

Bijaisoradat, M., & Kuhn, C. W. (1988). Detection of two viruses in peanut seeds by complementary DNA hybridization tests. *Plant Disease*, *72*, 956–959.

Biswas, K. K., Tarafdar, A., & Biswas, K. (2012). Viral diseases and its mixed infection in mungbean and urdbean: Major biotic constraints in production of food pulses in India. In Asha Sinha, B. K. Sharma, & Manisha Srivastava (Eds.), *Modern trends in microbial biodiversity of natural ecosystem* (pp. 301–317). New Delhi: Biotech Books.

Biswas, K. K., & Varma, A. (2000). Identification of variants of mungbean yellow mosaic geminivirus by host reaction and nucleic acid spot hybridization. *Indian Phytopathology*, *53*, 37–38.

Blair, M. W., Beaver, J. S., Nin, J. C., Prophete, E., & Singh, S. P. (2006). Registration of PR9745-232 and RMC-3 red-mottled dry bean germplasm lines with resistance to *Bean golden yellow mosaic virus*. *Crop Science*, *46*, 1000–1002.

Blair, M. W., Rodriguez, L. M., Pedraza, F., Morales, F., & Beebe, S. (2007). Genetic mapping of the bean golden yellow mosaic geminivirus resistance gene *bgm-1* and linkage with potyvirus resistance in common bean (*Phaseolus vulgaris* L.). *Theoretical and Applied Genetics*, *114*, 261–271.

Blok, V. C., Ziegler, A., Scott, K., Dangora, D. B., Robinson, D. J., & Murant, A. F. (1995). Detection of groundnut rosette umbravirus infections with radioactive and non-radioactive probes to its satellite RNA. *Annals of Applied Biology*, *127*, 321–328.

Bock, K. R., Murant, A. F., & Rajeswari, R. (1990). The nature of resistance in groundnut to rosette disease. *Annals of Applied Biology*, *117*, 379–384.

Bonfim, K., Faria, J. C., Nogueira, E. O., Mendes, E. A., & Aragao, F. J. L. (2007). RNAi-mediated resistance to *Bean golden mosaic virus* in genetically engineered common bean (*Phaseolus vulgaris*). *Molecular Plant-Microbe Interactions*, *20*, 717–726.

Boonham, N., Smith, P., Walsh, K., Tame, J., Morris, J., Spence, N., et al. (2002). The detection of *Tomato spotted wilt virus* (TSWV) in individual thrips using real time fluorescent RT-PCR (Taq Man). *Journal of Virological Methods, 101*, 37–48.

Borah, B. K., & Dasgupta, I. (2012). Begomovirus research in India: A critical appraisal and the way ahead. *Journal of Biosciences, 37*, 791–806.

Bos, L. (2008). Legume viruses. In W. J. M. Brian, & M. H. V. Van Regenmortel (Eds.), *Desk encylopedia of plant and fungal virology* (pp. 419–426). United Kingdom: Elsevier/Academic Press.

Boukar, O., Bhattacharjee, R., Fatokun, C., Kumar, P. L., & Gueye, B. (2013). Cowpea. In S. Mohar, H. D. Upadhyaya, & S. B. Iswari (Eds.), *Genetic and genomic resources of grain legume improvement* (pp. 137–156). Amsterdam: Elsevier.

Bragard, C., Doucet, D., Dieryck, B., & Delfosse, P. (2008). Pecluviruses. In G. P. Rao, P. L. Kumar, & R. J. Holguin-Peña (Eds.), *Vegetable and pulse crops: Vol. 3. Characterization, diagnosis and management of plant viruses* (pp. 125–140). Houston, Texas: Studium Press LLC.

Breyel, E., Casper, R., Ansa, O. A., Kuhn, C. W., Misari, S. M., & Demski, J. W. (1988). A simple procedure to detect a dsRNA associated with groundnut rosette. *Journal of Phytopathology, 121*, 118–124.

Bruun-Rasmussen, M., Møller, I. S., Tulinius, G., Hansen, J. K. R., Lund, O. S., & Johansen, I. E. (2007). The same allele of translation initiation factor 4E mediates resistance against two Potyvirus spp. in *Pisum sativum*. *Molecular Plant-Microbe Interactions, 20*, 1075–1082.

Cantonwine, E. G., Culbreath, A. K., Stevenson, K. L., Kemerait, R. C., Jr., Brenneman, T. B., Smith, N. B., et al. (2006). Integrated disease management of leaf spot and spotted wilt of peanut. *Plant Disease, 90*, 493–500.

Castillo-Urquiza, G. P., Maia, F. G., Carvalho, M. G., Pinto, C. M., & Zerbini, F. M. (2006). Characterization of a *Bean rugose mosaic virus* (BRMV) isolate from Minas Gerais, and yield loss estimate in beans upon single infection and double infection with BCMV. *Fitopatologia Brasileira, 31*, 455–461.

Chander Rao, S., Bhatnagar-Mathur, P., Kumar, P. L., Reddy, A. S., & Sharma, K. K. (2013). Pathogen-derived resistance using a viral nucleocapsid gene confers only partial non-durable protection in peanut against *Peanut bud necrosis virus*. *Archives of Virology, 158*, 133–143.

Chaudhury, D., Madanpotra, S., Jaiwal, R., Saini, R., Kumar, P. A., & Jaiwal, P. K. (2007). *Agrobacterium tumefaciens*-mediated high frequency genetic transformation of an Indian cowpea (*Vigna unguiculata* L.) cultivar and transmission of transgenes into progeny. *Plant Science, 172*, 692–700.

Chickpea Research Highlights (2009). *All India Coordinated Research Project on Chickpea* (p. 70). Kanpur, India: Indian Institute of Pulses Research.

Chilangane, L. A., Tryphone, G. M., Protas, D., Kweka, E., Kusolwa, P. M., & Nchimbi-Msolla, S. (2013). Incorporation of resistance to angular leaf spot and Bean common mosaic necrosis virus diseases into adapted common bean (*Phaseolus vulgaris* L.) genotype in Tanzania. *African Journal of Biotechnology, 12*, 4343–4350.

Cho, E. K., & Goodman, R. M. (1979). Strains of *Soybean mosaic virus*: Classification based on virulence in resistant soybean cultivars. *Phytopathology, 69*, 467–470.

Chomic, A., Pearson, M. N., Clover, G. R. G., Farreyrol, K., Saul, D., Hampton, J. G., et al. (2010). A generic RT-PCR assay for the detection of *Luteoviridae*. *Plant Pathology, 59*, 429–442.

Coco, D., Calil, I. P., Brustolini, O. J., Santos, A. A., Inoue-Nagata, A. K., & Fontes, E. P. (2013). Soybean chlorotic spot virus, a novel begomovirus infecting soybean in Brazil. *Archives of Virology, 158*, 457–462.

Cruz, A. R. R., & Aragao, F. J. L. (2014). RNAi-based enhanced resistance to *Cowpea severe mosaic virus* and *Cowpea aphid-borne mosaic* virus in transgenic cowpea. *Plant Pathology, 63*, 831–837. http://dx.doi.org/10.1111/ppa.12178.

Cui, X., Chen, X., & Wang, A. (2011). Detection, understanding and control of *Soybean mosaic virus*. In A. Sudaric (Ed.), *Soybean-molecular aspects of breeding* (pp. 335–354). Rijeka, Croatia: InTech. ISBN: 978-953-307-240-1.

Culbreath, A. K., Branch, W., Holbrook, C., & Tilman, B. L. (2009). Effect of seeding rate on spotted wilt incidence in new peanut cultivars and breeding lines. *Phytopathology, 99*, S197.

Culbreath, A. K., & Srinivasan, R. (2011). Epidemiology of spotted wilt disease of peanut caused by *Tomato spotted wilt virus* in the southeastern U.S. *Virus Research, 159*, 101–109.

Culbreath, A. K., Tillman, B. L., Gorbet, D. W., Holbrook, C. C., & Nischwitz, C. (2008). Response of new field-resistant peanut cultivars to twin-row pattern or in-furrow applications of phorate for management of spotted wilt. *Plant Disease, 92*, 1307–1312.

Culbreath, A. K., Todd, J. W., & Brown, S. L. (2003). Epidemiology and management of tomato spotted wilt in peanut. *Annual Review of Phytopathology, 41*, 53–75.

Culver, J. N., Sherwood, J. L., & Melouk, H. A. (1987). Resistance to *Peanut stripe virus* in *Arachis* germplasm. *Plant Disease, 71*, 1080–1082.

D'Arcy, C. J., Torrance, L., & Martin, R. R. (1989). Discrimination among luteoviruses and their strains by monoclonal antibodies and identification of common epitopes. *Phytopathology, 79*, 869–873.

Davies, J. C. (1975). Use of Menazon for control of rosette disease of in Uganda. *Tropical Agriculture (Trinidad), 52*, 389–501.

De Breuil, S., Giolitti, F., & Lenardon, S. (2005). Detection of *Cucumber mosaic virus* in peanut (*Arachis hypogaea* L.) in Argentina. *Journal of Phytopathology, 153*, 722–725.

Delfosse, P., Reddy, A. S., Legreve, A., Devi, P. S., Thirumala-Devi, K., Maraite, H., et al. (1999). *Indian peanut clump virus* infection on wheat and barley: Symptoms, yield loss and transmission through seed. *Plant Pathology, 48*, 273–282.

Demski, J. W., & Reddy, D. V. R. (1997). Diseases caused by viruses. In K. B. La, D. M. Porter, R. Rodriguez-kabana, D. H. Smith, & P. Subrahmanyam (Eds.), *Compendium on peanut disease* (pp. 53–59). St. Paul, Minnesota: American Phytopathological Society.

Demski, J. W., Reddy, D. V. R., Wongkaew, S., Xu, Z. Y., Kuhn, C. W., Cassidy, B. G., et al. (1993). Peanut stripe virus. *Information bulletin: Vol. 38*. Griffin: Peanut CRSP, University of Georgia, ICRISAT, Patancheru (AP), India, 16 pp.

Demski, J. W., & Sowell, G., Jr. (1981). Resistance to *Peanut mottle Virus* in *Arachis* spp. *Peanut Science, 8*, 43–44.

Deom, C. M., Naidu, R. A., Chiyembekeza, A. J., Ntare, B. R., & Subrahmanyam, P. (2000). Sequence diversity within the three agents of groundnut rosette disease. *Phytopathology, 90*, 214–219.

Dhole, V. J., & Reddy, K. S. (2012). Genetic analysis of resistance to *Mungbean yellow mosaic virus* in mungbean (*Vigna radiata*). *Plant Breeding, 131*, 414–417.

Dieryck, B., Delfosse, P., Reddy, A. S., & Bragard, C. (2010). Targeting highly conserved 3'-untranslated region of pecluviruses for sensitive broad-spectrum detection and quantitation by RT-PCR and assessment of phylogenetic relationships. *Journal of Virological Methods, 169*, 385–390.

Dietzgen, R. G., Callaghan, B., Higgins, C. M., Birch, R. G., Chen, K., & Xu, Z. (2001). Differentiation of peanut seed-borne potyviruses and cucumoviruses by RT-PCR. *Plant Disease, 85*, 989–992.

Dietzgen, R. G., Mitter, N., Higgins, C. M., Hall, R., Teycheney, P. Y., Cruickshank, A., et al. (2004). Harnessing RNA silencing to protect peanuts from stripe disease. In *Proceedings of the fourth international crop science congress, Brisbane, Australia.*

Dinon, A. Z., Brod, F. C., Mello, C. S., Oliveira, E. M., Faria, J. C., & Arisi, A. C. (2012). Primers and probes development for specific PCR detection of genetically modified common bean (*Phaseolus vulgaris*) Embrapa 5.1. *Journal of Agricultural and Food Chemistry, 60*, 4672–4677.

Diouf, D. (2011). Recent advances in cowpea (*Vigna unguiculata* (L.) Walp.) "omics" research for genetic improvement. *African Journal of Biotechnology, 10*, 2803–2810.

Dubey, S. C., & Singh, B. (2013). Integrated management of major diseases of mungbean by seed treatment and foliar application of insecticide, fungicides and bioagent. *Crop Protection, 47*, 55–60.

Dumet, D., Fatokun, C., Pasquet, R., Ehlers, J., Kumar, P. L., Hearne, S., et al. (2012). Sharing of responsibilities of cowpea and wild relatives in long term conservation. In O. Boukar, O. Coulibaly, C. A. Fatokun, K. Lopez, & M. Tamo (Eds.), *Innovation research along the cowpea value chain. Proceedings of the fifth world cowpea conference on improving livelihoods in the cowpea value chain through advancement in science* (pp. 56–65), Nigeria: IITA.

Dwivedi, S. L., Gurtu, S., Chandra, S., Upadhyaya, H. D., & Nigam, S. N. (2003). AFLP diversity among selected rosette resistant groundnut germplasm. *International Arachis Newsletter, 23*, 21–23.

Eapen, S. (2008). Advances in development of transgenic pulse crops. *Biotechnology Advances, 26*, 162–168.

Eduardo Vallejos, C., Astua-Mong, G., Jones, V., Plyler, T. R., Sakiyama, N. S., & Mackenzie, S. A. (2006). Genetic and molecular characterization of the '*I*' locus of *Phaselous vulgaris*. *Genetics, 172*, 1229–1242.

Ehlers, J. D., Diop, N.-N., Boukar, O., Muranaka, S., Wanamaker, S., Issa, D., et al. (2012). Modern approaches for cowpea breeding. In O. Boukar, O. Coulibaly, C. A. Fatokun, K. Lopez, & M. Tamo (Eds.), *Innovation research along the cowpea value chain. Proceedings of the fifth world cowpea conference on improving livelihoods in the cowpea value chain through advancement in science* (pp. 3–16). Nigeria: IITA.

Elbeaino, T., Digiaro, M., Uppala, M., & Sudini, H. (2014). Deep sequencing of pigeonpea sterility mosaic virus discloses five RNA segments related to emaraviruses. *Virus Research, 188*, 27–31.

Elbeaino, T., Whitefield, A., Sharma, M., & Digiaro, M. (2013). Emaravirus-specific degenerate PCR primers allowed the identification of partial RNA-dependent RNA polymerase sequences of *Maize red stripe virus* and *Pigeonpea sterility mosaic virus*. *Journal of Virological Methods, 188*, 37–40.

El-Beshehy, E. K. F., & Farag, A. G. (2013). Antiserum production and reverse transcription polymerase chain reaction (RT-PCR) for detection of bean leaf roll virus. *African Journal of Microbiology Research*, 7, 2853–2861.

El-Bramawy, M. A. S. A., & El-Beshehy, E. K. F. (2012). Inheritance of resistance to *Bean yellow mosaic virus* in faba bean plants. *International Journal of Virology, 8*, 98–105.

El-Kewey, S. A., Sidaros, S. A., Abdelkader, H. S., Emeran, A. A., & EL-Sharkawy, M. (2007). Molecular detection of *Broad bean stain comovirus* (BBSV) and *Cowpea aphid-borne potyvirus* (CABMV) in faba bean and cowpea plants. *Journal of Applied Sciences Research, 3*, 2013–2025.

EPPO. (1997). Tomato spotted wilt tospovirus. *EPPO data sheets on quarantine pests*. Wallingford, United Kingdom: CABI. https://www.eppo.int/QUARANTINE/virus/Tomato_spotted_wilt_virus/TSWV00_ds.pdf.

Erskine, W., Muehlbauer, F. J., Sarker, A., & Sharma, B. (2009). *The lentil: Botany, production and uses*. Wallingford, United Kingdom: CABI. ISBN: 9781845934873.

Fan, Q., Niroula, M., Fildstein, P. A., & Bruening, G. (2011). Participation of the *Cowpea mosaic virus* protease in eliciting extreme resistance. *Virology, 417*, 71–78.

FAOSTAT. (2012). *Statistics*. Rome: FAO: Food and Agriculture Organization of the United Nations. http://www.faostat.fao.org/site/567/DesktopDefault.aspx?PageID=567#ancor.

Fernandes, F. R., Cruz, A. R., Faria, J. C., Zerbini, F. M., & Aragao, F. J. (2009). Three distinct begomoviruses associated with soybean in Central Brazil. *Archives of Virology, 154*, 1567–1570.

Ferreira, J. J., Campa, A., Perez-Vega, E., Rodriguez-Suarez, C., & Giraldez, R. (2012). Introgression and pyramiding into common bean market class fabada of genes conferring resistance to anthracnose and potyvirus. *Theoretical and Applied Genetics, 124*, 777–788.

Fiallo-Olive, E., Marquez-Martin, B., Hassan, I., Chirinos, D. T., Geraud-Pouey, F., Navas-Castillo, J., et al. (2013). Complete genome sequences of two novel begomoviruses infecting common bean in Venezuela. *Archives of Virology, 158*, 723–727.

Freeman, A. J., Spackman, M. E., Aftab, M., McQueen, V., King, S., van Leur, J. A. G., et al. (2013). Comparison of tissue blot immunoassay and reverse transcription polymerase chain reaction assay for virus-testing pulse crops from a South-Eastern Australia survey. *Australasian Plant Pathology, 42*, 675–683.

Frison, E. A., Bos, L., Hamilton, R. I., Mathur, S. B., & Taylor, J. D. (1990). *FAO/IBPGR technical guidelines for the safe movement of legume germplasm*. Rome: Research Institute for Plant Protection. Food and Agriculture Organisation of the United Nations, International Board for Plant Genetic Resources, Rome.

Gagarinova, A. G., Babu, M., Stromvik, M. V., & Wang, A. (2008). Recombination analysis of *Soybean mosaic virus* sequences reveals evidence of RNA recombination between distinct pathotypes. *Virology Journal, 5*, 143.

Ganapathy, K. N., Byre, G. M., Ajay, B. C., Venkatesha, S. C., Gnanesh, B. N., Gomesha, S. S., et al. (2012). Inheritance studies of sterility mosaic disease (SMD) resistant in vegetable type pigeonpea (*Cajanus cajan* (L.) Millsp.). *Australian Journal of Crop Science, 6*, 1154–1158.

Ganapathy, K. N., Byre, G. M., Rama Chandra, R., Venkatesha, S. C., Gnanesh, B. N., & Girish, G. G. (2009). Identification of AFLP markers linked to sterility mosaic disease in pigeonpea *Cajnus cajan* (L.) Millsp. *International Journal of Integrative Biology, 7*, 145–149.

Gao, Z., Johansen, E., Eyers, S., Thomas, C. L., Noel Ellis, T. H., & Maule, A. J. (2004). The potyvirus recessive resistance gene, *sbm1*, identifies a novel role for translation initiation factor *eIF4E* in cell-to-cell trafficking. *The Plant Journal, 40*, 376–385.

Gazala, I. F. S., Sahoo, R. N., Pandey, R., Mandal, B., Gupta, V. K., Singh, R., et al. (2013). Spectral reflectance pattern in soybean for assessing yellow mosaic disease. *Indian Journal of Virology, 24*, 242–249.

Ghanekar, A. M., Sheila, V. K., Beniwal, S. P. S., Reddy, M. V., & Nene, Y. L. (1992). Sterility mosaic of pigeonpea. In U. S. Singh, A. N. Mukhopadhyay, J. Kumar, & H. S. Chaube (Eds.), *Diseases of cereals and pulses: Vol. 1. Plant diseases of international importance* (pp. 415–428). New Jersey: Prentice Hall.

Gillaspie, A. G., Jr. (2007). Attempts to improve the method for screening cowpea germplasm for resistance to *Cucumber mosaic virus* and *Black eye cowpea mosaic virus*. *Plant Pathology Journal, 6*, 202–205.

Gillaspie, A. G., Jr., Pittman, R. N., Pinnow, D. L., & Cassidy, B. G. (2000). Sensitive method for testing peanut seed lots for *Peanut stripe* and *Peanut mottle viruses* by immunocapture reverse transcription-polymerase chain reaction. *Plant Disease, 84*, 559–561.

Gillaspie, A. G., Jr., Wang, M. L., Pinnow, D. L., & Pittman, R. N. (2007). Polymerase chain reaction for detection of peanut mottle and peanut stripe viruses in *Arachis hypogaea* L. germplasm seed lots. *Plant Pathology Journal, 6*, 87–90.

Girish, K. R., & Usha, R. (2005). Molecular characterization of two soybean-infecting begomoviruses from India and evidence for recombination among legume-infecting begomoviruses from South-East Asia. *Virus Research, 108*, 167–176.

Gnanesh, B. N., Bohra, A., Sharma, M., Byregowda, M., Pande, S., Wesley, V., et al. (2011). Genetic mapping and quantitative trait locus analysis of resistance to sterility mosaic disease in pigeonpea [*Cajanus cajan* (L) Millsp.]. *Field Crops Research, 123*, 53–61.
Gnanesh, B. N., Ganapathy, K. N., Ajay, B. C., & Byregowda, M. (2011). Inheritance of sterility mosaic disease resistance to Bangalore and Patancheru isolates in pigeonpea [*Cajanus cajan* (L) Millsp.]. *Electronic Journal of Plant Breeding, 2*, 218–223.
Goenaga, R., Quiles, A., & Gillaspie, A. G. (2008). Assessing yield potential of cowpea genotypes grown under virus pressure. *HortScience, 43*, 673–676.
Gopal, K., Muniayappa, V., & Jagadeeshwar, R. (2011). Management of peanut bud necrosis disease in groundnut by botanical pesticides. *Archives of Phytopathology and Plant Protection, 44*, 1233–1237.
Gopal, K., Sreenivasulu, Y., Gopi, V., Subasini, P., Ahammed, S. K., Govindarajulu, B., et al. (2010). Resistant sources in groundnut germplasm lines against peanut bud necrosis tospovirus disease. *Archives of Phytopathology and Plant Protection, 43*, 501–506.
Grossi-de-Sa, M. F., Pelegrini, P. B., & Fragoso, R. R. (2011). Genetically modified soybean for insect-pests and disease control. *Soybean-molecular aspects of breeding 19*, (pp. 429–452). Rijeka, Croatia: InTech.
Gumedzoe, M. Y. D., Rossel, H. W., Thottappilly, G., Asselin, A., & Huguenot, C. (1998). Reaction of cowpea (*Vigna unguiculata* L. Walp.) to six isolates of *Blackeye cowpea mosaic virus* (BlCMV) and *Cowpea aphid-borne mosaic virus* (CABMV), two potyviruses infecting cowpea in Nigeria. *International Journal of Pest Management, 44*, 11–16.
Gunduz, I., Buss, G. R., Chen, P., & Tolin, S. A. (2004). Genetic and phenotypic analysis of *Soybean mosaic virus* resistance in PI 88788 soybean. *Phytopathology, 94*, 687–692.
Gunduz, I., Buss, G. R., Ma, G., Chen, P., & Tolin, S. A. (2001). Genetic analysis *of Soybean mosaic virus* in OX 670 and Harosoy soybean. *Crop Science, 41*, 1785–1791.
Gupta, S., Gupta, D. S., Anjum, T. K., Pratap, A., & Kumar, J. (2013). Inheritance and molecular tagging of MYMIV resistance gene in blackgram (*Vigna mungo* L. Hepper). *Euphytica, 193*, 27–37.
Habib, S., Shad, N., Javaid, A., & Iqbal, U. (2007). Screening of mungbean germplasm for resistance/tolerance against yellow mosaic disease. *Mycopathology, 5*, 89–94.
Hadfield, J., Thomas, J. E., Schwinghamer, M. W., Kraberger, S., Stainton, D., Dayaram, A., et al. (2012). Molecular characterization of dicot-infecting mastreviruses from Australia. *Virus Research, 166*, 13–22.
Hajimorad, M. R., Eggenberger, A. L., & Hill, J. H. (2006). Strain-specific P3 of *Soybean mosaic virus* elicits *Rsv1*-mediated extreme resistance, but absence of P3 elicitor function alone is insufficient for virulence on *Rsv1*-genotype soybean. *Virology, 345*, 156–166.
Hampton, R. O., & Thottappilly, G. (2003). Cowpea. In G. Loebenstein, & G. Thottappilly (Eds.), *Virus and virus-like diseases of major crops in developing countries* (pp. 355–376). Dordrecht: Kluwer Academic Publishers.
Haq, Q. M. I., Ali, A., & Malathi, V. G. (2010). Engineering resistance against *Mungbean yellow mosaic India virus* using antisense RNA. *Indian Journal of Virology, 21*, 82–85.
Hartman, G. L., Sinclair, J. B., & Rupe, J. C. (1999). Bean pod mottle virus. *Compendium of soybean diseases* (4th ed.). St. Paul, MN: The American Phytopathological Society.
Hema, M., Gogoi, N., Dasgupta, I., & Sreenivasulu, P. (2014). Genetically engineered pathogen resistant crops in Asia and Latin America. In D. V. R. Reddy, P. Ananda Kumar, G. Loebenstein, & P. L. Kumar (Eds.), *Genetically Engineered Crops in Developing Countries* (pp. 161–219). Houston, Texas: Studium Press LLC.
Herselman, L., Thwaites, R., Kimmins, F. M., Courtois, B., Van der Merwe, P. J. A., & Seal, S. E. (2004). Identification and mapping of AFLP markers linked to peanut (*Arachis hypogaea* L.) resistance to the aphid vector of groundnut rosette disease. *Theoretical and Applied Genetics, 109*, 1426–1433.

Higgins, C. M., Hall, R. M., Mitter, N., Cruickshank, A., & Dietzgen, R. G. (2004). Peanut stripe potyvirus resistance in peanut (*Arachis hypogaea L.*) plants carrying viral coat protein gene sequences. *Transgenic Research, 13*, 59–67.

Hill, J. H. (2003). Soybean. In G. Loebenstein, & G. Thottappilly (Eds.), *Viruses and virus-like diseases of major crops in developing countries* (pp. 377–395). Dordrecht: Kluwer Academic Publishers.

Hobbs, H. A., Reddy, D. V. R., Rajeshwari, R., & Reddy, A. S. (1987). Use of direct antigen coating and protein-A coating ELISA procedures for detection of three peanut viruses. *Plant Disease, 71*, 747–749.

Hooks, C. R. R., & Fereres, A. (2006). Protecting crops from non-persistently aphid-transmitted viruses: A review on the use of barrier plants as a management tool. *Virus Research, 120*, 1–16.

Hormozi-Nejad, M. H., Mozafari, J., & Rakhshandehroo, F. (2010). Elimination of *Bean common mosaic virus* using an electrotherapy technique. *Journal of Plant Diseases and Protection, 117*, 201–205.

Horn, N. M., & Reddy, S. V. (1996). Survey of chickpea (*Cicer ariettinum*) for chickpea stunt disease associated viruses in India and Pakistan. *Plant Disease, 80*, 286–290.

Horn, N. M., Reddy, S. V., & Reddy, D. V. R. (1995). Assessment of yield losses caused by *Chickpea chlorotic dwarf geminivirus* in chickpea (*Cicer ariettinum*) in India. *European Journal of Plant Pathology, 101*, 221–224.

Horn, N. M., Reddy, S. V., Roberts, I. M., & Reddy, D. V. R. (1993). *Chickpea chlorotic dwarf virus*, a new leafhopper transmitted geminivirus of chickpea in India. *Annals of Applied Biology, 122*, 467–479.

Horn, N. M., Reddy, S. V., van den Heuvel, J. F. J. M., & Reddy, D. V. R. (1996). Survey of chickpea (*Cicer ariettinum* L.) for chickpea stunt disease and associated viruses in India and Pakistan. *Plant Disease, 80*, 286–290.

Hughes, J., Naidu, R. A., & Shoyinka, S. A. (2001). Strategy for strengthening plant virus research in sub-Saharan Africa agriculture. In *Book of abstracts Plant virology in sub-Saharan Africa conference I, June 4–8*. Ibadan, Nigeria: IITA.

Hughes, J. d. 'A., & Shoyinka, S. A. (2003). Overview of viruses of legumes other than groundnut in Africa. In J. D. A. Hughes, & B. Odu (Eds.), *Plant virology in sub-Saharan Africa. Proceeding of plant virology* (pp. 553–568). Ibadan, Nigeria: IITA.

Huguenot, C., Givord, L., Sommermeyer, G., & Van Regenmortel, M. H. (1989). Differentiation of *Peanut clump virus* serotypes by monoclonal antibodies. *Virus Research, 140*, 87–102.

Huguenot, C., van den Dobbelsteen, G., de Haan, P., Wagemakers, C. A., Drost, G. A., Osterhaus, A. D., et al. (1990). Detection of *Tomato spotted wilt virus* using monoclonal antibodies and riboprobes. *Archives of Virology, 110*, 47–62.

Hwang, T. Y., Jeong, S. C., Kim, O., Park, H.-M., Lee, S.-K., Seo, M.-J., et al. (2011). Intra-host competition and interactions between *Soybean mosaic virus* (SMV) strains in mixed-infected soybean. *Australian Journal of Crop Science, 5*, 1379–1387.

Hwang, T. Y., Moon, J. K., Yu, S., Yang, K., Mohan Kumar, S., Yu, Y. H., et al. (2006). Application of comparative genomics in developing molecular markers tightly linked to the virus resistance gene *Rsv4* in soybean. *Genome, 49*, 380–388.

The International Committee on Taxonomy of Viruses (ICTV) (2012). The viruses. In A. M. Q. King, M. J. Adams, E. B. Carstens, & E. J. Lefkowitz (Eds.), *Virus taxonomy—Ninth report of the international committee on taxonomy of viruses* (pp. 21–1197). United Kingdom: Elsevier/Academic Press.

ICTVdB Management. (2006). *ICTVdB–the universal virus database, version 4*. New York: Columbia University.

Ilyas, M., Qazi, J., Mansoor, S., & Briddon, R. W. (2010). Genetic diversity and phylogeography of begomoviruses infecting legumes in Pakistan. *The Journal of General Virology, 91*, 2091–2101.

Islam, M. N., Sony, S. K., & Borna, R. S. (2012). Molecular characterization of *Mungbean yellow mosaic disease* and coat protein gene in mungbean varieties of Bangladesh. *Plant Tissue Culture and Biotechnology, 22*, 73–81.

Ivo, N. L., Nascimento, C. P., Vieira, L. S., Campos, F. A. P., & Aragao, F. J. L. (2008). Biolistic-mediated genetic transformation of cowpea (*Vigna unguiculata*) and stable Mendelian inheritance of transgenes. *Plant Cell Reports, 27*, 1475–1483.

Iwo, G. A., & Olorunju, P. E. (2009). Yield stability and resistance to leaf spot diseases and rosette in groundnut. *Czech Journal of Genetics and Plant Breeding, 45*, 18–25.

Jain, R. K., Pappu, S. S., Pappu, H. R., Culbreath, A. K., & Todd, J. W. (1998). Rapid molecular diagnosis of tomato spotted wilt tospovirus (TSWV) infection: Application to peanut and other TSWV susceptible greenhouse and row crops. *Plant Disease, 82*, 900–904.

Jain, R. K., Vemana, K., & Sudeep, B. (2008). *Tobacco streak virus*-an emerging virus in vegetable crops. In G. P. Rao, P. L. Kumar, & R. J. Holguin-Peña (Eds.), *Vegetable and pulse crops: Vol. 3. Characterization, diagnosis and management of plant viruses* (pp. 203–212). USA: Studium Press, LLC.

Jayan, P. K., & Maya, C. N. (2001). Studies on the germplasm collection of horse gram (*Macrotyloma uniflorum* (Lam.) Verdc.). *Indian Journal of Plant Genetic Resources, 14*, 43–47.

Jellis, G. J., Bond, D. A., & Boulton, R. E. (1998). Diseases of faba bean. In D. J. Allen, & J. M. Lenne (Eds.), *The pathology of food and pasture legumes* (pp. 371–422). Wallingford, United Kingdom: CABI International.

Jeong, S. C., Kristipati, S., Hayes, A. J., Maughan, P. J., Noffsinger, S. L., Gunduz, I., et al. (2002). Genetic and sequence analysis of markers tightly linked to the *Soybean mosaic virus* resistance gene *Rsv3*. *Crop Science, 42*, 265–270.

Jinendra, B., Tamaki, K., Kuroki, S., Vassileva, M., Yoshida, S., & Tsenkova, R. (2010). Near infrared spectroscopy and aquaphotomics: Novel approach for rapid in vivo diagnosis of virus infected soybean. *Biochemical and Biophysical Research Communications, 397*, 685–690.

Johansen, I. E., Keller, K. E., Dougherty, W. G., & Hampton, R. O. (1996). Biological and molecular properties of a pathotype P-1 and a pathotype P-4 isolate of *Pea seed-borne mosaic virus*. *The Journal of General Virology, 77*, 1329–1333.

Jones, R. A. (2009). Plant virus emergence and evolution: Origins, new encounter scenarios, factors driving emergence, effects of changing world conditions, and prospects for control. *Virus Research, 141*, 113–130.

Jones, R. A. C., & Barbetti, M. J. (2012). Influence of climate change on plant disease infections and epidemics caused by viruses and bacteria. *CAB Reviews: Perspectives in Agriculture, Veterinary Science, Nutrition and Natural Resources*, 7(22), 1–31.

Jones, A. T., Kumar, P. L., Saxena, K. B., Kulkarni, N. K., Muniyappa, V., & Waliyar, F. (2004). Sterility mosaic disease—The "green plague" of pigeonpea: Advances in understanding the etiology, transmission and control of a major virus disease. *Plant Disease, 88*, 436–445.

Jyothi, V., Nagaraju, N., Padmaja, A. S., & Rangaswamy, K. T. (2013). Transmission, detection and management of yellow mosaic virus in polebean (*Phaseolus vulgaris* L.). *Journal of Mycology and Plant Pathology, 43*, 229–236.

Kaiser, W. J. (1973). Biology of bean yellow mosaic and pea leaf roll viruses affecting *Vicia faba* in Iran. *Journal of Phytopathology, 78*, 253–263.

Kamaal, N., Akram, M., Pratap, A., & Yadav, P. (2013). Characterization of a new begomovirus and a beta satellite associated with the leaf curl disease of French bean in northern India. *Virus Genes, 46*, 120–127.

Kanakala, S., Sakhare, A., Verma, H. N., & Malathi, V. G. (2012). Infectivity and the phylogenetic relationship of a mastrevirus causing chickpea stunt disease in India. *European Journal of Plant Pathology, 135*, 429–438.

Kanakala, S., Verma, H. N., Vijay, P., Saxena, D. R., & Malathi, V. G. (2013). Response of chickpea genotypes to *Agrobacterium*-mediated delivery of *Chickpea chlorotic dwarf virus*

(CpCDV) genome and identification of resistance source. *Applied Microbiology and Biotechnology, 97*, 9491–9501.

Karkashian, J., Ramos-Reynoso, E. D., Maxwell, D. P., & Ramirez, P. (2011). Begomoviruses associated with bean golden mosaic disease in Nicaragua. *Plant Disease, 95*, 901–906.

Karthikeyan, A., Sudha, M., Pandiyan, M., Senthil, N., Shobana, V. G., & Nagarajan, P. (2011). Screening of MYMV resistant mungbean (*Vigna radiata* L. Wilczek) progenies through agroinoculaiton. *International Journal of Plant Pathology, 2*, 115–125.

Kasai, M., & Kanazawa, A. (2012). RNA silencing as a tool to uncover gene function and engineer novel traits in soybean. *Breeding Science, 61*, 468–479.

Kelly, J. D., Gepts, P., Miklas, P. N., & Coyne, D. P. (2003). Tagging and mapping of genes and QTL molecular marker-assisted selection for traits of economic importance in bean and cowpea. *Field Crops Research, 82*, 135–154.

Khan, M. H., Tyagi, S. D., & Dar, Z. A. (2013). Screening of soybean (*Glycine max* (L.) Merrill) genotypes for resistance to rust, yellow mosaic and pod shattering. In H. A. El-Shemy (Ed.), *Soybean-pest resistance*. Rijeka, Croatia: InTech. ISBN: 978-953-51-0978-5. http://dx.doi.org/10.5772/54697.

Kim, Y., Roh, J., Kim, M., Im, D., & Hur, I. (2000). Seasonal occurrence of aphids and selection of insecticides for controlling aphids transmitting *Soybean mosaic virus*. *Korean Journal of Crop Science, 45*, 353–355.

Kitsanachandee, R., Somta, P., Chatchawankanphanich, O., Akhtar, K. P., Shah, T. M., Nair, R. M., et al. (2013). Detection of quantitative trait loci for *Mungbean yellow mosaic India virus* (MYMIV) resistance in mungbean (*Vigna radiata* (L) Wilczek) in India and Pakistan. *Breeding Science, 63*, 367–373.

Kosaka, Y., & Fukunishi, T. (1994). Applications of cross-protection to the control of black soybean mosaic disease. *Plant Disease, 78*, 339–341.

Kraft, J. M., & Pfleger, F. L. (2001). *Compendium of pea diseases and pests* (2nd ed.). Saint Paul, MN: APS Press.

Kulkarni, N. K., Reddy, A. S., Kumar, P. L., Vijaynarasimha, J., Rangaswamy, K. T., Muniyappa, V., et al. (2003). Broad-based resistance to pigeonpea sterility mosaic disease in the accessions of *Cajanus scarabaeoides*. *Indian Journal of Plant Protection, 31*, 6–11.

Kumar, P. L., Jones, A. T., & Reddy, D. V. R. (2003). A novel mite-transmitted virus with a divided RNA genome closely associated with pigeonpea sterility mosaic disease. *Phytopathology, 93*, 71–81.

Kumar, P. L., Jones, A. T., & Waliyar, F. (2004). Biology, etiology and management of pigeonpea sterility mosaic disease. *Annual Review of Plant Pathology, 3*, 77–100.

Kumar, P. L., Jones, A. T., & Waliyar, F. (2008). Virus diseases of pegionpea. In G. P. Rao, P. L. Kumar, & R. J. Holguin-Peña (Eds.), *Vegetable and pulse crops: Vol. 3. Characterization, diganosis and management of plant viruses* (pp. 235–258). USA: Studium Press LLC.

Kumar, P. L., Jones, A. T., Waliyar, F., Sreenivasulu, P., Muniyappa, V., Latha, T. K. S., et al. (2007). Pigeonpea sterility mosaic disease. In P. L. Kumar, & F. Waliyar (Eds.), *Diagnosis and detection of viruses infecting ICRISAT mandate crops* (pp. 15–21). Patancheru, India: ICRISAT.

Kumar, P. L., Kumari, S. M. G., & Waliyar, F. (2008). Virus diseases of chickpea. In G. P. Rao, P. L. Kumar, & R. J. Holguin-Peña (Eds.), *Vegetable and pulse crops: Vol. 3. Characterization, diagnosis and management of plant viruses* (pp. 213–234). USA: Studium Press LLC.

Kumar, P. L., Latha, T. K. S., Kulkarni, N. K., Raghavendra, N., & Saxena, K. B. (2005). Broad based resistance to pigeonpea sterility mosaic disease in wild relatives of pigeonpea (*Cajanus: Phaseoleae*). *Annals of Applied Biology, 146*, 371–379.

Kumar, P. L., Prasada Rao, R. D. V. J., Reddy, A. S., Madhavi, K. J., Anitha, K., & Waliyar, F. (2008). Emergence and spread of *Tobacco streak virus* menace in Indian and control strategies. *Indian Journal of Plant Protection, 36*, 1–8.

Kumar, B., Talukdar, A., Verma, K., Girmilla, V., Bala, I., Lal, S. K., et al. (2014). Screening of soybean (*Glycine max* (L.)) genotypes for yellow mosaic virus (YMV) disease and their molecular characterization using RGA and SSRs markers. *Australian Journal of Crop Science, 8*, 27–34.
Kumari, S. G., Attar, N., Mustafayev, E., & Akparov, Z. (2009). First report of *Faba bean necrotic yellows virus* affecting legume crops in Azerbaijan. *Plant Disease, 93*, 1220.
Kumari, S. G., & Makkouk, K. M. (1995). Variability among twenty lentil genotypes in seed transmission rated and yield loss induced by pea seed-borne mosaic potyvirus infection. *Phytopathologia Mediterranea, 34*, 129–132.
Kumari, S. G., & Makkouk, K. M. (1996). Inactivation of broad bean stain comovirus in lentil seeds by dry heat treatment. *Phytopathologia Mediterranea, 35*, 124–126.
Kumari, S. G., & Makkouk, K. M. (2003). Differentiation among Bean leaf roll virus susceptible and resistant lentil and faba bean genotypes on the basis of virus movement and multiplication. *Journal of Phytopathology, 151*, 19–25.
Kumari, S. G., & Makkouk, K. M. (2007). Virus diseases of faba bean (*Vicia faba* L) in Asia and Africa. *Plant Viruses, 1*, 93–105.
Kumari, S. G., Makkouk, K. M., & Attar, N. (2006). An improved antiserum for sensitive serologic detection of chickpea chlorotic dwarf virus. *Journal of Phytopathology, 154*, 129–133.
Kumari, S. G., Makkouk, K. M., & Ismail, I. D. (1996). Variation among isolates of two viruses affecting lentils their affect on yield and seed transmissiblity. *Arab Journal of Plant Protection, 14*, 81–85.
Kumari, S. G., Makkouk, K. M., Loh, M. H., Negassi, K., Tsegay, S., Kidane, R., et al. (2008). Viral diseases affecting chickpea crops in Eritrea. *Phytopathologia Mediterranea, 47*, 42–49.
Kumari, S. G., Najar, A., Attar, N., Loh, M. H., & Vetten, H. J. (2010). First report of Beet mosaic virus infecting chickpea (*Cicer arietinum*) in Tunisia. *Plant Disease, 94*, 1068.
Lewis, G., Schrire, B., MacKinder, B., & Lock, M. (Eds.), (2005). *Legumes of the world*. UK: Royal Botanical Gardens.
Li, Z., Jarret, R. L., & Demski, J. W. (1997). Engineered resistance to tomato spotted wilt virus in transgenic peanut expressing the viral nucleocapsid gene. *Transgenic Research, 6*, 297–305.
Li, H. C., Zhi, H. J., Gai, J. Y., Guo, D. Q., Wang, Y. W., Li, K., et al. (2006). Inheritance and gene mapping of resistance to *Soybean mosaic virus* strain SC14 in soybean. *Journal of Integrative Plant Biology, 48*, 1466–1472.
Lim, S. M. (1985). Resistance to *Soybean mosaic virus* in soybeans. *Phytopathology, 75*, 199–201.
Loebenstein, G., & Thottappilly, G. (Eds.). (2003). *Virus and virus-like diseases of major crops in developing countries* (p. p. 800). Dordrecht: Kluwer Academic Publishers.
Lynch, R. E., Demski, J. W., Branch, W. D., Holbrook, C. C., & Morgan, L. W. (1988). Influence of *Peanut stripe virus* on growth, yield and quality of Florunner peanut. *Peanut Science, 15*, 47–52.
Ma, G., Chen, P., Buss, G. R., & Tolin, S. A. (2004). Genetics of resistance to two strains of *Soybean mosaic virus* in differential soybean genotypes. *The Journal of Heredity, 95*, 322–326.
Mahdy, A. M. M., Fawzy, R. N., Hafez, M. A., Mohamed, H. A. N., & Shahwan, E. S. M. (2007). Inducing systemic resistance against Bean yellow mosaic potyvirus using botanical extracts. *Egyptian Journal of Virology, 4*, 129–145.
Maiti, S., Basak, J., Kundagrami, S., Kundu, A., & Pal, A. (2011). Molecular marker-assisted genotyping of mungben yellow mosaic India virus resistant germplasm of mungbean and urdbean. *Molecular Biotechnology, 47*, 95–104.

Makkouk, K. M., Bos, L., Azzam, O. I., Kumari, S. G., & Rizkallah, A. (1988). Survey of viruses affecting faba bean in six Arab countries. *Arab Journal of Plant Protection, 6*, 53–61.

Makkouk, K. M., Bos, L., Rizkallah, A., Azzam, O. I., & Katul, L. (1988). Broad bean mottle virus: Identification, serology, host range and occurrence on faba bean (*Vicia faba*) in West Asia and North Africa. *Netherlands Journal of Plant Pathology, 94*, 195–212.

Makkouk, K. M., & Kumari, S. G. (1990). Variability among 19 lentil genotypes I seed transmission rates and yield loss induced by broad bean stain virus infection. *LENS Newsletter, 17*, 31–33.

Makkouk, K. M., & Kumari, S. G. (2001). Reduction of incidence of the three persistently transmitted aphid-borne viruses affecting legume crops by seed treatment with the insecticide imidacloprid (Gaucho). *Crop Protection, 20*, 433–437.

Makkouk, K. M., & Kumari, S. G. (2009). Epidemiology and integrated management of persistently transmitted aphid-borne viruses of legume and cereal crops in West Asia and North Africa. *Virus Research, 141*, 209–918.

Makkouk, K. M., Kumari, S. G., Huges, J. D. A., Muniyappa, V., & Kulkarni, N. K. (2003). Other legumes. In G. Loebenstein, & G. Thottappilly (Eds.), *Virus and virus-like disease of major crops in developing countries* (pp. 447–476). Dordercht: Kluwer Academic Publisher.

Makkouk, K. M., Kumari, S., Sarker, A., & Erskine, W. (2001). Registration of six lentil germplasm lines with combined resistance to viruses. *Crop Science, 41*, 931–932.

Makkouk, K. M., Kumari, S. G., & van Leur, J. A. G. (2002). Screening and selection of faba bean (*Vicia faba* L.) germplasm resistant to *Bean leaf roll virus*. *Australian Journal of Agricultural Research, 53*, 1077–1082.

Makkouk, K., Pappu, H., & Kumari, S. G. (2012). Virus diseases of peas, beans, and faba bean in the Mediterranean region. *Advances in Virus Research, 84*, 367–402.

Malapi-Nelson, M., Wen, R.-H., Ownley, B. H., & Hajimorad, M. R. (2009). Co-infection of soybean with *Soybean mosaic virus* and *Alfalfa mosaic virus* results in disease synergism and alteration in accumulation level of both viruses. *Plant Disease, 93*, 1259–1264.

Malathi, V. G., & John, P. (2008). Gemini viruses infecting legumes. In G. P. Rao, P. L. Kumar, & R. J. Holguin–Peña (Eds.), *Vegetable and pulse crops: Vol. 3. Characterization, diagnosis and management of plant viruses* (pp. 97–123). USA: Studium Press LLC.

Malek, M. A., Rahman, L., Raffi, M. Y., & Salam, M. A. (2013). Selection of a high yielding soybean variety, Binasoybean-2 from collected germplasm. *Journal of Food Agriculture and Environment, 11*, 545–547.

Mallikarjuna, N., Senapathy, S., Jadhav, D. R., Saxena, K. B., Sharma, H. C., Upadhyaya, H. D., et al. (2011). Progress in the utilization of *Cajanus platycarpus* (Benth.) Maesen in pigeonpea improvement. *Plant Breeding, 130*, 507–514.

Mandal, B., Jain, R. K., Krishnareddy, M., Krishna Kumar, N. K., Ravi, K. S., & Pappu, H. R. (2012). Emerging problems of Tospoviruses (*Bunyaviridae*) and their management in the Indian Subcontinent. *Plant Disease, 96*, 468–479.

Manohar, S. K., Dollet, M., Dubern, J., & Gargani, D. (1995). Studies on variability of *Peanut clump virus*: Symptomatology and serology. *Journal of Phytopathology, 143*, 233–238.

Maruthi, M. N., Manjunatha, B., Rekha, A. R., Govindappa, M. R., Colvin, J., & Muniyappa, V. (2006). *Dolichos yellow mosaic virus* belongs to a distinct lineage of old world begomoviruses; its biological and molecular properties. *Annals of Applied Biology, 149*, 187–195.

McDonald, D., Reddy, D. V. R., Sharma, S. B., Mehan, V. K., & Subrahmanayam, P. (1998). Diseases of groundnut. In *The pathology of food and pasture legumes* (pp. 63–124). Wallingford, United Kingdom: CAB International in Association with the International Crops Research Institute for the Semi-Arid Tropics (ICRISAT).

Mehta, R., Radhakrishnan, T., Kumar, A., Yadav, R., Dobaria, J. R., Thirumalaisamy, P. P., et al. (2013). Coat protein-mediated transgenic resistance of peanut (*Arachis hypogaea* L.) to peanut stem necrosis disease through *Agrobacterium*-mediated genetic transformation. *Indian Journal of Virology, 24*, 205–213.

Melgarejo, T. A., Lehtonen, M. T., Fribourg, C. E., Rannali, M., & Valkonen, J. P. T. (2007). Strains of BCMV and BCMNV characterized from lima bean plants affected by deforming mosaic disease in Peru. *Archives of Virology, 152*, 1941–1949.

Mielke-Ehret, N., & Mühlbach, H. P. (2012). Emaravirus: A novel genus of multipartite, negative strand RNA plant viruses. *Viruses, 4*, 1515–1536.

Miklas, P. N., Kelly, J. D., Beebe, S. E., & Blair, M. W. (2006). Common bean breeding for resistance against biotic and abiotic stresses: From classical MAS breeding. *Euphytica, 147*, 105–131.

Mikoshiba, Y., Honda, K., Kanematsu, S., & Fujisawa, I. (1994). Differentiation of *Soybean dwarf virus* strains by monoclonal antibodies. In *Tenth international congress of virology, Jerusalem, Israel* (abstracts, 210).

Mishra, A., John, P., & Mishra, A. A. (2008). Potyviruses infecting grain legumes. In G. P. Rao, P. L. Kumar, & R. J. Holguin-Peña (Eds.), *Vegetable and pulse crops: Vol. 3. Characterization, diagnosis and management of plant viruses* (pp. 63–95). USA: Studium Press LLC.

Mondol, M. E. A., Rahman, H., Rashid, M. H., Hossain, M. A., & Islam, M. M. (2013). Screening of mungbean germplasm for resistance to mungbean yellow mosaic virus. *International Journal of Sustainable Crop Production, 8*, 11–15.

Monfort, W. S., Culbreath, A. K., Stevenson, K. L., Brenneman, T. B., & Perry, C. D. (2007). Use of resistant cultivars and reduced fungicide inputs for disease management in strip-tillage and conventional tillage systems. *Plant Health Progress*. http://dx.doi.org/10.1094/PHP-2007-0614-01-RS.

Morales, F. J. (2003). Common bean. In G. Loebenstein, & G. Thottappilly (Eds.), *Virus and virus-like diseases of major crops in developing countries* (pp. 425–445). The Netherlands: Kluwer Academic Publishers.

Morales, F. J. (2006). History and current distribution of begomoviruses in Latin America. *Advances in Virus Research, 67*, 127–162.

Morroni, M., Thompson, J. R., & Tepfer, M. (2008). Twenty years of transgenic plants resistant to *Cucumber mosaic virus*. *Molecular Plant-Microbe Interactions, 21*, 675–684.

Muhire, B., Martin, D. P., Brown, J. K., Navas-Castillo, J., Moriones, E., Zerbini, M., et al. (2013). A genome-wide pairwise-identity-based proposal for the classification of viruses in the genus *Mastrevirus* (family *Geminiviridae*). *Archives of Virology, 158*, 1411–1424.

Mumtaz, H., Kumari, S. G., Mansoor, S., Martin, D. P., & Briddon, R. W. (2011). Analysis of the sequence of a dicot infecting mastrevirus (family *Geminiviridae*) originating from Syria. *Virus Genes, 42*, 422–428.

Muniyappa, V., Rajeshwari, R., Bharathan, N., Reddy, D. V. R., & Nolt, B. L. (1987). Isolation and characterization of a geminivirus causing yellow mosaic disease of horsegram (*Macrotyloma uniflorum* (Lam.) Verdc.) in India. *Indian Journal of Phytopathology, 119*, 81–87.

Muniyappa, V., Reddy, H. R., & Mustak Ali, T. M. (1978). Studies on the yellow mosaic disease of horsegram *Dolichos biflorus*, IV. Epidemiology of the disease. *Mysore Journal of Agricultural Science, 12*, 277–279.

Murant, A. F., Kumar, I. K., & Robinson, D. J. (1991). Current research on groundnut rosette at SCRI. In *Groundnut virus diseases in Africa. Summary proceedings of the consultative group meeting, Montpellier, France*. Patancheru, AP, India: ICRISAT.

Murphy, A. M., & Colucci, P. E. (1999). A tropical forage solution to poor quality ruminant diets: A review of *Lablab purpureus*. *Livestock Research for Rural Development, 11*. Cali, Colombia: CIPAV. http://www.cipav.org.co/Irrd/Irrd11/2/colu.htm (accessed June 2005).

Nahid, N., Amin, I., Briddon, R. W., & Mansoor, S. (2011). RNA interference-based resistance against a legume mastrevirus. *Virology Journal, 8*, 499.

Nahid, N., Amin, I., Mansoor, S., Rybicki, E. P., van der Walt, E., & Briddon, R. W. (2008). Two dicot-infecting mastreviruses (family *Geminiviridae*) occur in Pakistan. *Archives of Virology, 153*, 1441–1451.

Naidu, R. A., Kimmins, F. M., Deom, C. M., Subrahmanyam, P., Chiyembekeza, A. J., & van der Merwe, P. J. A. (1999). Groundnut rosette: A virus disease affecting groundnut production in Sub-Saharan Africa. *Plant Disease, 83*, 700–709.

Naidu, R. A., Miller, J. S., Mayo, M. A., Wesley, S. V., & Reddy, A. S. (2000). The nucleotide sequence of *Indian peanut clump virus* RNA 2: Sequence comparisons among pecluviruses. *Archives of Virology, 145*, 1857–1866.

Naidu, R. A., Robinson, D. J., & Kimmins, F. M. (1998). Detection of each of the causal agents of groundnut rosette disease in plants and vector aphids by RT-PCR. *Journal of Virological Methods, 76*, 9–18.

Naik, S. J. S., Gowda, M. B., Venkatesha, S. C., Ramappa, H. K., Pramila, C. K., Reena, G. A. M., et al. (2012). Molecular diversity among pigeonpea genotypes differing in response to pigeonpea sterility mosaic disease. *Journal of Food Legumes, 25*, 194–199.

Naimuddin, A. M. (2010). Detection of mixed infection of begomoviruses in cowpea and their molecular characterization based on CP gene sequences. *Journal of Food Legumes, 23*, 191–195.

Nair, R. M., Yang, R. Y., Easdown, W. J., Thavaraiah, D., Thavaraiah, P., Hughes, J. d. 'A., et al. (2013). Biofortification of mungbean (*Vigna radiata*) as a whole food to enhance human health. *Journal of the Science of Food and Agriculture, 93*, 1805–1813.

Nascimento, L. C. D., Pensuk, V., da Costa, N. P., de Assis Filho, F. M., Pio-Ribeiro, G., Deom, C. M., et al. (2006). Evaluation of peanut genotypes for resistance to *Tomato spotted wilt virus* by mechanical and thrips inoculation. *Pesquisa Agropecuária Brasileira, 41*, 937–942.

Navas-Castillo, J., Fiallo-Olive, E., & Sanchez-Campos, S. (2011). Emerging virus diseases transmitted by whiteflies. *Annual Review of Phytopathology, 49*, 219–248.

Nene, Y. L. (1995). Sterility mosaic of pigeonpea: The challenge continues. *Indian Journal of Mycology and Plant Pathology, 25*, 1–11.

Nene, Y. L., & Reddy, M. V. (1987). Chickpea disease and their control. In M. C. Saxena, & K. B. Singh (Eds.), *The chickpea, viral disease* (pp. 233–270). Wallingford, Oxon: CAB International, ICARD.

Nigam, S. N., & Bock, K. R. (1990). Inheritance of resistance to *Groundnut rosette virus* in groundnut (*Arachis hypogaea* L.). *Annals of Applied Biology, 117*, 553–560.

Nischwitz, C., Mullis, S. W., Gitaitis, R. D., & Csinos, A. S. (2006). First report of *Tomato spotted wilt virus* in soybean (*Glycine max*) in Georgia. *Plant Disease, 90*, 524.

Nolt, B. L., Rajeshwari, R., Reddy, D. V. R., Barathan, N., & Manohar, S. K. (1988). Indian *Peanut clump virus* isolates: Host range, symptomatology, serological relationships and some physical properties. *Phytopathology, 78*, 310–313.

Ntare, B. R., Olorunju, P. E., & Waliyar, F. (2002). Progress in combating groundnut rosette disease in West and Central Africa: An overview. In F. Waliyar, & M. Adomou (Eds.), *Summary proceedings of the seventh ICRISAT regional groundnut meeting for Western and Central Africa, December 6–8, 2000, Cotonou, Benin* (pp. 39–41). Patancheru, AP, India: International Crops Research Institute for the Semi-Arid Tropics.

Odedara, O. O., Hughes, J. D. A., Odebode, A. C., & Odu, B. O. (2008). Multiple virus infections of lablab [*Lablab purpureus*(L.) sweet] in Nigeria. *Journal of General Plant Pathology, 74*, 322–325.

Ogunsola, K. E., Fatokun, C. A., Boukar, O., Ilori, C. O., & Kumar, P. L. (2010). Characterizing genetics of resistance to multiple virus infections in cowpea (*Vigna unguiculata* L. Walp). In *Abstracts: fifth world cowpea conference: Improving livelihoods in the cowpea value chain through advancement of science, September 27–October 1, 2010, Saly, Senegal* (pp. 26–27), Nigeria: IITA.

Ojuederie, O. B., Odu, B. O., & Ilori, C. O. (2009). Serological detection of seed-borne viruses in cowpea regenerated germplasm using protein a sandwich enzyme linked immunosorbent assay. *African Crop Science Journal, 17*, 125–132.

Olorunju, P. E., Kuhn, C. W., Demski, J. W., Misari, S. M., & Ansa, O. A. (1991). Disease reactions and yield performance of peanut genotypes under groundnut rosette and rosette free field environments. *Plant Disease, 75*, 1269–1273.

Olorunju, P. E., Kuhn, C. W., Demski, J. W., Misari, S. M., & Ansa, O. A. (1992). Inheritance of resistance in peanut to mixed infections of *Groundnut rosette virus* (GRV) and groundnut rosette assistor virus and a single infection of GRV. *Plant Disease, 76*, 95–100.

Olorunju, P. E., & Ntare, B. R. (2003). Combatting viruses and virus diseases of groundnut through the use of resistant varieties: A case study of Nigeria. In J. d. 'A. Hughes, & B. Odu (Eds.), *Plant virology in sub Saharan Africa* (pp. 189–202). Ibadan, Nigeria: IITA.

Orawu, M., Melis, R., Laing, M., & Derera, J. (2013). Genetic inheritance of resistance to *Cowpea aphid-borne mosaic virus* in cowpea. *Euphytica, 189*, 191–201.

Ortiz, V., Castro, S., & Romero, J. (2005). Optimization of RT-PCR for the detection of *Bean leaf roll virus* in plant hosts and insect vectors. *Journal of Phytopathology, 153*, 68–72.

Palukaitis, P., & Garcı´a-Arenal, F. (2003). Cucumoviruses. *Advances in Virus Research, 62*, 241–323.

Pappu, H. R., Jones, R. A. C., & Jain, R. K. (2009). Global status of tospovirus epidemics in diverse cropping systems: Successes achieved and challenges ahead. *Virus Research, 141*, 219–236.

Paul, P. C., Biswas, M. K., Mandal, D., & Pal, P. (2013). Studies on host resistance of mungbean against mungbean yellow mosaic virus in the agro-ecological condition of lateritic zone of West Bengal. *Bioscan, 8*, 583–587.

Pedersen, P., Grau, C., Cullen, E., Koval, N., & Hill, J. H. (2007). Potential for integrated management of soybean virus disease. *Plant Disease, 91*, 1255–1259.

Pensuk, V., Daengpluang, N., Wongkaew, S., Jogloy, S., & Patanothai, A. (2002). Evaluation of screening procedures to identify peanut resistance to *Peanut bud necrosis virus* (PBNV). *Peanut Science, 29*, 47–51.

Pensuk, V., Jogloy, S., Wongkaew, S., & Patanothai, A. (2004). Generation means analysis of resistance to peanut bud necrosis caused by peanut bud necrosis tospovirus in peanut. *Plant Breeding, 123*, 90–92.

Popelka, J. C., Gollasch, S., Moore, A., Molvig, L., & Higgins, T. J. V. (2006). Genetic transformation of cowpea (*Vigna unguiculata* L.) and stable transmission of the transgenes to progeny. *Plant Cell Reports, 25*, 304–312.

Potter, J. L., Nakhla, M. K., Mejia, L., & Maxwell, D. P. (2003). PCR and DNA hybridization methods for specific detection of bean infecting begomoviruses in Americas and Caribbean. *Plant Disease, 87*, 1205–1212.

Prabhakar, M., Prasad, Y. G., Desai, S., Thirupathi, M., Gopika, K., Ramachandra Rao, G., et al. (2013). Hyperspectral remote sensing of yellow mosaic severity and associated pigment losses in *Vigna mungo* using multinomial logistic regression models. *Crop Protection, 45*, 132–140.

Prasada Rao, R. D. V. J., Reddy, A. S., Reddy, S. V., Thirumala-Devi, K., Chander Rao, S., Manoj Kumar, V., et al. (2003). The host range of *Tobacco streak virus* in India and transmission by thrips. *Annals of Applied Biology, 142*, 365–368.

Provvidenti, R., & Alconero, R. (1988). Inheritance of resistance to a third pathotype of *Pea seed-borne mosaic virus* in *Pisum sativum*. *The Journal of Heredity, 79*, 76–77.

Qazi, J., Ilyas, M., Mansoor, S., & Briddon, R. W. (2007). Legume yellow mosaic viruses: Genetically isolated begomoviruses. *Molecular Plant Pathology, 8*, 343–348.

Qin, H., Feng, S., Chen, C., Guo, Y., Knapp, S., Culbreath, A., et al. (2012). An integrated genetic linkage map of cultivated peanut (*Arachis hypogaea* L.) constructed from two RIL populations. *Theoretical and Applied Genetics, 124*, 653–664.

Raj, S. K., Aslam, N., Srivastava, K. M., & Singh, B. P. (1988). Associaton of geminivirus-like particles with yellow mosaic disease of *Dolichos lablab. Current Science, 58*, 813–814.

Raj, S. K., Khan, M. S., Snehi, S. K., Srivastava, S., & Singh, H. B. (2006). First report of *Tomato leaf curl Karnataka virus* infecting soybean in India. *Plant Pathology, 55*, 817.

Rajeswari, R., Murant, A. F., & Massalski, P. R. (1987). Use of monoclonal antibody to *Potato leaf roll virus* for detecting groundnut rosette assistor virus by ELISA. *Annals of Applied Biology, 111*, 353–358.

Rangaswamy, K. T., Muniyappa, V., Kumar, P. L., Saxena, K. B., Byregowda, M., Raghavendra, N., et al. (2005). ICP 7035—A sterility mosaic resistant vegetable and grain purpose pigeonpea variety. *Journal of SAT Agricultural Research, 1*, 1–3.

Rao, G. P., Kumar, P. L., & Holguin- Peña, R. J. H. (Eds.). (2008). In *Vegetable and pulse crops: Vol. 3. Characterization diagnosis and management of plant viruses*. USA: Studium Press LLC, 408 p.

Rathi, Y. P. S. (1997). Temik treatment of pigeonpea seed for prevention of sterility mosaic. *Acta Botanica Indica*, 7, 90–91.

Rebedeaux, P. F., Gaska, J. M., Kurtzweil, N. C., & Grau, C. R. (2004). Seasonal progression and agronomic impact of *Tobacco streak virus* on soybean in Wisconsin. *Plant Disease, 89*, 391–396.

Reddy, D. V. R. (1998). Control measures for the economically important peanut viruses. In A. Hadidi, R. K. Khetarpal, & A. Koganezawa (Eds.), *Plant virus disease control* (pp. 541–546). USA: American Phytopathological Society Press.

Reddy, K. S. (2009). A new mutant for yellow mosaic virus resistance in mungbean (*Vigna radiata* L. Wilczek) variety SML-668 by recurrent gamma-ray irradiation. In Q. Y. Shu (Ed.), *Induced plant mutation in the genomics era* (pp. 361–362). Rome: Food and Agriculture Organization of the United Nations.

Reddy, D. V. R., Amin, P. W., McDonald, D., & Ghanekar, A. M. (1983). Epidemiology and control of groundnut bud necrosis and other diseases of legumes crops in India caused by *Tomato spotted wilt virus*. In R. T. Plumb, & J. M. Thresh (Eds.), *Plant virus epidemiology* (pp. 93–102). Oxford, UK: Blackwell Scientific Publications.

Reddy, D. V. R., Bragard, C., Sreenivasulu, P., & Delfosee, P. (2008). Pecluviruses. In B. W. J. Mahy, & M. Van Regenmortel (Eds.), *Desk encylopedia of plant and fungal virology* (pp. 257–263). New York: Academic Press.

Reddy, D. V. R., Buiel, A. A. M., Satyanarayana, T., Dwivedi, S. L., Reddy, A. S., Ratna, A. S., et al. (1995). Peanut bud necrosis disease: An overview. In A. A. M. Buiel, J. E. Parlevliet, & J. M. Lenne (Eds.), *Recent studies on peanut bud necrosis disease* (pp. 3–7). Patancheru, AP, India: ICRISAT.

Reddy, S. V., & Kumar, P. L. (2004). Transmission and properties of a new luteovirus associated with chickpea stunt disease in India. *Current Science, 86*, 1157–1161.

Reddy, A. S., Prasada Rao, R. D. V. J., Thirumala-Devi, K., Reddy, S. V., Subrahmanayam, K., Satyanarayana, T., et al. (2001). First record of tobacco streak ilavirus occurrence on peanut in India. *Plant Disease, 86*, 173–178.

Reddy, M. V., Raju, T. N., & Lenne, J. M. (1998). Diseases of pigeonpea. In D. J. Allen, & J. M. Lenne (Eds.), *The pathology of food and pasture legumes* (pp. 517–558). Wallingford, UK: CAB International.

Reddy, M. V., Raju, T. N., Sharma, S. B., Nene, Y. L., McDonald, D., Pande, S., et al. (2012). *Handbook of pigeonpea diseases (revised). Information bulletin no. 42: Technical report.* Patancheru, AP, India: International Crops Research Institute for the Semi-Arid Tropics.

Reddy, A. S., Reddy, L. M., Mallikarjuna, N., Abdurahman, M. D., Reddy, Y. V., Bramel, P. J., et al. (2000). Identification of resistance to *Peanut bud necrosis virus* (PBNV) in wild *Arachis* germplasm. *Annals of Applied Biology*, *137*, 135–139.
Reddy, M. V., Sharma, S. B., & Nene, Y. L. (1990). Pigeonpea: Disease management. In Y. L. Nene, S. D. Hall, & V. K. Sheila (Eds.), *The pigeonpea* (pp. 303–347). Wallingford, UK: CAB International.
Reddy, M. V., Sheila, V. K., Murthy, A. K., & Padma, P. (1995). Mechanism of resistance to *Aceria cajani* in pigeonpea. *International Journal of Tropical Plant Diseases*, *13*, 51–57.
Reddy, D. V. R., Sudarshana, M. R., Fuchs, M., Rao, N. C., & Thottappilly, G. (2009). Genetically engineered virus-resistant plants in developing countries: Current status and future prospects. *Advances in Virus Research*, *75*, 185–220.
Reddy, D. V. R., & Thirumala-Devi, K. (2003). Peanuts. In G. Loebenstein, & G. Thottappilly (Eds.), *Viruses and virus-like diseases of major crops in developing countries* (pp. 397–423). Dordrecht: Kluwer Academic Publishers.
Reddy, C., Tonapi, V. A., Navi, S. S., & Jayaram, R. (2005). Influence of plant age on infection and symptomological studies on urdbean leaf crinkle virus in urdbean (*Vigna mungo*). *International Journal of Agricultural Science*, *1*, 1–6.
Resende, R. d. O., de Avila, A. C., Goldbach, R. W., & Peters, D. (1991). Comparison of polyclonal antisera in the detection of *Tomato spotted wilt virus* using double antibody sandwich and cocktail ELISA. *Journal of Phytopathology*, *132*, 46–56.
Rey, M. E. C., Ndunguru, J., Berrie, L. C., Paximadis, M., Berry, S., Cossa, N., et al. (2012). Diversity of dicotyledenous-infecting geminiviruses and their associated DNA molecules in Southern Africa, including the South-West Indian Ocean Islands. *Viruses*, *4*, 1753–1791.
Rice, D. J., German, T. L., Mau, R. F. L., & Fujimoto, F. M. (1990). Dot blot detection of tomato spotted wilt virus RNA in plant and thrips by cDNA clones. *Plant Disease*, *74*, 274–276.
Riley, D. G., Joseph, S. V., Srinivasan, R., & Diffie, S. (2011). Thrips vectors of tosposviruses. *Journal of Integrated Pest Management*, *1*, 1–10. http://dx.doi.org/10.1603/IPM10020.
Rodrigues, M. A., Santos, C. A. F., & Santana, J. R. F. (2012). Mapping of AFLP loci linked to tolerance to cowpea golden mosaic virus. *Genetics and Molecular Research*, *11*, 3789–3797.
Rodriguez-Pardina, P. E., Zerbini, F. M., & Ducasse, D. A. (2006). Genetic diversity of begomoviruses infecting soybean, bean and associated weeds in North-Western Argentina. *Fitopatologia Brasileira*, *31*, 342–348.
Rojas, M. R., & Gilbertson, R. L. (2008). Emerging plant viruses: A diversity of mechanisms and opportunities. In J. Marilyn Roossinck (Ed.), *Plant virus evolution* (pp. 27–51). Berlin: Springer (Chapter 3).
Roman, M. A., Castaneda, A. M., Sanchez, J. C. A., Munoz, C. G., & Beaver, J. S. (2004). Inheritance of normal pod development in bean golden yellow mosaic resistant common bean. *Journal of American Society for Horticultural Science*, *129*, 549–552.
Ronco, A. E., Dal Bo, E., Ghiringhelli, P. D., Medrano, C., Romanowski, V., Sarachu, A. N., et al. (1989). Cloned cDNA probes for the detection of *Tomato spotted wilt virus*. *Phytopathology*, *79*, 1309–1313.
Roy, A., & Malathi, V. G. (2004). Development of specific detection technique for *Cowpea golden mosaic virus*. *Journal of Plant Biochemistry and Biotechnology*, *13*, 131–133.
Salam, S. A., Patil, M. S., & Byadgi, A. S. (2009). IDM of mungbean yellow mosaic disease. *Annals of Plant Protection Sciences*, *17*, 157–160.
Salem, N. M., Ehlers, J. D., Roberts, P. A., & Ng, J. C. K. (2010). Biological and molecular diagnosis of seed-borne viruses in cowpea germplasm of geographically diverse sub-Saharan origins. *Plant Pathology*, *59*, 773–784.

Sandhu, T. S., Brar, J. S., Sandhu, S. S., & Verma, M. M. (1985). Inheritance of resistance to mungbean yellow mosaic virus in green gram. *Journal of Research Punjab Agricultural University*, *22*, 607–611.

Sastry, K. S. (2013). *Seed-borne plant virus diseases*. India: Springer, (327 p). http://dx.doi.org/10.1007/978-81-322-0813-6 2.

Sastry, K. S., & Zitter, T. A. (2014). Management of virus and viroid diseases of crops in the tropics. In *Plant virus and viroid diseases in the tropics: Vol. 1.* (pp. 149–480). Springer Science+Business Media B.V. (Chapter 2). http://dx.doi.org/10.1007/978-94-007-7820-7_2.

Satyanarayana, T., Gowda, S., Lakshminaryana Reddy, K., Mitchell, S. E., Dawson, W. O., & Reddy, D. V. R. (1998). *Peanut yellow spot virus* is a member of a new serogroup of tospovirus genus based on small (S) RNA sequence and organization. *Archives of Virology*, *143*, 353–364.

Sayasa, T., Iwasaki, M., & Yamamoto, T. (1993). Seed transmission of *Bean yellow mosaic virus* in broad bean (*Vicia faba*). *Annals of the Phytopathological Society of Japan*, *59*, 559–562.

Schmutz, J., Cannon, S. B., Schlueter, J., Ma, J., Mitros, T., Nelson, W., et al. (2010). Genome sequence of the palaeopolyploid soybean. *Nature*, *463*, 178–183.

Schwartz, H. F., Steadman, J. R., Hall, R., & Foster, R. L. (Eds.), (2005). *Compendium of bean diseases* (2nd ed.). St. Paul, MN, USA: American Phytopathological Society.

Schwinghamer, M., Knights, T., Breeder, C., & Moore, K. (2009). Virus control in chickpea-special considerations. *Australian Pulse Bulletin*. http://www.pulseaus.com.au/pdf/Virus%20Contol%20in%20Chickpea.pdf.

Seo, Y. S., Gepts, P., & Gilbertson, R. L. (2004). Genetics of resistance to the geminivirus, *Bean dwarf mosaic virus*, and the role of the hypersensitive response in common bean. *Theoretical and Applied Genetics*, *108*, 786–793.

Seo, J. K., Ohshima, K., Lee, H. G., Son, M., Choi, H. S., Lee, S. H., et al. (2009). Molecular variability and genetic structure of the population of *Soybean mosaic virus* based on the analysis of complete genome sequences. *Virology*, *393*, 91–103.

Seo, Y. S., Rojas, M. R., Lee, J. Y., Lee, S. W., Jeon, J. S., Ronald, P., et al. (2006). A viral resistance gene from common bean functions across plant families and is up-regulated in a non-specific manner. *Proceedings of the National Academy of Sciences of the United States of America*, *103*, 11856–11861.

Sethuraman, K., Manivannan, N., & Natarajan, S. (2001). Management of yellow mosaic disease of urdbean using neem products. *Legume Research*, *24*, 197–199.

Shad, N., Mughal, S. M., Farooq, K., & Bashir, M. (2006). Evaluation of mungbean germplasm for resistance against mungbean yellow mosaic begomovirus. *Pakistan Journal of Botany*, *38*, 449–457.

Shahid, M. S., Ikegami, M., & Natsuaki, K. T. (2012). First report of *Mungbean yellow mosaic India virus* on lima bean affected by yellow mosaic disease in Nepal. *Australasian Plant Disease Notes*, *7*, 85–89.

Sharma, S. R., & Varma, A. (1984). Effect of cultural practices on virus infection in cowpea. *Journal of Agronomy and Crop Science*, *153*, 23–31.

Sharma, P. N., Pathania, A., Kapil, R., Sharma, P., Sharma, O. P., Patial, M., et al. (2008). Resistance to *Bean common mosaic virus* strains and its inheritance in some Indian landraces of common bean. *Euphytica*, *164*, 173–180.

Sharma, M., Rameshwar, T., & Pande, S. (2013). Identification and validation of resistance to *Fusarium* wilt and sterility mosaic disease in pigeonpea. *Indian Journal of Plant Protection*, *41*, 141–146.

Sharma, M., Rathore, A., Mangala, U. N., Ghosh, R., Sharma, S., Upadhyay, H. D., et al. (2012). New sources of resistance to *Fusarium* wilt and sterility mosaic disease in a mini-core collection of pigeonpea germplasm. *European Journal of Plant Pathology*, *133*, 707–714.

Sharma, S., Upadhyaya, H. D., Varshney, R. K., & Gowda, C. L. (2013). Pre-breeding for diversification of primary gene pool and genetic enhancement of grain legumes. *Frontiers in Plant Science*, *4*, 309.

Sharma, K. K., Waliyar, F., Nigam, S. N., Lavanya, M., Reddy, A. S., & Swamy Krishna, T. (2006). Development and evaluation of transgenic peanuts for induced resistance to the *Indian peanut clump virus*. In *Proceedings of XVI annual convention and international symposium on management of vector-borne viruses, February 7–10, 2006*. Patancheru, India: ICRISAT.

Shi, A., Chen, P., Li, D., Zheng, C., Zhang, B., & Hou, A. (2009). Pyramiding multiple genes for resistance *to Soybean mosaic virus* in soybean using molecular markers. *Molecular Breeding*, *23*, 113–124.

Shi, A., Chen, P., Vierling, R., Zheng, C., Li, D., Dong, D., et al. (2011). Multiplex Single nucleotide polymorphism (SNP) assay for detection of *Soybean mosaic virus* resistance genes in soybean. *Theoretical and Applied Genetics*, *122*, 445–457.

Shivaprasad, P. V., Thillaichidambaram, P., Balaji, V., & Veluthambi, K. (2006). Expression of full-length and truncated Rep genes from *mungbean yellow mosaic virus-vigna* inhibits viral replication in transgenic tobacco. *Virus Genes*, *33*, 365–374.

Shivashankar, G., Kulkarni, R. S., Shashidhar, H. E., & Mahishi, D. M. (1993). Improvement of field bean. In K. L. Chadha, & G. Kallo (Eds.), *Advances in horticulture: Vol. 5.* (pp. 77–286). New Delhi.

Shukla, G. P., & Pandya, B. P. (1985). Resistance to yellow mosaic in greengram. *SABRAO Journal*, *17*, 165–171.

Sillero, J. C., Villegas-Fernandez, A. M., Thomas, J., Rojas-Molina, M. M., Emeran, A. A., Fernandez-Aparicio, M., et al. (2010). Faba bean breeding for disease resistance. *Field Crops Research*, *115*, 297–307.

Silva, S. J. C., Castillo-Urquiza, G. P., Hora-Junior, B. T., Assuncao, I. P., Lima, G. S. A., Pio-Ribeiro, G., et al. (2011). Species diversity, phylogeny and genetic variability of begomovirus populations infecting leguminous weeds in North-Eastern Brazil. *Plant Pathology*, *61*, 457–467.

Singh, S. K., Chakroborthy, S., Singh, A. K., & Pandey, P. K. (2006). Cloning, restriction mapping and phylogenetic relationship of genomic components of MYMIV from *Lablab purpureus*. *Bioresource Technology*, *97*, 1807–1814.

Singh, P. K., Kumar, A., Rai, N., & Singh, D. V. (2012). Identification of host plant resistant to *Dolichos yellow mosiac virus* (DYMV) in *Dolichos* bean (*Lablab purpureus*). *Journal of Plant Pathology and Microbiology*, *3*, 130.

Singh, G., Sharma, Y. R., Shanmugasundaram, S., Shih, S. L., & Green, S. K. (2003). Status of *Mungbean yellow mosaic virus* resistant breeding. In *Final workshop and planning meeting on mungbean* (pp. 204–212).

Smýkal, P., Aubert, G., Burstin, J., Coyne, C. J., Ellis, N. T. H., Flavell, A. J., et al. (2012). Pea (*Pisum sativum* L.) in the genomic era. *Agronomy*, *2*, 74–115.

Smýkal, P., Safarova, D., Navratil, M., & Dostalova, R. (2010). Marker assisted pea breeding: eIF4E allele specific markers to *Pea seed-borne mosaic virus* (PSbMV) resistance. *Molecular Breeding*, *26*, 425–438.

Souframanien, J., & Gopalakrishna, T. (2006). ISSR and SCAR markers linked to the *Mungbean yellow mosaic virus* (MYMV) resistance gene in blackgram [*Vigna mungo* (L.) Hepper]. *Plant Breeding*, *125*, 619–622.

Souleymane, A., Aken'Ova, M. E., Fatokun, C. A., & Alabi, O. Y. (2013). Screening for resistance to cowpea aphid (*Aphis craccivora* Koch) in wild and cultivated cowpea (*Vigna unguiculata* L. Walp.) accessions. *International Journal of Science, Environment and Technology*, *2*, 611–621.

Sreekanth, M., Sriramulu, M., Rao, R. D. V. J. P., Babu, B. S., & Babu, T. R. (2002). Evaluation of greengram genotypes for resistance to *Thrips palmi* Karny and *Peanut bud necrosis virus*. *Indian Journal of Plant Protection*, *30*, 109–114.

Sreekanth, M., Sriramulu, M., Rao, R. D. V. J. P., Babu, B. S., & Babu, T. R. (2004a). Effect of intercropping on *Thrips palmi* (Karny) population and *Peanut bud necrosis virus* incidence in mungbean. *Indian Journal of Plant Protection, 32*, 45–48.

Sreekanth, M., Sriramulu, M., Rao, R. D. V. J. P., Babu, B. S., & Babu, T. R. (2004b). Evaluation of certain new insecticides against *Thrips palmi* (Karny), the vector of *Peanut bud necrosis virus* (PBNV) on mungbean (*Vigna radiata* L. Wilczek). *International Pest Control, 46*, 315–317.

Sreenivasulu, P., Demski, J. W., Reddy, D. V. R., Misari, S. M., Olorunju, P. E., & Kuhn, C. W. (1988). *Tomato spotted wilt virus* (TSWV) and strains of *Peanut mottle virus* that mimic TSWV symptoms in peanut in Georgia. *Plant Disease, 72*, 546.

Sreenivasulu, P., Demski, J. W., Reddy, D. V. R., Naidu, R. A., & Ratna, A. S. (1991). Purification and some serological relationships of *Tomato spotted wilt virus* isolates occurring on peanut (*Arachis hypogaea*) in the USA. *Plant Pathology, 40*, 503–507.

Sreenivasulu, P., Subba Reddy, C., Ramesh, B., & Kumar, P. L. (2008). Groundnut viruses. In G. P. Rao, S. M. Paul Khurana, & S. L. Lenardon (Eds.), *Industrial crops: Vol. 1. Characterization, diagnosis and management of plant viruses* (pp. 47–97). USA: Studium Press LLC.

Srinivas, T., Reddy, M. V., Jain, K. C., & Reddy, M. S. S. (1997). Studies on inheritance of resistance and allelic relationships for strain 2 of pigeon pea sterility mosaic pathogen. *Annals of Applied Biology, 130*, 105–110.

Steinlage, T. A., Hill, J. H., & Nutter, F. W., Jr. (2002). Temporal and spatial spread of *Soybean mosaic virus* (SMV) in soybeans transformed with the coat protein gene of SMV. *Phytopathology, 92*, 478–486.

Subrahmanayam, P., Naidu, R. A., Reddy, L. J., Kumar, P. L., & Ferguson, M. (2001). Resistance to groundnut rosette disease in wild *Arachis* species. *Annals of Applied Biology, 139*, 45–50.

Subrahmanyam, P., Hildebrand, G. L., Naidu, R. A., Reddy, L. J., & Singh, A. K. (1998). Sources of resistance to groundnut rosette disease in global groundnut germplasm. *Annals of Applied Biology, 132*, 473–485.

Sudha, M., Karthikeyan, A., Anusuya, P., Ganesh, N. M., Pandiyan, M., Senthil, N., et al. (2013). Inheritance of resistance to *Mungbean yellow mosaic virus* (MYMV) in inter- and intra-specific crosses of mungbean (*Vigna radiata*). *American Journal of Plant Science, 4*, 1924–1927.

Sundaraj, S., Srinivasan, R., Culbreath, A. K., Riley, D. G., & Pappu, H. R. (2014). Host plant resistance against *Tomato spotted wilt virus* in peanut (*Arachis hypogaea*) and its impact on susceptibility to the virus, virus population genetics, and vector feeding behavior and survival. *Phytopathology, 104*, 202–210.

Sunitha, S., Shanmugapriya, G., Balamani, V., & Veluthambi, K. (2013). *Mungbean yellow mosaic virus* (MYMV) AC4 suppresses post-transcriptional gene silencing and an AC4 hairpin RNA gene reduces MYMV DNA accumulation in transgenic tobacco. *Virus Genes, 46*, 496–504.

Swanson, M. M., Varma, A., Muniyappa, V., & Harrison, B. D. (1992). Comparative epitope profiles of the particle proteins of whitefly-transmitted geminiviruses from nine crop legumes in India. *Annals of Applied Biology, 120*, 425–433.

Taliansky, M. E., Robinson, O. J., & Murant, A. F. (2000). Groundnut rosette disease virus complex: Biology and molecular biology. *Advances in Virus Research, 55*, 357–400.

Taliansky, M. E., Ryabov, E. V., & Robinson, D. J. (1998). Two distinct mechanisms of transgenic resistance mediated by *Groundnut rosette virus* satellite RNA sequences. *Molecular Plant-Microbe Interactions, 11*, 367–374.

Terauchi, H., Kanematsu, S., Honda, K., Mikoshiba, Y., Ishiguro, K., & Hidaka, S. (2001). Comparison of complete nucleotide sequences of genomic RNAs of four *Soybean dwarf virus* strains that differ in their vector specificity and symptom production. *Archives of Virology, 146*, 1885–1898.

Thein, H. X., Bhat, A. I., & Jain, R. K. (2003). Mungbean necrosis disease caused by a strain of *Groundnut bud necrosis virus*. *Indian Phytopathology, 56*, 54–60.
Thira, S. K., Cheema, S. S., & Kang, S. S. (2004). Pattern of bud necrosis disease development in groundnut crop in relation to different dates of sowing. *Plant Disease Research, 19*, 125–129.
Thomas, J. E., Parry, J. N., Schwinghamer, M. W., & Dann, E. K. (2010). Two novel mastreviruses from chickpea (*Cicer arietinum*) in Australia. *Archives of Virology, 155*, 1777–1788.
Thresh, J. M. (2003). Control of plant virus diseases in sub-Saharan Africa: The possibility and feasibility of an integrated approach. *The African Crop Science Journal, 11*, 199–223.
Todd, J. W., & Culbreath, A. K. (1995). Thrips populations and spotted wilt disease progress on resistance/susceptible cultivars treated with various insecticides. In *Proceedings of the American Peanut Research and Education Society*: Vol. 27 (p. 35).
Tollefson, J. (2011). Brazil cooks up transgenic bean. *Nature, 478*, 168.
Tougou, M., Furutani, N., Yamagishi, N., Shizukawa, Y., Takahata, Y., & Hidaka, S. (2006). Development of resistant transgenic soybeans with inverted repeat-coat protein genes of *Soybean dwarf virus*. *Plant Cell Reports, 25*, 1213–1218.
Tougou, M., Yamagishi, N., Furutani, N., Shizukawa, Y., Takahata, Y., & Hidaka, S. (2007). *Soybean dwarf virus*-resistant transgenic soybeans with the sense coat protein gene. *Plant Cell Reports, 26*, 1967–1975.
Tryphone, G. M., Chilagane, L. A., Protas, D., Kusolwa, P. M., Nchimbi-Msolla, S., Tryphone, G. M., et al. (2013). Marker assisted selection for common bean diseases improvement in Tanzania: Prospects and future needs. In S. B. Andersen (Ed.), *Agricultural and biological sciences. Plant breeding from laboratories to fields,* (298 p). Rijeka, Croatia: InTech. ISBN: 978-953-51-1090-3. Chapter 5.
Tsai, W. S., Shih, S. L., Rauf, A., Safitri, R., Hidayati, N., Huyen, B. T. T., et al. (2013). Genetic diversity of legume yellow mosaic begomoviruses in Indonesia and Vietnam. *Annals of Applied Biology, 163*, 367–377.
Tsompana, M., & Moyer, J. W. (2008). Tospovirus. In B. W. J. Mahy, & M. H. V. Van Regenmortel (Eds.), *Desk encyclopedia of plant and fungal virology* (pp. 353–359). New York: Academic Press.
Usharani, K., Surendranath, B., Haq, Q. M. R., & Malathi, V. G. (2004). Yellow mosaic virus infecting soybean in Northern India is distinct from the species infecting soybean in Southern and Western India. *Current Science, 86*, 845–850.
Usharani, K. S., Surendranath, B., Haq, Q. M. R., & Malathi, V. G. (2005). Infectivity analysis of a soybean isolate of *Mungbean yellow mosaic India virus* by agroinoculation. *Journal of General Plant Pathology, 71*, 230–237.
Varma, A., & Malathi, V. G. (2003). Emerging geminivirus problems: A serious threat to crop production. *Annals of Applied Biology, 142*, 145–164.
Varma, A., Mandal, B., & Singh, M. K. (2011). Global emergence and spread of whitefly (*Bemisia tabaci*) transmitted geminiviruses. In W. M. O. Thompson (Ed.), *The whitefly, Bemisia tabaci (Homoptera: Aleyrodidae) interaction with geminivirus-infected host plants* (pp. 205–292). The Netherlands: Springer.
Varshney, R. K., Chen, W., Li, Y., Bharti, A. K., Saxena, R. K., Schlueter, J. A., et al. (2012). Draft genome sequence of pigeonpea (*Cajanus cajan*), an orphan legume crop of resource-poor farmers. *Nature Biotechnology, 30*, 83–89.
Varshney, R. K., Mohan, S. M., Gaur, P. M., Gangarao, N. V. P. R., Pandey, M. K., Bohra, A., et al. (2013). Achievements and prospects of genomics-assisted breeding in three legume crops of the semi-arid tropics. *Biotechnology Advances, 31*, 1120–1134.
Varshney, R. K., Song, C., Saxena, R. K., Azam, S., Yu, S., Sharpe, A. G., et al. (2013). Draft genome sequence of chickpea (*Cicer ariettinum*) provides a resource for trait improvement. *Nature Biotechnology, 31*, 240–246.

Vemana, K., & Jain, R. K. (2013). Comparative reaction of popular high yielding spanish and virginia bunch groundnut varieties of *Tobacco streak virus*. *Indian Journal of Virology, 24*, 214–219.

Venkataravanappa, V., Swarnalatha, P., Lakshminarayana Reddy, C. N., Mahesh, B., Rai, A. B., & Krishna Reddy, M. (2012). Molecular evidence for association of *Tobacco curly shoot virus* and a betasatellite with curly shoot disease of common bean (*Phaseolus vulgaris* L.) from India. *Journal of Plant Pathology and Microbiology, 3*, 148.

Waliyar, F. (2005). *Characterisation of the causal virus of Pigeonpea sterility mosaic disease: A further step towards attaining sustainability of Pigeonpea production in the Indian subcontinent: Final technical report*. Patancheru, AP, India: International Crops Research Institute for the Semi-Arid Tropics, 78 pp.

Wang, X., Eggenberger, A. L., Nutter, F. W., Jr., & Hill, J. H. (2001). Pathogen-derived transgenic resistance to *Soybean mosaic virus* in soybean. *Molecular Breeding, 8*, 119–127.

Wang, X., Gai, J., & Pu, Z. (2003). Classification and distribution of strains of *Soybean mosaic virus* in middle and lower Huanghuai and Changjiang river valleys. *Soybean Science, 22*, 102–107.

Wang, D., Ma, Y., Liu, N., Yang, Z., Zheng, G., & Zhi, H. (2011). Fine mapping and identification of the soybean *Rsc4* resistance candidate gene to *Soybean mosaic virus*. *Plant Breeding, 130*, 653–659.

Wang, D.-G., Ma, Y., Liu, N., Zheng, G.-J., Yang, Z.-L., Yang, Y.-Q., et al. (2012). Inheritance of resistance to *Soybean mosaic virus* strains SC4 and SC8 in soybean. *Acta Agronomica Sinica, 38*, 202–209.

Wang, H., Pandey, M. K., Qiao, L., Qin, H., Culbreath, A. K., He, G., et al. (2013). Genetic mapping and QTL analysis for disease resistance using F_2 and F_5 generation based genetic maps derived from Tifrunner × GT–C20 in peanut (*Arachis hyopgaea* L.). *The Plant Genome, 6*, 1–10. http://dx.doi.org/10.3835/plantgenome2013.05.0018 (Special submission).

Wangai, A. W., Pappu, S. S., Pappu, H. R., Deom, C. M., & Naidu, R. A. (2001). Distribution and characteristics of groundnut rosette disease in Kenya. *Plant Disease, 85*, 470–474.

Weekes, R. J., Mumford, R. A., Barker, I., & Wood, K. R. (1996). Diagnosis of tospoviruses by RT-PCR. *Acta Horticulturae, 431*, 159–166.

Wightman, J. A., & Amin, P. W. (1988). Groundnut pests and their control in the semi-arid tropics. *Tropical Pest Management, 34*, 218–226.

Wrather, A., Shannon, G., Balardin, R., Carregal, L., Escobar, R., Gupta, G. K., et al. (2010). Effect of diseases on soybean yield in the top eight producing countries in 2006. *Plant Health Progress*. http://dx.doi.org/10.1094/PHP-2010-0125-01-RS.

Xu, Z., & Barnett, O. W. (1984). Identification of a *Cucumber mosaic virus* strain from naturally infected peanuts in China. *Plant Disease, 68*, 386–389.

Yadav, R. K., Shukla, R. K., & Chattopadhyay, D. (2009). Soybean cultivar resistant to *Mungbean yellow mosaic India virus* infection induces viral RNA degradation earlier than the susceptible cultivar. *Virus Research, 144*, 89–95.

Yamagishi, N., Terauchi, H., Honda, K., Kanematsu, S., & Hidaka, S. (2006). Discrimination of four Soybean dwarf virus strains by dot-blot hybridization with specific probes. *Journal of Virological Methods, 133*, 219–222.

Yamagishi, N., Terauchi, H., Kanematsu, S., & Hidaka, S. (2006). Biolistic inoculation of soybean plants with *Soybean dwarf virus*. *Journal of Virological Methods, 137*, 164–167.

Yi, W.-X., Chen, S.-S., & Yang, C. Y. (2011). One-tube real-time PCR assay for simultaneous detection of *Bean pod mottle virus* and *Tobacco ring spot virus* in soybean seeds. *Acta Phytopathology Sinica, 41*, 85–92.

Yin, X., Wang, J., Cheng, H., Wang, X., & Yu, D. (2013). Detection and evolutionary analysis of soybean miRNAs responsive to *Soybean mosaic virus*. *Planta, 237*, 1213–1225.

Yu, K. (2012). Bacterial artificial chromosome libraries of pulse crops: Characteristics and applications. *Journal of Biomedicine and Biotechnology, 2012,* Article ID 493186, 8 p.

Zein, H. S., Nakazawa, M., Ueda, M., & Miyatake, K. (2007). Development of serological procedures for rapid and reliable detection of *Cucumber mosaic virus* with dot-immunobinding assay. *World Journal of Agricultural Sciences, 3,* 430–439.

Zhang, C., Grosic, S., Whitham, S. A., & Hill, J. H. (2012). The requirement of multiple defense genes in soybean *Rsv1*-mediated extreme resistance to *Soybean mosaic virus*. *Molecular Plant-Microbe Interactions, 25,* 1307–1313.

Zhang, C., Hajimorad, M. R., Eggenberger, A. L., Tsang, S., Whitham, S. A., & Hill, J. H. (2009). Cytoplasmic inclusion cistron of *Soybean mosaic virus* serves as a virulence determinant on *Rsv3*-genotype soybean and a symptom determinant. *Virology, 391,* 240–248.

Zhang, X., Sato, S., Ye, X., Dorrance, A. E., Morris, T. J., Clemente, T. E., et al. (2011). Robust RNAi-based resistance to mixed infection of three viruses in Soybean plants expressing separate short hairpins from a single transgene. *Virology, 101,* 1264–1269.

Zheng, C., Chen, P., & Gergerich, R. (2005). Characterization of resistance to *Soybean mosaic virus* in diverse soybean germplasm. *Crop Science, 45,* 2503–2509.

INDEX

Note: Page numbers followed by "*f*" indicate figures and "*t*" indicate tables.

C

Q

R

S

U

Yehezkel Antignus, Figure 1.1 A demonstration of aphids phototaxis toward UV irradiation and the formation of the "two-compartment effect" in greenhouses with roof arches covered alternately with UV-absorbing (films with a bluish hue) and ordinary films (A). The massive immigration of aphids into greenhouse sections with a rich UV environment resulted in total infection with the aphid-borne nonpersistent *Zucchini yellow mosaic virus* (ZYMV) (B Left hand). All the stunted plants show the typical mosaic symptoms of the virus (C). None of the plants grown under the UV-absorbing film were infected (A Right hand).

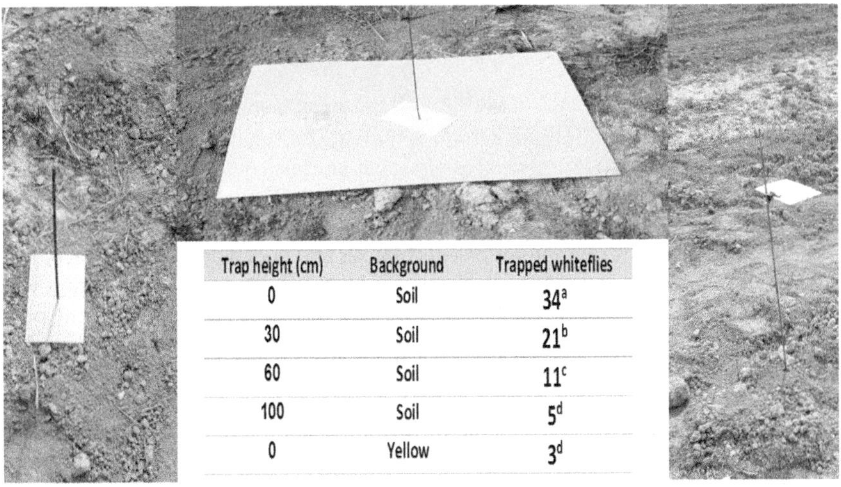

Trap height (cm)	Background	Trapped whiteflies
0	Soil	34[a]
30	Soil	21[b]
60	Soil	11[c]
100	Soil	5[d]
0	Yellow	3[d]

Yehezkel Antignus, Figure 1.9 The contrast effect on whiteflies (*Bemisia tabaci*) trapping by yellow-sticky cards. The highest trapping efficiency was obtained when traps were placed directly on bare soil. A gradient of numbers of trapped insects was formed according to the distance of the yellow cards' plane from ground level. Lower trapping numbers are correlated with lower levels of contrast between the yellow color and background formed by the brown soil. When the yellow cards were placed over a large yellow "poligal" plate trapping was zero or near zero, indicating again the importance of high contrast between the trap and its background.

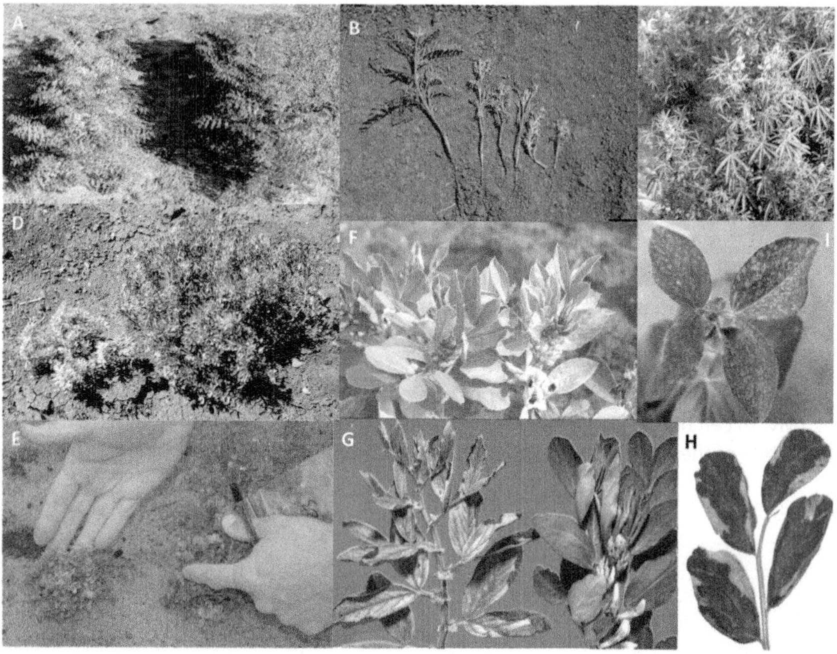

Khaled M. Makkouk *et al.*, Figure 4.1 (A) Symptoms caused by *Cucumber mosaic virus* (CMV) infection in chickpea: leaf chlorosis in kabuli chickpea (left), leaf reddening in desi chickpea (right). (B) Symptoms of leaf chlorosis and plant stunting (smaller shoots on right) caused by infection with *Chickpea chlorotic dwarf virus* (right), healthy shoot (single shoot on left). (C) Symptoms of leaf chlorosis and bunching of young leaves caused by CMV infection in a narrow-leafed lupin plant. (D) Marked plant stunting with chlorosis and bunching of young leaves caused by luteovirus infection in a lentil plant (left), a healthy plant (right). (E) Chlorosis and bunching of young leaves and marked plant stunting caused by *Bean yellow mosaic virus* (BYMV) infection in lentil plant (left), less severely affected plants (right). (F) Leaf symptoms of mild mosaic and chlorotic spotting caused by *Pea seed-borne mosaic virus* infection in a faba bean plant (right), healthy plant (left). (G) Leaf symptoms of stiffness, interveinal chlorosis, and marginal necrosis caused by infection with *Bean leafroll virus* in a plant of faba bean (left), healthy plant (right). (H) Chlorotic marginal leaf mottle caused by infection with *Broad bean stain virus* in a leaf of faba bean. (I) Chlorotic leaf spotting symptoms caused by *Broad bean mottle virus* infection in leaves of faba bean.

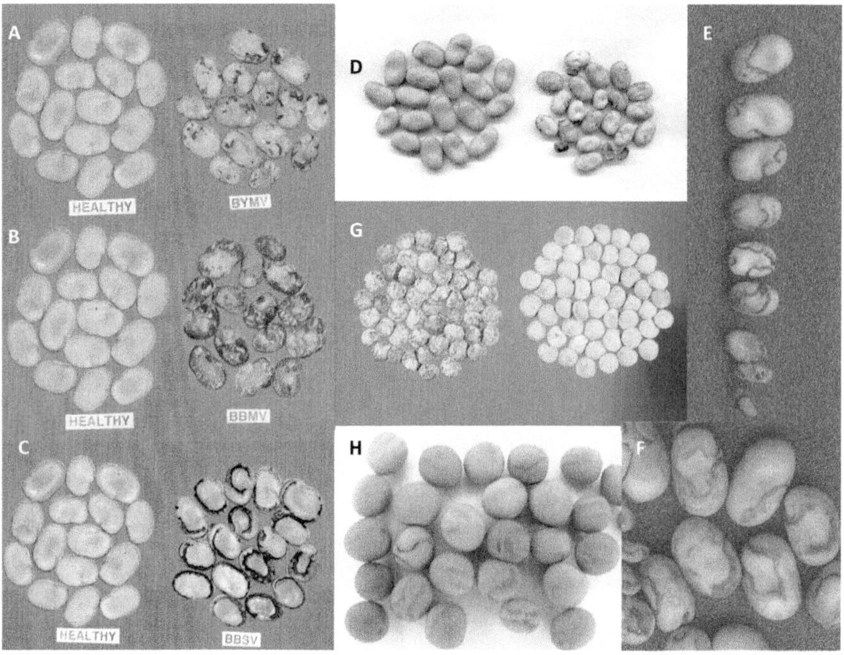

Khaled M. Makkouk *et al.*, Figure 4.2 Symptoms of necrosis, reduction in size, and malformation in seeds of faba bean, virus-infected (right) and healthy (left): causal viruses are *Bean yellow mosaic virus* (A), *Broad bean mottle virus* (B), and *Broad bean stain virus* (BBSV) (C). Symptoms of malformation, reduction in size and necrotic rings caused by *Pea seed-borne mosaic virus* (PSbMV) in faba bean seeds (D), virus-infected (right), healthy (left). Necrotic rings caused by PSbMV in broad bean seeds (E, F). Symptoms of necrosis in lentil seeds caused by BBSV (G), virus-infected (left), healthy (right). Necrotic rings caused by PSbMV in pea seed (H).

Hervé Lecoq and Nikolaos Katis, Figure 5.1 Different methods used to control cucurbit viruses. (A) Plastic mulches repel aphid and whitefly vectors and delay virus spread. (B) Row covers prevent insect vectors from reaching the plants, but they should be partially removed at flowering stage to allow pollination. (C) Cucumber plant grafted on *Cucurbita* rootstock (indicated by arrow). (D) Mild-strain protection against ZYMV in zucchini squash: left protected plants and right nonprotected plants. (E) Machine developed to inoculate seedlings in nurseries with ZYMV mild strain. (F) An example of *Cucurbita* genetic resources in which virus resistance can be looked for. (G) Susceptible squash cultivar inoculated by a severe ZYMV strain. (H) Same cultivar carrying a transgenic resistance to ZYMV, inoculated the same day with the same strain as the susceptible plant in (G).

Lawrence Kenyon *et al.*, Figure 6.1 (A) Remnant plant from previous season left growing at the edge of the field because it was still producing a few fruits, but it was acting as a potent source of PepYLCIV for the new pepper crop in the adjacent field, Yogyakarta area, Indonesia. (B) Young volunteer pepper plant severely infected with PepYLCV and acting as a source of infection for the new field of peppers adjacent, Yogyakarta area, Indonesia.

Lawrence Kenyon *et al.*, Figure 6.2 (See legend on next page.) *(Continued)*

Lawrence Kenyon *et al.*, Figure 6.3 Severe chlorosis and leaf curl caused by mixed infection of an unidentified begomovirus and ChiVMV, Khon Kaen area, Thailand.

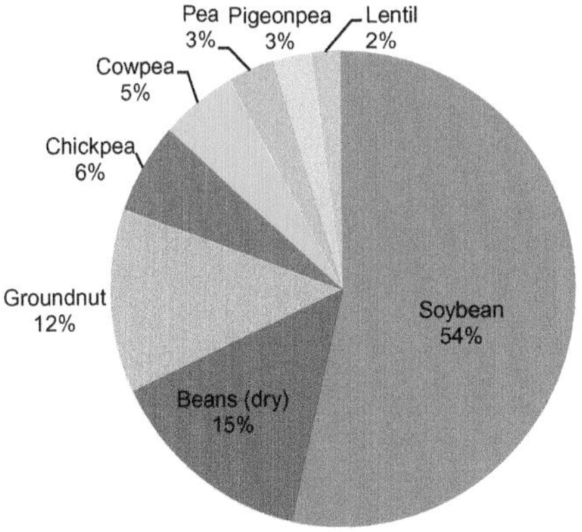

Crop	Area (ha)
soybean	106,625,241
beans (dry)	28,780,377
Groundnut	24,625,099
Chickpea	12,144,639
Cowpea	10,688,653
Pea	6,326,999
Pigeonpea	5,323,407
Lentil	4,249,725
Total	198,764,140

Masarapu Hema *et al.*, Figure 9.1 Percentage share of global production area of major food legumes. Total area = 198,764,140 ha. Source: *FAO Production Statistics of 2012.*

Lawrence Kenyon *et al.*, Figure 6.2—Cont'd (A) ChiVMV (Potyvirus) inducing relatively mild symptoms in pepper, Khon Kaen area, Thailand. (B) PVMV (Potyvirus) infecting pepper in Taiwan. (C) PeVYV (Polerovirus) infected plant with general chlorosis/yellowing that could easily be mistaken for nutrient deficiency or toxicity, Kamphaeng Saen area, Thailand.